D1811020

Introduction to Black Hole Physics

Introduction to Black Hole Physics

Valeri P. Frolov & Andrei Zelnikov

University of Alberta

OXFORD
UNIVERSITY PRESS

OXFORD

UNIVERSITY PRESS

Great Clarendon Street, Oxford OX2 6DP

Oxford University Press is a department of the University of Oxford.
It furthers the University's objective of excellence in research, scholarship,
and education by publishing worldwide in

Oxford New York

Auckland Cape Town Dar es Salaam Hong Kong Karachi
Kuala Lumpur Madrid Melbourne Mexico City Nairobi
New Delhi Shanghai Taipei Toronto

With offices in

Argentina Austria Brazil Chile Czech Republic France Greece
Guatemala Hungary Italy Japan Poland Portugal Singapore
South Korea Switzerland Thailand Turkey Ukraine Vietnam

Oxford is a registered trade mark of Oxford University Press
in the UK and in certain other countries

Published in the United States
by Oxford University Press Inc., New York

British Library Cataloguing in Publication Data

Data available

Library of Congress Cataloging in Publication Data

Data available

Typeset by SPI Publisher Services, Pondicherry, India
Printed in Great Britain
on acid-free paper by
CPI Antony Rowe, Chippenham, Wiltshire

ISBN 978–0–19–969229–3

1 3 5 7 9 10 8 6 4 2

To Werner and Inge Israel
with our deep gratitude and admiration

Preface

This book is about black holes, one of the most intriguing problems of modern theoretical Physics and Astrophysics. It is intended to provide a thorough introduction to this area and to serve as a text for graduate students, postdocs, and young researches.

The book is based on the lectures that during 30 years one of the authors (V.F.) has been reading at different universities. In 1981–1992 he gave lectures on general relativity for students of the Moscow Institute of Physics and Technology, chair of problems of Physics and Astrophysics. The founder and the head of the chair was academician Vitaly Ginzburg. This was a one-year course, and half of it was devoted to black holes. Starting from 1993 he has read 'Black hole physics' as a one-term course PHYS 696, for the PhD students at the University of Alberta. It is natural that during this long time the content of the course has changed considerably, however, its structure remained practically the same. This book is based on the lecture notes for this course and reflects its structure. The main part of the book (presented in Chapters 1–8) follows the lectures. This provides one with material sufficient for a one-term self-consistent course on black holes. The book also contains an additional material, e.g. concerning higher-dimensional gravity, black holes and extra dimensions, hidden symmetries, wormholes and 'time machines', which can be used for term papers related with the course (Chapters 9 and 10 and Appendices).

It took us more than a year to transform the lecture notes into the book. During this work we tried to include answers to many questions often asked by students. We included many assignments that may help to understand the material better. Most of the problems are given with their solutions. We use small letter text for this material. The same small letter text is also used for remarks, which contain examples illustrating the subject, or brief notes on possible generalizations. The book is rich with illustrations. In our experience proper illustrations are extremely helpful for better understanding of the physical and mathematical ideas. We follow the advice of Synge (1959): "It is a mistake to economize in relativistic diagrams. They form the bridge across which can pass those astronomers whose minds work in the great Greek tradition."

The history of black holes is quite dramatic. The black hole paradigm is based on the Einstein theory of general relativity. During many years theoreticians studied black hole solutions of the gravity equations, but until the beginning of the 1970s the 'astrophysical community' in general had a quite reluctant attitude to these ideas. Only after the discovery of pulsars and the first missions of the space-based X-ray telescopes, did this attitude slowly change and warm up. Now, 40 years later, we cannot imagine our world without black holes. They became an important element of the 'big picture' of our Universe. There are more and more pieces of evidence that their role in the universe evolution might be not only important but decisive.

For many years black holes have been considered as an interesting pure theoretical construction of general relativity with a number of amusing mathematical properties. Now, after the discovery of black holes, the Einstein gravity became an important tool for their study. Collection of data on the properties of the matter in strong gravitational field in the vicinity of black holes will also probe the validity of the Einstein theory itself in the strong-field regime. Recently, black holes have been widely discussed in connection with many fundamental problems of theoretical physics. It is worth also mentioning recent interest in the higher-dimensional gravity discussing the possibility of black hole creation in particle colliders.

In this book we tried to present the theory of black holes in the form that might be useful for students and young scientists who are interested in this area. Our book combines physical, mathematical, and astrophysical aspects of the black hole problem. This is a self-contained textbook, which covers both the most important established results and the most recent new developments. The book includes pedagogically presented 'standard' material on black holes and it also contains relatively new subjects such as the role of hidden symmetries in black hole physics and black holes in spacetimes with large extra dimensions. We address the book to those who would like to study black hole physics and its applications.

The name 'black holes', invented by John Wheeler in 1967, has now become very popular. This is not only because of their astrophysical importance, but also because black holes play the role of gates from our world, with known physics, to an interior abyss, where new physics will become important. This Planckian-scale physics is the Holy Grail of theoretical physics. Quantum gravity and string theory are the first attempts to understand it. Black holes, and also connected with them such subjects as wormholes and time machines, are now almost compulsory features of most science fiction books.

To illustrate how popular is the notion of the black hole we made a brief search on Goggle. The name 'black hole' has about 11,800,000 results, while for example 'neutron star' is used in about 3,120,000 documents. Even if we exclude such things as, e.g., 'black holes in economy', one can conclude that at least in the virtual reality black holes play an important role and are more 'popular' than say 'neutron stars'.

There are many nice popular books on black holes. In the more advanced books on general relativity this subject is often discussed as one of its applications. The wonderful books "Gravitation", by Misner, Thorne and Wheeler (1973) , "Gravitation and Cosmology: Principles and Applications of the General Theory of Relativity" by Weinberg (1972), "General Relativity", by Wald (1984) and "Large Scale Structure" by Hawking and Ellis (1973) are the most famous examples. More recently, there was published a textbook "A relativist's toolkit: the mathematics of black-hole mechanics" by Poisson (2007).

There exist a number of more specialized books that cover different aspects of black hole physics, for example mathematically oriented monographs like "The Mathematical Theory of Black Holes" by Chandrasekhar (1983), "Black Hole Uniqueness Theorems" by Heusler (1996), or "Gravity and Strings" by Ortín (2004), which focus mainly on mathematical aspects of black hole theory. In 1998 one of the authors (V.F.) and Novikov published a monograph "Black hole physics. Basics concepts and new developments". There are also more recent books that are oriented mostly towards the general public and reflect the contemporary state of black holes in observational astrophysics, e.g., "The Black hole at the Center of our Galaxy" and "High-Energy Astrophysics" by Melia (2009). Usually, such books are mostly descriptive and don't pay much attention to the mathematical component of the Einstein theory, and don't describe derivation of many important results from first principles.

Our book is aimed mainly at students and postdocs with the specialization in physics, mathematics and astrophysics. It will also attract the attention of those who would like to investigate the wonderful 'Black Hole Land' on their own, and to become equipped with mathematical tools and ideas that would allow them to find out answers to many of their questions.

Valeri P. Frolov
Andrei Zelnikov

Edmonton, February 2011

Acknowledgements

During almost 20 years of our work at the University of Alberta the authors have enjoyed the financial support of the Killam trust. We really appreciate this support that, in particular, helped this book to be written. Some of the material, which enters the book, is based on our own research, which was supported by the Natural Sciences and Engineering Council of Canada (NSERC). The thanks go to the NSERC for this support.

We wish to express our sincere thanks to our teachers Moisei Markov and Vitaly Ginzburg. Most of our knowledge and skills were developed during the time when we were their students and (later) collaborators. We enjoyed the unfailing support of our colleagues from the University of Alberta and Lebedev Physical Institute and thank them for their remarks concerning the material of the lectures used in this book. We are grateful to the Institute of Theoretical Physics and Department of Physics of the University of Alberta for friendly support.

This book is devoted to two wonderful persons who played an important role in our life: Werner and Inge Israel. We benefit a lot from their friendship and help. Many discussions and joint work with Werner, who is the world-wide recognized expert in black holes, clarified many aspects of black holes to us.

We wish to thank Craig Heinke for his comments on astrophysical aspects of black holes. We would specially thank Pavel Krtouš and Andrei Shoom, who carefully read the draft of the book and made many useful remarks.

This book would never have appeared if we had not received support and encouragements from members of our families. We owe the deepest debt of gratitude to them.

Contents

1 Black Holes: Big Picture   1
  1.1   Gravity and Black Holes   1
  1.2   Brief History of Black Holes   8
  1.3   'Dark Stars' vs. Black Holes   12
  1.4   Final State of Stellar Evolution   15
  1.5   Equilibrium of Gravitating Systems   19
  1.6   Important Notions of Astrophysics   22
  1.7   Black Holes in Astrophysics and Cosmology 26
  1.8   Stellar-Mass Black Holes   29
  1.9   Supermassive Black Holes   34
  1.10 Primordial Black Holes   40
  1.11 Black Holes in Theoretical Physics   41
  1.12 Black Holes and Extra Dimensions   44

2 Physics in a Uniformly Accelerated Frame   47
  2.1   Minkowski Spacetime and Its Symmetries   47
  2.2   Minkowski Spacetime in Curved Coordinates 50
  2.3   Uniformly Accelerated Reference Frame   57
  2.4   Homogeneous Gravitational Field   59
  2.5   Causal Structure   63
  2.6   Wick's Rotation in the Rindler Space   65

3 Riemannian Geometry   67
  3.1   Differential Manifold. Tensors   67
  3.2   Metric   74
  3.3   Covariant Derivative   79
  3.4   Lie and Fermi Transport   81
  3.5   Curvature Tensor   84
  3.6   Parallel Transport of a Vector   90
  3.7   Spacetime Symmetries   97
  3.8   Submanifold   102
  3.9   Integration   106

4 Particle Motion in Curved Spacetime   109
  4.1   Equations of Motion   109
  4.2   Phase Space   114
  4.3   Complete Integrability   120

5   Einstein Equations   127
    5.1   Einstein–Hilbert Action   127
    5.2   Einstein Equations   129
    5.3   Linearized Gravity   135
    5.4   Gravitational radiation   142
    5.5   Gravity in Higher-Dimensions   149

6   Spherically Symmetric Black Holes   162
    6.1   Spherically Symmetric Gravitational Field   162
    6.2   Schwarzschild–de Sitter Metric   166
    6.3   Global Structure of the Schwarzschild Spacetime   172
    6.4   Black Hole Interior   178
    6.5   Painlevè–Gullstrand Metric   180
    6.6   Eddington–Finkelstein Coordinates   181
    6.7   Charged Black Holes   181
    6.8   Higher-Dimensional Spherical Black Holes   182

7   Particles and Light Motion in Schwarzschild Spacetime   185
    7.1   Equations of Motion   185
    7.2   Particle Trajectories   188
    7.3   Kepler's Law   196
    7.4   Light Propagation   200
    7.5   Ray-Tracing in Schwarzschild Spacetime   211
    7.6   Black Hole as a Gravitational Lens   215
    7.7   Radiation from an Object Moving Around the Black Hole   223
    7.8   Equations of Motion in 'Tilted' Spherical Coordinates   230
    7.9   Magnetized Schwarzschild Black Hole   230
    7.10   Particle and Light Motion Near Higher-Dimensional Black Holes   236

8   Rotating Black Holes   242
    8.1   Kerr Spacetime   242
    8.2   Ergosphere. Horizon   246
    8.3   Particle and Light Motion in Equatorial Plane   257
    8.4   Spinning up the Black Hole   268
    8.5   Geodesics in Kerr Spacetime: General Case   272
    8.6   Light Propagation   276
    8.7   Hidden Symmetries of Kerr Spacetime   286
    8.8   Energy Extraction from a Rotating Black Hole   289
    8.9   Black Holes in External Magnetic Field   293

9   Classical and Quantum Fields near Black Holes   298
    9.1   Introduction   298
    9.2   Static Field in the Schwarzschild Spacetime   299
    9.3   Dimensional Reduction   302
    9.4   Quasinormal Modes   307
    9.5   Massless Fields in the Kerr Spacetime   314

9.6   Black Hole in a Thermal Bath   315
9.7   Hawking Effect   320
9.8   Quantum Fields in the Rindler Spacetime   325
9.9   Black Hole Thermodynamics   336
9.10   Higher-Dimensional Generalizations   340

10   Black Holes and All That Jazz   347
10.1   Asymptotically Flat Spacetimes   347
10.2   Black Holes: General Definition and Properties   355
10.3   Black Holes and Search for Gravitational Waves   362
10.4   'Black Holes' in Laboratories   370
10.5   Black Holes in Colliders?   376
10.6   Higher-Dimensional Black Holes   382
10.7   Wormholes   393
10.8   'Time Machine' Problem   399

Appendix A   Fundamental Constants and Units   408
A.1   Fundamental Constants   408
A.2   Planck Units   408
A.3   Conversion Factors   408
A.4   Various Scales of Masses   409
A.5   Milky Way Galaxy Observational Data   409
A.6   Universe Observational Data   409
A.7   Dimensionless Entropy (S/k_B) 409

Appendix B   Gauss–Codazzi Equations   410
B.1   Gauss–Codazzi Equations   410
B.2   Static Surface in a Static Spacetime   411

Appendix C   Conformal Transformations   412

Appendix D   Hidden Symmetries   414
D.1   Conformal Killing Tensor   414
D.2   Killing–Yano Tensors   414
D.3   Primary Killing Vector   417
D.4   Properties of the Primary Killing Vector   418
D.5   Secondary Killing Vector   418
D.6   Darboux Basis   420
D.7   Canonical Form of Metric   422
D.8   Separation of Variables in Canonical Coordinates   425
D.9   Higher-Dimensional Generalizations   425
D.10   Higher-Dimensional Kerr-NUT-(A)dS Metric   428

Appendix E   Boundary Term for the Einstein–Hilbert Action   431
E.1   An Example Illustrating the Problem   431
E.2   Boundary Term for the Einstein–Hilbert Action   432
E.3   Boundary Term for the Euclidean Einstein–Hilbert Action   434

Appendix F   Quantum Fields   435
   F.1   Classical Oscillator   435
   F.2   Quantum Oscillator   437
   F.3   Quantum Field in Flat Spacetime   442
   F.4   Quantum Theory in (1+1)-Spacetime   454

References   463

Index   477

1

Black Holes: Big Picture

1.1 Gravity and Black Holes

1.1.1 What is a black hole?

One of the most intriguing predictions of Einstein's theory of general relativity is the existence of black holes. A *black hole* is a spacetime region where the gravitational field is so strong that no information carrying objects and signals can escape it. A black hole is formed when the size of a gravitating object of mass M becomes smaller than its gravitational radius

$$r_S = \frac{2GM}{c^2}. \tag{1.1.1}$$

The boundary surface that restricts the no-return region is called the *event horizon*.

The properties of black holes are so unusual that even Einstein himself couldn't accept the idea that these weird objects may exist in reality in spite of the fact that they were the solutions of the relativity theory.[1] The last decades have been time of a real revolution in astronomy and astrophysics. Besides ground-based facilities of fantastically increased power, several space observatories have been launched, the best known of which is the Hubble Space Telescope. Now astronomers receive information in all possible frequencies of electromagnetic radiation, which besides visible light include gamma- and X-rays, ultraviolet and infrared radiation, and radio waves. The sensitivity of the observations also greatly increased. As a result, our vision of the universe changes. Black holes were discovered and now they have become an important component of the big picture of the universe.

Now we believe that black holes are 'everywhere'. Less powerful stellar mass black holes are observed in our galaxy, the Milky Way, and its vicinity. Much more powerful supermassive black holes were discovered in the centers of active galaxies. It is plausible that they are present in the centers of most of the galaxies. The black hole with a mass of about four million solar masses in the center of our own galaxy is the closest to us supermassive black hole.

The theoretical study of black holes and their properties is the subject of *black hole physics*. This is an exciting area of research. The discovery of astrophysical black holes made the subject really breathtaking. The discovery of black holes has importance going far beyond

---

[1] In his paper published in Annals of Mathematics, Vol. 40, No. 4 (1939), Einstein studied a realistic model of a star. He came to a conclusion that " 'Schwarzschild singularities' do not exist in physical reality ... The 'Schwarzschild singularity' does not appear for the reason that matter cannot be concentrated arbitrarily". Apparently, nature happened to be prone to more extreme conditions than Einstein had expected.

astrophysics. Black holes are different from all other known astrophysical objects. What makes black holes so special and unique? Normal stars are made of matter. Black holes are regions of strong gravitational field left after the matter collapses. The matter forming a black hole after its collapse disappears in the black hole abyss. What is left behind is an empty space with an extremely strong gravitational field. The black hole vicinity is the world (realm) of gravity. According to general relativity, gravity is simply the curvature of spacetime. In this sense a black hole is 'plenty of nothing', just self-supported empty curved spacetime. By studying black holes one probes physics in a very strong gravitational field. General relativity provides us with the required tools. The 'holy grail' of black hole physics and astrophysics is to demonstrate that our understanding of spacetime properties in a strong gravitational field is correct.

This book discusses black holes and their properties. We discuss the structure of black holes and physical processes in their vicinity. As any other gravitating object a black hole affects particle motion and field propagation. At a far distance these effects are small and can be well described by corrections to the Newtonian gravity. Such corrections are determined by the global characteristics of an object, such as its mass and angular momentum, and thus they are common for black holes and, e.g. for neutron stars. In this book we focus our attention on the effects in a strong gravitational field of a black hole, which distinguish black holes from stars. The study of black holes and their properties is the subject of black hole physics. This book is an introduction to this interesting area.

There arise many natural questions related to black holes. Here are some of them:

- What is a covariant (i.e. coordinate independent) definition of a black hole?
- What are the conditions required for black hole formation?
- Do black holes exist in the universe? How many? How to observe them?
- What is the most general solution of the Einstein equations describing an isolated stationary black hole?
- Is this solution stable?
- How to prove that we really observe a black hole?
- What is the role of black holes in structure formation in the early universe?
- How do particles and light propagate in the vicinity of black holes?
- How much energy can be released by matter falling into a black hole?
- Matter falling into a black hole and carrying a net angular momentum forms an *accretion disk*. What are its properties?
- How to 'measure' the mass of a black hole?
- What happens when a black hole collides with a star or another black hole?
- If a black hole newly formed in this process is at first non-stationary, what happens with it later?
- If black holes are stable classically, what can one say about their quantum stability? Is a physical vacuum stable in the presence of a black hole?
- Can one use black holes as probes of new physics? Can they, for example, serve as 'probes of extra dimensions'?

This list of questions can be easily continued. Many of these questions have already received quite complete answers, but some of them are still open. In the process of studying black holes several fundamental results were obtained. In particular, a deep relation has been established between black holes and what are at first glance quite different areas, such as quantum theory, thermodynamics, information theory, and quantum computing. One can speak now about *black hole physics* as a well-established scientific area, with a well-defined subject, developed tools of study, and exciting astrophysical applications.

The aim of this book is to provide an introduction into black hole physics. The book contains a brief summary of both mathematical and physical subjects required for understanding the theory of black holes. We restrict ourselves by presenting the main results and describing their possible applications. We do not give more complicated and detailed material. Instead, we provide many problems, that illustrate the material and allow one to test his/her understanding. We tried to focus on those aspects of the theory that are important both for astrophysical applications and for understanding deep theoretical problems connected with black holes.

1.1.2 The weakest interaction

Black holes are an amazing prediction of the gravitational theory. What makes gravity so special and which of its properties are 'responsible' for the black hole's existence? Let us discuss the general properties of gravity and compare it with other interactions.

Gravity is the weakest of four fundamental interactions in nature. Consider, e.g., two protons at a distance comparable to their size. Their electromagnetic interaction is at least $1/137$ times weaker than their nuclear (strong) interaction. The characteristic energy of their weak interaction is about 10^{-5} of that of the strong one, while their gravitational attraction is 10^{-38} weaker. And in spite of this, the gravitational force was 'discovered' first. Newton's law, describing this force, was used as a model for the formulation of the laws of the other interactions. The gravitational interaction plays the key role in the understanding the majority of phenomena in astrophysics and cosmology, including the large-scale structure formation in the universe, the existence of galaxies, stars, and planets, not to speak of our own existence. The reasons for this are the remarkable properties of gravity, which strongly 'amplify' its action at astronomical scales.

1.1.3 Long-range interaction

The role of different interactions at low energy is quite different. The fields responsible for the weak and the strong interactions are massive. That is why the corresponding forces are short range. This means that they are important at short distances, while at distances larger than the Compton wavelength $\hbar/(mc)$ connected with the mass m of the field their force falls exponentially. As a result, these short-range forces play an important role in the structure of matter. But when we are speaking about astrophysical objects, the short-range interactions can be effectively reduced to the contribution to the corresponding equation of state. The electromagnetic and gravitational interactions are transferred by photons and gravitons, respectively. Both of these quanta are massless and the corresponding forces are long range. They fall according to the power law $\sim 1/r^2$. That is why gravity and electromagnetism play an important role in astrophysics.

1.1.4   Hierarchy problem

The relative strength of the electromagnetic and gravitational interactions is quite different. The force of the electric repulsion of two protons at the distance r is

$$F_e = \frac{e^2}{r^2}.$$ (1.1.2)

The gravitational attraction of the same protons is

$$F_g = -\frac{Gm_p^2}{r^2}.$$ (1.1.3)

It is easy to find that the ratio of these two forces does not depend on the distance r and is

$$\frac{|F_g|}{F_e} = \frac{Gm_p^2}{e^2} = 0.8 \times 10^{-36}.$$ (1.1.4)

This ratio is extremely small. The problem of why it is so small is known as a *hierarchy problem*. This problem is one of the most fundamental problems of theoretical physics.

1.1.5   Attraction of gravity

There is another important difference between gravity and electromagnetism. There exist two kinds of electric charges, positive and negative. Opposite charges attract each other, while charges with the same sign repulse. In the presence of surrounding matter and in the absence of other forces, the processes in charged systems 'try' to minimize the potential energy, that is to reduce the net electric charge. As a result of this screening the net charge of macroscopic objects usually is not large.

For gravity the situation is quite different. For gravitating bodies the gravitational charges, i.e. their masses, always have the same 'sign', and they attract one another. This property is directly connected with the fact that gravitons have spin 2, while the photons, responsible for the electromagnetic interaction, have spin 1. The more massive a body is the more stable it is against disintegration. There exists a general mechanism of self-amplification of gravity: a massive body attracts surrounding matter that falls onto the body and increases its ability to attract. The gravitational forces, being extremely weak for a single elementary particle, are greatly amplified when such particles form a macroscopic body. At cosmic scales the gravitational forces can reach very large values, often becoming the leading force that determines the evolution of the system.

In other words, in astrophysics, the weakness of Newtonian coupling constant is compensated by the large value of the gravitational 'charge' (mass). The described mechanism of the self-amplification of gravity results in the domination of gravity at large scales. Under these conditions, a homogeneous distribution of matter becomes unstable and small perturbations grow, resulting in the formation of planets, stars, galaxies, and clusters of galaxies. This instability is known as the *Jeans instability*. The size of the compact astrophysical objects (stars) is determined by the condition of the equilibrium between gravitation and pressure of the matter. If the pressure and rotation are small, self-gravitating bodies collapse and form a black hole.

1.1.6 Universality

Gravitational interaction has one more, very important, distinguishing property: it is universal. For any other interaction there exist neutral (with respect to this interaction) particles. For gravity this is not so. The total energy of a system plays the role of the gravitational charge. Even particles with zero rest mass, like photons, interact gravitationally. Everything that exists in nature has energy, and hence 'participates' in the gravitational interaction. Gravity is a *universal interaction.*

The universality of gravity has another important aspect. For any other interaction one can use a neutral (with respect to this interaction) particle as a preferred *reference frame.* For example, charged particles in an electric field are moving differently from the neutral one. The relative acceleration of charged particles with respect to the neutral particle frame determines the strength of the field. This strength is a gradient of the field potential. For gravity the situation is quite different. There are no gravitationally 'neutral' particles, that is particles that are not affected by the gravitational field. Moreover, because the inertial mass m_i and the gravitational mass m_g are equal, the acceleration of all freely moving particles in the gravitational field is the same. If we choose one of these particles as an origin of the reference frame, the other particles, close to it, are moving without relative acceleration.

This property is valid locally. For an extended object, like a cloud of dust particles, spatially separated parts of the object have non-vanishing relative acceleration, proportional to the separation distance. This is a so-called *tidal force*, which can be used as a measure of the strength of the gravitational field. It is described by the second derivatives of the gravitational potential. In general relativity the object 'responsible' for this property of the gravitational field is the *spacetime curvature.*

1.1.7 Equivalence principle

The *equivalence principle*, which is one of the basic principles of general relativity, gives a concrete description of the universality of gravity. Consider a particle in a gravitational field. The motion of the particle depends on its initial position and velocity, but it is independent of other factors, such as its mass and its structure. It should be emphasized that this simple principle has far-reaching consequences. The reason is the following. If a particle has an inner structure, then its mass depends on it. For example, two atoms of different chemical elements have different contributions of electromagnetic and other interactions between their constituents to the rest mass of the nuclei. If they move in the gravitational field in exactly the same manner, the ratio of the contributions of different forces to the inertial and gravitational mass must be the same. This statement, known as the *Einstein principle of equivalence*, in fact means the universality of the gravitational interaction.

For a compact object its own gravitational field also modifies the mass of the object, as any other interaction. The *strong equivalence principle* suggests that the Einstein principle of equivalence is valid for such objects as well. Namely, *the outcome of any local (gravitational or non-gravitational) experiment in a freely falling laboratory is independent of the velocity of the laboratory and its location in spacetime.* This principle is satisfied in general relativity.

For a compact object its ability to attract depends on its mass. But this mass depends also on the gravitational interaction of the body constituents. As a result, the gravitational equations are essentially non-linear.

1.1.8   Geometrization

The equivalence principle makes it possible to 'geometrize' the theory of gravity. The dynamical problem of solving the equation of motion for a particle in a gravitational field can be reduced to a 'pure' geometric problem, finding a geodesic in the properly chosen metric describing the gravitational field. The Einstein equations allow one to find the corresponding metric for a given distribution of the matter. In the weak-field regime for non-relativistic sources the Einstein equations correctly reproduce the results of Newtonian gravity, and the geodesic equation automatically reproduces the Newtonian equation for particle motion. The Einstein theory also uniquely determines the corrections to this theory. Such lower-order corrections have been experimentally tested.

1.1.9   Gravity and causal structure of spacetime

Consider a 'box', inertially moving in a gravitational field, such that its interior is isolated from an external 'world'. Suppose an observer in such a 'laboratory' performs different experiments. If the size of the 'box' l is much smaller than the characteristic curvature radius L, the tidal effects are suppressed by the factor $(l/L)^2$. Thus, free particles in such a 'box' move practically without acceleration relative to one another. Local experiments with free particles, performed by the observer in such a 'laboratory' will give the same results as experiments performed by a similar observer in a similar box that is moving inertially in a flat spacetime. The *Einstein equivalence principle* implies that this conclusion is valid not only for particles, but for any other experiment testing the physical laws. In mathematical language it means that the metric of the gravitational field $g_{\mu\nu}$ can be locally reduced to the *Minkowski metric* $\eta_{\mu\nu}$ by coordinate transformations.

Choose a point O as an origin in the Minkowski spacetime, and consider straight lines passing through it. We say that a line with a tangent vector u^μ is time-like, null, or space-like, if $\eta_{\mu\nu}u^\mu u^\nu$ is negative, zero, or positive, respectively.[2] Time-like lines represent the motion of massive particles, while null lines describe the motion of massless particles, such as photons. Null lines emitted from the origin O form a *null cone*. Physical fields in a flat spacetime are described by hyperbolic partial differential equations. Their characteristics are null surfaces generated by null lines. In particular, this implies that a perturbation of the field at O always propagates within the null cone with the apex at O. This property is known as the *local causality* of the physical theory.

In the presence of the gravitational field $g_{\mu\nu}$ one can define local null cones. If dx^μ is a displacement from an original point x^μ then a *local null cone* at this point is defined by the condition $g_{\mu\nu}dx^\mu dx^\nu = 0$. According to the equivalence principle the field perturbations propagate within the local null cones. But there is a big difference between general relativity and physics in the Minkowski spacetime. Namely, in the Minkowski spacetime all local null cones can be obtained by a parallel transport from one another, so that a set of local null cones forms a 'rigid structure'. In a curved spacetime such a rigid structure is absent. As a result, the *global causal structure* of the spacetime in general relativity can be complicated. The global causal structure is determined by the metric of the spacetime. The metric is a solution of the Einstein equations and it depends on the matter distribution. Formation of a black hole in

---

[2]In the sign convention adopted in this book the metric has signature $(-,+,\ldots,+)$.

gravitational collapse is an example of a situation when the causal structure of the spacetime is non-trivial.

1.1.10 Local vs. non-local effects

When the gravitational field is weak one can use Newtonian theory to describe it quite accurately. General relativity is required when the gravity is strong. But what does 'a strong gravitational field' mean? In what sense is the gravity of a black hole strong?

Does it mean, for example, that the spacetime curvature, which characterizes the strength of the tidal forces, is large? Not necessarily. The curvature is proportional to the second derivatives of the gravitational potential: i.e. at the surface of the black hole it is of the order of

$$\mathcal{R} \sim r_S^{-2}. \tag{1.1.5}$$

So, for large black holes the curvature can be very small. This conclusion, based on the Newtonian approximation, is valid for general relativity as well. It means that an object of the size l and mass m freely falling into the black hole 'feels' at its surface a tidal force

$$F \sim \frac{lm}{r_S^2}. \tag{1.1.6}$$

This force is trying to stretch the object in the radial direction and to compress it in the transverse directions. The larger the mass of the black hole, the smaller the curvature at its surface. For very large masses the *local* curvature effects in the black hole exterior become negligibly small.

> *Consider an extended object falling in the gravitational field of a body of mass M. Two parts of the object separated by the distance l move with slightly different accelerations. When the body is at the radius r, the relative acceleration of its two parts is a $\sim$ GM l/r³. This effect is caled the tidal effect, and the corresponding force is called* tidal force. *Let us compare tidal forces on the surface of the Earth and on the surface of a supermassive black hole. The ratio of these forces is*
>
> $$\frac{F_\oplus}{F_{BH}} \sim \frac{M_\oplus}{M_{BH}} \left(\frac{r_S(M_{BH})}{r_\oplus} \right)^3. \tag{1.1.7}$$
>
> *Here, $M_\oplus$ and $r_\oplus$ are the mass of the Earth and its radius, and M_{BH} and $r_S(M_{BH})$ are the mass of the black hole and its gravitational radius. This ratio is equal to one when*
>
> $$r_S(M_{BH}) \sim \sqrt{\frac{r_\oplus^3}{r_S(M_\oplus)}}. \tag{1.1.8}$$
>
> *Thus, these tidal forces are equal if the black hole mass is about $6 \times 10^7 M_\odot$. It shows that for an observer, who is freely falling onto the supermassive black hole, the tidal effects at the moment when he/she crosses its horizon are rather small.*

However, even when the curvature in the black hole exterior is small, non-linearity of the Einstein equations is important for black hole solutions. In this sense the gravity near the black hole is strong. It is also easy to show that the difference $\Delta\Phi$ of the gravitational potentials between infinity and the gravitational radius, calculated in Newtonian gravity is of the order of c^2. Under these conditions one can expect that the global properties of such a compact object must differ from its Newtonian description. It is really so. For example, let us consider a test object of mass m in the gravitational field of a black hole. When it is at rest at infinity (initial state), its proper energy is $E_\infty = mc^2$. The energy of this body at rest at a finite radius r, calculated in general relativity, is

$$E = mc^2\sqrt{1 - r_S/r}. \tag{1.1.9}$$

In the limit, when a particle reaches the horizon, this energy vanishes. Thus, in such a hypothetical process one can extract 100% of the matter as proper energy. This certainly indicates that the field of the black hole is very strong.[3]

Modification of the expression for the energy of a particle moving in the gravitational field of the black hole, dictated by general relativity, implies that the orbital motion of particles near a black hole and light propagation in its vicinity also differs from the Newtonian case. The existence of the innermost stable circular Kepler orbit in the black hole spacetime is one of the examples. In all these cases the spacetime regions of interest are comparable to or larger than the gravitational radius. In this sense such effects are non-local.

1.2   Brief History of Black Holes

The term 'a black hole' was invented by Wheeler in 1967, although the possibility of the existence of such objects was discussed a long time before this. At the end of the eighteenth century Michell and Laplace independently came to the conclusion that if the mass of a star were large enough its gravity would not allow light to escape. Though this conclusion was based on the Newtonian theory the obtained result for the size of such 'dark stars' (the *gravitational radius*) coincides with the later prediction of Einstein's theory of gravity. (see, e.g., Barrow and Silk 1983; Israel 1987; Novikov 1990).

In 1916, within a year after general relativity had been developed, Schwarzschild obtained the first exact (spherically symmetric) solution of the *Einstein equations* in vacuum. This solution describes the gravitational field of spherical compact objects. In addition to a singularity at the center of symmetry (at $r = 0$), this solution had a singularity on the gravitational-radius surface (at $r = r_S$). Very soon, it was understood that the latter singularity is quite different from $r = 0$ – singularity. The nature of this Schwarzschild singularity was a mystery for many years.[4]

Many scientists contributed to the solution of this problem, see, e.g., (Flamm 1916; Weyl 1917; Eddington 1924; Lemaître 1933; Einstein and Rosen 1935), before the final complete solution was obtained (Synge 1950; Finkelstein 1958; Fronsdal 1959; Kruskal 1960; Szekeres 1960; Novikov 1963, 1964). The main lesson obtained during this study is that the spacetime manifold with the metric, representing the gravitational field, may have global properties, that

---

[3]It should be emphasized that this extreme efficiency is a global property of the black hole geometry. We shall see that it is directly related to the existence of the black hole horizon, which is also defined non-locally.

[4]In the famous paper by Einstein and Rosen the authors discussed global aspects of the Schwarzschild solution, and in particular, what was called later its *Einstein–Rosen bridge*. This paper contains the following discussion: "In this case it will not be possible to describe the whole field by means of a single coordinate system without introducing singularities. The simplest procedure appears to be to chose coordinate systems in the following way: (1) One coordinate system to describe one of the congruent sheets. With respect to this system the field will appear to be singular at every bridge. (2) One coordinate system for every bridge, to provide a description of the field at the bridge and in the neighborhood of the latter, which is free from singularities.

Between the coordinates of the sheet system and those of each bridge system there must exist outside of the hypersurfaces $g = 0$, a regular coordinate transformation with non-vanishing determinant."

differ from the Minkowski spacetime of special relativity. In the case of a black hole such global properties are connected with its topology and the causal structure.

Though a solution for a (non-rotating) black hole has been known for quite a long time, there was the general belief that nature could not admit a body with a size that would be comparable to its gravitational radius; this viewpoint was shared by the creator of general relativity himself (see footnote[1], see also Israel 1987) and references therein). Some interest in the properties of very compact gravitational systems was stimulated in the 1939s after Chandrasekhar's (1931) work on white dwarfs and the works of Landau (1932), Baade and Zwicky (1934), and Oppenheimer and Volkoff (1939) who showed that neutron stars are possible, with a radius only a few times that of the gravitational radius. The gravitational collapse of a massive star that produces a black hole was first described by Oppenheimer and Snyder (1939). This description[5] has remained practically unchanged till now.

For a long period of time black holes were considered as very exotic objects. The names 'frozen' or 'collapsed' stars had been used by specialists until the end of the 1960s for the description of these objects. These terms reflect the property of a black hole to serve as a 'grave of matter'. In 1963 Kerr discovered a solution of the Einstein equation, describing the gravitational field of a stationary rotating black hole. This solution also has the gravitational radius describing a position of the *event horizon*. The Kerr solution also has a real singularity, but if the angular momentum $J \leq J_* = GM^2/c$ then the latter is surrounded by the horizon and hence is invisible to an external observer.

A main new feature of the Kerr solution is the *dragging-into-rotation effect*, induced by the rotation of the black hole. This effect differs general relativity from the Newtonian theory of gravity. Near the horizon of a rotating black hole this dragging effect becomes so strong that any matter in this domain is necessarily corotating with the black hole. This domain is called an *ergosphere*. It surrounds the black hole horizon. The angular momentum J of the black hole cannot exceed the critical value $J_* = GM^2/c$. In the opposite case a solution describes a naked singularity. Later, it was proved that in the absence of matter the Kerr solution is the most general solution for stationary black holes.

The Kerr metric is quite complicated. It is stationary and axisymmetric. If one studies matter fields in this background these symmetries are not sufficient for the separation of variables in the Hamilton–Jacobi and field equations. It was quite a big surprise when Carter (1968) discovered a new type of integrals of motion in the Kerr metric, connected with its *hidden symmetries*. This discovery made it possible to use the separation of variables method to study particle motion and field propagation in the Kerr spacetime. Another important 'breakthrough' of the end of 1960s – beginning of the 1970s was application of a global geometrical approach to the black hole theory, which not only allowed one to give a covariant definition of a black hole but also to prove several fundamental theorems. The now-classic theorems stating that 'black holes have no hair' (that is, no external individual attributes except mass, angular momentum, and charge), that a black hole contains a singularity inside it, and that the black hole area cannot decrease were proved during this period. These and other

---

[5]They wrote: "When all thermonuclear sources of energy are exhausted a sufficiently heavy star will collapse. Unless fission due to rotation, the radiation of mass, or the blowing off of mass by radiation, reduce the star's mass to the order of that of the sun, this contraction will continue indefinitely. ... The total time of collapse for an observer comoving with the stellar matter is finite, and for this idealized case and typical stellar masses, of the order of a day; an external observer sees the star asymptotically shrinking to its gravitational radius."

results made it possible to construct a qualitative picture of the formation of a black hole, to describe its possible further evolution and its interaction with matter and other classical physical fields. Many of these results were summarized in the well-known monographs of Zel'dovich and Novikov (1971a), Misner *et al.* (1973), Hawking and Ellis (1973), Thorne *et al.* (1986), and Frolov and Novikov (1998).

In 1967 in his public lecture J.A. Wheeler coined the name 'black hole'. Soon after that this name was adopted enthusiastically by everybody. It reflects very picturesquely the remarkable properties of the object.

After pulsars (neutron stars) were discovered at the end of the 1960s, astrophysicists had to examine the prospects for the observational detection of black holes. Certainly, one cannot see what happens inside the gravitational radius. But the gravitational field of black holes is so strong and has several important distinguishing features that one can detect their existence by observing matter and fields propagation in their vicinity. For example, one can try to use lensing effect to identify a black hole by its effect on the light propagating in its vicinity. But the typical stellar mass black holes have such small gravitational radius (of order of 10 km) that the direct observation of black holes at the astronomical distances by this method is practically impossible. Nevertheless, black holes were discovered. What made it possible?

New astronomical techniques had to be invented to search for black holes. Luckily this was also a period of dramatic innovation in astrophysics; observational discoveries and theoretical interpretations together produced evidence for black holes and other compact stars. In 1962, Giacconi *et al.* (1962) identified the first X- ray source outside our solar system, named Sco X-1. Such a strong X-ray source (1000s of times stronger than expected) required a new and powerful method of generating X-rays. The next year, Maarten Schmidt discovered that the radio source 3C 273, associated with an optical source of unusual spectrum, was receding from us at high velocities (16% of light speed), and thus was located 2 billion light-years away. Such a distance required that it be 100s of times more luminous than any galaxy so far known. Schmidt termed 3C 273 a 'quasistellar' object or quasar.

New ideas for powering these objects were required. Simultaneously and independently, Salpeter (1964) and Zeldovich (1964) realized that massive objects such as black holes can liberate huge quantities of energy by shocks within gas falling into them, perhaps explaining the energetics of quasars. Lynden-Bell (1969) showed that the gas would form an *accretion disk* around a central massive object, which could explain the observed spectra of quasars. Guseinov and Zel'dovich (1966) showed that collapsed stars in binaries could pull gas from their companion stars that would liberate energy in a shock with temperatures of millions of degrees, producing an X-ray source. This idea was given weight by Sandage *et al.*'s (1966) discovery that Sco X-1 was associated with a blue star. Shklovsky (1967) produced a detailed theory to explain Sco X-1's radiation as a neutron star accreting gas from its companion.

The existence of neutron stars was proven by Hewish *et al.*'s (1968) discovery of pulsed radio emission, and some X-ray sources were proven to be neutron stars through the discovery of pulsations in their X-rays (Schreier *et al.* 1972). However, the non-detection of pulsations does not prove the existence of a black hole; more subtle methods were required to identify them. Zel'dovich and Novikov (1971a) showed that black holes do not themselves radiate in any special way, and that identifying an X-ray source as a black hole would require measurement of the mass of the compact object through observations of the velocity changes of the companion star. Such measurements of the companion star in the X-ray binary Cygnus

X-1 soon gave strong evidence for the existence of a black hole (Webster and Murdin 1972; Bolton 1972).

The existence of stellar mass black holes was predicted by the theory. The discovery of the other types of black holes of much greater mass (up to 10^6–10^9 solar masses) was quite unexpected. At first, such supermassive black holes were identified with the central engine of quasars. Now, such black holes are expected to exist in the centers of many (or even most of the) galaxies. After these discoveries the black hole paradigm became an important element of modern astrophysics. The gravitational field of a black hole is so strong that it may transform a considerable part of the proper mass of the in-falling matter into radiation. This is the next most powerful mechanism of the energy extraction from matter, which concedes only to the process of annihilation of matter and anti-matter. Naturally, black hole models are often used to explain powerful radiation produced by very compact objects. Another dramatic change of the astrophysical paradigm is the understanding that supermassive black holes in the center of galaxies may have a strong influence on their properties, and, for example, became important at the early stage of galaxy evolution and even during its first stages of formation.

The sensational 'news' of the possible discovery of a black hole in an X-ray binary (*Cygnus X-1*) had scarcely died down when a new unexpected result obtained by Hawking (1974, 1975) again focused physicists' attention on black holes. It was found that as a result of the instability of the vacuum in the strong gravitational field of a black hole, these objects are sources of quantum radiation. The most intriguing property of this radiation is that it happened to have a thermal spectrum. In other words, if one neglects the scattering of the radiation by the external gravitational field, a black hole radiates exactly like a heated black body. If the black hole mass is small (smaller than 10^{15} g), it decays over a time shorter than the age of the universe. Such small black holes, now called primordial black holes, may have been formed only at a very early stage of the Universe's evolution (Zel'dovich and Novikov 1967, 1971a; Hawking 1971). The discovery of primordial black holes or of their decay products would supply valuable information on the physical processes occurring in the universe at that period.

Hawking's discovery stimulated a large number of papers that analyzed specific features of quantum effects in black holes. In addition to a detailed description of the effects due to the creation of real particles escaping to infinity, substantial progress has been achieved in the understanding of the effect of vacuum polarization in the vicinity of a black hole. This effect is important for the construction of a complete quantum description of an 'evaporating' black hole.[6]

More recently, another aspect of black hole physics became very important for astrophysical applications. The collision of a black hole with a neutron star or coalescence of a pair of black holes in binary systems is a powerful source of gravitational radiation that might be strong enough to reach the Earth and be observed in a new generation of gravitational-wave experiments (LIGO, LISA, and others). The detection of gravitational waves from these sources requires a detailed description of the gravitational field of a black hole during the collision. In principle, gravitational astronomy opens up remarkable opportunities to test

---

[6]There were a number of review articles written in the 1970s and early 1980s that had summarized the main results obtained during this 'heroic' period of black hole physics. Here are references to some of them: (Penrose 1972; Carter 1973, 1976; Markov 1970, 1974; DeWitt 1975; Sciama 1976; Frolov 1976; Bekenstein 1980; Israel 1983; Dymnikova 1986). See also a remarkable review article by Israel (1987) that describes the history of evolution of the black hole idea.

gravitational field theory in the limit of very strong gravitational fields. In order to be able to do this, besides the construction of the gravitational antennas, it is also necessary to obtain the solution of the gravitational equations describing this type of situation. Until now, there exist no analytical tools that allow this to be done. Under these conditions one of the important tasks is to study colliding black holes numerically.

A half a century ago black holes were considered as highly exotic objects, and the general attitude in the wider physical and astrophysical communities (i.e. among the scientists who were not working on this subject) to these objects was quite cautious. Now the situation has changed drastically. It happened both because of numerous new astrophysical data and the considerable development of the theory.

1.3 'Dark Stars' vs. Black Holes

1.3.1 Newtonian potential

Before we start discussion of black holes (our main subject), let us make some remarks concerning properties of compact gravitating objects in Newtonian gravity. The *Newtonian potential* φ for a mass distribution μ obeys the *Poisson equation*

$$\Box \varphi = 4\pi G \rho. \tag{1.3.1}$$

Here, G is the *Newtonian coupling constant*. This equation is similar to the equation for the *electrostatic potential*, with the only difference being the sign in the right-hand side. This sign difference reflects the fact that gravitational force is attractive, while the electrostatic force between two equal charges is repulsive. For a spherical distribution of matter we have

$$\frac{1}{r^2}(r^2\varphi')' = 4\pi G\rho, \qquad (\ldots)' = d(\ldots)/dr. \tag{1.3.2}$$

Integrating Eq. (1.3.2) over r from 0 to some radius r we get

$$\varphi' = \frac{GM(r)}{r^2}, \qquad M(r) = 4\pi \int_0^r dr r^2 \rho. \tag{1.3.3}$$

Here, $M(r)$ is the mass within the sphere of radius r. Choosing $\varphi(\infty) = 0$ and assuming that beyond some radius $r > r_0$ the mass M is constant we have

$$\varphi = -\frac{GM}{r}, \text{ for } r > r_0. \tag{1.3.4}$$

We define the gravitational radius by the relation

$$r_S = \frac{2GM}{c^2}. \tag{1.3.5}$$

Notice that at the gravitational radius one has $\varphi/c^2 = -1/2$.

The same potential written in the *Cartesian coordinates* (X_1, X_2, X_3) is

$$\varphi = -\frac{GM}{\sqrt{X_1^2 + X_2^2 + X_3^2}}. \tag{1.3.6}$$

Gravitational force strength, F_i is determined as $F_i = -\varphi_{,i}$. Here, $(\ldots)_{,i} = \partial(\ldots)/\partial X_i$.

Consider two closely located points $\mathbf{r}$ and $\mathbf{r} + \Delta\mathbf{r}$. The difference of gravitational forces acting at these points on a unit mass is

$$\Delta F_i = -\varphi_{,i}(\mathbf{r} + \Delta\mathbf{r}) + \varphi_{,i}(\mathbf{r}) = -\sum_j \varphi_{,ij}\Delta r_j. \tag{1.3.7}$$

The quantity $\varphi_{,ij}$ determines *tidal forces* acting on an extended body. It is called a *tensor of tidal forces*.

Problem 1.1: *Using expression Eq. (1.3.6) calculate the tensor of tidal forces at a point $(X = Y = 0, Z = r)$. Estimate an absolute value of the tidal force at the gravitational radius Eq. (1.3.5) for 3 different masses: (a) $10M_\odot$; (b) $10^6 M_\odot$; and (c) $10^9 M_\odot$.*

1.3.2   Dark stars

In Newtonian theory there exists an analog of a black hole, that we call a *dark star*. A 'dark star' by definition is a star that does not radiate its light to infinity because of its strong gravitational field. Let M be the mass of such a star, what is its 'gravitational radius'? In Newtonian gravity the escape velocity v from a surface of an object of radius r is determined by the equation

$$\frac{GMm}{r} = \frac{mv^2}{2}. \tag{1.3.8}$$

The mass of the particle m, which enters both sides of this equation, can be cancelled. This equation shows that if a particle emitted from the surface r of the body has a velocity less than or equal to v it cannot escape to infinity if $r < r_v$, where

$$r_v = \frac{2GM}{v^2}. \tag{1.3.9}$$

If the speed of light c is the 'limiting velocity' then the escape velocity cannot exceed it. If $r \le r_S$

$$r_S = \frac{2GM}{c^2}, \tag{1.3.10}$$

then no light or matter can escape from the surface of the radius r_S to infinity. Such a star would be dark.[7]

For the Sun mass $M_\odot = 2 \times 10^{33}$ g the gravitational radius of a dark star is about 3 km.[8] For an arbitrary mass M we have

$$r_S \simeq 3 \left(\frac{M}{M_\odot}\right) \text{ km}. \tag{1.3.11}$$

[7]Certainly this 'derivation' is not accurate. It contains two 'mistakes'. First, for the object of the radius r_S the gravitational potential is $\varphi \sim c^2$, and one must use general relativity instead of the Newtonian theory. Secondly, the light quanta are ultrarelativistic 'particles', and for their energy one must use the relativistic formula. By neglecting these two factors, one nevertheless not only arrives at the qualitatively correct conclusion, but 'miraculously' obtains the exact value for the gravitational radius, which follows from general relativity.

[8]The numerical value of the Newtonian coupling constant G, the speed of light c, as well as other useful constants can be found in Appendix A.

Problem 1.2: *Calculate the gravitational radius of black holes with the masses: (a) $10M_\odot$; (b) $10^6 M_\odot$; (c) $10^9 M_\odot$. Compare the latter radius with the Earth–Sun distance $1 A.U. = 1.5 \times 10^{13} cm$.*

Problem 1.3: *What are the masses of the objects with the gravitational radii equal to: (a) the nuclear size $\sim 10^{-13}$ cm; (b) 1 cm; (c) 1 km; (d) 1 A.U? What are the effective "mass densities" M/r_S^3 of such objects?*

1.3.3 Black holes vs. dark stars: Similarities and differences

By definition, a *black hole* is a spacetime region from where no information-carrying signals can escape to infinity. The words 'information-carrying' means signals that propagate with a velocity smaller or equal to the speed of light.

Are 'dark stars' black holes? No, they are not. The big difference between the Newtonian theory and general relativity is that in the former there exists an *information-mining process*, which allows one to obtain information from the surface of a 'dark star' and even from the regions located deep inside it.

Consider the following *chain process*. Let a particle move with the velocity c ('photon') radially in the gravitational field of mass M. If it is emitted at the surface of the radius r_0 it can reach the radius r_1 determined by the relation[9]

$$-\frac{GM}{r_1} = -\frac{GM}{r_0} + \frac{c^2}{2}, \tag{1.3.12}$$

or, equivalently,

$$\frac{1}{r_1} = \frac{1}{r_0} - \frac{1}{r_S}. \tag{1.3.13}$$

If $r_0 < r_S$ the photon does not reach infinity but it can be used to send information from r_0 to points up to the radius r_1. One can use another 'photon' to transfer the information further to larger values of r up to the radius r_2. It is easy to see that after n steps the information reaches the radius r_n, obeying the relation

$$\frac{1}{r_n} = \frac{1}{r_0} - \frac{n}{r_S}. \tag{1.3.14}$$

The condition that after n steps the light reaches infinity is $n \geq r_S/r_0$. Denote by n_0 the minimal value of n when it happens and by $[A]$ an integer part of A, then

$$n_0 = [r_S/r_0]. \tag{1.3.15}$$

This result shows that after a finite number of steps (equal to n_0) the information from the surface of radius r_0 can be transmitted to infinity.

In general relativity such a chain process is forbidden. In the presence of a gravitational field the geometry of spacetime is modified. As a result, gravity affects the structure of the local null cones. Light emitted inside the gravitational radius moves to the smaller value of the radius r from the very beginning (see Figure 1.1). Later, we discuss this question in detail.

---

[9]Once again we use here the 'wrong' non-relativistic equations.

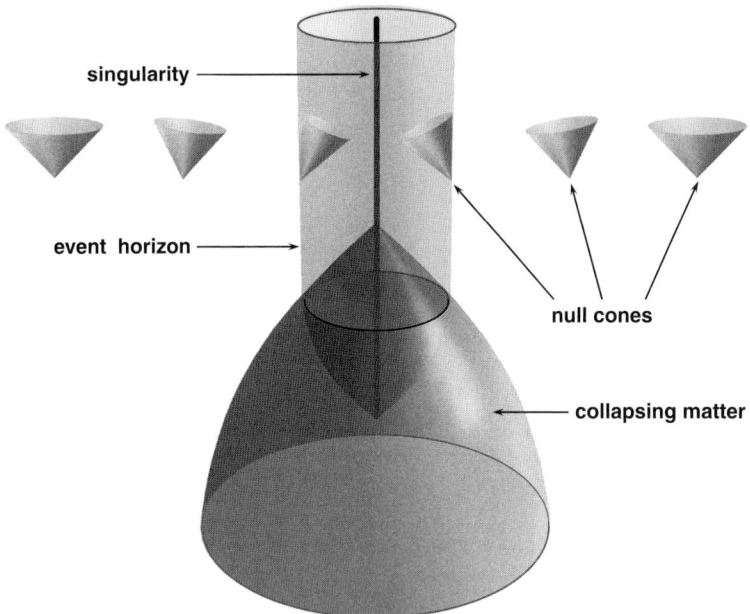

singularity

event horizon

null cones

collapsing matter

Fig. 1.1 Schematic picture illustrating black hole formation. Time goes in the vertical direction. Collapse of the matter results in the creation of the event horizon. It is formed some time before the surface of the collapsing body crosses the gravitational radius. Soon after the formation the horizon becomes stationary. Future directed local null cones are shown. Inside the horizon these cones are strongly tilted, so that motion with a velocity less than or equal to the speed of light brings a particle closer to the singularity.

1.4   Final State of Stellar Evolution

1.4.1   What is a 'fate' of stars?

The first theoretical prediction of the existence of astrophysical black holes was made for the *stellar mass black holes*. Namely, for these objects we understand better the mechanism of their formation. Indeed, if one asks what might be a good candidate for a black hole, the natural place to look is the world of 'dead stars'. During the long phase of its active life a star is in a state of equilibrium. The pressure of the hot matter balances the gravitational attraction. The condition of the equilibrium determines the main parameters of the star.

A natural question is what happens after the nuclear reactions, supplying the energy keeping the star hot, end. When a star cools down unbalanced gravity forces it to shrink. This process (the *gravitational collapse*) may be stopped if at some stage (i.e. at some density of the matter) the equation of state becomes rigid enough. The theory predicts three types of final configurations of the star evolution: (i) white dwarfs, (ii) neutron stars; and (iii) black holes. White dwarfs and neutron stars are formed when the mass of a collapsing star core is less than some critical value. For the higher mass of the collapsing star core a black hole is formed. In order to explain this conclusion let us consider the properties of matter at high densities.

1.4.2 Matter at high density

In Nature there exist two kind of particles, bosons and fermions. Their properties at low temperature are quite different. Let us assume that there is no interaction between particles. Then, at zero temperature all bosons are at the same lowest-energy state. They form a *Bose–Einstein condensate*.

The behavior of fermions is quite different. Because of the Pauli principle any two Fermi particles must be in a different quantum state. Electrons, neutrons, protons are the Fermi particles with spin 1/2. For the given momentum there may be only two such identical particles with opposite spins. As a result, even at zero temperature fermions are moving and have momenta up to the *Fermi momentum* p_F. Motion of fermions produces non-vanishing pressure. For a 'cold' gravitating star[10] this is the main force that makes possible their equilibrium. For this reason the problem of finding a final state of a star is greatly simplified and reduces to the study of stable gravitating configurations of a system of the *degenerate Fermi gas*.

In the case of stars there are two realistic options. First, when there exists a degenerate gas of electrons, and secondly, when this is a gas of neutrons. An equilibrium state of a 'cold' star in the former case is a *white dwarf*, while in the latter case it is a *neutron star*. For each of these two configurations there exists an upper limit of mass. Above this critical mass the equilibrium becomes impossible. For the white dwarfs this critical limit is known as the *Chandrasekhar mass* and it is about $1.4M_\odot$. For neutron stars this limit is about $3M_\odot$ or slightly higher. For higher mass, the core of a star at the final state of its evolution forms a *black hole*. In this section we discuss briefly the problem of equilibrium of 'cold' stars.

When the matter density is very high and volume per unit atom is less than the atomic volume, then atoms are squeezed so much that electrons and nuclei are mixed together in one 'hot soup'. Denote by Z a number of nuclei per unit electron. Denote by n_e the number density of electrons. If

$$n_e \gg \left(\frac{m_e e^2}{\hbar^2} \right)^3 Z^2, \qquad (1.4.1)$$

then the electromagnetic interaction of electrons and nuclei is not important. It happens when $\rho \gg 20Z^2$ g/cm^3. For even higher density, the Fermi energy of electrons becomes larger than the mass difference of a neutron and a proton

$$\Delta m = m_n - (m_p + m_e) \approx 1.4 \cdot 10^{-27} \text{ g}, \quad \Delta m\,c^2 \approx 7.8 \cdot 10^5 \text{ eV} \approx 9.1 \cdot 10^9 \text{ K} \qquad (1.4.2)$$

If the Fermi energy of electrons becomes greater than Δmc^2 then the matter becomes 'neutronized'. When $\rho \sim 3 \times 10^{11}$ g/cm^3 neutrons dominate over electrons. At the density $\rho \sim 10^{12}$ g/cm^3 their pressure dominates as well.

At the final state of stellar evolutions the pressure of the 'cold' degenerate Fermi gas of electrons and neutrons plays the main role in supporting a star's equilibrium.

---

[10]In fact 'cold' means only that the temperature of the object Θ is much less than the Fermi energy.

1.4.3 Degenerate Fermi gas

Let us derive the equation of state of the *degenerate Fermi gas*. Let V be a space volume, then the number of fermions with momentum from p to $p + dp$ is

$$dN = \frac{\text{(number of spin states)} \times \text{(phase space volume)}}{\text{(phase space volume per one state)}}$$

$$= 2\frac{4\pi p^2\, dp\, V}{(2\pi\hbar)^3} = V\frac{p^2\, dp}{\pi^2\hbar^3}. \tag{1.4.3}$$

We consider fermions with spin $1/2$ and include a factor 2, number of spin states. Integrating over p we obtain a total number of fermions

$$N = \frac{V}{\pi^2\hbar^3}\int_0^{p_F} p^2\, dp = \frac{Vp_F^3}{3\pi^2\hbar^3}. \tag{1.4.4}$$

Since $N/V = n$ is the density of fermions, we have

$$p_F = (3\pi^2)^{1/3}n^{1/3}\hbar, \qquad n = \frac{p_F^3}{3\pi^2\hbar^3}. \tag{1.4.5}$$

The energy of the Fermi gas is

$$\mathcal{E} = \frac{V}{\pi^2\hbar^3}\int_0^{p_F} \varepsilon(p)\, p^2\, dp, \tag{1.4.6}$$

where

$$\varepsilon(p) = c\sqrt{m^2c^2 + p^2} \tag{1.4.7}$$

is the energy of a particle with momentum p.

There are two different regimes. The Fermi gas is non-relativistic when $p_F \ll mc$, and it is relativistic when $p_F \gg mc$. A transition between these two regimes occurs at $p_F \sim mc$. Using expression (1.4.5) we can find that the corresponding density of particles, when the change in regimes occurs, is

$$n_t \sim \frac{1}{3\pi^2}\frac{1}{\lambda_m^3}. \tag{1.4.8}$$

Here, $\lambda_m = \hbar/(mc)$ is the *Compton wavelength* of the Fermi particle. For electrons and neutrons one has, respectively,

$$\lambda_e = 3.86 \times 10^{-11}\text{ cm}, \qquad n_t^{(e)} \sim 5.86 \times 10^{29}\ \frac{1}{\text{cm}^3}, \tag{1.4.9}$$

$$\lambda_n = \frac{\hbar}{m_nc} = 2.1 \times 10^{-14}\text{ cm}, \qquad n_t^{(n)} \sim 3.6 \times 10^{39}\ \frac{1}{\text{cm}^3}.$$

The mass density of matter at this transition is $\rho_t = m'n_t$, where m' is the mass per single Fermi particle. One has approximately

$$\rho_t^{(e)} \sim 10^6 \, \frac{g}{cm^3}, \qquad \rho_t^{(n)} \sim 6 \times 10^{15} \, \frac{g}{cm^3}. \qquad (1.4.10)$$

The neutronization starts when for the electron gas $p_F \sim \Delta m \, c$. The corresponding 'Compton wavelength' and density $n_{\Delta m}$ are

$$\lambda_{\Delta m} = \frac{\hbar}{\Delta mc} \sim 0.24 \times 10^{-10} \, cm, \quad n_{\Delta m} = (3\pi^2 \lambda_{\Delta m}^3)^{-1} \sim 0.25 \times 10^{31} \, \frac{1}{cm^3}. \quad (1.4.11)$$

Hence, the electron gas becomes relativistic before neutronization starts.

Let us summarize. The electron gas becomes relativistic at the density of matter $\rho \sim 10^6$ g/cm^3. This gas is degenerate if $\Theta < \Theta_e \sim 6 \times 10^9$ K. The corresponding density and temperature for the neutron gas are $\rho \sim 6 \times 10^{15}$ g/cm^3 and $\Theta_n \sim 10^{13}$ K, respectively.

1.4.4   Non-relativistic gas

Let us derive and compare the equations of state of the Fermi gas for two limiting cases, of non-relativistic and relativistic particles. For non-relativistic particles $\varepsilon(p) = mc^2 + p^2/(2m)$. Hence,

$$\mathcal{E} = E_0 + E, \qquad E_0 = Nmc^2,$$

$$E = \frac{Vp_F^5}{10m\pi^2\hbar^3} = \frac{3(3\pi^2)^{2/3}\hbar^2}{10m} \frac{N^{5/3}}{V^{2/3}}. \qquad (1.4.12)$$

For the non-relativistic gas one has

$$P = \frac{2}{3}\frac{E}{V} = \frac{(3\pi^2)^{2/3}}{5} \frac{\hbar^2}{m} n^{5/3}. \qquad (1.4.13)$$

The condition that the non-relativistic Fermi gas is degenerate is

$$\Theta \ll \varepsilon_F, \qquad \varepsilon_F = \frac{p_F^2}{2m} \sim \frac{\hbar^2}{m} n^{2/3}. \qquad (1.4.14)$$

1.4.5   Relativistic gas

For ultrarelativistic particles $\varepsilon \sim pc$ and one has

$$E = \frac{cV}{\pi^2\hbar^3} \int_0^{p_F} p^3 \, dp = \frac{Vcp_F^4}{4\pi^2\hbar^3} = \frac{3(3\pi^2)^{1/3}\hbar c}{4} \frac{N^{4/3}}{V^{1/3}}. \qquad (1.4.15)$$

For Fermi (as well as for Bose) relativistic gases one has

$$P = \frac{E}{3V} = \frac{(3\pi^2)^{1/3}}{4} \hbar c n^{4/3}. \qquad (1.4.16)$$

The condition that the relativistic Fermi gas is degenerate is

$$\Theta \ll \varepsilon_F, \qquad \varepsilon_F = p_F c = (3\pi^2)^{1/3} \hbar c n^{1/3}. \qquad (1.4.17)$$

1.5   Equilibrium of Gravitating Systems

1.5.1   Equilibrium condition

The pressure of the Fermi degenerate gas depends on its density as $P \sim n^\gamma$. For the relativistic gas $\gamma = 4/3$, while for the non-relativistic one $\gamma = 5/3$. That is, the equation of state for the relativistic gas is *softer* than for the non-relativistic one. This has a simple explanation. In the non-relativistic case there are 2 factors that result in increasing pressure with increase of the density: both the velocity of particles and their (Fermi) momentum, increase. For the relativistic gas one of the factors (increase of the particle velocities) does not work since all the particles have a velocity practically equal to the speed of light.

The effective 'softening' of the equation of state results in an instability of gravitating systems with the mass higher than some critical value M_*. Let us discuss the problem of equilibrium of gravitating systems in more detail. In order to estimate the critical mass M_* and other parameters of the system we omit all the numerical factors of order of 1. For example, for a star of mass M and radius R we use the following expression for the density

$$\rho \sim M/R^3. \tag{1.5.1}$$

Similarly, the gradient of the pressure can be estimated as $\nabla P \sim P/R$. In this approximation the condition for the equilibrium

$$\frac{GM\rho}{R^2} \sim \nabla P \tag{1.5.2}$$

takes the form

$$GM^2 \sim PR^4. \tag{1.5.3}$$

Notice that the mass M and the number N of the Fermi particles, responsible for pressure, are related as follows

$$M = m'N, \tag{1.5.4}$$

where m' is the mass per single Fermi particle. For example, for the electron gas

$$m' = (\text{number of nuclei per 1 electron}) \times (\text{nuclear mass}). \tag{1.5.5}$$

1.5.2   Gravitating spheres of a non-relativistic Fermi gas

Using the equation of state Eq. (1.4.13) one has

$$P \sim \frac{\hbar^2}{m} n^{5/3} \sim \frac{\hbar^2}{m} \cdot \frac{N^{5/3}}{R^5}, \tag{1.5.6}$$

and the equilibrium condition Eq. (1.5.3) takes the form

$$GM^2 \sim \frac{\hbar^2}{m} \frac{N^{5/3}}{R}, \tag{1.5.7}$$

or, using Eq. (1.5.4)

$$R \sim \frac{\hbar^2}{Gmm'^{5/3}} \frac{1}{M^{1/3}}.$$ (1.5.8)

The larger the mass M, the smaller is the radius R of the sphere of the gravitating matter. At the same time, the density of the matter grows with mass. As a result, at a large enough value of the mass the gas becomes relativistic. This happens when the particle density

$$n \sim \frac{M^3}{m'R^3} \sim \frac{M^2 c^3 m^3 m'^4}{\hbar^3 m_{\text{Pl}}^6}$$ (1.5.9)

becomes of the order of λ_m^{-3}. This gives the following condition for the value of mass M_* when the gas becomes relativistic

$$M \sim M_* = \frac{m_{\text{Pl}}^3}{m'^2}.$$ (1.5.10)

Here, m_{Pl} is the Planck mass, $m_{\text{Pl}} = (\hbar c/G)^{1/2} \approx 2.176 \times 10^{-5}$ g.

1.5.3   Gravitating spheres of a relativistic Fermi gas

Using the relativistic equation of state Eq. (1.4.16)

$$P \sim \hbar c n^{4/3},$$ (1.5.11)

we write the equilibrium equation Eq. (1.5.3) in the form

$$GM^2 \sim \hbar c N^{4/3}.$$ (1.5.12)

For the non-relativistic gas this equation determines the equilibrium radius of the gravitating system. Now, the radius R disappears from this relation and the equation uniquely fixes the mass of the system

$$M \sim M_* = \frac{(\hbar c/G)^{3/2}}{m'^2}.$$ (1.5.13)

Here, as earlier, m' is a mass per unit Fermi particle, so that the critical mass M_* is

$$M_* \sim \frac{m_{\text{Pl}}^3}{m'^2},$$ (1.5.14)

and it coincides with Eq. (1.5.10). This means that after the gas becomes relativistic the mass of the star cannot be greater that M_*. This value is the upper limit for the equilibrium star mass.

Since for both the electron and neutron degenerate gases $m' \sim m_{\text{n}} \approx 10^{-24}$ g, the critical masses in both cases are close to one another. They are of the order of

$$M_* \sim 10^{33} \text{ g}$$ (1.5.15)

that is of the order of the solar mass $M_\odot \approx 2 \times 10^{33}$ g. At the same time the critical radii for these two configurations are quite different. Indeed,

$$R \sim R_* = \frac{\lambda_m \lambda_{m'}}{l_{\text{Pl}}}, \qquad (1.5.16)$$

where $\lambda_m = \hbar/(mc)$ and $\lambda_{m'} = \hbar/(m'c)$. One also has

$$R_* = \frac{m_{\text{Pl}}^2}{mm'} l_{\text{Pl}}. \qquad (1.5.17)$$

For the *white dwarf* $m = m_e$ and we have

$$R_* = 4.5 \times 10^8 \text{ cm} = 4500 \text{ km}. \qquad (1.5.18)$$

This means that the radius of a write dwarf is close to the radius of the Earth, while the size of a neutron star is smaller and is about 10 km.

Let us note that the expression Eq. (1.5.10) for the critical mass contains the Planckian mass. It does not mean that the *quantum gravity* effects are important for our problem. We have considered the *gravitational equilibrium* of a *quantum* system. This explains why the constants G and $\hbar$ appear in the formula. The limiting mass exists because of the softening of the equation of state, which occurs when the fermions become relativistic. This is why the speed of light c enters the expression for the limiting mass of the gravitating Fermi gas system.

> **Problem 1.4:** *There is strong evidence that the neutrinos have non-zero mass. Assume that the electron neutrino mass is 0.01 eV. What are the corresponding critical mass M_* and the size R_* of a 'neutrino star' that is made of such Fermi particles?*

1.5.4   White dwarfs and neutron stars

A *white dwarf* is the final state of evolution of a star whose mass is not too high. The pressure of the degenerate electron gas provides its equilibrium. The critical mass of a white dwarf (the *Chandrasekhar limit*) is about $1.44 M_\odot$. The size of the white dwarfs is comparable with the size of the Earth. Their density is about 10^6 g/cm^3. Over 97% of the stars in our Galaxy are or will become white dwarfs.

Neutron stars are remnants of massive stars after their gravitational collapse. Neutron stars are created during supernova events. They are composed almost entirely of neutrons. A typical mass of a neutron star is between 1.35 and about 2.1 solar masses. Their corresponding radii are between 15 and 8 km. The heaviest known neutron star has a mass of about 2 solar mass. Neutron stars have an average density of order $8.4 \times 10^{13} - 1 \times 10^{15}$ g/cm^3. This density is comparable with the nuclear density $\sim 3 \times 10^{14}$ g/cm^3. The density is higher at the center where it becomes above $\sim 6 - 8 \times 10^{14}$ g/cm^3. At the surface (at the crust) the density is about $\sim 1 \times 10^6$ g/cm^3.

During neutron star formation most of the initial angular momentum of the core is conserved. Since its moment of inertia is sharply reduced, a neutron star is formed with very high rotation speed. Then, the angular velocity of a newly formed isolated neutron star gradually slows down. For a neutron star in a close binary the matter flow from a companion star can speed up this rotation again or slow it down (depending on the flow rate and the magnetic field). Neutron stars are known to have rotation periods between about 1.4 ms to 2.74 h. The neutron stars are so compact that the surface gravity is very high. An escape velocity from a surface of a neutron star is around 100 000 km/s, about 30% of the speed of light.

During the contraction, the magnetic field of the collapsing star, which is 'frozen' in the plasma, increases. The rotating neutron stars with a strong magnetic field are known as *pulsars* and *magnetars*. The value of the magnetic field on the surface of a pulsar can reach 10^{12} – 10^{15} G. The total number of known neutron stars in the Milky Way and the Magellanic Clouds is about 2000. Most of them have been detected as radio pulsars. About 5% of the neutron stars enter binary systems.

The critical mass of the neutron stars, a so-called *Tolman–Oppenheimer–Volkoff limit*, is approximately $2 - 3$ solar masses. There exists an uncertainty in the calculation of the critical mass of a neutron star. It arises because we do not know exactly the equation of state of the matter with nuclear and higher densities. Also, rapid rotation of the neutron star might be important for these calculations. It is usually excepted that the Tolman–Oppenheimer–Volkov critical mass is about $3M_\odot$, or a bit less.

There exists a hypothetical possibility that for the slightly higher mass a *quark star* or *strange star* might be created. It may happen if the degenerate Fermi particles are quarks, however, this option is quite speculative and uncertain. Gravitational collapse will always occur on any star core over 3–5 solar masses, inevitably producing a black hole.

1.6 Important Notions of Astrophysics

To identify an astrophysical object as a black hole candidate one needs first to 'measure' its mass. Such a measurement is possible, for example, when a black hole enters a binary system. Approximately 1/3 of the star systems in the Milky Way are binary or multiple. The remaining 2/3 are single stars. A heavier companion in the stellar binary has a faster evolution than the lighter one. If the mass of the heavy star is large enough a system may become a *black hole binary*.

Most of the known candidates for stellar mass black holes were discovered in such systems. The motion of a star companion in the binary can be detected. This gives information about the mass of the black hole candidate. At the same time, the star companion provides matter that falls onto the black hole and becomes a source of intensive radiation. Both the motion of companions in the binary system and the main properties of the accreting matter can be studied in the framework of the Newtonian theory. In this sense, most of the preliminary information concerning a black hole candidate, and especially its mass, can be obtained by the same methods as used by astronomers to study 'usual' binaries. For massive and supermassive black holes, which we consider later, the situation is similar. To estimate the mass of such a black hole one uses information about the velocity of accreting gas and stars motion near it. This can be done in the framework of the standard Newtonian theory. To 'prove' that a compact massive object is a black hole one needs additional observations testing effects specific for general relativity.

In this section we describe some of the 'standard' tools based on Newtonian gravity adopted in the astrophysics.

1.6.1 Mass function for a binary system

Let us first discuss how to determine the mass of objects in the binary system. Consider a non-relativistic two-body problem with the Lagrangian

$$L = \frac{1}{2}M_1\dot{\mathbf{r}}_1^2 + \frac{1}{2}M_2\dot{\mathbf{r}}_2^2 - U(|\mathbf{r}_1 - \mathbf{r}_2|). \tag{1.6.1}$$

Let us denote $\mathbf{r} = \mathbf{r}_1 - \mathbf{r}_2$, and choose the *center-of-mass* coordinate system

$$M_1\mathbf{r}_1 + M_2\mathbf{r}_2 = 0, \qquad \mathbf{r}_1 = \frac{M_2}{M_1 + M_2}\mathbf{r}, \qquad \mathbf{r}_2 = -\frac{M_1}{M_1 + M_2}\mathbf{r}. \tag{1.6.2}$$

Then, the Lagrangian (1.6.1) reduces to

$$L = \frac{1}{2}\mu\dot{\mathbf{r}}^2 - U(|\mathbf{r}|), \qquad \mu = \frac{M_1 M_2}{M_1 + M_2}. \tag{1.6.3}$$

Here, μ is the *reduced mass* of the binary system. For the gravitational interaction

$$U = \frac{\alpha}{r}, \qquad \alpha = GM_1 M_2. \tag{1.6.4}$$

Solving the Kepler problem one can show that

- An orbit is always planar, i.e. lies within a plane.
- An orbit with negative energy E is a closed ellipse. Such a motion is periodic.
- The major semi-axis is $a = \alpha/2|E|$.
- The period is $P = \pi\alpha\sqrt{\mu/(2|E|^3)}$.

The relation between the period P and the size of the system, a, is given by Kepler's law

$$\frac{P^2}{a^3} = \frac{4\pi^2\mu}{\alpha} = \frac{4\pi^2}{G(M_1 + M_2)}. \tag{1.6.5}$$

Let M_1 be the mass of a 'visible' star. Since

$$a = \frac{M_1 + M_2}{M_2}a_1, \tag{1.6.6}$$

one has

$$\frac{4\pi^2(a_1 \sin i)^3}{GP^2} = \frac{M_2^3 \sin^3 i}{(M_1 + M_2)^2}. \tag{1.6.7}$$

Denote $v_1 = 2\pi a_1/P$ the effective velocity of a visible star (for a circular orbit) and denote the maximal line-of-sight Doppler velocity of the visible star by $K = v_1 \sin i$, where i is the *inclination angle* of the binary orbit. Then, Eq. (1.6.7) can be rewritten as

$$\frac{K^3 P}{2\pi G} = f(M_2, i), \tag{1.6.8}$$

where

$$f(M_2, i) = \frac{M_2 \sin^3 i}{(1 + M_2/M_1)^2} \qquad (1.6.9)$$

is the *mass function*. Observing the visible star one can determine parameters K and P and, hence, find the mass function. It is easy to see that by measuring the mass function $f(M_2, i)$ one obtains the *lower limit* for the mass M_2. Indeed,

$$M_2 = (1 + \frac{M_2}{M_1})^2 (\sin i)^{-3} f(M_2, i) > f(M_2, i). \qquad (1.6.10)$$

For the elliptic motion (with an eccentricity e) the mass function can be written as follows (see, e.g., Cherepashchuk (1996))

$$f(M_2, i) = 1.038 \times 10^{-7} (1 - e^2)^{3/2} K^3 P_0. \qquad (1.6.11)$$

Here, M_1, M_2 and f are given in the solar mass $M_\odot$, P_0 is the period in days, and K is the amplitude of the velocity shift for the star companion measured in km/s.

1.6.2 Roche lobe

In a binary system an important notion is the *Roche lobe*. For a circular motion of the elements of the binary it is convenient to use a corotating frame. In such a coordinate system with its center at the center-of-mass, the potential energy is the sum of the gravitational potential energy of the stars and the effective potential of the centrifugal force

$$U = -\frac{GM_1}{r_1} - \frac{GM_2}{r_2} - \frac{1}{2}\omega^2 (X^2 + Y^2). \qquad (1.6.12)$$

We choose Z so that $Z = 0$ is the equation of the rotation plane and coordinates X, Y are within this plane. The axis X is directed from the first star to the second one, while Y is orthogonal to both X and Z. Here M_1 and M_2 are masses of the stars, X_1 and X_2 are distances to the stars from the center of mass, and

$$r_i = \sqrt{(X - X_i)^2 + Y^2 + Z^2}. \qquad (1.6.13)$$

We also denote by ω the frequency of rotation

$$\omega^2 = \frac{G(M_1 + M_2)}{|X_2 - X_1|^3}. \qquad (1.6.14)$$

An example of the equipotential surfaces at $Z = 0$-plane for the potential Eq. (1.6.12) is shown in Figure 1.2.

Equipotential surfaces in the vicinity of each of the stars look like a deformed sphere surrounding the star. At larger distance they have a different topology. The change of topology occurs at the critical value of the potential. The critical equipotential surface, which plays the role of the separatrix, intersects itself at the *Lagrange point* 1 and forms a two-lobe figure-of-eight. Each of the stars is at the center of its own lobe (see Figure 1.2).

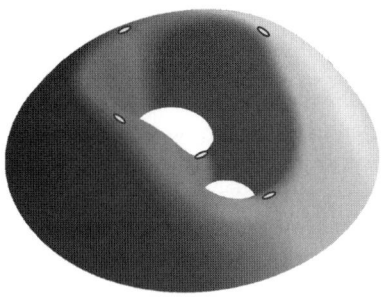
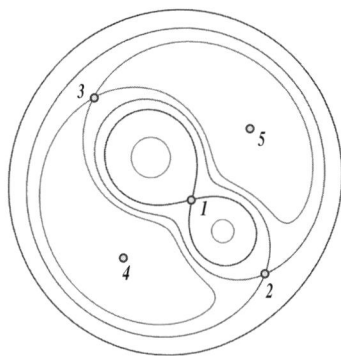

Fig. 1.2 Effective potential U at the slice $Z = 0$. Points 1, 2, 3, 4, 5 are critical points of the potential where its gradient vanishes. Points 1, 2, and 3 are located on the X-axis, while the Y-axis is orthogonal to this direction. The right plot shows the equipotential surfaces (at $Z = 0$). Small circles between 1 and 2, and 1 and 3, show the equipotential surfaces close to the points of the location of the first and second mass, respectively. The topology of the equipotential surfaces changes when the value of the potential passes through the critical value. To construct this particular plot we put $G = 1$ and choose $M_1 = 2$, $M_2 = 1$ and $X_2 = 1$. For these values of the parameters one has $U_4 = U_5 = -25/8$, $U_1 = -3.94557$, $U_2 = -3.54474$, and $U_2 = -3.32144$.

If the masses of the stars are M_1 and M_2, the characteristic size R_1 of the Roche lobe around the first mass, M_1, can be approximated as

$$\frac{R_1}{R} = \begin{cases} 0.38 + 0.2\ln(M_1/M_2), & \text{for } 0.3 < M_1/M_2 < 20; \\ 0.46224[M_1/(M_1 + M_2)]^{1/3}, & \text{for } M_1/M_2 < 0.8. \end{cases} \tag{1.6.15}$$

Here, R is semi-major axis of the system. The accuracy of this formula is about 2%. If one of the stars fills its *Roche lobe*, its matter can flow to the second companion.

> *Consider a function $U(\xi_1, \xi_2)$ of two variables ξ_1 and ξ_2 and denote $U_{,i} = \partial U/\partial \xi^i$. A point where the gradient of U vanishes, $U_{,i} = 0$, is called a* critical point. *If $det(U_{,ij})$ at the critical point does not vanish, the critical point is called a regular, or* Morse point. *The* Morse index *is the number of negative eigenvalues of $U_{,ij}$ at the Morse point. There are 3 types of critical points of U, minima, saddles, and maxima, where the Morse index is 0, 1, and 2, respectively. The topology of the equipotential lines $U = U_0 =$const changes, when U_0 passes through a critical point. The character of the change of the topology is described by the* Morse theorem. *The potential for a system with the Roche lobe is a very simple special case of the general* Morse theory *(see, e.g., (Dubrovin et al. 1990)). For the gravitational potential Eq. (1.6.12) the* Lagrange points *coincide with the critical points. Points 4 and 5 are maxima, while points 1, 2, and 3 are saddle points.*

1.6.3 Eddington luminosity

Consider a non-relativistic motion of a charged particle with the mass m_e and charge e. Under the action of electromagnetic radiation such a particle is accelerated. It moves in the direction of the oscillating field, and becomes a source of the dipole electromagnetic radiation. As a result, the initial electromagnetic radiation is scattered by the charged particle. The corresponding *Thomson scattering cross-section* for an electron is

$$\sigma_{\rm T} = \left(\frac{e^2}{m_e c^2}\right)^2 \approx 7.9407875 \times 10^{-26}~{\rm cm}^2. \qquad (1.6.16)$$

For the ionized plasma the radiation acts both on the electrons and ions. Since the mass of ions is much larger than the electron mass the resulting force of the radiation on plasma is determined by the Thomson scattering cross-section for the electron.

Consider now *spherical accretion* of the matter on a massive radiating body. Denote by L the total luminosity of the object, and by M its mass. We assume that the accreting matter is fully ionized gas and μ is the molecular weight per electron. If the density of the matter is ρ, the number density of electrons is $\rho/(\mu m_p)$. The outward force of radiation on the matter at the sphere of the radius r is

$$F_{\rm rad} = \frac{\sigma_T \rho}{\mu m_p c} \frac{L}{4\pi r^2}. \qquad (1.6.17)$$

It is equal to the gravitational attraction when the luminosity is

$$L_{\rm Edd} = \frac{4\pi G M \mu m_p c}{\sigma_T} = 1.3 \times 10^{38}~{\rm erg/s} \times \mu \frac{M}{M_\odot}. \qquad (1.6.18)$$

This is the so-called *Eddington luminosity*. For the Eddington luminosity there is an equilibrium between the gravitational attraction of the accreting matter to the body and its repulsion by the radiation of the body .

For accreting black holes their luminosity is determined by the rate of accretion of the matter on it. Only part of the produced energy is emitted, while the other part is absorbed by the black hole. Although in many cases the accretion is not spherical, nevertheless the Eddington luminosity plays an important role in black hole physics, allowing one to obtain restrictions on the mass of black holes.

1.7   Black Holes in Astrophysics and Cosmology

1.7.1   Black hole classification

The *black hole paradigm* proved to be very fruitful and rich not only in astrophysics, but also in theoretical physics. 40 years ago not so many astronomers believed in the existence of black holes. The black holes were considered as quite exotic objects and the general opinion was that they require for their formation quite unusual conditions. Many believed that even if black hole were formed, it would be very difficult, if even possible, to observe them. Now, 40 years later the situation has changed dramatically. Black holes are considered now as important elements of the universe. Black hole models are often used to explain phenomena where huge energy is emitted by a small-size compact object. Quite often it is difficult (or practically impossible) to explain observations without attracting such models.

Let us briefly discuss the modern status of black holes. The main parameter that determines black hole properties is its mass M. In principle, black holes can have a wide spectrum of masses. Black holes are usually classified as follows:

- stellar-mass black holes with $M \sim 3 - 30 M_\odot$;
- (super)massive black holes with $M \sim 10^5 - 10^9 M_\odot$;

- intermediate-mass black holes with $M \sim 10^3 M_\odot$;
- primordial black holes with mass up to $M_\odot$;
- micro-black holes.

Astrophysical observations confirm the existence of supermassive and stellar black holes. There is much less known about the existence of the intermediate-mass black holes, which might be logical precursors of the supermassive black holes. In principle, they can be formed by the matter accretion on a stellar mass black hole, or as a result of the collision of such black holes. Primordial black holes can be formed in the early universe from the initial inhomogeneities. Their formation may occur if during the universe evolution there exist stages when the equation of state of the matter becomes 'soft', for example, 'dust-like'. If the pressure is low, the *Jeans instability* may create compact overdense regions, where matter collapses and forms a black hole.

Micro-black holes are hypothetical objects with a very small mass. The minimal mass of the black hole can be determined as follows. Consider an object of mass μ_* for which the gravitational radius r_S coincides with its *Compton wavelength*, $\hbar/(\mu_* c)$. In Einstein gravity $r_S \sim G\mu_*/c^2$, and hence for such an object

$$\mu_* \sim \mu_{Pl} = \sqrt{\frac{\hbar c}{G}} \approx 10^{-5}\,\text{g}. \tag{1.7.1}$$

Black holes of smaller mass do not exist, at least in the standard classical sense. The quantum gravity effects become important for black holes with $\mu \gtrsim \mu_{Pl}$. Black holes of the Planckian mass are called *maximons* (Markov 1965) or *elementary black holes*.

Recent interest in micro-black holes was stimulated by models with *large extra dimensions*. In the presence of large (up to 0.1 mm) extra dimensions, the effective higher-dimensional gravitational constant is large, so that the elementary black hole mass μ_* is much smaller that μ_{Pl}. The new fundamental scale and, hence, the minimal mass of micro-black holes is usually adopted to be of the order of TeV. (We discuss these models in Section 10.5.)

1.7.2   Search for astrophysical black holes

To identify an astrophysical object as a black hole candidate one needs first to 'measure' its mass. This is true for both the stellar mass black holes and the supermassive ones. At this initial stage it is often sufficient to use the Newtonian gravitational law. For example, known stellar mass black hole candidates usually belong to a binary system. Observing the motion of the black hole companion, which is a normal star, it is possible to get information about the mass of the black hole candidate. If this mass is greater than, say, $3M_\odot$, it cannot be a white dwarf or a neutron star. Using this argument, based on general relativity, one concludes that most likely this is a black hole.

Similarly, for the objects in the center of galaxies, one can estimate their mass by observing the velocity of the gas and stars around it. If the radiation from the central object comes from a small-size region containing a large mass, one can suspect that the central massive object is a black hole. Combination of such properties as small size, large mass, large power of radiation emission in different bands of the electromagnetic spectrum are characteristic for black holes.

After a black hole candidate is found, a lot of work is to be done in order to confirm that this is really a black hole. Ideally, one needs to prove the existence of the black hole horizon. But the size of the horizon (gravitational radius) is so small that it is so far impossible to resolve a region of this size at the astrophysical distance. Future observations of the gravitational waves in the black hole coalescence, would give more direct information about the regions close to the horizon. At the moment, astrophysicists use different, more indirect ways to confirm that a found candidate is a black hole. Usually one assumes that general relativity is a valid theory of gravity in the strong-field regime and uses general relativity as tools in this study. Practically all the information about the black hole properties is obtained by studying the electromagnetic radiation emitted by the matter in the vicinity of the black hole and reaching us. Combination of radio, optical, infrared, X-ray, and gamma-ray observations allows one to obtain very rich information about the black hole candidates. Observation of black holes and the study of their properties provide a good test of predictions of the Einstein gravity.

There are several robust predictions of general relativity that can be used as tests for black holes:

- An astrophysical black hole is uniquely specified by two parameters: the black hole mass M and its angular momentum, J. Instead of J one can use the dimensionless rotation parameter $\alpha = cJ/(GM^2)$. This parameter vanishes for a non-rotating black hole and is 1 for an extremely rotating one. For $\alpha > 1$ a black hole does not exist. The corresponding formal solution in this case has a *naked singularity*.

- A black hole has no rigid boundary surface. Any matter, falling into a black hole, disappears into its abyss. In the case of a neutron star the interaction of the accreting matter with its surface produces specific radiation. There is no such (classical) emission from the black hole horizon.

- The gravitational force grows infinitely at the black hole surface. As a result, there exists an *innermost stable circular orbit*. For a non-rotating black hole ($\alpha = 0$) it lies at $6GM/c^2$, while for the extremely rotating one ($\alpha = 1$) it lies at GM/c^2.

- The characteristic orbital frequencies are 2200 Hz$(M/M_\odot)^{-1}$ and 16 50 Hz$(M/M_\odot)^{-1}$ for the non-rotating and extremely rotating black holes, respectively.

- All the mass and rotation multipole moments of the gravitational field of the rotating black hole are uniquely determined by its mass M and the rotation parameter α.

Besides these general robust tests there exists a variety of other predictions of general relativity, that can be used to identify a candidate with a black hole. We consider them in more detail later.

1.7.3   Black hole coalescence

A close binary system containing a black hole (for example, a black-hole–black-hole (BH–BH) or black-hole–neutron-star (BH–NS) binary) emits gravitational waves. As a result it loses its angular momentum and energy and the components of the system move closer to each other. The revolution frequency of the system becomes higher. A final stage of such a system is the formation of a larger-size single black hole. The gravitational waves emitted during this process, and especially in its final stage of coalescence, are strong enough to be detected on Earth.

Today, a set of gravitational-wave detectors is operational, including laser interferometers LIGO, VIRGO, GEO600 and TAMA. These detectors can register the gravitational radiation from close binary coalescence, provided the source of the radiation is not too far. A future, space-based observatory LISA, would be able to register lower-frequency gravitational waves from massive and supermassive black hole collisions. These and other existing and future experiments for observation of the gravitational waves from black holes will give us direct information about spacetime properties in the vicinity of the black hole horizons. They will test both general relativity itself and its prediction of black holes.

To study the gravitational waves produced by the black hole coalescence and by other compact relativistic sources the calculations based on general relativity in the strong-field regime are required. Recently, remarkable progress was achieved in solving numerically the gravity equations. Combination of numerical and analytical results now gives quite complete understanding of these processes (see Section 10.3).

1.8   Stellar-Mass Black Holes

1.8.1   Black holes in binary systems

A stellar mass black hole is the final stage of the evolution of a massive star. We know that the upper limit of the white dwarf mass (the *Chandrasekhar limit*) is about $1.4M_\odot$. For a neutron star, the critical mass (*Tolman–Oppenheimer–Volkoff limit*) about $3M_\odot$, or a bit less.

If after the collapse of a progenitor massive star, the mass of the central collapsing core is larger than these limits, one expects a black hole formation. Stars may have initial mass up to about $100M_\odot$, or, in the distant past, even higher. There are uncertainties in the description of the evolution of such massive stars. A considerable part or even most of the mass of such stars can be ejected during the final state and the collapse of the central core. One can roughly estimate the upper limit of the stellar mass black hole at about $30M_\odot$. This mass may increase by further matter accretion and collision with stars and other black holes.

All the best-known candidates of stellar mass black holes are in X-ray compact binary systems (see Table 1.1). Such close binaries contain a black hole and a normal star moving around their common center of mass. This makes these systems convenient for observations. A normal star companion is the source of the matter falling into the black hole. The matter has an angular momentum and moves mainly in the plane of rotation of the system. It forms an accretion disk around the black hole. The matter in the disk cannot fall into the black hole until it loses its angular momentum. In order for the material to fall into the black hole, it must transfer its angular momentum to the outer parts of the accretion disk, and it must radiate away much of its energy. The details of this process are quite complicated. The properties of the accretion disk is a subject of active research. The energy of matter at the inner orbits is smaller than the energy at the outer orbits. The released energy is transformed into the thermal energy of the accretion disk. The inner part of the disk is so hot that it emits X-rays. As a result the luminosity of the disk in X-rays is extremely high and such binaries are bright X-ray sources in the sky. By using optical observations it is possible to register the companion star. Periodic *Doppler shift* of its frequency can be used to confirm that both the black hole and the star are elements of the same binary system. Using the information concerning the motion of the star companion, one can find the *mass function* for the binary and use it for the estimation of the mass of the black hole. When its mass is larger than $3M_\odot$ one concludes that the companion of the star in the binary is a candidate for a black hole. This is a general scheme of selection of the *black hole binary* candidates. Further, more detailed study of these objects is required to prove that these objects are really black holes.

1.8.2 How many black holes are in the Galaxy?

Most of the black hole candidates have been observed in close binaries in our Milky Way (see Table 1.1). Only four well-studied black hole candidates are extragalactic (LMC X1, LMC X3, M33, IC10). The exact number of stellar-mass black holes in our Galaxy is not known. Estimates give this number up to $10^8 - 10^9$ (Remillard and McClintock 2006). A similar or larger number of neutron stars exist in the Galaxy.

Since the total number of stars in our Galaxy is about 10^{11}, the approximate ratio of the number of black holes to the number of stars can be as high as 10^{-2}–10^{-3}. These estimations are based on counting the number of very massive stars that exist now and have existed at earlier stages of the universe. These numbers are not very precise and there is also uncertainty in the exact value of the critical mass of a neutron star. Nevertheless, the main conclusion that the abundance of stellar mass black holes in our Galaxy and in the universe is not extremely small is quite robust.

1.8.3 Black hole binaries

The first discovered black hole binary candidate was Cygnus X-1 (abbreviated Cyg X-1). This is a galactic X-ray source in the constellation Cygnus. It was discovered in 1964 during a rocket flight. It is one of the strongest X-ray sources seen from Earth. This X-ray binary has been studied in detail. It is persistently bright in X-rays but shows variability on timescales of seconds. Its companion was identified as a blue supergiant variable star designated HDE 226868. The size of this binary is about 0.2 AU.[11]

Following Cyg X-1, the second black hole binary candidate discovered was LMC X-3. This is one of the brightest X-ray sources in the Large Magellanic Cloud, with luminosity $> 10^{38}$ erg/s. This object is also a bright persistent X-ray source, with a companion star being massive O/B-type star.[12] The third one discovered in 1975 candidate, A 0620 − 00, is quite different. It was discovered as an X-ray nova, when its emission becomes suddenly so high that it was the brightest non-solar X-ray source in the sky. Then, during the year its luminosity reduced. Its companion was identified as a K-dwarf, and the X-ray source was shown to have a mass $> 3M_\odot$.

The launch of the first X-ray satellite UHURU in 1971 resulted in the discovery of about a hundred stellar mass X-ray binaries. New data from the Chandra X-ray Observatory has revealed similar objects in many distant galaxies.

Table 1.1 contains a list of the 22 best-known candidates for the black hole binaries. It is a slightly updated version of the table given in the review by Ronald A. Remillard and Jeffrey E. McClintock (2006). We add only some additional data that can be found in (Orosz *et al.* 2007, 2009; Silverman and Filippenko 2008).

1.8.4 Transient vs. persistent black holes

Three of the stellar mass black hole candidates, LMC X-1, LMC X-3 and Syg X-1 are so-called *persistent black holes*. They emit X-rays continuously. Most of the other candidates in Table 1.1 are so-called *transient black holes*. One observes bright X-ray emission from these

---

[11] AU stands for the abbreviation of the astronomical unit. This is a unit of length equal to the distance from the Sun to the Earth. It is about 149, 597, 871 kilometers.

[12] Star classification uses the letters O, B, A, F, G, K and M for different types of stars. The letter O denotes the hottest stars. The other letters of this sequence denote successively cooler stars. M stars are the coolest ones.

Table 1.1 Stellar mass black holes candidates in binary systems (From Remillard and McClintock 2006)

Coordinate	Name	Year	Spec.	P_{orb} (h)	$f(M)/M_\odot$	M_{BH}
0422+32	(GRO J)	1992/1	M2V	5.1	1.19 ±0.02	3.7–5.0
0538–641	LMC X–3	–	B3V	40.9	2.3 ±0.3	5.9–9.2
0540–697	LMC X–1	–	O7III	93.82008	0.13 ±0.05	10.91 ±1.41
0620–003	(A)	1975/1	K4V	7.8	2.72 ±0.06	8.7–12.9
1009–45	(GRS)	1993/1	K7/M0V	6.8	3.17 ±0.12	3.6–4.7
1118+480	(XTE J)	2000/2	K5/M0V	4.1	6.1 ±0.3	6.5–7.2
1124–684	Nova Mus 91	1991/1	K3/K5V	10.4	3.01 ±0.15	6.5–8.2
1354–64	(GS)	1987/2	GIV	61.1^g	5.75 ±0.30	–
1543–475	(4U)	1971/4	A2V	26.8	0.25 ±0.01	8.4–10.4
1550–564	(XTE J)	1998/5	G8/K8IV	37.0	6.86 ±0.71	8.4–10.8
1650–500	(XTE J)	2001/1	K4V	7.7	2.73 ±0.56	–
1655–40	(GRO J)	1994/3	F3/F5IV	62.9	2.73 ±0.09	6.0–6.6
1659–487	GX 339–4	1972/10	–	42.1	5.8 ±0.5	–
1705–250	Nova Oph 77	1977/1	K3/7V	12.5	4.86 ±0.13	5.6–8.3
1819.3–2525	V4641 Sgr	1999/4	B9III	67.6	3.13 ±0.13	6.8–7.4
1859+226	(XTE J)	1999/1	–	9.2	7.4 ±1.1	7.6–12.d
1915+105	(GRS)	1992/Q	K/MIII	804.0	9.5 ±3.0	10.0–18.0
1956+350	Cyg X–1	–	O9.7I	134.4	0.244 ±0.005	6.8–13.3
2000+251	(GS)	1988/1	K3/K7V	8.3	5.01 ±0.12	7.1–7.8
2023+338	V404 Cyg	1989/1	K0III	155.3	6.08 ±0.06	10.1–13.4
	M33 X-7		O7-O8	82.80	0.46 ±0.08	15.65 ±1.45
	IC 10 X-1		WR	34.40	7.64 ±1.26	23–33

This table contains 22 confirmed black hole binary candidates. Column 1 gives coordinates of the object. Column 2 gives either its common name, or the prefix identifying the discovering mission. 'J' indicates that the coordinate epoch is J2000. The year of discovery and the number of observed cycles of X-ray activity is in column 3. The spectral class of a star companion is given in column 4. The orbital period of the binary, its mass function and the mass of the black hole candidate is given in columns 5, 6, and 7, respectively. This table is based on Remillard and McClintock 2006. We also included in it updated parameters for LMC X-1 from (Orosz 2007), and new data about the sources M33 X-7 (Orosz 2009) and IC 10 X-1 (Silverman and Filippenko 2008).

objects only during some rather short time, while for the most of the time they are not active. During the X-ray outburst the X-ray luminosity changes from $10^{30} - 10^{33}$ erg/s (in the calm state) to $10^{37} - 10^{39}$ erg/s in the maximum. This increase occurs in a few days. After the maximum the radiation falls (often exponentially) during several months.

Two candidates, GRS 1915+105 and GX 339-4, are quite special. GRS 1915+105 has remained bright for more than a decade since 1992 when it was discovered. *GX* 339 − 4 demonstrated frequent outbursts followed by very quiet periods. Most of the other black hole candidates are transient.

The difference between persistent and transient black holes might be explained as follows. The companions of the persistent sources are large stars, filling their *Roche lobes*. They provide a permanent flow of accreting matter on the black hole. For the transient sources the size of a star component is smaller than its Roche lobe size. The accretion is non-stationary and the matter supply from such stars occurs only during periods of time when a star companion is active.

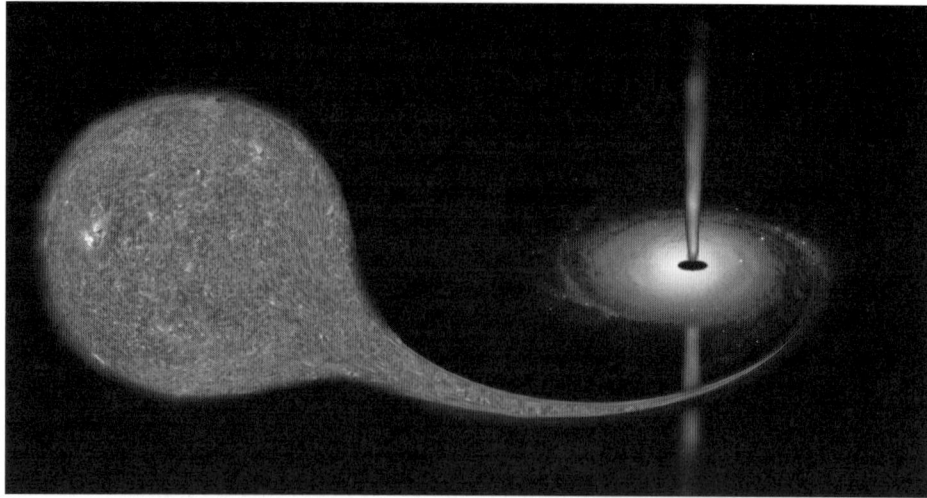

Fig. 1.3 A black hole in a close binary system. A star companion emits matter that is falling into a black hole. The picture shows a case when the size of the star is comparable with the size of the *Roche lobe*. The accreting matter has the angular momentum and forms an *accretion disk* around the black hole. Matter in the disk is rotating around the black hole at the Keplerian orbits. Transition from the outer to inner orbits occurs as a result of friction. As a result of this the released energy is partly transformed into the thermal energy that makes the disk hot and emitting X-rays. Such a system can be observed as the X-ray binary. The picture also shows the jet, which is a collimated flux of the relativistic plasma, emitted by some of the black holes.

1.8.5   Iron X-ray line broadening

Important information about stellar mass black holes can be obtained by observations of the broadening of the Fe Kα line. This line is associated with iron emission at 6.4 keV. At low temperature the line is very narrow. Its width is less than 100 eV. This line was observed in the spectrum of many astrophysical objects. It is explained by the Fe K fluorescence generated by the irradiation of the weakly ionized disk by a source of hard X-rays. It was also found in Cyg X-1 and other X-ray binaries. For the emission from the black hole binaries the observed spectrum differs from the sharp original line. It becomes much broader. The form of the broadened line can be explained if one takes into account that a line emitted by a matter in the inner part of the rotating accretion disk is modified by the Doppler effect, gravitational redshift and relativistic beaming. By fitting the disk model to observations one can find the radius of the Keplerian orbit of iron ions emitting these lines, as well as the spin of the central black hole. For some sources, e.g. for GX 339-4, the radiation region is likely extended to $(2-3)r_S$. In the most extreme case of XTE J1650-500 the inner edge may be located at $1.2r_S$ (Miller *et al.* 2009). This confirms the black hole model of the central body and also suggests that this particular black hole is rotating with nearly maximal spin.

1.8.6   Measuring black hole spin

While at large distances the gravitational force in Newtonian theory and general relativity is practically the same, in the vicinity of the black hole Einstein gravity is effectively much more strongly than the Newtonian one. At the same time, the effects connected with rotation of a massive body do not exist in Newtonian gravity. These effects in the weak-field limit exhibit themselves as the *Lense–Thirring force*. Close to a rapidly rotating massive body this effect of dragging spacetime into rotation becomes much more profound. For this reason study of possible observational consequences of the rotation of the black hole can give a rather clear indication on the validity of the Einstein theory and help to identify the objects as black holes. In the black hole binaries there exists a simple mechanism that speeds up the rotation of the black hole. Matter in the accretion disk, surrounding the black hole, preserves its angular momentum. It changes slowly. The rate of the change is determined by the viscosity effects, which is relatively small. The slow change of the angular momentum of an element of the matter in the disk results in the decrease of its Keplerian orbit. It comes closer and closer to the black hole until it reaches the *innermost stable circular orbit*. After this, matter falls into the black hole almost freely. Such falling matter brings with it not only energy, but also angular momentum. Because of this effect one can expect that the rotation parameter of the black hole in such systems may be quite high, or even close to its extremal value $\alpha = 1$ (see Section 8.4).

The main effect, connected with the rotation of a black hole, is the position of the innermost stable circular orbit. For a non-rotating (Schwarzschild) black hole of mass M this orbit is at $6GM/c^2$, while for an *extremely rotating Kerr black hole* a corotating circular orbit can be as close as GM/c^2 (see Chapters 6 and 7). Particles after reaching this orbit are practically in a regime of free fall into the black hole. The energy of the particles of the accretion disk, which can be transformed into the radiation before they reach the critical orbit, is about 5.7% of their proper energy mc^2, for a non-rotating black hole, and is about 42.4% for the extremely rotating one (see Chapter 8). Part of the radiation, emitted by the disk, is absorbed by the black hole. The black hole rotation can also affect the shape of the inner part of the accretion disk, and deflect the rays emitted by the disk. All these effects can be used to probe the properties of the spacetime in the close vicinity of the black hole. To give the quantitative estimations of their role one must use the 'machinery' of general relativity.

The method used to measure the angular momentum of a black hole is based on the study of the spectrum of the accretion disk. Since the inner parts of the accretion disk give a considerable contribution to the flux, the geometry of the inner parts of the disk, and in particular how close it is to the black hole, has an impact on the observations. At present there are estimations of the rotation parameters for 7 black hole binaries (LMC X-3, XTE J1550-564, GRO J1655-40,M33 X-7, 4U1543-47, LMC X-1, GRS 1915+105). The last two of these black holes are very rapidly rotating: $\alpha = 0.85 - 0.97$ and $\alpha = 0.98 - 1.0$ for LMC X-1 and GRS 1915+10, respectively (Shaffi *et al.* 2006; McClintock *et al.* 2006, 2009; Gou *et al.* 2009).

1.8.7 Microquasars

Some of the black hole (as well as neutron star) X-ray binaries have strong and variable radio emission. This radio emission is produced by *relativistic jets*. The term 'jet' is used to denote a relativistic outflow of plasma with high degree of collimation. These objects are called *microquasars*.[13] The jets are formed close to the compact object. Similar, but more powerful, jets are observed in quasars. They are connected with supermassive black holes in quasars (see later). The jets from stellar mass black holes are smaller-scale 'relatives' of the quasar jets.

---

[13] Some of the jets show the apparent superluminal motion, i.e. motion of 'blobs' across the sky at speeds that appear to exceed the speed of light.

Examples of black hole microquasars are Cygnus X-1 and GRS 1915+105, with very high jet velocity. In the case of the X-ray black hole binary XTE J1550-564 the powerful microquasar was observed a few years after the ejection event (for a general review, see, e.g., Gallo (2010)).

The mechanism of the jet formation is not known. It is believed that the energy for the jets is provided by the rotating black hole surrounded by the accreting matter, while the collimation of the jets is the result of the presence of the magnetic field. There are two most popular models explaining how the energy is transferred from a black hole to the jet: (i) the Blandford–Znajek process, and (ii) the Penrose mechanism. These general relativistic effects are discussed in Chapter 8.

1.8.8 Stellar black holes vs. neutron stars

Black hole candidates in the X-ray binaries usually have considerably higher mass than the neutron stars. However, if the mass of the black hole is small it is difficult to distinguish it from a neutron star. The main characteristic features of the neutron stars that can be used for this purpose are:

- existence of a solid surface;
- strong magnetic field leading to pulsations in *X*-ray emission;
- explosions (thermonuclear bursts) of the matter accumulated on the surface of the neutron stars.

1.8.9 Gamma-ray bursts

Gamma-ray bursts are flashes of gamma rays emitted by explosions of compact extra galactic objects. It is commonly believed that at least some of these bursts are produced during the collapse of high-mass rapidly rotating stars that produce black holes. The distribution of the gamma-ray bursts over the sky is very isotropic, with no concentration to the plane of the Milky Way. This indicates their extragalactic origin. The sources of most of the gamma-ray bursts are very far from the Earth, up to distances of a billion light years. This means that the energy emitted during such flashes, which are very short of about a few seconds duration, is very large. This makes a model of a black hole, as an engine of this energy production, quite natural and attractive (see Section 10.3.8).

1.9 Supermassive Black Holes

Stellar mass black holes were predicted by the theory. The discovery of supermassive black hole in quasars and at the center of galaxies was more surprising. Some supermassive black hole candidates are collected in Table 1.2.[14]

---

[14]Column 1 contains the galaxy name. The prefix M stands for the objects in the *Messier catalogue* published by the French astronomer Charles Messier in 1794. Its final version contains 110 objects. NGC (The New General Catalogue) is a well-known catalog of deep-sky objects compiled in the 1880s. It contains 7840 objects. IC (1 and 2) stands for the Index Catalog published by Dreyer in 1895. It contains 5386 objects. 3C is the Third Cambridge Catalog of Radio Sources. It was published in 1959. The prefix 3C is followed by the entry number in this catalog. The prefix Mrk stands for the Markarian catalog. PG is the Palomar-Gree catalog of ultraviolet-excess objects that was published in 1986. It contains 1874 objects. The prefix PG is followed by the coordinates of the object. Column 2 gives the distance of the object from the Earth, while column 3 gives the estimate of the black hole mass.

Table 1.2 Supermassive black hole candidates (From (Ho 2002))

Galaxy	D (Mpc)	$M/M_\odot$.	Galaxy	D (Mpc)	$M/M_\odot$.
3C 120	137.8	2.3×10^7	NGC 4258	7.3	4.1×10^7
3C 390.3	241.2	3.4×10^8	NGC 4261	31.6	5.2×10^8
Ark 120	134.6	1.84×10^8	NGC 4291	26.2	1.5×10^8
Arp 102B	99.7	2.2×10^8	NGC 4342	16.8	3.4×10^8
Circinus	4.0	1.3×10^6	NGC 4374	18.4	1.6×10^9
Fairall 9	199.8	8.0×10^7	NGC 4395	3.6	$< 1.1 \times 10^5$
IC 342	1.8	$< 5.0 \times 10^5$	NGC 4459	16.1	6.5×10^7
IC 1459	29.2	3.7×10^8	NGC 4473	15.7	1.0×10^8
IC 4329A	65.5	5.0×10^6	NGC 4486	16.1	3.4×10^9
Milky Way	0.008	$\sim 4 \times 10^6$	NGC 4486B	16.1	6.0×10^8
Mrk 79	91.3	5.2×10^7	NGC 4564	15.0	5.7×10^7
Mrk 110	147.7	5.6×10^6	NGC 4593	39.5	8.1×10^6
Mrk 279	126.6	4.2×10^7	NGC 4594	9.8	1.1×10^9
Mrk 335	106.6	6.3×10^6	NGC 4596	16.8	5.8×10^7
Mrk 509	143.8	5.78×10^7	NGC 4649	16.8	2.0×10^9
Mrk 590	109.2	1.78×10^7	NGC 4697	11.7	1.2×10^8
Mrk 817	131.0	4.4×10^7	NGC 4945	4.2	1.1×10^6
NGC 205	0.74	$< 9.3 \times 10^4$	NGC 5548	70.2	1.23×10^8
NGC 221	0.81	3.9×10^6	NGC 5845	25.9	3.2×10^8
NGC 224	0.76	3.3×10^7	NGC 6251	94.8	5.4×10^8
NGC 598	0.87	$< 1.5 \times 10^3$	NGC 7052	63.6	3.6×10^8
NGC 821	24.1	5.0×10^7	NGC 7457	13.2	3.4×10^6
NGC 1023	11.4	3.9×10^7	NGC 7469	66.6	6.5×10^6
NGC 1068	14.4	1.6×10^7	PG 0026+129	627.4	5.4×10^7
NGC 2778	22.9	2.0×10^7	PG 0052+251	690.4	2.2×10^8
NGC 2787	7.5	3.9×10^7	PG 0804+761	429.9	1.89×10^8
NGC 3031	3.9	6.3×10^7	PG 0844+349	268.4	2.16×10^7
NGC 3115	9.7	9.1×10^8	PG 0953+414	1118	1.84×10^8
NGC 3227	20.6	3.9×10^7	PG 1211+143	361.7	4.05×10^7
NGC 3245	20.9	2.1×10^8	PG 1226+023	705.1	5.5×10^8
NGC 3377	11.2	1.0×10^8	PG 1229+204	268.4	7.5×10^7
NGC 3379	10.6	1.0×10^8	PG 1307+085	690.4	2.8×10^8
NGC 3384	11.6	1.8×10^7	PG 1351+640	370.7	4.6×10^7
NGC 3516	38.9	2.3×10^7	PG 1411+442	379.8	8.0×10^7
NGC 3608	22.9	1.1×10^8	PG 1426+015	366.2	4.7×10^8
NGC 3783	38.5	9.4×10^6	PG 1613+658)	565.3	2.41×10^8
NGC 3998	14.1	5.6×10^8	PG 1617+175	494.7	2.73×10^8
NGC 4051	17.0	1.3×10^6	PG 1700+518	1406	6.0×10^7
NGC 4151	20.3	1.53×10^7	PG 1704+608	1857	3.7×10^7
NGC 4203	14.1	$< 1.2 \times 10^7$	PG 2130+ 099	255.3	1.44×10^8

1.9.1 Black holes in the center of galaxies

Most of the galaxies have compact central regions with dense gas and star populations. They are called the *galactic nuclei*. Such nuclei are usually well seen in the spiral galaxies. A small fraction (approximately 1%) of galaxies have *active galactic nuclei*. Physical processes in active galactic nuclei generate powerful electromagnetic radiation in the X-ray, ultraviolet and radio bands. One of the types of such galaxies are *quasars*. Quasars are the most powerful sources of the radiation in the universe. Their total luminosity reaches 10^{47}–10^{48} erg/s. This is about 3 orders of magnitude higher than the optical luminosity of their parent galaxy. The first quasars were discovered at the end of the 1950s in radio telescope observations. More than 200 000 quasars are known now. Fast variation of the radiation from some of the quasars allowed one to prove that this radiation comes from a small-size compact region. Study of quasars demonstrated that the most probable model explaining their activity is a model of a central supermassive black hole.

Now, a standard model for the nuclear activity in galaxies assumes the existence of black holes with mass 10^6–$10^9 M_\odot$. in their center, which acts as a central engine producing radio emission. For many of the galaxies with active nuclei there was confirmed the existence of such black holes. Now it is believed that the centers of most of the galaxies, including the Milky Way, contain a massive black hole. Usually a black hole has a mass about 10^{-3} of the mass of the galaxy and can play an important role in its evolution. The mechanism of formation of supermassive black holes is not known. There have been proposed several models for the explanation of the creation of black holes of such a big mass. In some of these models a black hole is formed in the already existing galaxy, either as a result of slow matter accretion on an original stellar mass black hole, or a collapse of a dense stellar cluster into a black hole without supernova explosions. Other models discuss formation of the large-mass primordial black holes, which later serve as seeds for the galaxy formation. The main problem is to explain how the large-mass was concentrated in a small enough volume. This process requires a mechanism that transfers the initial large angular momentum outwards. There exists a gap in the observed mass spectrum of black holes. At the moment there are no certain confirmations of the intermediate-mass black hole existence in the range between around $50 M_\odot$ (possible largest mass of stellar black holes) and up to $10^4 M_\odot$.

Black holes have one dimensional parameter, their mass M, that uniquely determines the scale. Since the mass of a supermassive black hole is 10^6–$10^{10} M_\odot$, its gravitational radius is $\sim 10^5$–10^9 times bigger than the gravitational radius of a stellar mass black hole (with mass $\sim 10 M_\odot$). The main differences in the properties of supermassive black holes and their stellar-mass cousins are determined by this scale difference. Another important difference is that for the stellar-mass black holes in binaries there is a permanent source of matter, accreting into it, i.e. the normal star companion. For the black holes in the galactic nuclei the matter falling into the black holes is the gas and stars 'living' in the nuclei. Thus, the rate of the accretion is directly connected with the properties of the galactic nuclei. Matter falling into a black hole and having non-vanishing angular momentum usually forms an *accretion disk*. Its plane orientation may be determined by the plane of the galaxy or by the spin direction of the central black hole.

1.9.2   Mass estimation

To identify a compact object in the center of a galaxy with a black hole candidate one first needs to 'measure' its mass. There are several robust estimations that can be used for this purpose.

Consider an accretion of matter into a black hole. If the accretion is spherical the luminosity of its radiation is restricted by the *Eddington luminosity*, Eq. (1.6.18),

$$L_{\rm Edd} = (1.3 \times 10^{38} \,{\rm erg/s}) \mu \left(\frac{M}{M_\odot} \right). \tag{1.9.1}$$

Here, μ is the molecular weight per electron. $L_{\rm Edd}$ is the luminosity at which the radiation pressure exactly balances the gravitational attraction of mass M (for a fully ionized plasma). If $L \ll L_{\rm Edd}$, gravity dominates, and photon pressure is not important. If one assumes that the luminosity of the object L is not larger than the $L_{\rm Edd}$ one can obtain from Eq. (1.9.1) the lower limit of the corresponding M.

The reliable way to detect black holes in the galactic nuclei is analogous to the case of black holes in binaries. Namely, one must prove that there is a large dark mass in a small volume, and that it can be nothing other than a black hole. In order to obtain such a proof, we can use arguments based on both stellar kinematics and surface photometry of the galactic nuclei. The mass M inside radius r can be inferred from observed data through the formula (Kormendy 1993):

$$M(r) = \frac{v^2 r}{G} + \frac{\sigma_r^2 r}{G} \left[-\frac{d \ln I}{d \ln r} - \frac{d \ln \sigma_r^2}{d \ln r} - \left(1 - \frac{\sigma_\theta^2}{\sigma_r^2} \right) - \left(1 - \frac{\sigma_\phi^2}{\sigma_r^2} \right) \right]. \tag{1.9.2}$$

Here, I is the brightness, v is the rotation velocity, σ_r, σ_θ, σ_ϕ are the radial and the two tangential components of the velocity dispersion. These values must all be obtained from observations. More complex formulas are used for more sophisticated models.

1.9.3   Relativistic jets

Relativistic jets are extremely powerful streams of relativistic plasma that are observed in some active galaxies and quasars. They are similar to the jets in microquasars, but have much bigger scale and power. In some of these objects, the visible length of jets reaches hundred and thousands of light years. For example, for the quasar 3C 273, which was the first quasar discovered, the jet size is ~ 62 kpc.

The mechanism of the jet formation is not known. It is believed that the energy for the jets is provided by the rotating black hole surrounded by the accreting matter. The magnetic field trapped inside the ionized accreting matter plays an important role in this process. It also provides the collimation of the jets. To explain this process, better understanding of the magnetohydrodynamics in a strong gravitational field is required.

1.9.4   Flashes from the accretion disk and iron line broadening

Quasars emit electromagnetic waves in all the bands: radio, optical, X-rays and gamma-rays. Because of their high luminosity they are visible at very large distances. From 5 to 25% of quasars are 'radio loud' with powerful radio emission produced by jets.

The emitted electromagnetic radiation comes from the different regions of the accretion disk surrounding a black hole. The closer to the black hole the part of the disk, the higher is the effective temperature of the radiation. After reaching the innermost stable (circular) orbit the matter of the disk freely falls into the black hole. As we have already mentioned, the radius of this innermost stable orbit is determined by the general relativistic effects. In order to calculate it one studies the *relativistic Kepler problem* (see Chapters 6 and 7). The radius of the innermost stable circular orbit depends on the angular momentum of the black hole. It is different for co- and counter-rotating particles (see Chapter 7). Flashes of the light produced by inhomogeneities of the accretion disk may give information about the rotation angular velocity of the radiating parts of the accretion disk. The existence of the innermost stable circular orbits restricts the characteristic timescales of the visible fluctuations.

Another important observation is the discovery of a strong broad line in the spectrum of X-ray emission. This is a direct analog of the effect we have already discussed for stellar-mass black holes. The broadening spectrum profile allows one to determine the Doppler velocity of the emitting ions, as well as the gravitational potential in the origin of the emission. In principle, for rapidly rotating black holes this may give information about the spacetime properties close to the horizon.

1.9.5   Multiple images

The strong gravitational field of a black hole deflects the light. The light rays passing near a black hole with the impact parameter close to the critical one have a large bending angle. As a result, the black hole will create more than one image of a distant object. This strong-field *lensing effect* can be used to identify and study black holes. The quantitative study of this and similar effects requires the development of the *ray-tracing* methods in the gravitational field of a black hole. The ray-tracing and its applications are discussed in Chapter 7. Rotating black holes as gravitational lenses are discussed in Section 8.6.

1.9.6   Star disruption

The gas accretion onto a black hole and black hole mergers are two mechanisms of the black hole mass growing that are often discussed in the literature. There is another interesting process that results in the mass increase of the central black hole. This is a *stellar capture* and *stellar disruption*. In the Newtonian theory a condition of tidal disruption of a star of mass m and radius b passing at a distance R from the mass M follows from the relation

$$\frac{GMb}{R^3} \sim \frac{Gm}{b^2}. \tag{1.9.3}$$

The term in the left-hand side is the tidal force (the force difference) acting on the object of size b in the gravitational field of mass M. The right-hand side is a self-gravity force on the

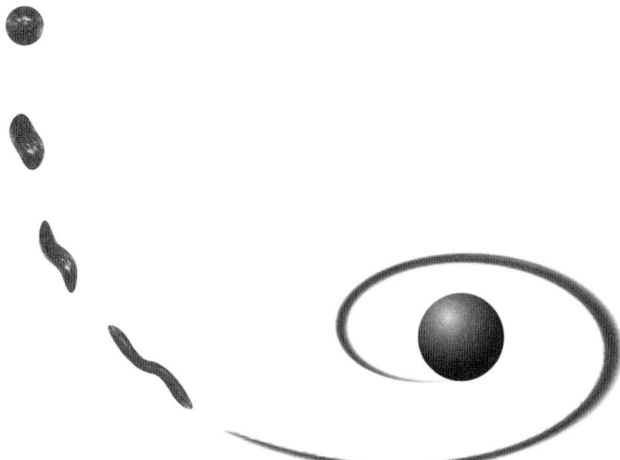

Fig. 1.4 Illustration of the tidal disruption of a star at its close encounter with a massive black hole. The star is shown at different times moments. At early time, when the star is far from the black hole, it has a spherical form. Tidal forces deform the star. The distortion may become so high that the star is disrupted. The matter of the star falls into the black hole.

surface of the star of size b produced by its mass m. This relation shows that if a star in its motion comes closer to the black hole of mass M than the distance

$$R = \left(\frac{M}{m} \right)^{1/3} b, \tag{1.9.4}$$

it will be disrupted by the tidal forces. For the close encounter it is necessary to use general relativity in order to describe a stars motion and its disruption.

This problem can be solved as follows. First, one finds a trajectory of a star in the gravitational field of the black hole, by integrating the geodesic equation in the corresponding metric. Secondly, one solves the equation of the parallel transport of a frame along such a geodesic and calculates the component of the curvature tensor in this frame. And finally, one solves the star evolution equation in a comoving frame under the action of the known tidal forces. It is interesting that there exist several observations of luminous flares from the supermassive black holes, which were explained as a result of the tidal disruption of stars, captured by the black hole, followed by accretion of its debris onto the black hole (Gezari 2009, Burrows *et al.* 2011).

1.9.7 Maser effect

Astrophysical *masers* are well-known phenomena. After their first discovery in 1965, they were found in comets, planetary atmospheres, star-forming regions, supernova remnants and in some extragalactic sources. *Maser radiation* is similar to the laser. It is generated when a coherent beam of electromagnetic radiation passes through media with a pumped population inversion. Such a beam induces transitions from upper occupied energy levels to a lower-energy one and gain in the energy of this process. There are many types of molecules that can produce the maser effect in the astrophysical environments. Observations confirm the existence of water vapor and other molecules that can produce stimulated microwave emission in the accretion disks around supermassive black holes.

Maser radiation from quasars is used to establish parameters of the central black hole engine. For example, the observation of the maser line emission allows one to test the mass of the black hole with a very high accuracy. The spectral lines emitted by masers are very bright and narrow. These lines are observed as broadened by the *Doppler effect* because of the Keplerian motion of the matter of the disk, emitting and amplifying these lines. The broad spectral lines are used as probes of this motion and give additional information about the black hole mass.

1.9.8   Black hole in Milky Way

This is the case of the best-known black hole. The object, known as 'Sagittarius A*', is located in the center of our Galaxy, which is about 27 000 light-years from the Earth. In a 16-year-long study of this object there were found very strong indications of the presence there of a black hole with the mass about 4 million solar masses. Basically, the method used for the estimation of the mass of the 'invisible' central object is based on finding the mass function from observations of 'visible' stars moving around it. Very accurate estimation of the black hole mass was obtained by watching the motion of about 30 stars in the central region. These observations also allowed astronomers to pinpoint the center of the Galaxy with great precision. One particular star, known as S2, orbits the Milky Way's center so fast that it has completed already one full revolution within the 16-year period of the study. The better resolution of the central, close to the black hole, region will be obtained by combining the light from the four 8.2-m VLT unit telescopes and using the interferometry technique.

1.10   Primordial Black Holes

Primordial black holes are hypothetical objects that might be created during the Early evolution of the universe. Such black holes might arise from overdense regions of the expanding matter. Suppose a black hole of mass M is formed. Then, the characteristic average mass density in the region of the size $r_S \sim GM/c^2$ occupied by the black hole is

$$\rho_g \sim M/r_S^3 \sim \frac{c^6}{G^3 M^2}. \tag{1.10.1}$$

One can expect that such black holes might be formed from the inhomogeneities of the matter of the universe preferably at the time when the average mass density is $\rho \sim \rho_g$. At the *radiation-dominated stage* the mass density of the matter decreases with time as $\rho \sim a^{-4}$, where a is a universe scale factor. The mass of primordial black holes created at the stage when this factor is a is of the order of $M \sim a^2$. This means the earlier such black holes are created the smaller mass they have.

In principle, the mass of primordial black holes can be as small as the Planckian mass. Primordial black holes of mass less than about 5×10^{14}g would decay by the present time as a result of *Hawking radiation* (see Chapter 8). Thus, only primordial black holes with mass larger than 5×10^{14}g could exist at the present time. The problem of the primordial black holes will be discussed in Section 10.2.6.

1.11   Black Holes in Theoretical Physics

1.11.1   Black hole uniqueness

A remarkable property of a black hole is that its gravitational field is completely determined by two parameters, its mass M and angular momentum J. When a black hole has an electric charge, Q, it enters as an additional third parameter. For a given set of the parameters (M, J, Q) a black hole exists only when these parameters obey an inequality

$$M^2 - (J/M)^2 - Q^2 \geq 0. \tag{1.11.1}$$

Here, we set $G = c = 1$. For the parameters obeying Eq. (1.11.1) the black hole solution of the Einstein–Maxwell equations is unique.[15]

Consider a neutral black hole. If the angular momentum of a black hole vanishes, its geometry is spherically symmetric. In the presence of rotation, the geometry is axially symmetric. The corresponding gravitational field of such a black hole at the large distance is similar to the field of an extended body. One can use the standard multipole decomposition to define the mass, μ_ℓ, and current, j_ℓ, multipole coefficients ($\ell \geq 0$). In the lowest order these coefficients are

$$\mu_0 = M, \qquad j_1 = J = aM. \tag{1.11.2}$$

The gravitational field of a normal astrophysical object at far distance can also be described in terms of the multipole coefficients. These coefficients depend on the structure of the body, that is on the matter and angular momentum distribution inside it. What singles out a black hole, is that higher-order multipole coefficients for its gravitational field are not arbitrary, but are uniquely determined by its mass and angular momentum. It is possible to show that these multiple moments obey the following simple relation (Hansen 1974)

$$\mu_\ell + i j_\ell = M(ia)^\ell. \tag{1.11.3}$$

Thus, for black holes with the same mass and angular momentum all the details of the gravitational field are identical.

What is the physical reason for this universality? The gravitational field of a normal astrophysical object depends on its matter distribution. It determines the multipole moments and, at the same time, 'supports' them. A black hole is an empty space. The matter, which created the black hole, has fallen into its abyss. The matter behind the *event horizon* cannot affect the exterior. Only those characteristics of the field that are connected with globally conserved integrals, such as the mass and angular momentum, survive. No-hair theorems, which have been proved for stationary black holes, provide us with an accurate description of this result.

One can describe this process of 'baldening' of the black hole after its formation as follows. In a general case, a newly born black hole is non-stationary. Its gravitational field is time dependent. The propagation of the excitation modes results in the following two effects: Radiation of the gravitational waves to infinity, and absorption of the other part of the excitations by the black hole. In the absence of stationary matter, supporting the excitation

---

[15]If magnetic monopoles exist in Nature, a black hole can have a magnetic charge P. In this case the inequality Eq. (1.11.1) is modified by substituting $Q^2 + P^2$ instead of Q^2.

modes, the final result of this process is a stationary configuration. It has a very special property: the spacetime is regular near the event horizon. But the gravitational field at the horizon, as measured by a stationary observer, is infinitely strong. The regularity condition, working as the boundary condition for the stationary gravitational field, singles out only those configurations that are uniquely determined by the globally conserved quantities, the mass and the angular momentum.

1.11.2 'Miraculous properties' of black holes

The *Einstein equations* is a set of 10 coupled highly non-linear partial differential equations for the metric $g_{\mu\nu}$. As we have already mentioned, a subclass of asymptotically flat vacuum stationary solutions of these equations is quite narrow, when one assumes that a solution is regular and it possesses a regular horizon. This result follows from the uniqueness theorems, discussed in the previous subsection. The *Kerr metric*, describing the most general vacuum stationary black hole, written explicitly, looks quite complicated. Nevertheless, its properties are remarkably simple.

For example, geodesic equations for particle and light motion in this metric are completely integrable (see Chapter 8 and Appendix D). This is quite a rare case when a physically interesting problem can be solved in quadratures. This property means that the corresponding motion is 'regular' (not chaotic). The spacetime symmetries of the Kerr metric, generated by Killing vectors, are not sufficient for explanation of this 'miracle'. This 'miraculous' property of the Kerr metric is a result of its *hidden symmetry*. This symmetry also implies that the massless field equations in the Kerr spacetime can be decoupled and reduced to a single scalar *master equation*. The latter equation is completely separable. The recent development demonstrated that the hidden symmetries is a generic property of black holes not only in four-dimensions but also in the higher-dimensional gravity. Why black hole solutions have this fundamental property is still an open question.

1.11.3 Quantum effects

Hawking radiation

The possible role of quantum effects in *black hole physics* depends on the black hole mass M, or, more exactly, on the natural dimensionless parameter M/m_{Pl}. When this parameter is large quantum effects are usually small and not important at short time scale. However, even if the quantum effects do not modify the black hole metric, they may produce observable effects and change the parameters of the black hole after long enough time. The *Hawking effect* is the most famous example.

In 1974 Hawking (Hawking 1974, 1975) theoretically discovered that the vacuum in the presence of a black hole is unstable. The quantum vacuum decay generates particles. A fraction of the particles created by the black hole reaches an observer at infinity and forms the *Hawking radiation*. This radiation has a thermal distribution over energies and the corresponding temperature (as measured at infinity) is known as the *Hawking temperature*. For a black hole of mass M it is (see Section 9.7)

$$\Theta_{\mathrm{H}} = \frac{\hbar c^3}{8\pi GkM}. \tag{1.11.4}$$

As a result of this process the black hole loses its mass. The rate of mass loss is

$$\dot{M} = -C(m_{Pl}/M)^2. \qquad (1.11.5)$$

Here, C is a dimensionless coefficient that depends of the number and properties of the emitted (massless) fields. The lifetime of the black hole with respect to the *Hawking evaporation* is

$$T \sim t_{Pl}(M/m_{Pl})^3. \qquad (1.11.6)$$

This effect might be important for the small-mass primordial black hole (see Section 10.2.6).

Quantum gravity effects

In the calculation of the Hawking radiation, the black hole is considered as a classical object. Its metric is not quantized. One quantizes physical fields in the background of a stationary black hole. These fields may also include gravitational perturbations. This approach is called the *semiclassical approximation*. For black holes with $M \gg m_{Pl}$ it is very accurate. The tools required for such calculation are well developed.

The *quantum gravity* is a regime of the gravitational theory when quantum effects modify the classical theory essentially. For example, when the curvature is large, the Einstein action must be modified by including the quantum corrections into it. These corrections contain both higher in curvature local terms, and non-local contributions. The quantum gravity effects are important for black holes of small mass that is comparable with the Planckian one. If the mass M of the black hole is of the order of the Planckian mass, the spacetime curvature at its surface is comparable with the Planckian curvature,

$$\mathcal{R} \sim l_{Pl}^{-4}. \qquad (1.11.7)$$

In this case, one cannot neglect the quantum gravity corrections that are of the order of $\sim (m_{Pl}/M)^4$. There exists another quantum gravity effect, the quantum fluctuations of the spacetime metric. For black holes with a mass comparable with the Planckian mass one can expect these fluctuations to be large, so that the meaning of the classical metric becomes not well defined.

There are indications that the gravitational field itself is an *emergent phenomenon*. One can use a metric of the spacetime in low-energy regime, when the energies are much smaller than the Planckian one. In this approach the gravitational field is not a fundamental one, but it is effectively a collective variable for a more fundamental theory. The string theory, which is considered as the most mathematically developed version of the quantum theory, is a good example. These problems go far beyond the format of the present book. For their discussion see, e.g., Barceló *et al.* (2005).

1.11.4 Black hole thermodynamics and information loss

A black hole radiates as a hot *black body*. Consider a black hole in a thermostat. If its temperature is the same as the Hawking temperature of the black hole, the complete system, that is the black hole and the radiation surrounding it, is in equilibrium.[16] Such a system can be studied by using standard thermodynamical laws, provided the black hole subsystem

---

[16]For a large-sized thermostat this equilibrium is unstable.

has such characteristics as the energy, entropy and so on. The basic ideas of the black hole thermodynamics were formulated by Bekenstein and Hawking in 1975 (see Section 8.7). According to this analogy a non-rotating black hole of mass M has the entropy

$$S_\mathrm{H} = \frac{\mathcal{A}_H}{4 l_\mathrm{Pl}^2},\tag{1.11.8}$$

where $\mathcal{A}_\mathrm{H} = 4\pi r_\mathrm{S}^2 = 16\pi\, G^2 M^2/c^4$ is the surface area of the black hole.

There are two fundamental problems connected with the *black hole entropy*. The first one is: What is the microscopic origin of this entropy? We know from *statistical mechanics* that the entropy is defined as the amount of information needed to exactly specify the state of the system with given macroscopic parameters. What are the *constituents* responsible for encoding this information and, hence, for the black hole entropy?

The second problem is often formulated as the *information-loss paradox*. In the process of the *black hole evaporation*, its energy is radiated away and its mass reduces. During the Hawking process the initially *pure quantum state* (initial vacuum), is transformed into an *entangled pure state*. In this state some of the particles, forming the *Hawking radiation* are located outside the black hole, while the other fraction of the particles is located inside the black hole. There exists a strong correlation between these two subsystems of spatially separated particles, which makes a state of a complete system, formed by these two subsystems, pure. After the complete evaporation of the black hole it disappears. In such a process the initial pure quantum-mechanical system would be transformed into a mixed state of outer particles, described by a (non-pure) density matrix. In spite of a lot of progress and many interesting proposals, these two problems, the origin of the black hole entropy and the information loss paradox, still remain open.

1.12 Black Holes and Extra Dimensions

1.12.1 Black holes and string theory

The idea that the spacetime can have more than 4 dimensions is quite old. Nordström (1914), Kaluza (1921) and Klein (1926) used this idea trying to unify gravity and electromagnetism. Suppose one has a metric in $4 + k$ dimensional space. This metric can be naturally decomposed into a 4-dimensional metric and a set of k 4-dimensional vectors and k^2 scalar fields. Thus, the gravitational theory in the higher-dimensional spacetime generates a set of 4-dimensional fields, and automatically unifies 4-dimensional gravity and a set of scalar and vector fields.

This idea has received a far-reaching generalization in the *string theory*. This theory is the most developed scheme unifying physical interactions. In particular, it includes quantum gravity. The conditions of the consistency of string theory impose the restriction on the allowed dimensionality of the spacetime. In the presence of fermions, this number must be equal to 10. Namely, for this number of dimensions the *conformal anomalies* vanish. The conformal anomalies that are present for the other numbers of dimensions reflect the existence of unphysical degrees of freedom that make the theory inconsistent.

Since our world has (3+1) dimensions, it is usually assumed that extra dimensions are invisible because they are compact and the size of these compact extra dimensions is small. This makes them undetectable in our low-energy experiments. In string theory the natural size of the compactification is determined by the Planck length. In the low-energy limit string theory generates the *theory of supergravity*. This theory contains besides the gravitational field $g_{\mu\nu}$, also a scalar dilaton field, the antisymmetric tensor fields and a set of Fermi fields. The total number of degrees of freedom of bosons is equal to the number of degrees of freedom of fermions. This is one of the consequences of *supersymmetry*, that is the symmetry that relates bosons and fermions. This property implies that after a natural reduction to the lower-dimensional theory, one is dealing with a theory that has a negative *cosmological constant*.

The interesting property of such theories, known as the *Maldacena conjecture*, is the existence of the highly non-trivial relations between the properties of the theory in the bulk of such a spacetime, and the theory on the boundary which is one dimension less. Black hole solutions with the anti-de Sitter asymptotic is the subject of very high interest. String theory has made many important contributions to the black hole physics. In particular, strings are natural fundamental constituents. Gravity in the string theory is an *emergent phenomenon*. The standard Einstein theory is an effective theory valid at low (with respect to the Planckian) energy scales. In a number of cases it was demonstrated that counting the string states gives the correct value of the black hole entropy. Unfortunately, these and some other results of string theory essentially rely on the use of supersymmetry, the property that is apparently violated in our real world, at least at the present time.

1.12.2   Black holes and large extra dimensions

String theory describes the very high energy phenomena. The characteristic scale of this theory is the Planckian one. This is also a natural scale for the size of extra dimensions. As a result, effects predicted by the string theory and that can be used to test this theory are extremely small. This makes the experimental verification of string theory extremely difficult, if even possible. Recently, it became popular to consider models where the size of extra dimensions is not microscopically small. These are so-called *models with large extra dimensions*. There are two main types of such models. In the first type of models, usually called *ADD models*[17] in which it is assumed that extra dimensions are compact but flat, while in the second one, so-called *Randall–Sundrum models* (*Randall–Sundrum* (1999)), the bulk space of extra dimensions is AdS, that is curved with a constant negative cosmological constant.

In the ADD models one usually assumes that the size L of the compactified extra dimensions is of the order of fraction of a millimeter. The dimensional reduction to the 4-dimensional theory in this model generates a *Kaluza–Klein tower* of particles, which differs by the mass gaps of the order of $(\Delta m)^2 \sim (\hbar/Lc)^2$. For $L \sim 0.1$ mm the existence of the Kaluza–Klein tower of partners of the known particles is experimentally excluded. In order to make such models consistent one needs to assume additionally that all particles and fields are 'living' within a 4D brane, representing our world. Gravity can live in the bulk. The choice of the scale of L less than 0.1–0.05 mm also implies that one does not observe the violation of *Newtonian law* in the experiments performed till now (Kapner 2007).

The original motivation of the models with large extra dimensions was to provide a natural 'solution' of the long-standing fundamental problems, such as the *cosmological constant*

---

[17] ADD stands for the abbreviation of the names of the authors, Arkani-Hamed, Dimopoulos, and Dvali (1998, 1999).

problem and the *hierarchy problem*. An important property of such models is that the gravitational coupling constant in the bulk space can be much larger than the 4D Newtonian constant. In other words, the higher-dimensional gravity at small distances is much stronger than the 4D gravity. This makes it possible that already at the scales of the order of TeV gravity becomes as strong as the other interactions. (This formally solves the hierarchy problem.) For example, in the ADD models this happens for 2 extra dimensions of the size of the order of 0.1 mm. At the TeV and higher scales the gravitational interaction would become the leading one. This makes possible such 'exotic' processes as emission of gravitons to the bulk with a visible non-conservation of the 4D energy within the brane, and formation of micro-black holes in the collision of two ultrarelativistic particles. The possibility to test predictions of the models with large extra dimensions for chosen parameters in the future collider experiments makes them quite interesting (see Section 10.6).

It should be emphasized that the proposed models with large extra dimensions do not follow from a fundamental theory and, thus, have a 'phenomenological' nature. Nevertheless, these ideas, even being quite speculative, have generated a lot of research and created very high interest in the higher-dimensional generalizations of 4D Einstein gravity. Among other reasons explaining the high interest in such models, one can single out the following two. First, in spite of the fact that the probability that these models are valid (at least at their present form) is very close to zero, the importance of possible discovery of black holes in the high-energy experiments is extremely high. This, at least in principle, makes the theoretical study of such models attractive. Secondly, study of the higher-dimensional theory of gravity allows one to understand better which results of the 4D Einstein theory are generic and valid in any number of dimensions, and which results are specific for 4D gravity only. This approach has already brought a number of interesting results, especially concerning the properties of the higher-dimensional black holes. A review of models with large extra dimensions and black holes in such theories can be found in, e.g., (Emparan and Reall 2008). Can mini-black holes provide us with a window into extra dimensions? This is an interesting open question.

To summarize, the subject of black holes and their interaction with matter is a fast-developing area of research. Black holes have 'many faces'. For astrophysicists, they are a new kind of astrophysical objects, which are 'responsible' for many important phenomena in the universe. For scientists working in gravitational physics black holes are the solutions of the Einstein equations, which have remarkable properties. More generally, for theoretical physicists, black holes provide an interesting and unique opportunity to test many fundamental ideas. They play the role of the 'Rosetta stone', which allows us to establish relations between different areas of physics, such as gravity, thermodynamics, statistical mechanics, and quantum theory. For philosophers, the existence of black holes poses a fundamental question: Are black holes really the 'graves of matter'? That is, does the formation of a black hole mean the end of matter evolution? Black holes are a challenge to our imagination. They are a favorite subject of science fiction books and movies.

2

Physics in a Uniformly Accelerated Frame

In this chapter we study physical effects in a *uniformly accelerated frame*. This is important for two reasons. First, according to the *equivalence principle* the physical laws in a uniformly accelerated frame are locally identical to the physical laws in a *homogeneous gravitational field*. This principle plays the role of the 'Rosetta stone' of gravitational physics. Thus, study of physics in a uniformly accelerated frame is a natural first step to gravitational physics. Secondly, the case of the homogeneous gravitation field is a special limiting case of the near-horizon geometry of the black hole, when its mass M infinitely grows. In this limit the curvature of the spacetime $\sim 1/r_S^2$ vanishes, while an observer located at a finite distance from the black hole horizon becomes a *Rindler observer*, that is a uniformly accelerated one. As we shall see, a lot of important properties of black holes can be understood better if one understands properties of the flat spacetime in a uniformly accelerated frame.

2.1 Minkowski Spacetime and Its Symmetries

2.1.1 Minkowski spacetime

Consider the *Minkowski spacetime* and denote by $X^0 = cT$ and X^i $(i = 1, 2, 3)$ the *Cartesian coordinates* in it. There is a natural map between these coordinates and an *inertial reference frame*. The inertial frame has an origin, O, in which an observer is located. This observer has a *standard clock*, which measures the proper time T. One can also imagine that there exist 3 mutually orthogonal ideal rods, which play the role of rulers and allow one to define the position of any point in the space by giving 3 numbers, (X^1, X^2, X^3), distances of the orthogonal projections to the rods measured by a standard ruler. Such an inertial observer characterizes an event in his frame by four Cartesian coordinates X^μ, $(\mu, \nu, \ldots = 0, 1, 2, 3)$. As a result, the spacetime, considered as a set of events, is mapped onto a four-dimensional manifold $\mathbb{R}^4$. Since X^μ cover all the spacetime, we call these coordinates *global*. The defined global frame is *rigid*. This means that a distance between any two points X^i and $\tilde{X}^i$ measured at time X^0 does not depend on X^0.

In a curved spacetime such global and rigid frames usually do not exist. Instead of them one uses *local frames*. They also are connected with some observer, but now an observer can move non-inertially. This frame is also provided with ideal solid rods and an ideal clock, but now the rods have finite size. The clocks also may be working only during some finite interval of time. We shall give an exact definition of the local frames and describe their geometric meaning later.

The *Minkowski metric* in the Cartesian coordinates is

$$ds^2 = \eta_{\mu\nu} \, dX^\mu \, dX^\nu, \qquad \eta_{\mu\nu} = \mathrm{diag}(-1, 1, 1, 1). \tag{2.1.1}$$

This metric determines an *interval* between two close events with coordinates X^μ and $X^\mu + dX^\mu$. In particular, $ds = 0$ is an equation of a *local null cone*. Two events are time-like separated if $ds^2 < 0$, and space-like separated if $ds^2 > 0$.

It should be emphasized that not every coordinate system can be related to some reference frame. To illustrate this let us consider the following example. Let us introduce new coordinates Y^μ, connected with the Cartesian coordinates X^μ as follows

$$Y^0 = \alpha X^0 + \beta X^1, \quad Y^1 = \gamma X^0 + \delta X^1, \quad Y^2 = X^2, \quad Y^3 = X^3.$$

In these coordinates the line element Eq. (2.1.1) takes the form

$$ds^2 = \Delta^{-2}[(\gamma^2 - \delta^2)dY^{0^2} - 2(\alpha\gamma - \beta\delta)dY^0 dY^1 + (\alpha^2 - \beta^2)dY^{1^2}]$$
$$+ dY^{2^2} + dY^{3^2}, \qquad \Delta = \alpha\delta - \beta\gamma.$$

Coordinate lines are the curves along which all but one of the coordinates of the system are constant. Choose for example, $\alpha = \gamma = 1$, $\delta = -\beta$, $|\beta| < 1$. Then, the coordinate lines of Y^0 and Y^1, as well as the coordinate line of Y^2 and Y^3, are space-like. But the existence of the reference frame assumes that one of the coordinate lines is time-like.

2.1.2   Poincaré group

Linear transformations of the coordinates that preserve the form of the interval Eq. (2.1.1) form the *Poincaré group* of symmetries. Consider a linear transformation

$$X^\mu = X_0^\mu + \Lambda^\mu_{\ \nu} \tilde{X}^\nu. \tag{2.1.2}$$

For $\Lambda^\mu_{\ \nu} = 0$ the vector X_0^μ defines a finite shift of the origin of the frame $O \to \tilde{O}$. These transformations form a subgroup of *translations*. Evidently the translations do not change the form of the interval. Let us consider now transformations $X^\mu = \Lambda^\mu_{\ \nu} \tilde{X}^\nu$, which result in

$$ds^2 = \Lambda^\alpha_{\ \mu} \Lambda^\beta_{\ \nu} \eta_{\alpha\beta} \, d\tilde{X}^\mu \, d\tilde{X}^\nu. \tag{2.1.3}$$

The condition of the form invariance of the interval is

$$\eta_{\mu\nu} = \Lambda^\alpha_{\ \mu} \Lambda^\beta_{\ \nu} \eta_{\alpha\beta}. \tag{2.1.4}$$

This condition determines the matrix of the *Lorentz transformation* $\Lambda^\alpha_{\ \mu}$. The matrices $\Lambda^\alpha_{\ \mu}$ with the natural matrix product form the *Lorentz group* $O(1, 3)$.

2.1.3   Generators of the symmetry transformations

For infinitely small Lorentz transformation one has

$$\Lambda^\alpha_{\ \mu} = \delta^\alpha_{\ \mu} + \zeta^\alpha_{\ \mu}, \tag{2.1.5}$$

where $\zeta^{\alpha}{}_{\mu}$ are the generators of this transformation. Denote $\zeta_{\nu\mu} = \eta_{\nu\alpha}\zeta^{\alpha}{}_{\mu}$, then the condition Eq. (2.1.4) implies

$$\zeta_{\mu\nu} + \zeta_{\nu\mu} = 0. \tag{2.1.6}$$

Thus, a generator of the Lorentz transformations $\zeta_{\mu\nu}$ is a (4×4)-antisymmetric matrix. The total number of independent parameters that define such a matrix is $4 \times (4 - 1)/2 = 6$. 3 of these 6 parameters, ζ_{ij}, are responsible for spatial rotations, while the other 3 parameters, ζ_{0i}, determine the *boost transformations*. Together with 4 generators of translations one has 10 parameters. Thus, the group of transformations that preserve the form of the interval in the Minkowski spacetime has 10 parameters. This group is called the *Poincaré group*.

2.1.4 Killing equation

Consider now how spacetime points are moved under the symmetry transformation. For an infinitely small transformation one has

$$\begin{aligned}
X^{\mu} &\to X^{\mu} + \xi^{\mu}, \\
\xi^{\mu} &= \varepsilon^{\mu} + \eta^{\mu\alpha}\zeta_{\alpha\nu}X^{\nu}.
\end{aligned} \tag{2.1.7}$$

Here, $\xi^{\mu}(X)$ is a generator of the symmetry transformations. It is called a *Killing vector*. It depends on 10 parameters ε^{μ} and $\zeta_{\alpha\nu}$. It is easy to check that it obeys the following *Killing equation*.

$$\xi_{\mu,\nu} + \xi_{\nu,\mu} = 0. \tag{2.1.8}$$

Let us show that Eq. (2.1.7) is the most general solution of this equation. For this purpose we write a general Taylor expansion of an arbitrary vector field ξ^{μ} near $X^{\mu} = 0$

$$\xi_{\mu}(X) = \sum_{J=0}^{\infty} \zeta_{\mu\alpha_1\ldots\alpha_J}X^{\alpha_1}\ldots X^{\alpha_J}, \qquad \zeta_{\mu\alpha_1\ldots\alpha_J} = \zeta_{\mu(\alpha_1\ldots\alpha_J)}. \tag{2.1.9}$$

For $J = 0$ the equation Eq. (2.1.8) does not impose any conditions, while for $J \geq 1$ one has

$$\zeta_{(\mu\nu)\ldots\alpha_J} = 0. \tag{2.1.10}$$

For $J = 1$ it means that $\zeta_{\mu\nu}$ is an antisymmetric matrix. For $J > 1$ one has

$$\zeta_{\mu\alpha_1\alpha_2\ldots\alpha_J} = -\zeta_{\alpha_1\mu\alpha_2\ldots\alpha_J} = -\zeta_{\alpha_1\alpha_2\ldots\alpha_J\mu} = 0. \tag{2.1.11}$$

The last equality follows from the fact that $\zeta_{\alpha_1\alpha_2\ldots\alpha_J\mu}$ must be simultaneously antisymmetric with respect α_1 and α_2, as follows from Eq. (2.1.10), and symmetric with respect to the same indices, as follows from Eq. (2.1.9).

In Cartesian coordinates the vectors of the symmetry generators can be written as a linear combination with constant coefficients of the following vectors

$$\xi^{\hat{\alpha}}_{(\hat{\mu})}\partial_{\hat{\alpha}} = \partial_{\hat{\mu}}, \qquad \lambda^{\hat{\alpha}}_{(\hat{\mu})(\hat{\nu})}\partial_{\hat{\alpha}} = X_{\hat{\mu}}\partial_{\hat{\nu}} - X_{\hat{\nu}}\partial_{\hat{\mu}}. \tag{2.1.12}$$

Here, the symbols $(\hat{\mu})$ and $(\hat{\nu})$ are used to enumerate these generators.

2.2   Minkowski Spacetime in Curved Coordinates

2.2.1   Metric

Working in Minkowski spacetime, one is not restricted by using only the Cartesian coordinates. In fact, the choice of the coordinates is a matter of convenience. Let x^α be coordinates in a spacetime domain V related to the Cartesian coordinates as follows $x^\alpha = x^\alpha(X^\mu)$. We assume that $\det(\partial X^\mu/\partial x^\alpha) \neq 0$ in V, so that there exists an inverse transformation $X^\mu = X^\mu(x^\alpha)$. Since

$$dX^\mu = \frac{\partial X^\mu}{\partial x^\alpha}\, dx^\alpha, \tag{2.2.1}$$

one can rewrite Eq. (2.1.1) in the form

$$ds^2 = g_{\alpha\beta}(x)\, dx^\alpha\, dx^\beta, \tag{2.2.2}$$

where

$$g_{\alpha\beta}(x) = \eta_{\mu\nu}\, \frac{\partial X^\mu}{\partial x^\alpha}\, \frac{\partial X^\nu}{\partial x^\beta} \tag{2.2.3}$$

is the metric of the Minkowski spacetime in the curved coordinates x^α.

Consider two sets of the curved coordinates x^α (acting in a domain) V and $x'^{\lambda'}$ (acting in a domain V'). In the intersection of these domains $V \cap V'$ the curved coordinates x^α and $x'^{\lambda'}$ are related to one another $x^\alpha = x^\alpha(x'^{\lambda'})$. In these coordinates the flat spacetime metric is $g_{\alpha\beta}(x)$ and $g'_{\lambda'\rho'}(x')$, respectively. It is easy to show that

$$g'_{\lambda'\rho'}(x') = \frac{\partial x^\alpha}{\partial x'^{\lambda'}}\, \frac{\partial x^\beta}{\partial x'^{\rho'}}\, g_{\alpha\beta}(x). \tag{2.2.4}$$

In other words the metric $g_{\alpha\beta}(x)$ is a tensor of the second order, which is symmetric and has the signature $(-, +, +, +)$ (see Section 3.2).

There are several standard adopted rules that allow one to write this and similar relations in a short form. We have already used one of them, namely the *Einstein summation rule*. In accordance with this rule we omit a symbol of summation and always assume that for any two repeated indices, one in the upper position and the other in the down position, there is summation. We shall also use indices with primes to refer to the 'new' coordinates and use 'comma' for partial derivatives. So we write $x^{\lambda'} \equiv x'^{\lambda'}$ and $\partial x^\alpha/\partial x'^{\lambda'} = x^\alpha_{,\lambda'}$. Using these notations we can rewrite Eq. (2.2.4) as

$$g_{\lambda'\rho'} = x^\alpha_{,\lambda'} x^\beta_{,\rho'} g_{\alpha\beta}. \tag{2.2.5}$$

One has

$$x^\alpha_{,\lambda'} x^{\lambda'}_{,\beta} = \delta^\alpha_\beta, \qquad x^{\lambda'}_{,\alpha} x^\alpha_{,\nu'} = \delta^{\lambda'}_{\nu'}. \tag{2.2.6}$$

Determine the inverse metric $g^{\alpha\beta}$ by the relation

$$g^{\alpha\gamma} g_{\gamma\beta} = \delta^\alpha_\beta, \tag{2.2.7}$$

then its transformation law is

$$g^{\lambda' \rho'} = x^{\lambda'}_{,\alpha} x^{\rho'}_{,\beta} g^{\alpha \beta}. \tag{2.2.8}$$

▌ Problem 2.1: *Prove the relations Eq. (2.2.6) and Eq. (2.2.8).*

Suppose one has a metric $g_{\alpha \beta}(x)$ determined in some domain V. A natural question is: Is it possible to find such a coordinate transformation $X^{\mu}(x^{\alpha})$ so that in the new coordinates X^{μ} the metric takes the form Eq. (2.1.1). In a general case the answer is negative. Relations Eq. (2.2.3) form a set of first-order partial differential equations for $X^{\mu}(x^{\alpha})$. The existence of a solution requires fulfillment of what is called the integrability conditions. As we shall see later, these integrability conditions are equivalent to the requirement that the curvature of spacetime vanishes.

According to the adopted notations, in order to indicate that we are dealing with objects in a 'new' coordinate system, we put primes over its indices. Thus, looking at the indices one may easily conclude which coordinates are used. We also use 'bold-faced' letters as symbols of vectors, tensors, and other geometrical objects. For the components of such an object in some coordinates we use the regular-font version of the same letter. For example $g_{\mu\nu}$ are components of the metric tensor **g**. In a flat spacetime there exist preferable (inertial) reference frames and connected with them preferable Cartesian coordinates. Quite often, to indicate that the components of the object are given in these special coordinates we shall use the 'hat' over the corresponding indices. For example, $A^{\hat{\mu}}$ are the components of a vector **A** in the Cartesian coordinates. Later, we shall use hats similarly, for the components of an object in a *local orthonormal frame*.

2.2.2 Observables in curved coordinates

In special relativity the Cartesian coordinates is a natural and convenient choice of coordinates. However, one can use any other (curved) coordinates. The result of calculations of the physical observables does not depend on this choice. This invariance becomes more transparent if the calculations are done in the covariant form.

Tensors

First, any tensor **T**, which in the Cartesian coordinates $X^{\hat{\mu}}$ has components $T^{\hat{\nu}_1...\hat{\nu}_m}_{\hat{\mu}_1...\hat{\mu}_n}$, in curved coordinates x^{α} has the components

$$T^{\beta_1...\beta_m}_{\alpha_1...\alpha_n} = X^{\hat{\mu}_1}_{,\alpha_1} \ldots X^{\hat{\mu}_n}_{,\alpha_n} x^{\beta_1}_{,\hat{\nu}_1} \ldots x^{\beta_m}_{,\hat{\nu}_m} T^{\hat{\nu}_1...\hat{\nu}_m}_{\hat{\mu}_1...\hat{\mu}_n}. \tag{2.2.9}$$

Here, $X^{\hat{\mu}}_{,\alpha} = \partial X^{\hat{\mu}}/\partial x^{\alpha}$ and $x^{\beta}_{,\hat{\nu}} = \partial x^{\beta}/\partial X^{\hat{\nu}}$.

▌ Problem 2.2: *Suppose x^{α} and $x^{\alpha'}$ are two curved coordinate systems in a flat spacetime. Prove that the components of the tensor **T** determined in these coordinates by Eq. (2.2.9) are related as follows*

$$T^{\beta'_1...\beta'_m}_{\alpha'_1...\alpha'_n} = x^{\alpha_1}_{,\alpha'_1} \ldots x^{\alpha_n}_{,\alpha'_n} x^{\beta'_1}_{,\beta_1} \ldots x^{\beta'_m}_{,\beta_m} T^{\beta_1...\beta_m}_{\alpha_1...\alpha_n}. \tag{2.2.10}$$

Covariant derivatives

Physical laws formulated in a flat spacetime contain usually two operations, differentiation and integration, which have a standard definition in Cartesian coordinates. As soon as one starts working in curved coordinates one needs to properly modify the definition of these operations.

Let us consider first the operation of the partial derivative. Suppose, for example, **A** is a covector and its components in the Cartesian coordinates are $A_{\hat{\mu}}$. Let

$$B_{\hat{\mu}\hat{v}} = \frac{\partial A_{\hat{\mu}}}{\partial X^{\hat{v}}}. \tag{2.2.11}$$

Both of the objects, **A** and **B** can be written in curved coordinates

$$A_\alpha = X^{\hat{\mu}}_{,\alpha} A_{\hat{\mu}}, \qquad B_{\alpha\beta} = X^{\hat{\mu}}_{,\alpha} X^{\hat{v}}_{,\beta} B_{\hat{\mu}\hat{v}}. \tag{2.2.12}$$

It is evident that $B_{\alpha\beta} \neq A_{\alpha,\beta}$. In fact, they are related as follows

$$B_{\alpha\beta} = \nabla_\beta A_\alpha \equiv \frac{\partial A_\alpha}{\partial x^\beta} - \Gamma^\gamma_{\alpha\beta} A_\gamma. \tag{2.2.13}$$

Here,

$$\Gamma^\gamma_{\;\alpha\beta} = \Gamma^\gamma_{\;(\alpha\beta)} = -\frac{\partial X^\mu}{\partial x^\alpha} \frac{\partial X^\nu}{\partial x^\beta} \frac{\partial^2 x^\gamma}{\partial X^\nu \partial X^\mu},$$

$$\Gamma^\gamma_{\;\alpha\beta} = g^{\gamma\delta} \Gamma_{\delta\alpha\beta}, \qquad \Gamma_{\delta\alpha\beta} = \frac{1}{2} \left(\partial_\alpha g_{\beta\delta} + \partial_\beta g_{\alpha\delta} - \partial_\delta g_{\alpha\beta} \right). \tag{2.2.14}$$

In the presence of the metric the operator ∇_α, called the *covariant derivative*, can be easily defined for an arbitrary tensor. If ∇_α acts on a scalar function, it coincides with a partial derivative ∂_α, while acting on a tensor it gives a new tensor with one additional index α in the low position.

Problem 2.3: *Suppose we define the covariant derivative so that the following relation*

$$\nabla_\beta (A_\alpha B^\alpha) = (\nabla_\beta A_\alpha) B^\alpha + A_\alpha \nabla_\beta B^\alpha \tag{2.2.15}$$

is valid for arbitrary **A** *and* **B**. *Using Eq. (2.2.13) prove that*

$$\nabla_\beta B^\alpha = \partial_\beta B^\alpha + \Gamma^\alpha_{\beta\gamma} B^\gamma. \tag{2.2.16}$$

Using a covariant derivative, the Killing equation Eq. (3.7.4) can be written as

$$\nabla_{(\alpha} \xi_{\beta)} = 0. \tag{2.2.17}$$

Integration

Similarly, the integration over the Cartesian coordinates can be rewritten in the covariant manner, that is in the form that does not depend on a particular choice of the coordinates. Suppose φ is a scalar function. Consider an integral

$$J = \int \varphi(X) \, d^4 X. \tag{2.2.18}$$

Using curved coordinates x^α we write it in the form

$$J = \int \varphi(X(x)) \, |\det(X^{\hat{\mu}}_{,\alpha})| d^4 x. \tag{2.2.19}$$

Taking the determinant of both sides of the relation

$$g_{\alpha\beta} = X^{\hat{\mu}}_{,\alpha} X^{\hat{\nu}}_{,\beta} \eta_{\hat{\mu}\hat{\nu}} \tag{2.2.20}$$

one obtains

$$-g \equiv -\det g = |\det X^{\hat{\mu}}_{,\alpha}|^2. \tag{2.2.21}$$

Thus, one has

$$J = \int \varphi(X(x)) \sqrt{-g(x)} \, d^4x. \tag{2.2.22}$$

Hence, $\sqrt{-g} d^4x$ is an *invariant volume element*.

2.2.3 Covariance principle and physics in curved spacetime

For physical laws formulated in the Minkowski spacetime there is a simple way to 'upgrade' them in order to include the gravitational interaction. This universal method is based on the *covariance principle*, which states that the physical laws must be valid for an arbitrary choice of the coordinates. This principle is satisfied when the corresponding equations are written in the tensorial form $\mathbf{T} = 0$. In fact if the components of the tensor vanish in some coordinate system, it is also true for any other coordinate system. If a physical system is described by an action, the covariance principle implies that the action must be invariant under coordinate transformations. Physical laws written in the Minkowski spacetime in the Cartesian coordinates can be easily 'upgraded' to satisfy the covariance principle.

For this purpose it is sufficient to do the following:

- Instead of a tensor describing physical fields in the Cartesian coordinates, say $T^{\hat{\nu}}_{\hat{\mu}}$, use its value in a curved coordinates, say T^{ν}_{μ}.
- Change $\eta_{\hat{\mu}\hat{\nu}}$ by the metric $g_{\mu\nu}$.
- Change the partial derivative $\partial/\partial_{\hat{\mu}}$ by the covariant derivative ∇_{μ}.
- Use the invariant volume element $\sqrt{-g(x)} \, d^4x$ instead of d^4X.
- After this is done, use the obtained expressions for an arbitrary (not necessarily flat) metric g.

By applying this procedure one obtains equations for the physical system in an arbitrary gravitational field, described by the metric $g_{\alpha\beta}$.

Strictly speaking, the 'covariantization' procedure is not uniquely determined. The problem may arise, for example, if an expression contains several covariant derivatives. Since the covariant derivatives do not commute in a curved spacetime (see later), their different ordering might give different answers. The resulting difference is proportional to the spacetime curvature. In a general case in order to control such terms, describing direct interaction of the matter with the curvature, additional information based on the observations is required.

Particle motion

To give a simple example of how the *covariance principle* works, let us consider the motion of a particle. The action of a particle in the Minkowski spacetime is

$$S[X^{\hat{\mu}}(\lambda)] = \int_1^2 d\lambda \sqrt{-\eta_{\hat{\mu}\hat{\nu}} \frac{dX^{\hat{\mu}}}{d\lambda} \frac{dX^{\hat{\nu}}}{d\lambda}}. \tag{2.2.23}$$

In accordance with the covariance principle we rewrite it in the form

$$S[x^{\alpha}(\lambda)] = \int_1^2 d\lambda \sqrt{-g_{\alpha\beta} \frac{dx^{\alpha}}{d\lambda} \frac{dx^{\beta}}{d\lambda}}. \tag{2.2.24}$$

Here, $g_{\mu\nu}$ is the flat metric in curved coordinates. Let us skip this restriction, and allow the metric to be arbitrary. The same action Eq. (2.2.24) now describes the motion of a particle in the gravitational field $g_{\alpha\beta}$. Let us note that both of the actions, Eq. (2.2.23) and Eq. (2.2.24), remain invariant under a change of the parametrization $\lambda \to \tilde{\lambda}(\lambda)$ of the curve.

The equations of motion for a particle can be obtained by varying the action Eq. (2.2.24) with the fixed boundary points

$$x^{\mu}(\lambda_1) = x_1^{\mu}, \qquad x^{\mu}(\lambda_2) = x_2^{\mu}. \tag{2.2.25}$$

By performing the variations, and after imposing the following *gauge-fixing condition*

$$g_{\alpha\beta} \frac{dx^{\alpha}}{d\lambda} \frac{dx^{\beta}}{d\lambda} = -1, \tag{2.2.26}$$

one obtains

$$\frac{d^2 x^{\alpha}}{d\lambda^2} + \Gamma_{\beta\gamma}^{\alpha} \frac{dx^{\beta}}{d\lambda} \frac{dx^{\gamma}}{d\lambda} = 0. \tag{2.2.27}$$

This is a *geodesic equation*. The condition Eq. (2.2.26) implies that λ coincides with the proper time along the corresponding worldline of the particle.

▌ **Problem 2.4:** *Derive the geodesic equation Eq. (2.2.27) by the variation of the action Eq. (2.2.24).*

Twin paradox

Consider two points p_1 and p_2 in the Minkowski spacetime connected by a time-like geodesic γ_0. This is a straight line. It is always possible to find a reference frame in which the equation of this line is $(T, 0, 0, 0)$. The proper time between p_1 and p_2 along the geodesic is

$$\Delta T = T_2 - T_1. \tag{2.2.28}$$

Consider now another time-like curve γ connecting the same pair of points. Let $\Delta\tau$ be a proper time between p_1 and p_2 along γ. Let us show that

$$\Delta\tau \leq \Delta T, \tag{2.2.29}$$

and the equality takes place only when γ coincides with γ_0.

This result is a general formulation of the so-called *twin paradox*. Namely, consider twins (say Alice and Bob). Assume that Bob moves inertially (along γ_0), while Alice starts her non-inertial motion (along γ) at the point p_1 and sometime later meets Bob again at the point p_2. The relation Eq. (2.2.29) means that at the moment of their second meeting Alice will be *younger* than Bob.

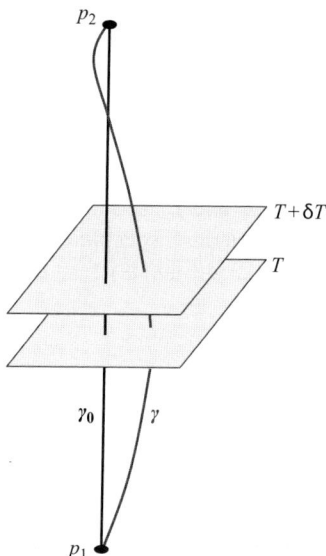

To prove this we consider 2 planes, orthogonal to γ_0 and described by the equations $T = \text{const}$ and $T + \delta T = \text{const}$ (see Figure 2.1). The geodesic γ_0 crosses these planes at the points A and C. Let $\boldsymbol{u}$ be a vector connecting these points. One has $\boldsymbol{u}^2 = -(\Delta T)^2$. (See Figure 2.2). Similarly, B and D are the points where the line γ crosses the planes. Denote by $\boldsymbol{v}$ a vector connecting these points. It is easy to see that

$$\boldsymbol{v} = \boldsymbol{u} + \boldsymbol{s}, \tag{2.2.30}$$

where $\boldsymbol{s}$ is a space-like vector in the plane $T + \Delta T$. For the square of the proper time $\Delta \tau$ between the planes along γ one has

$$(\Delta \tau)^2 \equiv -\boldsymbol{v}^2 = -\boldsymbol{u}^2 - \boldsymbol{s}^2 \leq (\Delta T)^2. \tag{2.2.31}$$

Hence, for any element of the lines γ_0 and γ between the planes T and $T + \Delta T$ one has

$$\Delta \tau \leq \Delta T. \tag{2.2.32}$$

The equality takes place only when $\boldsymbol{s} = 0$. Summing Eq. (2.2.32) over all the segments, one concludes that the proper time between p_1 and p_2 measured in the reference frame of Alice is always shorter (or equal), than the proper time in the Bob's frame. The equality of their proper times is achieved only in the case when for each of the slices the vector $\boldsymbol{s}$ vanishes, that is, when γ coincides with γ_0. For γ close to γ_0 this result can be easily obtained from the variation procedure described above.

Fig. 2.1 Illustration of the 'twin paradox'. Points p_1 and p_2 show two events when Alice and Bob meet one another. A straight line connecting these points is the worldline of Bob, who is moving inertially. In a chosen frame, where he is at rest, this line coincides with the line of time T. The curved line connecting p_1 and p_2 is the worldline of Alice. Two slices $T = \text{const}$ and $T + \delta T = \text{const}$ are shown.

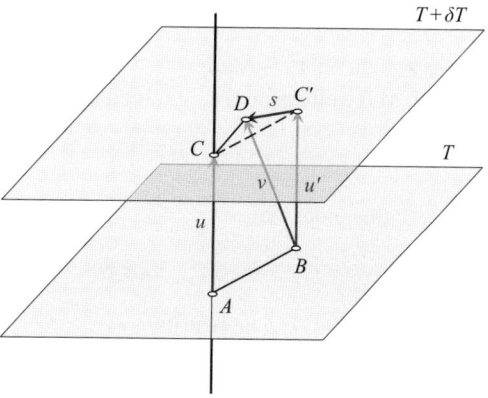

Fig. 2.2 This is a 'zoomed in' version of the previous figure. A segment AC is a segment of Bob's worldline between planes of T and $T + \delta T$. A similar segment for Alice's line is BD. The interval between B and D is always smaller than between A and C. It is equal to the latter only if Alice is at rest with respect to Bob.

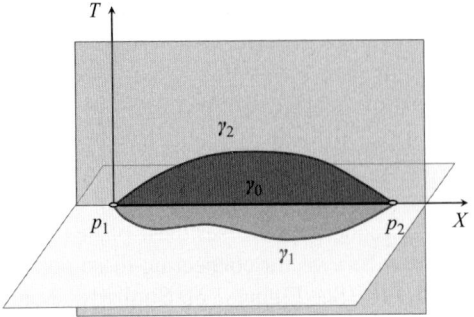

Fig. 2.3 Let γ_0 be a space-like geodesic connecting two spatially separated points p_1 and p_2 and its length be L. Let γ_1 be a space-like 'deformation' of γ_0 in the plane $T = $ const, and γ_2 be a similar time-like 'deformation' of γ_0 in the $(T - X)$-plane. The length of γ_1 is larger than L, while the length of γ_2 is smaller than L.

It is interesting that a space-like geodesic between two points p_1 and p_2 is neither the shortest non the longest line between these points. Figure 2.3 illustrates this. To explain why it is so let us use a frame in which a geodesic γ_0 is given by the equation $(0, X, 0, 0)$, so that the geodesic distance between these points is

$$L = X_2 - X_1. \tag{2.2.33}$$

Consider two space-like curves connecting p_1 with p_2. Let one of them, γ_1 be at the plane $T = $ const, while the other γ_2 be in the $(T - X)$-plane. The proper length of γ_1 is greater than L and the proper length of γ_2 is smaller than L.

■ **Problem 2.5:** *Prove this result.*

Maxwell equations

Let us derive *Maxwell's equations* in an external gravitational field. The action for the electromagnetic field in the flat spacetime is

$$S[A_{\hat{\mu}}(X)] = -\frac{1}{16\pi} \int d^4 X \eta^{\hat{\mu}\hat{\lambda}} \eta^{\hat{\nu}\hat{\rho}} F_{\hat{\mu}\hat{\nu}} F_{\hat{\lambda}\hat{\rho}}, \tag{2.2.34}$$

where $F_{\hat{\mu}\hat{\nu}} = A_{\hat{\nu},\hat{\mu}} - A_{\hat{\mu},\hat{\nu}}$. Its covariant generalization is

$$S[A_\alpha(x)] = -\frac{1}{16\pi} \int d^4 x \sqrt{-g}\, g^{\alpha\beta} g^{\gamma\delta} F_{\alpha\gamma} F_{\beta\delta}, \tag{2.2.35}$$

where $F_{\alpha\beta} = \nabla_\alpha A_\beta - \nabla_\beta A_\alpha$.

The Maxwell equations, obtained by the variation of this action, are

$$\left(\sqrt{-g} F^{\alpha\beta}\right)_{,\beta} = 0, \qquad F^{\alpha\beta} = g^{\alpha\gamma} g^{\beta\delta} F_{\gamma\delta}. \tag{2.2.36}$$

To prove this, note that

$$\nabla_\alpha A_\beta - \nabla_\beta A_\alpha = A_{\beta,\alpha} - A_{\alpha,\beta}. \tag{2.2.37}$$

Equation (2.2.36) can be rewritten as

$$F^{\alpha\beta}{}_{;\beta} = 0. \tag{2.2.38}$$

Here and later, we use a semicolon as another notation for the covariant derivative, $(\ldots)_{;\alpha} = \nabla_\alpha(\ldots)$. It is easy to see that Eq. (2.2.38) is the covariant form of the standard Maxwell equations in the flat spacetime

$$F^{\hat\mu\hat\nu}_{,\hat\nu} = 0. \tag{2.2.39}$$

2.3 Uniformly Accelerated Reference Frame

2.3.1 Uniformly accelerated motion

Consider a worldline in the Minkowski spacetime given by the following equation

$$X^{\hat\mu}(\tau) = (a^{-1}\sinh(a\tau),\ a^{-1}\cosh(a\tau),\ 0,\ 0), \tag{2.3.1}$$

where τ is a parameter along the trajectory. The line $X^{\hat\mu}(\tau)$ is a hyperbola. At $\tau = 0$, when a particle is at $X^0 = 0$, it is at the distance $l = a^{-1}$ from the origin O. (In dimensionful units $l = c^2 a^{-1}$.) It is easy to show that the interval between the point of the worldline and the origin remains the same and equal to $l = a^{-1}$ for any moment of time τ.

Simple calculations give the velocity u and acceleration w for such a motion

$$u^{\hat\mu} = \frac{dX^{\hat\mu}}{d\tau} = (\cosh(a\tau),\ \sinh(a\tau),\ 0,\ 0),$$

$$w^{\hat\mu} = \frac{du^{\hat\mu}}{d\tau} = (a\sinh(a\tau),\ a\cosh(a\tau),\ 0,\ 0), \tag{2.3.2}$$

$$u^2 = u_{\hat\mu} u^{\hat\mu} = -1, \qquad w^2 = w_{\hat\mu} w^{\hat\mu} = a^2. \tag{2.3.3}$$

The first of the relations Eq. (2.3.3) implies that τ is the proper time along the trajectory, while the second one means that the acceleration is constant. Such a motion, when both the direction and the value of the acceleration remain constant, is called a *uniformly accelerated motion* (see Figure 2.4).

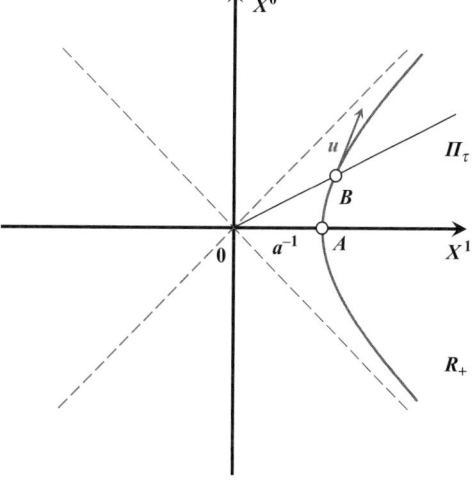

Fig. 2.4 Uniformly accelerated motion. Particle velocity and its acceleration are in the (X^0-X^1)-plane. The particle worldline is shown by a solid line. It is a hyperbola. At the moment $X^0 = 0$ it crosses the X^1 line at a point A, located at the distance a^{-1} from the center of the frame O. Planes orthogonal to the 4-velocity vector u, such as Π_τ, pass through the origin O.

Problem 2.6:   *Consider a charged (with charge e) particle of mass m that is initially at rest. Suppose at the moment of time $X^0 = 0$ one switches on a constant electric field $F_{\hat{\mu}\hat{\nu}} = -2E\,\delta^0_{[\hat{\mu}}\delta^1_{\hat{\nu}]}$. Prove that the particle moves with a constant acceleration a. Find its value. Use the equation of the motion*

$$\frac{du^{\hat{\mu}}}{d\tau} = \varepsilon F^{\hat{\mu}}{}_{\nu}\,u^{\nu}, \qquad \varepsilon = \frac{e}{m}. \qquad (2.3.4)$$

2.3.2   Uniformly accelerated frame

For the accelerated observer events simultaneous with $X^{\mu}(\tau)$ at a given moment of time τ form the hyperplane Π_{τ} orthogonal to the velocity $u^{\mu}(\tau)$. Since $w_{\mu}\,u^{\mu} = 0$, the vector of the acceleration at the moment of time τ belongs to the plane Π_{τ}. The points of this plane can be parametrized by 3 coordinates (ρ, X^2, X^3) and the plane equation is

$$X^{\hat{\mu}} = (\rho\sinh(a\tau_0),\ \rho\cosh(a\tau_0),\ X^2,\ X^3). \qquad (2.3.5)$$

A one-parameter family of planes Π_{τ} for $\tau \in (-\infty, \infty)$ sweeps the right wedge R_+ of the Minkowski spacetime (see Figure 2.5). In this wedge one can use (τ, ρ, X^2, X^3) as new curved coordinates. A line with a fixed value of (ρ, X^2, X^3) is a worldline of a uniformly accelerated observer with the acceleration ρ^{-1}. The closer a line is to the origin O the higher is the acceleration. The right wedge R_+ is covered by a 3-parameter family of trajectories of uniformly accelerated observers. This foliation is parametrized by 3 parameters (ρ, X^2, X^3).

All hyperplanes Π_{τ} intersect at the two-dimensional plane $X^0 = X^1 = 0$. Thus, the uniformly accelerated frame constructed cannot be extended to the global one beyond the region $X^0 - X^1 < 0,\ X^0 + X^1 > 0$ (the wedge R_+). Inside the wedge R_+ a spacetime point p can be parametrized by 4 coordinates (τ, ρ, X^2, X^3). We choose one of the accelerated observers to serve as the origin of the accelerated frame. If the particle acceleration is a, the distance of this observer from the origin O of the Minkowski frame is $l = c^2/a$. Note that this distance, as measured by an accelerated observer, remains constant.

One can introduce what we call a *local reference frame*. We define it as a set of 4 mutually orthogonal (in the Minkowski metric) unit vectors $e^{\alpha}_{\hat{\nu}}$. The local reference frame for a uniformly accelerated observer $u^{\hat{\mu}}$ is formed by 4 mutually orthogonal unit vectors

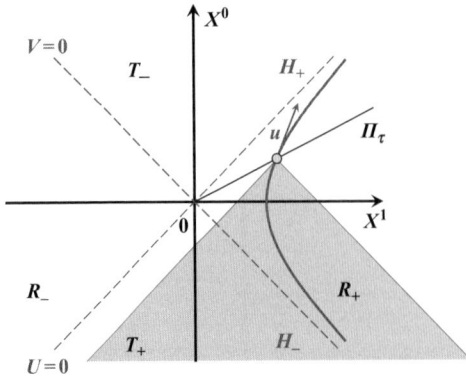

Fig. 2.5 Uniformly accelerated observer is shown by a solid line. A shadowed 'triangle' region shows the set of events that such an observer can see at the moment τ. In the limit $\tau \to \infty$ this region tends to the half-space located below H_+. The events above H_+ would have never be seen by this observer. Thus, a null plane H_+ plays the role of the event horizon. The future event horizon, H_+, and the past event horizon, H_-, divide the spacetime into 4 domains: In R_+: $X^1 > |X^0|$; in R_-: $-X^1 > |X^0|$; in T_-: $-X^0 > |X^1|$; and in T_+: $X^0 > |X^1|$.

$$e^\mu_{\hat{1}} = u^{\hat{\mu}}, \quad e^\mu_{\hat{1}} = \frac{1}{a} w^\mu, \quad e^\mu_{\hat{2}} = \delta^\mu_2, \quad e^\mu_{\hat{3}} = \delta^\mu_3. \qquad (2.3.6)$$

2.4 Homogeneous Gravitational Field

2.4.1 Metric

We defined 'new' coordinates $x^\alpha = (\tau, \rho, X^2, X^3)$ connected with a uniformly accelerated frame. Let us calculate the line element in these 'new' coordinates. Using the following expressions for the coordinate differentials

$$dX^0 = \sinh(a\tau)\,d\rho + a\rho\cosh(a\tau)\,d\tau,$$
$$dX^1 = \cosh(a\tau)\,d\rho + a\rho\sinh(a\tau)\,d\tau, \qquad (2.4.1)$$

one obtains

$$-(dX^0)^2 + (dX^1)^2 = -a^2\,\rho^2\,d\tau^2 + d\rho^2. \qquad (2.4.2)$$

We denote $y = X^2$ and $z = X^3$. Then,

$$ds^2 = -a^2\,\rho^2\,d\tau^2 + d\rho^2 + dy^2 + dz^2. \qquad (2.4.3)$$

The frame of reference determined by these coordinates is rigid. This means that the distance between any two points with fixed values of the spatial coordinates (ρ, y, z) does not depend on the moment of time τ when it is measured.

An observer that is at rest with respect to this frame, that is that has constant values of its (ρ, y, z) coordinates, is equivalent to the observer moving with a constant accelerated in the Minkowski spacetime with respect to its inertial frame. The metric Eq. (2.4.3) is often called the *Rindler metric*, and the spacetime with such a metric is called a *Rindler spacetime*. One can interpret the metric Eq. (2.4.3) as a metric describing a *'homogeneous gravitational field'*.

> *Strictly speaking, this gravitational field is constant in time and homogeneous in y-and z-directions, but depends on the ρ coordinate. This dependence of the acceleration on the ρ is a price for the requirement of rigidity of the frame. Nevertheless, traditionally in the literature this metric is still presented as describing a 'homogeneous gravitational field', since in the local vicinity of the accelerated observer the inhomogeneity is small and the Rindler metric describes the system in a constant gravitational filed, which is the closest analog of the Einstein lift.*

According to the *equivalence principle*, physical laws in the homogeneous gravitational field are locally equivalent to the physical laws in a uniformly accelerated frame provided its acceleration $\vec{a}$ is equal to the acceleration $-\vec{g}$ of the free fall.

2.4.2 Motion of particles

The same problem in the Minkowski spacetime can be analyzed either by using the Cartesian coordinates or the Rindler frame. Certainly the answers will be the same. Let us compare these two kind of calculations for a free particle.

A free-particle motion in the Minkowski spacetime has a straight-line trajectory. In the Cartesian coordinates its equation is

$$X^{\hat{\mu}} = X_0^{\hat{\mu}} + v^{\hat{\mu}}\,t. \qquad (2.4.4)$$

$\mathbf{v}^2 = -1$ for the particle and $\mathbf{v}^2 = 0$ for the light. For the massive particle the parameter t is the proper time along the particle trajectory, while for the light t is the *affine parameter*. For the motion in the $X^2 = X^3 = 0$ plane one has

$$X^{\hat{\mu}} = b\,\delta_1^{\hat{\mu}} + v^{\hat{\mu}}\,t, \qquad v^{\hat{\mu}} = (\gamma, \beta\gamma, 0, 0), \qquad (2.4.5)$$

where $\gamma = 1/\sqrt{1 - \beta^2}$ is the *Lorentz gamma factor*.

To perform similar calculations for a free-particle motion in the *Rindler frame* one needs to solve a *geodesic equation*

$$\frac{d^2 x^\alpha}{d\lambda^2} + \Gamma^\alpha_{\beta\gamma} \frac{dx^\beta}{d\lambda} \frac{dx^\gamma}{d\lambda} = 0. \qquad (2.4.6)$$

Instead of solving these equations directly one can simply rewrite relations Eq. (2.4.5) in the Rindler coordinates

$$\rho \sinh(a\tau) = \gamma t, \qquad \rho \cosh(a\tau) = b + \gamma\beta t. \qquad (2.4.7)$$

Let us define τ_0 by the relation

$$\tanh(a\tau_0) = \beta. \qquad (2.4.8)$$

Then, one has

$$\rho = \frac{b \cosh(a\tau_0)}{\cosh[a(\tau - \tau_0)]}. \qquad (2.4.9)$$

We assume $b > 0$, so that the particle cross the R_+ region. It is easy to check that Eq. (2.4.9) is a required solution of the geodesic equations Eq. (2.4.6) in the *Rindler metric*.

Problem 2.7:   *Check that the relation Eq. (2.4.9) is a solution of the geodesic equations Eq. (2.4.6) in the Rindler metric.*

Let us define null coordinates U and V by relations

$$U = \frac{X^0 - X^1}{\sqrt{2}}; \qquad V = \frac{X^0 + X^1}{\sqrt{2}}. \qquad (2.4.10)$$

Conditions $U = 0$ and $V = 0$ determine two null planes. These planes are called future and past event horizons, respectively. In the null coordinates the region R_+ is defined by the equations $U < 0$ and $V > 0$.

Thus, the future and past horizons are boundaries of R_+ in the future and in the past, respectively. In its motion, the particle first crosses the past horizon and then the future one. Proper time Δt (in the particle's reference frame), which is required for the particle to propagate between the horizons, is

$$\Delta t = 2b\gamma. \qquad (2.4.11)$$

The corresponding Rindler time τ is infinite. In the Rindler frame the particle motion can be described as follows. At $\tau = -\infty$ the particle starts its motion at the horizon $\rho = 0$. It reaches the maximal distance from the horizon $\rho = \rho_0 = \gamma b$ at $\tau = \tau_0$, and after this falls back to the horizon. Its motion, as seen by the Rindler observer, is slowed down, so that it takes an infinite time τ for the particle to reach the horizon.

2.4.3 Electric field of a point-like charge

As another example, consider a problem of electrostatics in a homogeneous gravitational field. Namely, suppose that an electric point-like charge q is located at $\vec{r} = (\rho = a^{-1}, 0, 0)$ in the Rindler space. Denote by φ its electric potential in the Rindler coordinates. One has

$$A_\alpha = (-\varphi, 0, 0, 0), \qquad F_{\alpha\beta} = \nabla_\alpha A_\beta - \nabla_\beta A_\alpha = -2\delta^i_{[\alpha}\delta^\tau_{\beta]}\varphi_{,i}. \tag{2.4.12}$$

Since $\sqrt{-g} = a\rho$, it is easy to check that the Maxwell equation

$$F^{\alpha\beta}{}_{;\beta} = 4\pi J^\alpha \tag{2.4.13}$$

outside the charge is equivalent to the following equation

$$\varphi_{,\rho\rho} - \frac{1}{\rho}\varphi_{,\rho} + \varphi_{,yy} + \varphi_{,zz} = 0. \tag{2.4.14}$$

A required solution of this equation is

$$\varphi = qa \frac{\rho^2 + y^2 + z^2 + a^{-2}}{\sqrt{(\rho^2 + y^2 + z^2 + a^{-2})^2 - 4\rho^2 a^{-2}}}. \tag{2.4.15}$$

The potential φ satisfies Eq. (2.4.14) outside the source. This can be checked by direct substitution. To demonstrate that q is the charge of the source for this solution we write $\rho = a^{-1} + x$ and expand the function φ, near the position of the charge $x = y = z = 0$. One has

$$\varphi \approx \frac{q}{\sqrt{x^2 + y^2 + z^2}}. \tag{2.4.16}$$

This relation shows that q is the electric charge.

The Rindler reference frame is defined by static observers located at constant (ρ, y, z). Their 4-velocity $u^\alpha = ((a\rho)^{-1}, 0, 0)$. Now, we can calculate the electric field defined in the Rindler reference frame as

$$E_\alpha = F_{\alpha\beta}\, u^\beta. \tag{2.4.17}$$

Evidently, it is orthogonal to u^α and, hence, has only three (spatial) components

$$\vec{E} = -(a\rho)^{-1}\vec{\nabla}\varphi. \tag{2.4.18}$$

The electric-field lines are shown in Figure 2.6. Near the charge they are practically radially directed straight lines. They bend outside the charge and enter the horizon perpendicularly. The horizon resembles a conducting surface.

Problem 2.8: *A point charge that is at rest in the Rindler frame moves with a constant acceleration in the Minkowski spacetime. Using the expression for the Liénard-Wiechert potential for such a moving charge, obtain the expression Eq. (2.4.15).*

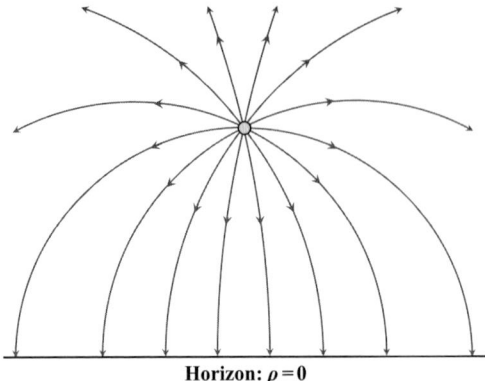

Horizon: $\rho = 0$

Fig. 2.6 The electric-field strength lines for a point-like charge at rest in a homogeneous gravitational field.

2.4.4   Field propagation in the Rindler spacetime

As a last example, let us consider a massless field propagation in the Rindler spacetime. For simplicity, we discuss a massless scalar field φ obeying the equation

$$\Box \varphi = 0. \tag{2.4.19}$$

One can think about such a field as a 'poor man's version of the Maxwell field. The main difference between these two cases is that the electromagnetic field has a spin. When spin effects are not important, spinless scalar field propagation in many aspects is similar to the propagation of the electromagnetic waves.

A solution of Eq. (2.4.19) in the Cartesian coordinates can be decomposed into modes

$$\varphi = \varphi_0 e^{i\Phi}, \qquad \Phi = k_{\hat{\mu}} X^{\hat{\mu}} = -\omega T + \vec{k}\vec{X}. \tag{2.4.20}$$

These modes are plane monochromatic waves with the *dispersion relation* $\omega = |\vec{k}|$. Four-vector k has components $(\omega, \vec{k})$, where ω is the frequency, and $\vec{k}$ is the wave vector. The quantity Φ is a phase function.

Consider a plane wave of frequency ω propagating in the X^1-direction

$$\Phi = -\omega(X^0 - X^1). \tag{2.4.21}$$

In the Rindler coordinates one has

$$\Phi \equiv -\frac{\omega}{a}[\sinh(a\tau) - \cosh(a\tau)] \equiv \frac{\omega}{a} e^{-a\tau}. \tag{2.4.22}$$

A uniformly accelerated observer is moving with respect to the Minkowski frame. That is why the frequency of the radiation, measured in such a frame, is different from ω due to the *Doppler shift*. The measured frequency is not constant since the velocity of the accelerated observer changes with time. The frequency $\tilde{\omega}$ in the uniformly accelerated frame is

$$\tilde{\omega} = -\frac{d\Phi}{d\tau} = -\Phi_{,\hat{\mu}} u^{\hat{\mu}}. \tag{2.4.23}$$

Using the relation Eq. (2.4.22) one obtains

$$\tilde{\omega} = \omega \, e^{-a\tau}, \tag{2.4.24}$$

where τ is the moment of observation.

The expression Eq. (2.4.21) is valid for waves emitted by an observer who is at rest with respect to the inertial frame connected with the worldline $(X^0 = t, X^1 = \text{const}, 0, 0)$. At the moment when such an 'emitter' crosses the horizon, that is when $\tau \to \infty$, the frequency detected by the uniformly accelerated observer tends to zero, $\tilde{\omega} \to 0$. From the point of view of such an observer the radiation emitted by an inertially moving particle crossing the event horizon experiences an infinite redshift effect. In other words, in the Rindler frame a particle falling into the horizon slows down its motion and becomes frozen near the horizon. Its radiation becomes more and more dark. If one takes into account that the radiation is quantized and during the emitter motion to the horizon only a finite number of quanta is emitted one comes to the conclusion that the horizon is black, or, in other words, it is a surface of a black hole.

Note that the frequency $\tilde{\omega}_0$ measured by an observer at the moment τ_0 corresponding to entering the emitter into R_+ (crossing the H^-) is finite. An explanation of this asymmetry of the properties of the future and past event horizons is the following. The radiation emitted near H^+ is affected by the gravitational redshift and the Doppler redshift. For the radiation emitted near H^- these two effects 'work' in the opposite direction. As a result of this combination, the observed frequency $\tilde{\omega}_0$ remains finite for light emitted at the past horizon.

2.5 Causal Structure

2.5.1 Horizons

The R_+ region, which is covered by the Rindler frame, is only a part of the complete spacetime. It takes a finite proper time for an inertially moving particle to cross R_+. This is evidence of the *geodesic incompleteness* of the Rindler frame. In fact, the complete Minkowski spacetime consists of 4 regions: $R_\pm$ and $T_\pm$ (see Figure 2.6). In the null coordinates

$$U = \frac{X^0 - X^1}{\sqrt{2}} \; ; \qquad V = \frac{X^0 + X^1}{\sqrt{2}}, \tag{2.5.1}$$

these regions are defined as follows

$$R_\epsilon : -\epsilon U > 0, \quad \epsilon V > 0; \qquad T_\epsilon : \epsilon U > 0, \quad \epsilon V > 0. \tag{2.5.2}$$

A null plane H_+, defined by the equation $U = 0$, is called a *future event horizon*. A null plane H_- defined by the equation $V = 0$ is called a *past event horizon*. A uniformly accelerated observer in R_+ cannot see events beyond the future event horizon. The adopted names for $H_\pm$ reflect this property: the future event horizon separates the domain of the spacetime visible by the uniformly accelerated observer from the 'invisible' one. The Rindler observer can see events in T_-. In the terminology adopted in black hole physics R_+ is the *black hole exterior*; T_+ is the *black hole interior*; T_- is the *white hole region* region.

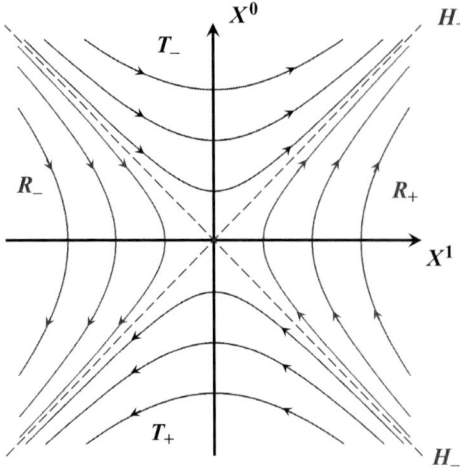

Fig. 2.7 Integral lines of the boost Killing vector in the $(X^0 - X^1)$-plane. These lines are future directed in R_+ and are past directed in R_-. In $T_\pm$ they are space-like. These integral lines are tangent to the horizons $H_\pm$ and coincide there with the horizon generators.

The horizon H_+ is a 3D null surface formed by 2D family of null rays (generators). A generator has the equation

$$U = 0, \qquad y = y_0 = \text{const}, \qquad z = z_0 = \text{const}. \qquad (2.5.3)$$

Problem 2.9: *Show that the distance of the null generator Eq. (2.5.3) from the Rindler observer, as measured in his/her reference frame always remains the same.*

2.5.2 Global structure of the boost generators

The Rindler metric Eq. (2.4.3) is static, that is its coefficients do not depend on time τ. The vector $\xi^\alpha = \delta^\alpha_\tau$ generating this symmetry transformations is a *Killing vector*. In fact, it is easy to check that it obeys the Killing equation Eq. (2.2.17). In the Minkowski spacetime, a complete set of Killing vectors is known (see section 2.1.4). Hence, $\boldsymbol{\xi}$ must be a special linear combination of them. In fact, the Killing vector ξ^α coincides with the generator of the boost transformations in the $(X^0\text{-}X^1)$-plane. Let us consider the properties of the boost transformations in more detail.

Let $\xi^{\hat\mu}$ is a vector in the Minkowski spacetime that has the following components in the Cartesian coordinates

$$\xi^{\hat\mu} = (aX^1, aX^0, 0, 0), \qquad \xi_{\hat\mu} = (-aX^1, aX^0, 0, 0). \qquad (2.5.4)$$

One has

$$\xi_{\hat\mu,\hat\nu} = a(-\delta^0_{\hat\mu}\,\delta^1_{\hat\nu} + \delta^1_{\hat\mu}\,\delta^0_{\hat\nu}) = -\xi_{\hat\nu,\hat\mu}. \qquad (2.5.5)$$

Hence, $\xi^{\hat\mu}$ is a *Killing vector*.
    In the Rindler coordinates one has

$$\xi_\alpha = a\,X^{\hat\mu}_{,\alpha}\,\xi_{\hat\mu} = a(-X^1 X^0_{,\alpha} + X^0 X^1_{,\alpha}). \qquad (2.5.6)$$

Simple calculations give

$$\xi_0 = -a^2 \rho^2, \quad \xi_1 = \xi_2 = \xi_3 = 0, \quad \xi^\alpha = \delta_\tau^\alpha. \tag{2.5.7}$$

Thus, ξ^α is a generator of the displacement along the Rindler time τ (in R_+).
If we use the null coordinates (U, V, X^2, X^3) the flat metric takes the form

$$ds^2 = -2\, dU\, dV + (dX^2)^2 + (dX^3)^2, \tag{2.5.8}$$

while the boost Killing vector is

$$\xi = \xi^\alpha \partial_\alpha = a(-U\,\partial_U + V\,\partial_V). \tag{2.5.9}$$

At $H^\pm$ the Killing vector is null and its integral lines

$$\frac{dx^\mu}{dt} = \xi^\mu \tag{2.5.10}$$

are generators of the future (H^+) and past (H^-) horizons. One also has

$$V = e^{at}, \quad (\text{at} \quad H^+); \qquad U = -e^{-at}, \quad (\text{at} \quad H^-). \tag{2.5.11}$$

2.6 Wick's Rotation in the Rindler Space

After the complex transformation

$$X^0 = iX^4 \tag{2.6.1}$$

the Minkowski metric

$$ds^2 = -(dX^0)^2 + (dX^1)^2 + (dX^2)^2 + (dX^3)^2 \tag{2.6.2}$$

takes the form

$$ds_E^2 = (dX^4)^2 + (dX^1)^2 + (dX^2)^2 + (dX^3)^2. \tag{2.6.3}$$

The transformation Eq. (2.6.1), known as *Wick's rotation*, is often helpful, e.g., for formulation of the quantum field theory.

A similar Wick's rotation can be made in the Rindler space. Under the transformation

$$\tau = i\tau_E \tag{2.6.4}$$

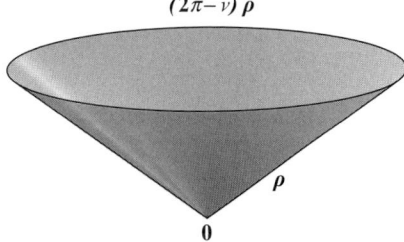

Fig. 2.8 Geometry of the $(\tau\text{-}\rho)$-sector of the space Eq. (2.6.6). The figure shows the embedding diagram for this sector in the 3D flat pace for the positive *angle deficit* ν. For $\nu \neq 0$ the surface has a *conical singularity* at $\rho = 0$.

the Rindler metric

$$ds^2 = -a^2 \rho^2 \, d\tau^2 + d\rho^2 + dy^2 + dz^2, \tag{2.6.5}$$

takes the form

$$ds_{\mathrm{E}}^2 = a^2 \rho^2 \, d\tau_{\mathrm{E}}^2 + d\rho^2 + dy^2 + dz^2. \tag{2.6.6}$$

The metric Eq. (2.6.6) has the Euclidean signature $(+, +, +, +)$ and it is flat everywhere outside $\rho = 0$. At $\rho = 0$ the metric may have a *conical singularity*. This singularity vanishes when τ_{E} is periodic with the period $2\pi/a$. To demonstrate this let us consider a two-dimensional sector $(\tau_{\mathrm{E}}, \rho)$ of the metric Eq. (2.6.6)

$$dl^2 = a^2 \rho^2 \, d\tau_{\mathrm{E}}^2 + d\rho^2. \tag{2.6.7}$$

Denote $\phi = a\tau_{\mathrm{E}}$, then this metric becomes

$$dl^2 = \rho^2 \, d\phi^2 + d\rho^2. \tag{2.6.8}$$

If the angle ϕ has the period 2π, this is a metric of a two-dimensional plane in the polar coordinates. For the different period there is a cone-like singularity at $\rho = 0$. The manifold with the period of ϕ less than 2π can be obtained from the plane by cutting the angle $2\pi - \nu$ and gluing the boundaries of the obtained wedge. The parameter ν is the *angle deficit*. The embedding diagram of this space with the positive angle deficit is shown in Figure 2.8.

> *In quantum field theory the Wick's rotation is often used as a method of finding a solution in the Minkowski spacetime, by solving a similar problem in the Euclidean space. In particular, the Wick's rotation connects statistical mechanics at the temperature Θ to the quantum mechanics of a system, which satisfies the property of periodicity in the imaginary time with the period Θ^{-1}. Using this method one can conclude that a Rindler observer moving with the acceleration a in the Minkowski vacuum would observe the thermal radiation with the temperature $\Theta = a/2\pi$. This effect, known as the* Unruh effect, *really exists and, in principle, it can be observed in experiments (see Section 9.8.5).*

3

Riemannian Geometry

3.1   Differential Manifold. Tensors

3.1.1   Introduction

The theory of gravity gives answers to the following two connected questions:

- How do matter and fields propagate in a given gravitational field?
- What is the gravitational field in the presence of matter and fields?

The covariance principle allows us to answer the first question. The spacetime in the presence of the gravitational field is curved, and the metric g describes this field. In general relativity the gravitational field obeys the Einstein equations. Schematically these equations have the form

$$\partial^2 g + \ldots \sim T. \tag{3.1.1}$$

The stress-energy tensor for the matter distribution T is on the right-hand side of this equation. The left-hand side contains second derivatives of the metric. The dots stand for the terms with fewer derivatives of g. In order to have a well-defined sense, these equations must be valid in arbitrary coordinates. This means that this equation must be a tensor equation. The covariance of the theory imposes severe restrictions on the form of the gravity equations, and, practically determines them uniquely (see Chapter 5).

   If the spacetime is curved, we no longer require that it has the same structure as the Minkowski space. The first fundamental question is: What is a mathematical model of the physical spacetime in general relativity? The adopted model is a *differential manifold* with the metric g. Each point of the manifold represents an *event*. The manifold has a differential structure, which makes it possible to use differential equations to describe physical systems. The metric determines the interval between close events. The differential manifold with the metric is called the *Riemannian geometry*. The matter and fields are described by the tensor fields. The tools developed in Riemannian geometry allow one to write equations of physics in a curved spacetime in a covariant form.

   This paradigm of general relativity has new important features that differentiates it from the physics in the flat spacetime.

   1. Global structure (topology) of the spacetime may be different from the flat one.

> *To illustrate this problem consider a two-dimensional space with constant positive curvature. Such a surface is a round sphere S^2. The topology of this space is different from the topology of a flat*

space R^2. One can also obtain spaces with more complicated topologies by factorizing the sphere S^2 over a discrete subgroup of the group $O(3)$ of rotations.

2. The *causal structure* of the spacetime can also be non-trivial.

The metric determines local null cones at each point. Locally causal curves are curves that at a given point are time-like or null. In the Minkowski spacetime the structure of the local null cones is 'rigid': Each of the local null cone can be obtained from another one by a parallel transport. In a curved spacetime such a 'rigid structure' is absent. As a result, the global causal structure of a spacetime can be very complicated. A black hole is an example of such a non-trivial spacetime.

This chapter contains a brief summary of the main concepts of the differential and Riemannian geometry. Since in this book we shall also speak about models with large extra dimensions, in this mathematical introduction we do not restrict ourselves to four dimensions. We assume that the number of spacetime dimensions is arbitrary. We denote this number by D.

3.1.2 Differential manifold

Definition: *A D-dimensional, C^m, real manifold M^D is a set of points together with a collection of subsets $\{O_\alpha\}$ satisfying the following properties:*

1. *Each point $p \in M^D$ lies in at least one O_α , i.e. the set $\{O_\alpha\}$ covers M^D;*

2. *For each α, there is a one-to-one, onto, map $\Psi_\alpha: O_\alpha \to U_\alpha$, where U_α is an open subset of $\mathbb{R}^D$ (linear D-dimensional space);*

3. *If any two sets O_α and O_β overlap, we can consider the map $\Psi_\beta \circ \Psi_\alpha^{-1}$ ($\circ$ denotes composition) that takes points in $\Psi_\alpha[O_\alpha \cap O_\beta] \subset U_\alpha \subset \mathbb{R}^D$ to points in $\Psi_\beta[O_\alpha \cap O_\beta] \subset U_\beta \subset \mathbb{R}^D$. We require this map to be C^m, i.e. m-times continuously differentiable.*

This definition is illustrated in Figure 3.1. A map Ψ_α is generally called a *chart* by mathematicians and a *coordinate system* by physicists. The collection of all charts is called an *atlas*. The map $\Psi_\beta \circ \Psi_\alpha^{-1}$ describes the transition from one ('old') coordinate system to

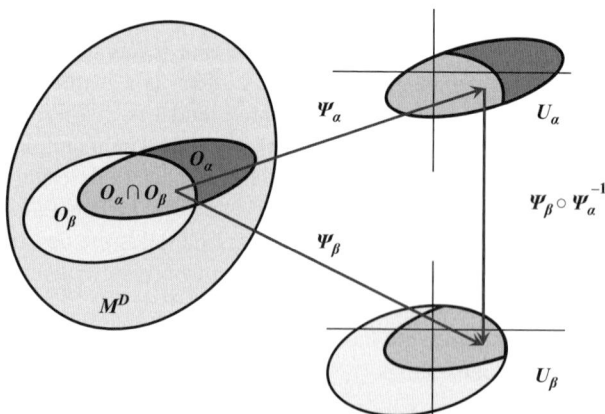

Fig. 3.1 Illustration of the differential manifold definition. M^D is the manifold, O_α and O_β are its two subsets. Their maps to $\mathbb{R}^D$ are Ψ_α and Ψ_β, respectively.

another ('new') one. If x^μ are D numbers labelling points in U_α, and $x^{\mu'}$ are D numbers labelling points in U_β, then the transition from one coordinate system (chart) to another one is given by functions $x^{\mu'} = x^{\mu'}(x^\mu)$[1]. The requirement that this transition is given by smooth functions allows one to use differential equations to describe the physical laws. For simplicity, we assume that $m = \infty$.

Let M^D and N^D be differential manifolds. A map $f : M^D \to N^D$ is said to be a *diffeomorphism* if it is one-to-one, onto, and it as well as its inverse is C^∞.

3.1.3 Tensors

Let p be a point of M^D and x^μ be a coordinate system that covers it. In such a coordinate system a k-times covariant and l-times contravariant tensor $\mathbf{T}$ is described by D^{k+l} numbers $T^{\mu_1\dots\mu_l}_{\nu_1\dots\nu_k}$ (its components), provided the components for any two such coordinate systems x^μ and $x^{\mu'}$ are related as follows

$$T^{\mu_1'\dots\mu_l'}_{\nu_1'\dots\nu_k'} = x^{\mu_1'}_{,\mu_1}\cdots x^{\mu_l'}_{,\mu_l}\, x^{\nu_1}_{,\nu_1'}\cdots x^{\nu_k}_{,\nu_k'}\, T^{\mu_1\dots\mu_l}_{\nu_1\dots\nu_k}. \tag{3.1.2}$$

We call (k, l) a *rank of the tensor*. We also say that it is k-times covariant and l-times contravariant. If all the components of the tensor vanish in one coordinate system, then they vanish in any other coordinates.

> The Kronecker delta symbol δ^ν_μ is *an example of the tensor of the rank* $(1, 1)$. *This follows from the following relation*

$$x^\alpha_{,\alpha'}x^{\beta'}_{,\beta}\delta^{\alpha'}_{\beta'} = \delta^\alpha_\beta. \tag{3.1.3}$$

If tensors $T^{\dots}_{\dots}$ and $P^{\dots}_{\dots}$ have the same rank (k, l), it is possible to define their linear combination: $\alpha\,T^{\dots}_{\dots} + \beta\,P^{\dots}_{\dots}$, which is again a tensor of rank (k, l).

Symmetrization $T_{(\mu_1\dots\mu_p)}$ and *antisymmetrization* $T_{[\mu_1\dots\mu_p]}$ of a tensor $T_{\mu_1\dots\mu_p}$ are defined as follows

$$T_{(\mu_1\dots\mu_p)} = \frac{1}{p!}\sum_{\substack{\text{over all}\\\text{permutations}\\(\mu_1,\dots,\mu_p)}} T_{\mu_1\dots\mu_p},$$

$$T_{[\mu_1\dots\mu_p]} = \frac{1}{p!}\sum_{\substack{\text{over all}\\\text{permutations}\\(\mu_1,\dots,\mu_p)}} (-1)^J\, T_{\mu_1\dots\mu_p}. \tag{3.1.4}$$

Here, $J = 0$ if the permutation is even, and $J = 1$ otherwise. Symmetrization and antisymmetrization over any subset of covariant or contravariant indices are defined similarly.

Contraction $T^{\dots\alpha\dots}_{\dots\alpha\dots}$ of a tensor of the rank (k, l) is a tensor of rank $(k - 1, l - 1)$. Indeed,

$$T^{\dots\alpha'\dots}_{\dots\alpha'\dots} = \delta^{\alpha'}_{\beta'}T^{\dots\beta'\dots}_{\dots\alpha'\dots} = \delta^{\alpha'}_{\beta'}\dots x^\alpha_{,\alpha'}x^{\beta'}_{,\beta}\dots T^{\dots\beta\dots}_{\dots\alpha\dots}. \tag{3.1.5}$$

---

[1] The Greek indices used for coordinates and tensor components take values $(0, 1, \dots, D - 1)$.

Statement: *Suppose the following relation*

$$P^{\alpha\ldots\beta\ldots}_{\gamma\ldots\delta\ldots}\, T^{\delta\ldots\mu\ldots}_{\beta\ldots\nu\ldots} = Q^{\alpha\ldots\mu\ldots}_{\gamma\ldots\nu\ldots} \tag{3.1.6}$$

is valid for arbitrary tensors **P** *and* **Q** *and for an arbitrary coordinate system. Then* **T** *is also a tensor.*

Proof: To prove that **T** has a correct tensorial transformation law we change the 'old' coordinates x^{α} to the 'new' one $x^{\alpha'}$. Suppose the components of the object **T** in the new coordinates are $\widetilde{T}^{\delta'\ldots\mu'\ldots}_{\beta'\ldots\nu'\ldots}$, so that in the new coordinates the relation Eq. (3.1.6) takes the form

$$P^{\alpha'\ldots\beta'\ldots}_{\gamma'\ldots\delta'\ldots}\, \widetilde{T}^{\delta'\ldots\mu'\ldots}_{\beta'\ldots\nu'\ldots} = Q^{\alpha'\ldots\mu'\ldots}_{\gamma'\ldots\nu'\ldots}. \tag{3.1.7}$$

On the other hand, for tensors **P** and **Q** one has

$$P^{\alpha\ldots\beta\ldots}_{\gamma\ldots\delta\ldots} = x^{\gamma'}_{,\gamma}\ldots x^{\delta'}_{,\delta}\ldots x^{\alpha}_{,\alpha'}\ldots x^{\beta}_{,\beta'}\ldots P^{\alpha'\ldots\beta'\ldots}_{\gamma'\ldots\delta'\ldots},$$
$$Q^{\alpha\ldots\mu\ldots}_{\gamma\ldots\nu\ldots} = x^{\gamma'}_{,\gamma}\ldots x^{\nu'}_{,\nu}\ldots x^{\alpha}_{,\alpha'}\ldots x^{\mu}_{,\mu'}\ldots Q^{\alpha'\ldots\mu'\ldots}_{\gamma'\ldots\nu'\ldots}. \tag{3.1.8}$$

Substituting these relations into Eq. (3.1.6) and using the relations

$$x^{\alpha}_{,\alpha'} x^{\alpha'}_{,\beta} = \delta^{\alpha}_{\beta}, \qquad x^{\alpha'}_{,\alpha} x^{\alpha}_{,\beta'} = \delta^{\alpha'}_{\beta'} \tag{3.1.9}$$

we can write Eq. (3.1.6) in the form

$$P^{\alpha'\ldots\beta'\ldots}_{\gamma'\ldots\delta'\ldots}\, x^{\beta}_{,\beta'}\ldots x^{\nu}_{,\nu'}\ldots x^{\delta'}_{,\delta}\ldots x^{\mu'}_{,\mu}\, T^{\delta\ldots\mu\ldots}_{\beta\ldots\nu\ldots} = Q^{\alpha'\ldots\mu'\ldots}_{\gamma'\ldots\nu'\ldots}. \tag{3.1.10}$$

Since both the obtained equation Eq. (3.1.10) and Eq. (3.1.7), are valid for arbitrary **P** and **Q**, one concludes that $\widetilde{T}^{\delta'\ldots\mu'\ldots}_{\beta'\ldots\nu'\ldots}$ are simply the components of **T** in a new coordinate system that obey the tensorial law. In other words, if the components of two tensors **P** and **Q** in any coordinates are related as Eq. (3.1.6), the coefficients **T** form a tensor, so that this is a tensorial equation.

Let x^{μ} be 'old' coordinates and $x^{\mu'}$ be new ones. At a fixed point $A^{\mu}_{\mu'} = x^{\mu}_{,\mu'}$ is $(D \times D)$ non-degenerate matrix. Such matrices with the standard matrix multiplication form a general linear group denoted by $GL(D, R)$. R indicates that the matrices have real components. Consider a tensor $T_{\mu_1\ldots\mu_n}$ of rank $(n, 0)$. Such tensors form a linear space R^N. Its dimension is $N = D^n$. The transformation law

$$T_{\mu_1'\ldots\mu_n'} = A^{\mu_1}_{\mu_1'}\ldots A^{\mu_n}_{\mu_n'} T_{\mu_1\ldots\mu_n} \tag{3.1.11}$$

is a linear transformation of the tensor space induced by the matrix $A^{\mu}_{\mu'}$. Such transformations are known as a representation of the linear group $GL(D, R)$.

By using the operations of symmetrization and antisymmetrization with respect to a selected subset of indices one generates from $T_{\mu_1\ldots\mu_n}$ new tensors of the same rank. Because of the imposed restrictions it has, in general, smaller number of components. The subset of tensors with adopted symmetry properties is again a linear space $\mathbb{R}^K$ with $K < N$. The linear transformations Eq. (3.1.11) respect the symmetry properties. Their action in the subspace of tensors with given symmetry property is again a linear representation of the linear group $GL(D, R)$. This representation is called irreducible if any further application of the operations of symmetrization and antisymmetrization applied to a tensor from $\mathbb{R}^K$ gives either zero or reproduces the initial tensor.

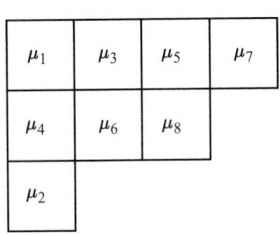

Fig. 3.2 An example of the Young tableau of the shape (4, 3, 1) for $n = 8$.

An elegant way to construct and classify irreducible tensors was proposed by Young (see, e.g., (Penrose and Rindler 1987; Fulton 1997)). First, impose symmetry conditions on some subsets of the indices and then 'saturate' the obtained tensor by the antisymmetrizations. To describe this procedure more accurately one uses what is called the Young tableau. For a tensor of rank $(n, 0)$ we consider a collection of n cells arranged in the left-justified rows. Each of the rows contains equal or fewer cells than the upper one. Put in each of the cells one of the numbers from 1 to n so that the numbers in different cells are different.

An example of the Young tableau of the shape (4, 3, 1) for $n = 8$ is shown in Figure 3.2. According to this tableau let us perform the following operations on the tensor $T_{\mu_1 \dots \mu_8}$.

First, symmetrize it with respect to the groups of indices $\{\mu_1, \mu_3, \mu_5, \mu_7\}$, $\{\mu_4, \mu_6, \mu_8\}$ and $\{\mu_2\}$. (The last operation for our example is trivial.). Then, antisymmetrize the obtained the tensor with respect to the indices $\{\mu_1, \mu_4, \mu_2\}$, $\{\mu_3, \mu_6\}$ and $\{\mu_5, \mu_8\}$. The tensor obtained the as a result of these operations is 'saturated' by symmetries and it is irreducible. This procedure allows one to decompose any tensor into its irreducible components.

The Young tableau allows one to calculate the dimension K of the linear space of irreducible tensors corresponding to it. Figure 3.3 contains two Young diagrams of the same shape as in Figure 3.2. But instead of the indices we fill them with numbers. The left diagram contains numbers corresponding to the dimension D. The second one contains what is known as a hook length. For any cell of the Young diagram this quantity is the number of cells in the same row to the right of it plus the number of cells in the same column below it, plus one. The number K of the independent components of the irreducible tensor for a given Young diagram is the product of all the numbers in the left diagram divided by the product of all the numbers in the right diagram. For our particular example one has

$$K = \frac{(D-2) \times (D-1) \times D^2 \times (D+1)^2 \times (D+2) \times (D+3)}{6 \times 4^2 \times 3 \times 2 \times 1^3}. \tag{3.1.12}$$

The Young tableau for the totally symmetric and totally antisymmetric tensor of rank n are shown in Figure 3.4.

Problem 3.1: *Using these diagrams show that the dimensions of spaces K_S and K_A of such irreducible totally symmetric and antisymmetric tensors, respectively, are*

$$K_S = \frac{D(D+1) \dots (D+n-1)}{n!}, \qquad K_A = \frac{D(D-1) \dots (D-n+1)}{n!}. \tag{3.1.13}$$

For or $n = 2$, $K_S = D(D+1)/2$, and $K_A = D(D-1)/2$ give the number of independent elements of symmetric and antisymmetric $(D \times D)$ matrices, respectively.

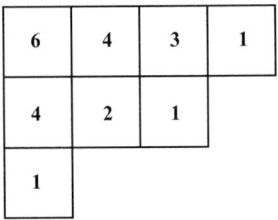

Fig. 3.3 The Young diagrams of the shape (4, 3, 1) necessary for the calculation of the dimension K of the linear space of irreducible tensors with these symmetries.

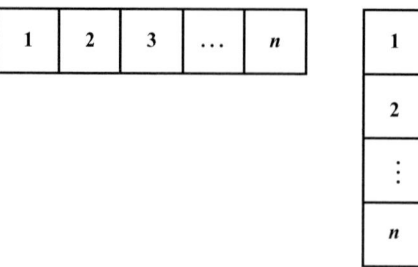

Fig. 3.4 The Young tableau for the totally symmetric and totally antisymmetric tensor of rank n.

3.1.4 Examples of tensors

Scalars

Scalars are the simplest tensors that have no indices and remain invariant under coordinate transformations. Their rank is (0,0).

Tangent space

A tensor of the rank $(0, 1)$ is called a *vector*. In a coordinate frame x^α a vector **A** is determined by its components A^α.

Consider a smooth curve γ on the manifold M^D. It is defined as a smooth map of an interval $(\lambda_1 < \lambda < \lambda_1)$ into M^D. In the coordinates x^α an equation of γ is $x^\alpha = x^\alpha(\lambda)$. According to this definition, two geometrically identical lines in M^D with different parametrization are considered as different curves. Choose a point p on the curve and let λ_0 be the corresponding value of the curve parameter. A tangent vector to γ at p is defined as

$$\xi^\alpha = \left.\frac{dx^\alpha}{d\lambda}\right|_{\lambda_0}. \tag{3.1.14}$$

Consider two close points p and p' with coordinates x^μ and $x^\mu + dx^\mu$, respectively. Under the change of the coordinates dx^μ transforms as follows

$$dx^{\mu'} = x^{\mu'}_{,\nu} dx^\nu. \tag{3.1.15}$$

This means that dx^μ is a vector. Thus, ξ^α defined by Eq. (3.1.14) is a vector as well.

Let the coordinates of p be x_0^α. A coordinate line $\gamma_{\check{\alpha}}$ passing through p is defined as a curve for which all the coordinates, except one, $x^{\check{\alpha}}$, remain constant. We use the index $\check{\alpha} = 0, \ldots, D-1$ to enumerate different coordinate lines passing through p. Evidently, the vectors tangent to the coordinate lines have the following components

$$\xi_{\check{\alpha}}^\alpha = \delta_{\check{\alpha}}^\alpha. \tag{3.1.16}$$

A tangent vector to any smooth curve passing through p can be presented as a linear combination of $\xi_{\check{\alpha}}$. This means that these vectors form a linear D-dimensional space. This space, which is defined at each point of the manifold, is called a *tangent space* and is denoted by T_pM. The basis $\{\xi_{\check{\alpha}}\}$ in T_pM, that is a set of $\xi_{\check{\alpha}}^\mu$ with $\check{\alpha} = 0, \ldots, D-1$, is called a *holonomic basis* associated with the coordinate frame x^α.

A *tangent vector* ξ defines the differential operator

$$\xi = \xi^\mu \, \partial_\mu, \tag{3.1.17}$$

acting on scalar functions on the manifold

$$\varphi \rightarrow \xi^\mu \, \partial_\mu \varphi. \tag{3.1.18}$$

Cotangent space

A tensor of rank $(1,0)$ is called a *covector*. Consider a scalar function φ on the manifold M^D. Its *gradient* in coordinates x^α is $\varphi_{,\alpha}(x)$. It is easy to see that

$$\varphi_{,\alpha'} = x^\alpha_{,\alpha'} \, \varphi_{,\alpha}. \tag{3.1.19}$$

Thus, $\varphi_{,\alpha}(x)$ transforms as the rank $(1,0)$ tensor, or a *covector*.

Consider a point p and a coordinate frame x^α in its vicinity. Choose one of the coordinates and denote it $x^{\check{\alpha}}$ ($\check{\alpha} = 0, \ldots, D-1$). Since $x^{\check{\alpha}}$ maps points in the vicinity of p to numbers, it is a scalar function. Its gradient at p is

$$\omega^{\check{\alpha}}_\alpha = \delta^{\check{\alpha}}_\alpha. \tag{3.1.20}$$

It is evident that any covector can be presented as a linear combination of $\omega^{\check{\alpha}}_\alpha$. This means that covectors form D-dimensional space, and $\omega^{\check{\alpha}}_\alpha$ is a *holonomic basis* associated with the coordinate frame x^α.

For any vector A^α and covector B_α one can define the scalar $\boldsymbol{B}(\boldsymbol{A}) = A^\alpha B_\alpha$. Thus, the *cotangent space* can be identified with a linear space of maps of the *tangent space* to real numbers. For the *holonomic bases* one has

$$\omega^{\check{\alpha}}(\xi_{\check{\beta}}) = \delta^{\check{\alpha}}_{\check{\beta}}. \tag{3.1.21}$$

Bases (not necessarily holonomic) in the tangent and cotangent spaces related by the conditions Eq. (3.1.21) are said to be dual to each other.

Differential forms

Consider antisymmetric tensors of rank $(p,0)$. These tensors form a linear space $\mathbb{R}^{N_p}$. The number N_p of its dimensions is (see Eq. (3.1.13))

$$N_p = \frac{D!}{(D-p)! \, p!}. \tag{3.1.22}$$

Elements of this space are called *p-forms*. In the space of forms two additional operations are defined: i) an *exterior product* (or *wedge product*) $\wedge$ and ii) an *exterior derivative* d. The wedge product $\gamma = \alpha \wedge \beta$ of a *p*-form α and of a *q*-form β is a $(p+q)$ form that has the following components

$$\gamma_{\mu_1 \ldots \mu_p \nu_1 \ldots \nu_q} = \frac{(p+q)!}{p! \, q!} \, \alpha_{[\mu_1 \ldots \mu_p} \beta_{\nu_1 \ldots \nu_q]}. \tag{3.1.23}$$

Here, *antisymmetrization brackets* are defined so that $\alpha_{\mu_1 \ldots \mu_p} = \alpha_{[\mu_1 \ldots \mu_p]}$ and, e.g., $\alpha_{\mu_1 \mu_2} = \alpha_{[\mu_1 \mu_2]} \equiv (\alpha_{\mu_1 \mu_2} - \alpha_{\mu_2 \mu_1})/2!$. The exterior derivative of the *p*-form α is a $(p+1)$-form $d\alpha$ with components

$$d\alpha_{\mu_1...\mu_p\mu_{p+1}} = (p+1)\,\alpha_{[\mu_1...\mu_p,\mu_{p+1}]}. \tag{3.1.24}$$

In D dimensions a completely skew-symmetric object the (Levi-Civita symbol) *is defined as*

$$\varepsilon_{\alpha_1...\alpha_D} = \begin{cases} 1 & - \{\alpha_1\alpha_2...\alpha_D\} = \{0\,1...(D-1)\} \text{ or even permutation}; \\ -1 & - \text{ odd permutation}; \\ 0 & - \text{ all other cases.} \end{cases} \tag{3.1.25}$$

This object is not a tensor. In the 'new' coordinates $x^{\alpha'}$ *it has the same components* $\varepsilon_{\alpha'_1...\alpha'_D}$, *while the tensorial transformation gives*

$$\tilde{\varepsilon}_{\alpha'_1...\alpha'_D} \equiv x^{\beta_1}_{\alpha'_1} \cdots x^{\beta_D}_{\alpha'_D}\,\varepsilon_{\beta_1...\beta_D} = det\left(x^\beta_{\alpha'}\right)\varepsilon_{\alpha'_1...\alpha'_D}. \tag{3.1.26}$$

Thus, the transformation of $\varepsilon_{\alpha_1...\alpha_D}$ *contains an additional factor* $det\left(x^\beta_{\alpha'}\right)$. *Such an object is called a tensor density.*

Let f be a *diffeomorphism* of a manifold M to N. If Ψ_α is a chart in the vicinity of a point $p \in M$, then $\Psi_\alpha \circ f^{-1}$ is a corresponding ('dragged along') chart in the vicinity of a point $p' = f(p) \in N$. A tensor T^* at p' is said to be 'dragged along' by the map f ($T^* \equiv f^* T$) if its components in the corresponding 'dragged along' chart are the same as the components of the original tensor T at the initial point p in the original chart. If $f : M \to M$ is a diffeomorphism, and T is a tensor field on M, one can compare T with $T^* \equiv f^* T$. If $T = T^*$, then the tensor field is said to be invariant with respect to f.

3.2 Metric

3.2.1 Metric

A model of a spacetime in general relativity is a *differential manifold* with a metric g on it. Points of such a manifold correspond to different physical *events*. A spacetime *metric* $g_{\mu\nu}$ is a non-degenerate symmetric covariant tensor field of rank $(2,0)$, which has signature[2] $(-,+,...,+)$. In standard general relativity the spacetime manifold has 4 dimensions. There are also modifications of the Einstein theory where the number of spacetime dimensions is higher. The *Riemannian geometry* studies the geometry of a *Riemannian space*, that is a differential manifold with a metric on it. The presence of a metric provides a rather rich structure. In particular, it makes it possible to define in a covariant way the operation of differentiation and integration in the Riemannian spaces.

Quite often, especially in the mathematical literature, the name 'Riemannian' is used for the spaces with a positive definite metric, while a space where the metric is not positive definite is called 'pseudo-Riemannian'. We use 'slang' adopted in the physical literature and keep the name 'Riemannian' for the spacetime geometry. In special cases, when the positive signature of the metric is important, we refer to it as to the Euclidean metric.

---

[2] At any given point $g_{\mu\nu}dx^\mu dx^{nu}$ is a quadratic form of the differentials dx^μ. By changing the coordinates the coefficients of this form can be transformed to a diagonal form $\eta_{\mu\nu}$. The signature of the metric shows the number of positive and negative components of this diagonal form. These numbers are evidently invariant under coordinate transformations.

Consider two vectors ζ and η, then a *scalar product* of these vectors is defined as

$$(\zeta, \eta) = g(\zeta, \eta) = g_{\alpha\beta}\zeta^\alpha\eta^\beta. \tag{3.2.1}$$

It is evident that a scalar product is really a scalar since the last expression on the right-hand side of Eq. (3.2.1) does not depend on the choice of the coordinates.

'Juggling with indices'

Define the metric with upper indices $g^{\mu\nu}$ by the relation

$$g^{\mu\nu}g_{\nu\lambda} = \delta^\mu_\lambda. \tag{3.2.2}$$

We assumed the metric to be non-degenerate. This means that the *inverse metric* $g^{\mu\nu}$ exists, and the relation Eq. (3.2.2) determines it in an arbitrary coordinate system. Because δ^μ_ν and $g_{\nu\lambda}$ are tensors, $g^{\mu\nu}$ is also a tensor.

In a *Riemannian space* the tensors $g_{\mu\nu}$ and $g^{\mu\nu}$ can be used to shuffle the indices between up and down positions. It is convenient to reserve a special position for each of the indices of the tensor. For example, in accordance with this agreement the following components describe the same tensor T

$$T\vert^\alpha\vert_\beta\vert^\gamma\vert \leftrightarrow T\vert_\alpha\vert_\beta\vert^\gamma\vert \leftrightarrow T_{\alpha\beta\gamma} \leftrightarrow T^{\alpha\beta\gamma}. \tag{3.2.3}$$

The order and position of the indices are important.

3.2.2   Completely antisymmetric tensor

Using the transformation law of the metric

$$g_{\mu'\nu'} = \frac{\partial x^\mu}{\partial x^{\mu'}}\frac{\partial x^\nu}{\partial x^{\nu'}}g_{\mu\nu} \tag{3.2.4}$$

and taking the determints of both its sides one gets

$$g' \equiv \det(g'_{\alpha'\beta'}) = \vert x^\mu_{,\mu'}\vert^2\det(g_{\alpha\beta}) \equiv \vert x^\mu_{,\mu'}\vert^2 g. \tag{3.2.5}$$

This means that the determinant of the metric is not a scalar, but a scalar density. Combining it with the *Levi-Civita symbol* $\epsilon_{\alpha_1...\alpha_D}$ one can define a completely antisymmetric *Levi-Civita tensor*

$$e_{\alpha_1...\alpha_D} = \sqrt{-g}\,\epsilon_{\alpha_1...\alpha_D}, \qquad e^{\alpha_1...\alpha_D} = -\frac{1}{\sqrt{-g}}\,\epsilon_{\alpha_1...\alpha_D}. \tag{3.2.6}$$

Using the Levi-Civita tensor one can construct the Hodge dual *tensor tensor* ***T** *from any antisymmetric tensor* **T**

$$*T_{\alpha_1...\alpha_{D-k}} = \frac{1}{k!}T^{\mu_1...\mu_k}e_{\mu_1...\mu_k\alpha_1...\alpha_{D-k}}. \tag{3.2.7}$$

The duality transformation relates antisymmetric tensors of rank k to antisymmetric tensors of rank D − k. Dual tensors contain precisely the same information. This becomes evident if one applies the duality transformation twice. One can show that, up to a sign, double-duality transformation restores the original tensor

$$**T = (-1)^{k(D-k)} \, \mathbf{s} \, T. \tag{3.2.8}$$

Here, k is the rank of the antisymmetric tensor and **s** *is the sign of the determinant of the metric. Thus, for spacetimes with the Minkowskian signature* $\mathbf{s} = -1$, *while in the Euclidean spaces* $\mathbf{s} = 1$.

3.2.3 Geodesics

The quantity $ds^2 = g_{\mu\nu} \, dx^\mu \, dx^\nu$ gives the *interval* between two close events x^μ and $x^\mu + dx^\mu$. A smooth curve $x^\mu(\lambda)$ is said to be space-like, time-like, or light-like (null) at a point $\lambda = \lambda_0$ if the vector $u^\mu = dx^\mu/d\lambda$ tangent to it at this point satisfies the condition $u^\mu u_\mu > 0$, $u^\mu u_\mu < 0$, or $u^\mu u_\mu = 0$, respectively. For a line $x^\mu(\lambda)$ the proper interval between the points $x^\mu(\lambda_1)$ and $x^\mu(\lambda_2)$ is

$$s = \int_{\lambda_1}^{\lambda_2} \sqrt{|g_{\mu\nu} \, (dx^\mu/d\lambda)(dx^\nu/d\lambda)|} \, d\lambda. \tag{3.2.9}$$

For a time-like curve the quantity s is the *proper time*, for a space-like curve s is a *proper distance*.

A *geodesic* is a line that provides a stationary point to the action Eq. (3.2.9). We have already considered a similar problem in Section 2.2.3 for a flat spacetime. But, in fact, in the derivation of the result we did not use the property of the metric to be flat. Repeating these calculations and imposing the following gauge-fixing condition

$$g_{\alpha\beta} \frac{dx^\alpha}{d\lambda} \frac{dx^\beta}{d\lambda} = \varepsilon, \tag{3.2.10}$$

one obtains the following equation

$$\frac{d^2 x^\alpha}{d\lambda^2} + \Gamma^\alpha_{\beta\gamma} \frac{dx^\beta}{d\lambda} \frac{dx^\gamma}{d\lambda} = 0. \tag{3.2.11}$$

A solution of this equation is a *geodesic*. For $\varepsilon = -1$ this is a time-like geodesic, while for $\varepsilon = 1$ it is a space-like one. Time-like geodesics describe the free motion of a particle in a given gravitational field g. The condition Eq. (3.2.10) implies that λ is *proper time* or *proper length* for $\varepsilon = -1$ and $\varepsilon = 1$, respectively. For $\varepsilon = 0$ Eq. (2.2.27) is a *null geodesic* that describes a propagation of a *null ray*. The corresponding parameter λ is called an *affine parameter*.

Problem 3.2: *Prove that the variational principle*

$$\delta \left[\int_{s_1}^{s_2} g_{\mu\nu} \, (dx^\mu/ds)(dx^\nu/ds) \, ds \right] = 0 \tag{3.2.12}$$

gives the same geodesic equation Eq. (3.2.11), and s is a proper time (distance), or an affine parameter.
  Denote

$$z = g_{\mu\nu} \, (dx^\mu/ds)(dx^\nu/ds), \tag{3.2.13}$$

and $f(z)$ is an arbitrary smooth monotonic function of z. Prove that the stationary point of the functional $\int_{s_1}^{s_2} f(z) \, ds$ is a geodesic line. For $f(z) = z$ one reproduces the action Eq. (3.2.12). This choice is especially convenient since it allows one to consider time-like, space-like, and null geodesics simultaneously.

A time-like geodesic connecting two points maximizes the functional Eq. (3.2.9).[3] For a space-like geodesic the variation of the functional Eq. (3.2.9) also vanishes, but it is neither maximum nor minimum. (See discussion in Section 2.2.)

Let us assume that two twins (Alice and Bob) meet one another at a point p and after this they move along different worldlines until they meet again at p'. Let Bob's worldline be a sole time-like geodesic γ_0 connecting these points. In other words, his motion is inertial (without an acceleration). As for Alice, her motion between these events is a time-like but non-geodesic curve in the vicinity of Bob's worldline. In other words, she moves with an acceleration. The proper time between p and p' as measured by Alice is always shorter than the proper time as measured by Bob. This is the *twin paradox* formulation for the motion in an external gravitational field.

Problem 3.3: *Consider an* ultrastatic spacetime, *that is a spacetime with the metric*

$$ds^2 = -dt^2 + h_{ij}dx^i dx^j, \qquad \partial_t h_{ij} = 0. \tag{3.2.14}$$

Show that a geodesic in this metric can be written in the form $x^\mu = (t, x^i(t))$, *where* $x^i(t)$ *is a geodesic in the metric* h_{ij}.

Proof: All the Christoffel coefficients for the metric Eq. (3.2.14) that contain the index t vanish. Thus, the geodesic equation Eq. (3.2.11) takes the form

$$\frac{d^2 t}{d\lambda^2} = 0, \qquad \frac{d^2 x^i}{d\lambda^2} + \Gamma^i{}_{jk} \frac{dx^j}{d\lambda} \frac{dx^k}{d\lambda} = 0. \tag{3.2.15}$$

Here, $\Gamma^i{}_{jk}$ are the Christoffel symbols for the metric h_{ij}. The first of the equations shows that one can take t as the affine parameter λ. The second equation is simply the geodesic equation in the metric h_{ij}.

3.2.4 Simultaneity and projectors

Let γ be a worldline of an observer. Suppose at a point $p \in \gamma$ he/she has the velocity u^μ. Events in the vicinity of p that are simultaneous with it in the reference frame of this observer, form a plane Π passing through p and orthogonal to u^μ. Denote

$$h^\mu_\nu = \delta^\mu_\nu - \frac{u^\mu u_\nu}{u^2}. \tag{3.2.16}$$

It is easy to see that $h_{\mu\nu} = g_{\mu\alpha} h^\alpha_\nu$ is a symmetric tensor. The tensor $\boldsymbol{h}$ annihilates $\boldsymbol{u}$, $h^\mu_\nu u^\nu = 0$, and it obeys the relation

$$h^\mu_\alpha h^\alpha_\nu = h^\mu_\nu. \tag{3.2.17}$$

Hence, $\boldsymbol{h}$ is the projector onto the plane Π orthogonal to u^μ. Any vector ζ^μ can be decomposed as follows

$$\zeta = \zeta_\parallel + \zeta_\perp,$$
$$\zeta_\parallel = -(\zeta, u)u, \qquad \zeta_\perp = h(\zeta). \tag{3.2.18}$$

---

[3] Strictly speaking, these extrema are local, since there may exist more than one geodesic connecting the same two points.

The vector $\zeta_\perp^\mu = h_\nu^\mu \zeta^\nu$ is a projection of $\boldsymbol{\zeta}$ on Π, and $\boldsymbol{\zeta}_\parallel$ is a vector orthogonal to Π. Consider an event p' that has a coordinate separation dx^μ from p. Then, the spatial distance between these two events, as measured by the observer $\boldsymbol{u}$, is

$$dl^2 = g_{\mu\nu}\, h_\lambda^\mu\, dx^\lambda\, h_\rho^\nu\, dx^\rho = h_{\lambda\rho}\, dx^\lambda\, dx^\rho. \tag{3.2.19}$$

To obtain the last relation we used the property Eq. (3.2.17) and symmetry of $h_{\mu\nu}$. Thus, $h_{\lambda\rho}$ determines a metric in the $(D-1)$-dimensional space Π. For a special case, when a particle worldline coincides with the time coordinate line x^0, one has

$$h_{\lambda\rho} \doteq g_{\lambda\rho} - \frac{g_{\lambda 0}\, g_{\rho 0}}{g_{00}}. \tag{3.2.20}$$

Here and later, we use a symbol $\doteq$ to indicate that a relation is valid only in a special coordinate frame.

3.2.5   Causal structure

Local null cones

Consider a point p and vectors dx^μ in the tangent space at this point that obey the condition

$$g_{\mu\nu} dx^\mu dx^\nu = 0. \tag{3.2.21}$$

Such vectors form a null surface that is a double-cone. The upper-cone and lower-cone are called the future and the past local null cones, respectively.

Take a time-like vector $\boldsymbol{\xi}_p$ inside the future null cone at p. Let q be another point and let γ be an arbitrary continuous curve connecting p and q. Consider a non-vanishing time-like vector field $\boldsymbol{\xi}$ that coincides with $\boldsymbol{\xi}_p$ at p and is continuous along γ. This vector field uniquely singles out a family of future local null cones along γ and uniquely determines a future local null cone at q. We call this procedure a 'synchronization' of null cones. We say a spacetime is *time orientable* if for any pair of its points p and q the adopted synchronization procedure is consistent. It is possible to show that the spacetime is *time orientable* if there exists a globally defined and nowhere vanishing continuous time-like vector field. In what follows we shall assume that a spacetime is time orientable, so that the definition of the future and the past local null cones has a well-defined meaning.

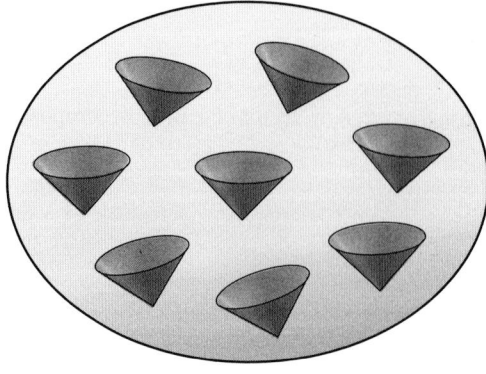

Fig. 3.5 Future-directed local null cones in a time-orientable spacetime.

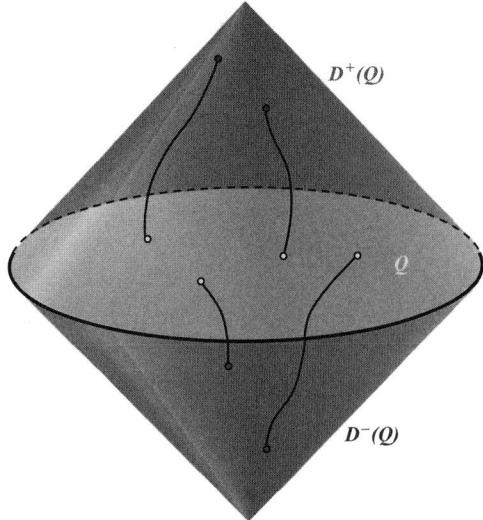

Fig. 3.6 Illustration of the future and past Cauchy domains for the set Q.

Causal sets

The *chronological future* $I^+(Q)$ (*chronological past* $I^-(Q)$) of a set Q is the set of points for each of which there is a past-directed (future-directed) time-like curve that intersects Q.

The curve $x^\mu(\lambda)$ is said to be causal (or *non-space-like*) if its tangent vector $u^\mu = dx^\mu/d\lambda$ obeys the condition $u^\mu u_\mu \le 0$ at each of its points. A non-space-like (causal) curve between two points, which is not a null geodesic, can be deformed into a time-like curve connecting these points.

The *causal future* $J^+(Q)$ (*causal past* $J^-(Q)$) of a set Q is the set of points for each of which there is a past-directed (future-directed) causal curve that intersects Q. The *future Cauchy domain* $D^+(Q)$ (*past Cauchy domain* $D^-(Q)$) of a set Q is the set of points such that any past-directed (future-directed) causal curve passing through it intersects Q (see Figure 3.6).

A surface Σ is called space-like, time-like, or null, if its normal vector is time-like, space-like, or null, respectively.[4] A *global Cauchy surface* in a spacetime M is a non-time-like hypersurface that is intersected by each causal curve exactly once.

3.3 Covariant Derivative

3.3.1 Definition of a covariant derivative

A partial derivative $\varphi_{,\mu}$ of a scalar function φ is a covector. But a second partial derivative $\varphi_{,\mu\nu}$ of φ is not a tensor. Indeed, one has

$$\varphi_{,\mu'\nu'} = (\varphi_{,\mu} x^\mu_{,\mu'})_{,\nu'} = \varphi_{,\mu\nu} x^\mu_{,\mu'} x^\nu_{,\nu'} + \varphi_{,\mu} x^\mu_{,\mu'\nu'}. \tag{3.3.1}$$

---

[4]If a surface Σ is locally defined by a relation $\varphi(x^\mu) = $ const, a normal vector to it is $n^\nu = g^{\mu\nu}\varphi_{,\nu}$

The second term in the last expression violates the tensorial law of transformation of $\varphi_{,\mu\nu}$. This is a general problem: *a partial derivative of a tensor is not a tensor*. In particular, the partial derivatives of the metric, $g_{\mu\nu,\alpha}$, do not form a tensor. In particular, one can always find such coordinates in which all the components of $g_{\mu\nu,\lambda}$ vanish at a given point.

For partial derivatives of a tensor the terms that violate the tensorial law have the standard structure $x^{\mu}_{,\nu'\lambda'}$. In the presence of the metric, this property can be used to cure the 'disease'. Partial derivatives of a tensor T can always be combined with the partial derivatives of the metric g in such a way that the terms violating the tensorial law in both objects cancel each other. This allows one to 'upgrade' the partial derivative and to define a *covariant derivative*.[5]

Let us consider special linear combinations of $g_{\mu\nu,\lambda}$ of the form

$$\Gamma^{\mu}{}_{\alpha\beta} = g^{\mu\nu}\,\Gamma_{\nu\alpha\beta}, \qquad \Gamma_{\nu\alpha\beta} = \frac{1}{2}\left(g_{\nu\alpha,\beta} + g_{\nu\beta,\alpha} - g_{\alpha\beta,\nu}\right). \tag{3.3.2}$$

The objects $\Gamma^{\mu}{}_{\alpha\beta}$ and $\Gamma_{\nu\alpha\beta}$ are known as the *Christoffel symbols*. The Christoffel symbols are symmetric with respect to the indices α and β.

The Christoffel symbols are not tensors. Under the coordinate transformations they transform as follows

$$\Gamma^{\rho'}{}_{\mu'\nu'} = x^{\rho'}_{,\rho}\,x^{\mu}_{,\mu'}x^{\nu}_{,\nu'}\,\Gamma^{\rho}{}_{\mu\nu} + x^{\rho'}_{,\rho}\,x^{\rho}_{,\mu'\nu'}. \tag{3.3.3}$$

At a given point the number of independent components of $x^{\rho}_{,\mu'\nu'}$ is $D^2(D+1)/2$. The Christoffel symbols have the same number of independent components. One can always find such coordinates near p in which $\Gamma^{\rho}{}_{\mu\nu}(p) \doteq 0$. Such coordinates are called Riemann normal coordinates. *Fermi demonstrated that there exist coordinates in which the components of the Christoffel symbols vanish along a geodesic line. These coordinates are called the* Fermi coordinates *(see Section 3.5.6).*

For an arbitrary tensor T the following combination of its partial derivatives and the Christoffel symbols

$$\nabla_{\lambda}\,T_{\alpha\ldots}{}^{\beta\ldots} = \partial_{\lambda}T_{\alpha\ldots}{}^{\beta\ldots} - \Gamma^{\rho}{}_{\alpha\lambda}\,T_{\rho\ldots}{}^{\beta\ldots} + \cdots + \Gamma^{\beta}{}_{\lambda\rho}\,T_{\alpha\ldots}{}^{\lambda\ldots} + \ldots\,. \tag{3.3.4}$$

forms a tensor. This tensor is called the *covariant derivative* of T. We shall also use another notation for the covariant derivative: $\nabla_{\alpha}(\ldots) \equiv (\ldots)_{;\alpha}$. The covariant derivative of the metric vanishes. Indeed,

$$\nabla_{\mu}\,g_{\alpha\beta} = g_{\alpha\beta,\mu} - \Gamma^{\lambda}{}_{\mu\alpha}\,g_{\lambda\beta} - \Gamma^{\lambda}{}_{\mu\beta}\,g_{\alpha\lambda} = g_{\alpha\beta,\mu} - \Gamma_{\beta\mu\alpha} - \Gamma_{\alpha\mu\beta} = 0. \tag{3.3.5}$$

3.3.2 Properties of the covariant derivative

The covariant derivative possesses the following properties:

1. *Linearity*: For constants a and b one has $\nabla_{\mu}(aA^{\ldots}_{\ldots} + bB^{\ldots}_{\ldots}) = a\nabla_{\mu}A^{\ldots}_{\ldots} + b\nabla_{\mu}B^{\ldots}_{\ldots}$.
2. *Leibnitz rule*: $\nabla_{\mu}(A^{\ldots}_{\ldots}B^{\ldots}_{\ldots}) = \nabla_{\mu}(A^{\ldots}_{\ldots})B^{\ldots}_{\ldots} + A^{\ldots}_{\ldots}\nabla_{\mu}(B^{\ldots}_{\ldots})$.

---

[5]To define an invariant differential operator one can use another tensorial object instead of the metric. For example, one can construct such an operator, called the *Lie derivative*, by using a vector field. (See Section 3.4.1.)

3. Commutativity with contraction: $\nabla_\mu(A^{...\beta...}_{...\beta...}) = (\nabla_\mu A)^{...\beta...}_{...\beta...}$.

4. For a scalar field: $\nabla_\mu\varphi = \varphi_{,\mu}$.

5. Torsion free: $\nabla_\mu\nabla_\nu\varphi = \nabla_\nu\nabla_\mu\varphi$.

6. $\nabla_\mu g_{\alpha\beta} = 0$.

These properties unambiguously determine the covariant derivative and hence they can be used as its definition.

Consider a line γ and let $\boldsymbol{\xi}$ be its tangent vector. A tensor field $A^{\alpha\cdots}_{\ \ \beta\cdots}$ is called parallel transported along γ if

$$\xi^\mu \nabla_\mu A^{\alpha\cdots}_{\ \ \beta\cdots} = 0. \tag{3.3.6}$$

3.4 Lie and Fermi Transport

3.4.1 Lie derivative

As we have already mentioned, one can 'upgrade' a partial derivative of a tensor to a covariant operation by using, e.g., a vector field rather than the metric. The corresponding operation is called the *Lie derivative*.

Consider a congruence of lines $x^\mu(\lambda, y^i)$, where y^i 'enumerates' lines, and λ is a parameter along a line. For this congruence $u^\mu = \partial x^\mu/\partial\lambda$ is a tangent vector to a line, and $v^\mu = (\partial x^\mu/\partial y^i)\,\delta y^i$ is a vector connecting two lines of the congruence with the same value of λ and with the parameters y^i and $y^i + \delta y^i$, respectively. It is easy to check that the following relations are valid:

$$\begin{aligned}
[u, v]^\mu &\equiv u^\alpha v^\mu_{\ ,\alpha} - v^\alpha u^\mu_{\ ,\alpha} = \frac{\partial x^\alpha}{\partial\lambda}\frac{\partial v^\mu}{\partial x^\alpha} - \delta y^i \frac{\partial x^\alpha}{\partial y^i}\frac{\partial u^\mu}{\partial x^\alpha} \\
&= \frac{\partial v^\mu}{\partial\lambda} - \delta y^i \frac{\partial u^\mu}{\partial y^i} = \left(\frac{\partial^2 x^\mu}{\partial\lambda\,\partial y^i} - \frac{\partial^2 x^\mu}{\partial y^i\,\partial\lambda}\right)\delta y^i = 0.
\end{aligned} \tag{3.4.1}$$

The quantity standing on the left-hand side of the first relation in Eq. (3.4.1) is called a *commutator of vector fields*. This is a special case of the *Lie derivative* of the vector $\boldsymbol{v}$ along

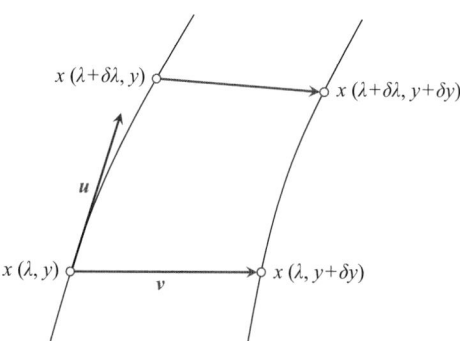

Fig. 3.7 Illustration of the Lie transport of a vector.

the vector u that can be defined for any tensor field (see later). The Lie derivative of the vector v coincides with the commutator

$$(\mathcal{L}_u v)^\mu = [u, v]^\mu. \tag{3.4.2}$$

Equation (3.4.1) shows that the Lie derivative $\mathcal{L}_u v$ of a vector field v, which is 'frozen' into the congruence generated by u vanishes.

3.4.2   Lie transport

One can generalize the above definition of the Lie derivative to the case of an arbitrary tensor field. Let M be a differential manifold, and let f_λ be a one-parameter group of diffeomorphisms of M onto M. Denote $p_\lambda = f_\lambda(p)$. For a chosen chart we define a function $x^\mu(\lambda; p) \equiv x^\mu(p_\lambda)$. Denote by

$$\xi^\mu(p) = \left. \frac{dx^\mu(\lambda; p)}{d\lambda} \right|_{\lambda=0} \tag{3.4.3}$$

a vector field generating a one-parameter group of diffeomorphisms. The Lie derivative $\mathcal{L}_\xi A^{\alpha\cdots}{}_{\beta\cdots}$ of a tensor field $A^{\alpha\cdots}{}_{\beta\cdots}$ along ξ^μ is defined by the relation

$$\mathcal{L}_\xi A^{\alpha\cdots}{}_{\beta\cdots} = \lim_{\lambda\to 0}\left[\frac{f^*_{-\lambda} A^{\alpha\cdots}{}_{\beta\cdots} - A^{\alpha\cdots}{}_{\beta\cdots}}{\lambda} \right]. \tag{3.4.4}$$

The following explicit expressions are valid for the *Lie derivative*

$$\begin{aligned}
\mathcal{L}_\xi A^{\alpha\cdots}{}_{\beta\cdots} &= \xi^\mu \, \partial_\mu A^{\alpha\cdots}{}_{\beta\cdots} - \partial_\mu \xi^\alpha \, A^{\mu\cdots}{}_{\beta\cdots} - \cdots + \partial_\beta \xi^\mu \, A^{\alpha\cdots}{}_{\mu\cdots} + \cdots \\
&= \xi^\mu \, \nabla_\mu A^{\alpha\cdots}{}_{\beta\cdots} - \nabla_\mu \xi^\alpha \, A^{\mu\cdots}{}_{\beta\cdots} - \cdots + \nabla_\beta \xi^\mu \, A^{\alpha\cdots}{}_{\mu\cdots} + \cdots .
\end{aligned} \tag{3.4.5}$$

The first line is a definition of the Lie derivative valid for an arbitrary manifold. In the presence of the metric the same expression can be identically rewritten by changing the partial derivations by the covariant ones (see the second line in Eq. (3.4.5)). The Lie derivative of a vector field η is

$$\mathcal{L}_\xi \eta^\alpha \equiv [\xi, \eta]^\alpha = \xi^\mu \, \partial_\mu \eta^\alpha - \eta^\mu \, \partial_\mu \xi^\alpha. \tag{3.4.6}$$

The Lie derivative obeys the following relations

1. $\mathcal{L}_\xi \xi = 0$;
2. $\mathcal{L}_\xi(\alpha A^{\cdots} + \beta B^{\cdots}) = \alpha \mathcal{L}_\xi A^{\cdots} + \beta \mathcal{L}_\xi B^{\cdots}$ (α and β are arbitrary constants);
3. $\mathcal{L}_\xi(A^{\cdots} B^{\cdots}) = \left(\mathcal{L}_\xi A^{\cdots}\right) B^{\cdots} + A^{\cdots} \mathcal{L}_\xi B^{\cdots}$;
4. $\mathcal{L}_\xi \mathcal{L}_\eta - \mathcal{L}_\eta \mathcal{L}_\xi = \mathcal{L}_{[\xi, \eta]}$.

$$\tag{3.4.7}$$

A tensor field $A^{\alpha\cdots}{}_{\beta\cdots}$ is said to be *Lie-transported* along ξ if

$$\mathcal{L}_\xi A^{\alpha\cdots}{}_{\beta\cdots} = 0. \tag{3.4.8}$$

Let ζ and η be two arbitrary Lie-transported along ξ vector fields. Their scalar product is constant along ξ if the metric obeys the condition

$$\mathcal{L}_\xi g_{\mu\nu} = 0. \tag{3.4.9}$$

3.4.3 Fermi transport

When both the vector field ξ and the metric g are defined on the manifold one can construct a new useful operation called the *Fermi transport*. Denote by w a vector

$$w^\alpha = \xi^\beta \, \xi^\alpha{}_{;\beta}, \qquad \varepsilon(\xi) = \text{sign}(-\xi^\mu \, \xi_\mu). \tag{3.4.10}$$

Then, the following antisymmetric tensor

$$\mathcal{F}^\alpha{}_\sigma = (w^\alpha \, \xi_\sigma - \xi^\alpha \, w_\sigma) \, \varepsilon(\xi), \tag{3.4.11}$$

is a generator of transformations in the (ξ, w)-plane. When ξ is a time-like vector, this is a *boost transformation*.

Using this tensor one can construct a new operation, called a *Fermi derivative*. The Fermi derivative $\mathcal{F}_\xi A^{\alpha\cdots}{}_{\beta\cdots}$ of a tensor field $A^{\alpha\cdots}{}_{\beta\cdots}$ along a vector field ξ $\;(\xi^\mu \, \xi_\mu \neq 0)$ is defined as:

$$\mathcal{F}_\xi A^{\alpha\cdots}{}_{\beta\cdots} = \xi^\mu \, \nabla_\mu A^{\alpha\cdots}{}_{\beta\cdots} + \mathcal{F}^\alpha{}_\sigma \, A^{\sigma\cdots}{}_{\beta\cdots} + \cdots - \mathcal{F}^\sigma{}_\beta \, A^{\alpha\cdots}{}_{\sigma\cdots} + \cdots. \tag{3.4.12}$$

A tensor field $A^{\alpha\cdots}{}_{\beta\cdots}$ is said to be Fermi-transported along ξ if

$$\mathcal{F}_\xi A^{\alpha\cdots}{}_{\beta\cdots} = 0. \tag{3.4.13}$$

It is easy to check that

$$\mathcal{F}_\xi g_{\mu\nu} = 0. \tag{3.4.14}$$

For a geodesic congruence the tensor $\mathbf{F}$ vanishes and the Fermi derivative $\mathcal{F}_\xi$ coincides with the covariant one $\xi^\mu \nabla_\mu$.

Let us consider a special case of a congruence of time-like curves, where τ is the proper time parameter. Then, $\xi^\mu = u^\mu = \partial x^\mu / \partial\tau$ is the velocity ($u_\mu \, u^\mu = -1$), and $w^\alpha = u^\beta \, u^\alpha{}_{;\beta}$ is the acceleration ($u^\mu \, w_\mu = 0$) vector. One has

$$\mathcal{F}_u u = 0. \tag{3.4.15}$$

If ζ and η are two Fermi-transported along u vector fields then their scalar product is constant along u.

Consider a particle with the velocity u. Consider a plane Π_0 orthogonal to u at a point p, and let e_i be mutually orthogonal unit vectors in Π_0. The vectors $\{u, e_i\}$ form a basis at p. This basis can be uniquely transferred along the integral line of u by assuming that its vectors are Fermi-propagated along it. Such a basis remains orthonormal and it possesses a very special property: its vectors e_i are not rotating.

3.5   Curvature Tensor

3.5.1   Commutator of covariant derivatives

In a flat spacetime two covariant derivatives commute. Non-commutativity of the covariant derivatives implies that the spacetime is curved. Using the definition of the covariant derivative Eq. (3.3.4), straightforward calculations give

$$V_{\beta;\mu\nu} - V_{\beta;\nu\mu} = V_\alpha R^\alpha{}_{\beta\mu\nu} = V^\alpha R_{\alpha\beta\mu\nu}, \tag{3.5.1}$$

where

$$R^\alpha{}_{\beta\mu\nu} = \partial_\mu \Gamma^\alpha{}_{\beta\nu} - \partial_\nu \Gamma^\alpha{}_{\beta\mu} + \Gamma^\alpha{}_{\sigma\mu} \Gamma^\sigma{}_{\beta\nu} - \Gamma^\alpha{}_{\sigma\nu} \Gamma^\sigma{}_{\beta\mu}. \tag{3.5.2}$$

Since V and its second covariant derivatives in a given point are arbitrary tensors, the object $R^\alpha{}_{\beta\mu\nu}$ standing on the right-hand side of Eq. (3.5.1) is a tensor. This tensor is called the *Riemann curvature*.

▌ **Problem 3.4:**   *Prove the relations Eqs. (3.5.1) and (3.5.2).*

Consider a scalar product $V_\alpha U^\alpha$ of two vectors V and U. Covariant derivatives acting on a scalar function commute. Hence,

$$0 = (V_\alpha U^\alpha)_{;\mu\nu} - (V_\alpha U^\alpha)_{;\nu\mu} = (V_{\alpha;\mu\nu} - V_{\alpha;\nu\mu}) U^\alpha + V_\alpha (U^\alpha{}_{;\mu\nu} - U^\alpha{}_{;\nu\mu}). \tag{3.5.3}$$

Using Eq. (3.5.1) one gets

$$V_\alpha (U^\alpha{}_{;\mu\nu} - U^\alpha{}_{;\nu\mu}) = -V_\alpha U^\beta R^\alpha{}_{\beta\mu\nu}. \tag{3.5.4}$$

Since this relation is valid for any V_α one has

$$U^\alpha{}_{;\mu\nu} - U^\alpha{}_{;\nu\mu} = -U^\beta R^\alpha{}_{\beta\mu\nu}. \tag{3.5.5}$$

In the general case:

$$\left[\nabla_\mu \nabla_\nu - \nabla_\nu \nabla_\mu\right] B^{\alpha\cdots}{}_{\beta\ldots} = R^\alpha{}_{\sigma\mu\nu} B^{\sigma\cdots}{}_{\beta\ldots} + \cdots - R^\epsilon{}_{\beta\mu\nu} B^{\alpha\cdots}{}_{\epsilon\ldots} - \cdots . \tag{3.5.6}$$

3.5.2   Symmetries of the curvature tensor

The curvature tensor $R_{\alpha\beta\gamma\delta}$ can be written explicitly

$$R_{\alpha\beta\gamma\delta} = \frac{1}{2} \left[g_{\alpha\delta,\beta\gamma} + g_{\beta\gamma,\alpha\delta} - g_{\alpha\gamma,\beta\delta} - g_{\beta\delta,\alpha\gamma} \right]$$
$$+ g_{\epsilon\sigma} \left(\Gamma^\epsilon{}_{\beta\gamma} \Gamma^\sigma{}_{\alpha\delta} - \Gamma^\epsilon{}_{\beta\delta} \Gamma^\sigma{}_{\alpha\gamma} \right). \tag{3.5.7}$$

Since $R_{\alpha\beta\gamma\delta}$ is a tensor it is sufficient to check this relation in any suitable coordinate system. It is convenient to use the *Riemann normal coordinates* where the Christoffel symbols vanish at a chosen point. The symmetry properties of the tensor Eq. (3.5.7) can be found by using this expression. Namely, in these coordinates it is sufficient to establish the symmetries of the expression inside the square brackets on the right-hand side of Eq. (3.5.7). This can be easily done since $g_{\mu\nu}$ is symmetric with respect to its indices, and the partial derivatives commute.

The curvature tensor $R_{\alpha\beta\gamma\delta}$ obeys the following symmetry properties:[6]

$$R_{\alpha\beta\gamma\delta} = R_{\alpha\beta[\gamma\delta]} = R_{[\alpha\beta]\gamma\delta}, \tag{3.5.8}$$

$$R_{\alpha\beta\gamma\delta} = R_{\gamma\delta\alpha\beta}, \tag{3.5.9}$$

$$R_{\mu[\nu\alpha\beta]} = 0. \tag{3.5.10}$$

These relations give a complete set of symmetries of the curvature tensor. It is easy to show that Eq. (3.5.10) can also be rewritten in the following two forms:

$$R_{[\mu\nu\alpha\beta]} = 0, \qquad e^{\mu_1\dots\mu_{D-4}\nu_1\nu_2\nu_3\nu_4} R_{\nu_1\nu_2\nu_3\nu_4} = 0. \tag{3.5.11}$$

3.5.3   Independent components of the curvature tensor

Let us calculate how many independent components the curvature tensor has. For D-dimensional spacetime we denote this number by $N_D[\mathcal{R}]$. To calculate $N_D[\mathcal{R}]$ it is convenient to introduce collective indices $A = [\mu\nu]$ and $B = [\alpha\beta]$, and to write the curvature tensor $R_{\mu\nu\alpha\beta}$ as R_{AB}. Each of the collective indices has $K = D(D-1)/2$ independent components. The property Eq. (3.5.9) implies that $R_{AB} = R_{BA}$. Hence, the number of components of this object is $K(K+1)/2$. Consider Eq. (3.5.11). The total number of independent conditions imposed by this relation coincides with the number of independent values of $[\mu_1 \dots \mu_{D-4}]$, which is equal to $D!/[(D-4)!4!] = \frac{1}{24}D(D-1)(D-2)(D-3)$.
    Subtracting this number from $K(K+1)/2$ we get

$$N_D[\mathcal{R}] = \frac{1}{12}D^2(D^2-1). \tag{3.5.12}$$

Problem 3.5:   *For a given tensor $R_{\delta\alpha\beta\gamma}$, obeying the symmetries Eqs. (3.5.8)–(3.5.10), we define new tensors A and S by the following relations*

$$A_{\alpha\beta\gamma\delta} = R_{\delta[\alpha\beta]\gamma}, \qquad S_{\alpha\beta\gamma\delta} = R_{\delta(\alpha\beta)\gamma}. \tag{3.5.13}$$

Show that

$$R_{\delta\alpha\beta\gamma} = 4A_{\alpha[\beta\gamma]\delta} = \frac{4}{3}S_{\alpha[\beta\gamma]\delta}. \tag{3.5.14}$$

Problem 3.6:   *For a given tensor $R_{\delta\alpha\beta\gamma}$, obeying the symmetries Eqs. (3.5.8)–(3.5.10), we define new tensors by the following relations*

$$B_{\alpha\gamma\beta\delta} = R_{\alpha\beta\gamma\delta}, \qquad \tilde{B}_{\alpha\beta\gamma\delta} = B_{(\alpha\gamma)(\beta\delta)}, \qquad \tilde{\tilde{B}}_{\alpha\beta\gamma\delta} = \tilde{B}_{[\alpha\beta][\gamma\delta]}. \tag{3.5.15}$$

Show that

$$\tilde{\tilde{B}}_{\alpha\beta\gamma\delta} = \frac{3}{4}R_{\alpha\beta\gamma\delta}. \tag{3.5.16}$$

Equations (3.5.14) show that operations of symmetrization and antisymmetrizations with respect to a pair of indices do not reduce the curvature tensor to new 'simpler' tensors. Relation

---

[6]The first of the equalities in Eq. (3.5.8) directly follows from the curvature definition Eq. (3.5.1).

α	γ
β	δ

D	$D{+}1$
$D{-}1$	D

3	2
2	1

Fig. 3.8 The Young diagrams of a shape $(2, 2)$ necessary for calculation of the dimension of tensors with symmetries of the Riemann curvature tensor.

Eq. (3.5.16) allows one to understand why it happens. In fact, it relates the curvature tensor to the tensor $\widetilde{B}$ that has the properties described by the following *Young tableau* shown at Figure 3.8. In this tableau we put indices themselves rather than the index numbers. This implies that the curvature tensor obeying the symmetry properties Eqs. (3.5.8)–(3.5.10) is, in fact, irreducible. It is saturated by these symmetries, and any attempts to 'simplify' it by further symmetrization or/and antisymmetrization either reproduce it or give a zero result. In other words, the relations Eqs. (3.5.8)–(3.5.10) form a complete set. Two other Young diagrams shown in Figure 3.8 allow one to show that the total number of independent components of this tensor is $D^2(D^2 - 1)/12$, which coincides with the result Eq. (3.5.12). Let us emphasize that these results are valid for any rank $(4, 0)$ obeying the symmetries Eqs. (3.5.8)–(3.5.10). (For more details, see, e.g., (Penrose and Rindler 1987).)

3.5.4   Ricci tensor, Ricci scalar, and Weyl tensor

One can obtain new tensors from $R_{\alpha\beta\gamma\delta}$ by contracting it with the metric tensor. The *Ricci tensor* is obtained from the curvature tensor by contraction

$$R_{\mu\nu} = g^{\alpha\beta} R_{\alpha\mu\beta\nu} = R^{\alpha}{}_{\mu\alpha\nu}. \tag{3.5.17}$$

The Ricci tensor $R_{\mu\nu}$ is symmetric and, hence, the total number of its independent components is

$$N_D[R] = \frac{1}{2}D(D + 1). \tag{3.5.18}$$

The *Ricci scalar* (or *scalar curvature*) is defined as follows:

$$R = g^{\mu\nu} R_{\mu\nu}. \tag{3.5.19}$$

Using the Ricci tensor and the Ricci scalar one defines the *Einstein tensor*

$$G_{\mu\nu} = R_{\mu\nu} - \frac{1}{2} g_{\mu\nu} R. \tag{3.5.20}$$

As we show later, this tensor has a special property $G^{\mu\nu}{}_{;\nu} = 0$ and it plays a key role in the formulation of the Einstein equations.

Another important tensor is the *Weyl tensor*

$$C_{\mu\nu\sigma\tau} = R_{\mu\nu\sigma\tau} - \frac{1}{D-2} \left(g_{\mu\sigma} R_{\nu\tau} + g_{\nu\tau} R_{\mu\sigma} - g_{\mu\tau} R_{\nu\sigma} - g_{\nu\sigma} R_{\mu\tau} \right)$$
$$+ \frac{1}{(D-1)(D-2)} \left(g_{\mu\sigma} g_{\nu\tau} - g_{\mu\tau} g_{\nu\sigma} \right) R. \tag{3.5.21}$$

The Weyl tensor has the same symmetries as the Riemann tensor and additionally the property $C^{\alpha}{}_{\mu\alpha\nu} = 0$, i.e. it is a traceless part of the Riemann tensor. This property imposes $D(D + 1)/2$

constraints. Subtracting this quantity from $N_D[\mathcal{R}]$ we obtain that for $D \geq 3$ the total number of independent components of the Weyl tensor is

$$N_D[\mathcal{C}] = \frac{1}{12}(D-3)D(D+1)(D+2). \tag{3.5.22}$$

In a four-dimensional spacetime

$$N_4[\mathcal{R}] = 20, \qquad N_4[R] = 10, \qquad N_4[\mathcal{C}] = 10. \tag{3.5.23}$$

In three dimensions one has

$$N_3[\mathcal{R}] = 6, \qquad N_3[R] = 6, \qquad N_3[\mathcal{C}] = 0. \tag{3.5.24}$$

This means that in three dimensions the curvature tensor can be expressed in terms of the Ricci tensor and the Ricci scalar.

Problem 3.7: *Show that in a 3-dimensional spacetime the Riemann tensor can be expressed in terms of the Ricci tensor, the Ricci scalar, and the metric as follows:*

$$R_{\alpha\beta\gamma\delta} = \left(g_{\alpha\gamma} R_{\beta\delta} + g_{\beta\delta} R_{\alpha\gamma} - g_{\alpha\delta} R_{\beta\gamma} - g_{\beta\gamma} R_{\alpha\delta} \right)$$
$$- \frac{1}{2} \left(g_{\alpha\gamma} g_{\beta\delta} - g_{\alpha\delta} g_{\beta\gamma} \right) R. \tag{3.5.25}$$

Problem 3.8: *Show that in a 2-dimensional spacetime the curvature tensor can be expressed in terms of the Ricci scalar and the metric only.*

$$R_{\alpha\beta\gamma\delta} = \frac{1}{2} \left(g_{\alpha\gamma} g_{\beta\delta} - g_{\alpha\delta} g_{\beta\gamma} \right) R. \tag{3.5.26}$$

The curvature tensor in the D-dimensional spacetime can be written as follows

$$R_{\alpha\beta\gamma\delta} = C_{\alpha\beta\gamma\delta} + E_{\alpha\beta\gamma\delta} + S_{\alpha\beta\gamma\delta}. \tag{3.5.27}$$

The three pieces in this decomposition are

1. Fully traceless part (Weyl tensor) $C_{\alpha\beta\gamma\delta}$.

$$C_{\alpha\beta\gamma\delta} \, g^{\alpha\gamma} = 0. \tag{3.5.28}$$

2. Semi-traceless part $E_{\alpha\beta\gamma\delta}$.

$$E_{\alpha\beta\gamma\delta} \, g^{\alpha\gamma} g^{\beta\delta} = 0. \tag{3.5.29}$$

3. Scalar part $S_{\alpha\beta\gamma\delta}$.

$$S_{\alpha\beta\gamma\delta} \, g^{\alpha\gamma} g^{\beta\delta} = R. \tag{3.5.30}$$

For the latter two components of the decomposition one has

$$E_{\alpha\beta\gamma\delta} = \frac{1}{D-2} \left(g_{\alpha\gamma} S_{\beta\delta} + g_{\beta\delta} S_{\alpha\gamma} - g_{\alpha\delta} S_{\beta\gamma} - g_{\beta\gamma} S_{\alpha\delta} \right),$$

$$S_{\alpha\beta} = R_{\alpha\beta} - \frac{1}{D} g_{\alpha\beta} R, \qquad S_{\alpha\beta} g^{\alpha\beta} = 0, \qquad (3.5.31)$$

$$S_{\alpha\beta\gamma\delta} = \frac{1}{D(D-1)} \left(g_{\alpha\gamma} g_{\beta\delta} - g_{\alpha\delta} g_{\beta\gamma} \right) R.$$

We know that the linear space of tensors with the same symmetry properties as the curvature tensor is irreducible with respect to the group GL(D, R). To obtain its decomposition Eq. (3.5.27) one uses the contraction, that is an operation that involves the metric. At a given point p one can introduce coordinates in which the metric takes the form $\mathbf{g} = diag(-1, 1, \ldots, 1)$. *The matrix* $\Lambda^\mu_{\ \mu'} = x^\mu_{\ ,\mu'}$ *of the transformations, which preserve this form of the metric, form the Lorentz group* $O(1, D-1)$. *Equation (3.5.27) gives the decomposition of the curvature tensor into the irreducible tensors with respect to this group. Since* $O(1, D-1) \subset GL(D, R)$, *this classification is more 'refined'.*

Problem 3.9:   *Consider a 2-dimensional Euclidean metric*

$$ds^2 = \frac{1}{f(z)} dz^2 + f(z) d\phi^2. \qquad (3.5.32)$$

Show that the Ricci scalar for this geometry is

$$R = -\frac{d^2}{dz^2} f(z). \qquad (3.5.33)$$

3.5.5   'Counting components'

The curvature tensor is constructed from the metric and its first and second derivatives. Can one construct another independent tensor by using $g_{\mu\nu}$, $g_{\mu\nu,\alpha}$, and $g_{\mu\nu,\alpha\beta}$? The answer is 'no'. To prove this let us count how many independent components the metric and its derivatives have.

Consider metric $g_{\mu\nu}$ at a given point p. In chosen coordinates it is a $D \times D$ symmetric matrix, and thus it has $D(D+1)/2$ independent components. Coordinate transformations at a given point p are described by a $D \times D$ matrix $x^\mu_{\ ,\mu'}$. Consider $D(D+1)/2$ equations

$$\eta_{\mu'\nu'} = x^\mu_{\ ,\mu'} x^\nu_{\ ,\nu'} g_{\mu\nu}. \qquad (3.5.34)$$

One can solve this system for $x^\mu_{\ ,\mu'}$ and there remain $D(D-1)/2$ free parameters.

These special transformations preserve the form of the flat metric $\eta_{\mu\nu}$. We already discussed such transformations in a flat spacetime and know that they form a D-dimensional Lorentz group that has $D(D-1)/2$ parameters. This is exactly the freedom in the choice of a solution of Eq. (3.5.34).

Consider now objects constructed from the metric and its derivatives up to the order n, $g_{\mu\nu,\alpha_1\ldots\alpha_n}$. Let us count the number of their independent components. This number is a product of $D(D+1)/2$ (the number of components of $g_{\mu\nu}$) and the number of possible components of n partial derivatives. The partial derivatives of $g_{\mu\nu}$ are symmetric with respect to their indices. Hence, they form an object with n elements and each of them takes D distinguished values. The corresponding total number of n combinations with replacements is

$$C^R(D,n) = \frac{(D+n-1)!}{n!\,(D-1)!}. \tag{3.5.35}$$

Therefore, the total number of different components of $g_{\mu\nu,\alpha_1...\alpha_n}$ is

$$Z_1 = \frac{(D+n-1)!}{n!\,(D-1)!}\,\frac{D(D+1)}{2}. \tag{3.5.36}$$

Transformations of $g_{\mu\nu,\alpha_1...\alpha_n}$ to the new coordinates $x^{\mu'}$ involve $x^{\alpha}_{,\mu'_1...\mu'_{n+1}}$. The total number of different components of this object is

$$Z_2 = \frac{(D+n)!}{(n+1)!\,(D-1)!}\,D. \tag{3.5.37}$$

The difference $Z(n,D) = Z_1 - Z_2$ is

$$Z(n,D) = \frac{D(n-1)}{2(n+1)!\,(D-2)!}\,(D+n-1)!. \tag{3.5.38}$$

This is the total number of independent components of $g_{\mu\nu,\alpha_1...\alpha_n}$ after the coordinate transformation freedom is used.

Consider special cases. For $n = 1$ one has $Z(1,D) = 0$. This means that all the components of $g_{\mu\nu,\alpha}$ can be put equal to 0 by proper coordinate transformation. In particular, this implies that there exist coordinates in which all the components of the Christoffel symbol at a given point vanish. Such coordinates, as we already know, are called the *Riemann normal coordinates*.

For $n = 2$ one has $Z(2,D) = \frac{1}{12}D^2(D^2 - 1)$. This is a number of independent components of $\partial^2 g$ that cannot be 'transformed away' by coordinate transformations. This number coincides with the total number of independent components of the curvature tensor $N_D[\mathcal{R}]$, Eq. (3.5.12). Thus, the only non-trivial tensor constructed from the metric and its derivatives up to the second order is the Riemann tensor.

The following general result is valid: *Any tensor constructed from the metric and its derivatives up to $n + 2$ can be constructed from the curvature tensor and its covariant derivatives up to the order n.*

3.5.6 Fermi coordinates

Consider a time-like curve γ and use the proper time τ as a parameter along it (see Figure 3.9). Let u^μ be the unit tangent vector. Choose three orthonormal vectors $e^\mu_{\hat{i}}(\tau_0)$ in a subspace orthogonal to $e^\mu_{\hat{0}} \equiv u^\mu$ at some moment of time τ_0. Define the vectors $\{e^\mu_{\hat{0}}, e^\mu_{\hat{i}}\}$ to be Fermi-propagated along γ. Points in the vicinity of γ can be parametrized as follows. A geodesic orthogonal to u at the moment of proper time τ is uniquely determined by the unit vector n^i ($i = 1, 2, 3$). If a point p lies on such a geodesic at the geodesic distance s, its Fermi coordinates are (τ, s, n^i) ($n_i n^i = 1$). A metric in the Fermi coordinates is

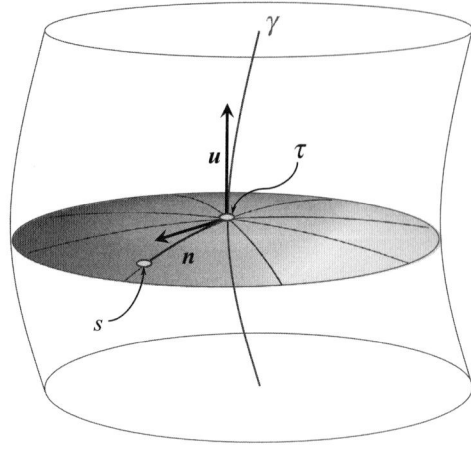

Fig. 3.9 Fermi coordinates $(\tau, s, \boldsymbol{n})$. Here, τ is a proper time along the curve γ, s is the geodesic distance to the point p, and n^i is a unit vector in the direction of p. This unit spatial vector $\boldsymbol{n}$ may be considered as a set of angles that describe a point on a $(D-2)$-dimensional unit sphere.

$$ds^2 \doteq -\left[1 + 2w_i\,x^i + (w_i\,x^i)^2 + R_{\alpha i \beta j}\,u^\alpha x^i\,u^\beta\,x^j\right](dx^0)^2$$

$$-\frac{4}{3}\,R_{\alpha k i l}\,u^\alpha\,x^k\,x^l\,dx^0\,dx^i \tag{3.5.39}$$

$$+\left(\delta_{ij} - \frac{1}{3}\,R_{ikjl}\,x^k\,x^l\right)dx^i\,dx^j + O(s^3).$$

Here, $w^\alpha = u^\mu \nabla_\mu u^\alpha$ is the vector of acceleration. For a geodesic line γ, $w^\alpha = 0$, and hence in the Fermi coordinates along this line $g_{\alpha\beta} \doteq \eta_{\alpha\beta}$ and $\Gamma^\lambda{}_{\alpha\beta} \doteq 0$.

3.6 Parallel Transport of a Vector

A flat spacetime possesses the following property. Any vector after its parallel transport along a closed contour coincides with itself. This allows one to define uniquely vectors over all the spacetime that are parallel to a chosen one. This property is called *absolute parallelism* of the spacetime. In a general case, in the presence of the curvature this property is not valid.

Consider a one-parameter family of smooth curves $x^\mu(\lambda, \sigma)$, let λ be a parameter along curves and σ be a parameter enumerating the curves. We denote $u^\mu = dx^\mu/d\lambda$ and $v^\mu = dx^\mu/d\sigma$. Consider two close curves with the parameters σ and $\sigma + \delta\sigma$. Denote $x_0^\mu = x^\mu(\lambda, \sigma)$. Consider a parallelogram with the apices at the points (see Figure 3.10)

$$p \leftrightarrow x_0^\mu, \quad p_u \leftrightarrow x_0^\mu + u^\mu \delta\lambda, \quad p_v \leftrightarrow x_0^\mu + v^\mu \delta\sigma, \quad p' \leftrightarrow x_0^\mu + u^\mu \delta\lambda + v^\mu \delta\sigma. \tag{3.6.1}$$

Let $\boldsymbol{A}$ be a vector at the point p. We propagate this vector along sides of the parallelogram. We denote the result of the parallel transform from p to p_u and p_v by $\boldsymbol{A}_u$ and $\boldsymbol{A}_v$, respectively. We denote a result of the parallel transport of $\boldsymbol{A}$ from p along the paths $p \rightarrow p_u \rightarrow p'$ and $p \rightarrow p_v \rightarrow p'$ by $\boldsymbol{A}_{uv}$ and $\boldsymbol{A}_{vu}$, respectively. Denote by $\delta\boldsymbol{A} = \boldsymbol{A}_{uv} - \boldsymbol{A}_{vu}$. This quantity depends linearly on $\boldsymbol{A}$, $\boldsymbol{u}$ and $\boldsymbol{v}$ and one has

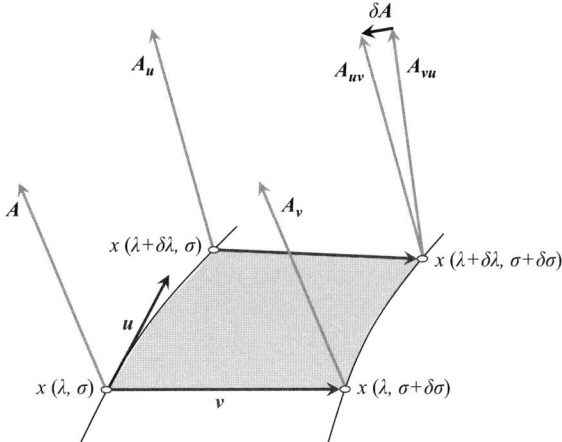

Fig. 3.10 The result of a parallel transport of a vector A depends on a path. Two vectors obtained from A by the parallel transport from an initial point p to a final point p' along 'upper' (p, p_u, p') and 'lower' (p, p_v, p') paths differs by the vector δA.

$$\delta A^\alpha = Q^\alpha{}_{\beta\gamma\delta} A^\beta u^\gamma v^\delta \, \delta\lambda\delta\sigma. \tag{3.6.2}$$

Since δA^α is a vector and Eq. (3.6.2) is valid for arbitrary A, u and v, $Q^\alpha{}_{\beta\gamma\delta}$ is a tensor. To find this tensor we use the Riemann normal coordinates at the point p in which

$$\Gamma^\alpha{}_{\beta\gamma}(p) \doteq 0, \quad \Gamma^\alpha{}_{\beta\gamma}(p_u) \doteq \Gamma^\alpha{}_{\beta\gamma,\delta}(p)u^\delta \, \delta\lambda, \quad \Gamma^\alpha{}_{\beta\gamma}(p_v) \doteq \Gamma^\alpha{}_{\beta\gamma,\delta}(p)v^\delta \, \delta\sigma. \tag{3.6.3}$$

The equation of a parallel transport gives

$$\delta_u A^\alpha = u^\mu \frac{\partial A^\alpha}{\partial x^\mu} = -\Gamma^\alpha{}_{\mu\nu} u^\mu A^\nu. \tag{3.6.4}$$

Since $\Gamma^\alpha{}_{\mu\nu}(p) \doteq 0$ one obtains $\delta_u A^\alpha$ and hence $A^\alpha(p_u) \doteq A^\alpha(p)$. Using Eq. (3.6.4) for the parallel transport of A from p_u to p_{uv} one gets

$$A^\alpha_{uv}(p_{uv}) = A^\alpha(p_u) - \Gamma^\alpha{}_{\beta\gamma}(p_u) A^\beta(p_u) v^\gamma \, \delta\sigma$$
$$\doteq A^\alpha(p) - \Gamma^\alpha{}_{\beta\gamma,\delta}(p) A^\beta(p) u^\delta v^\gamma \, \delta\lambda\delta\sigma. \tag{3.6.5}$$

Similarly, for the path $p \to p_v \to p_{vu}$ one has

$$A^\alpha_{vu}(p_{vu}) \doteq A^\alpha(p) - \Gamma^\alpha{}_{\beta\gamma,\delta}(p) A^\beta(p) u^\gamma v^\delta \, \delta\lambda\delta\sigma. \tag{3.6.6}$$

The difference of Eq. (3.6.5) and Eq. (3.6.6) has the form of Eq. (3.6.2) with

$$Q^\alpha{}_{\beta\gamma\delta} \doteq -(\Gamma^\alpha{}_{\beta\delta,\gamma} - \Gamma^\alpha{}_{\beta\gamma,\delta}). \tag{3.6.7}$$

Since in the Riemann coordinates the quantity in the brackets in the right-hand side of Eq. (3.6.7) coincides with the Riemann tensor, one has

$$\delta A^\alpha = -R^\alpha{}_{\beta\gamma\delta} A^\beta u^\gamma v^\delta \, \delta\lambda\delta\sigma. \tag{3.6.8}$$

This relation explicitly demonstrates that the presence of curvature creates an 'obstacle' for absolute parallelism.

3.6.1 Flatness and conformal flatness of the metric in a domain

Curvature and flatness

A spacetime region is called *simply connected* if any closed loop in it can be shrunk to a point by a continuous transformation.

Theorem: *The metric $g_{\mu\nu}$ in a simply connected region V of the Riemannian manifold M^D is flat (i.e. there exist flat coordinates $X^{\hat{\mu}}$ in which $g_{\hat{\mu}\hat{\nu}} = \eta_{\hat{\mu}\hat{\nu}}$) if and only if the Riemannian curvature tensor vanishes in this region.*

Proof: The direct statement is trivial: The curvature of a flat spacetime vanishes. To prove the second part of the theorem we assume that the curvature vanishes in some simply connected domain V. Let us first show that for any two points p and p' in this domain the result of the parallel transport from p to p' of a vector does not depend on the path connecting these points. Consider two paths γ and γ' connecting p and p'. In a simply connected region γ' can be obtained by a continuous transformation from γ (see Figure 3.11). This means that there exists a one-parameter family of curves $x^\mu(\lambda, \sigma)$ connecting p and p', where λ is a parameter along a curve with a fixed value of σ, while σ 'enumerates' curves of the family.

 We choose the parametrization of the curves so that for $\lambda = 0$ all the curves pass through p and for $\lambda = 1$ they pass through p'. (See Figure 3.11(a).) We also choose the parameter σ so that at $\sigma = 0$ a curve of the family coincides with γ, while at $\sigma = 1$ it coincides with γ'. Denote $\sigma_i = i\delta\sigma$, $\delta\sigma = 1/n$ ($i = 0, 1, \ldots, n$) and choose the number n to be large enough. Denote a curve with σ_i by γ_i. One has $\gamma_0 = \gamma$ and $\gamma_n = \gamma'$. Denote by V_i a domain restricted by curves γ_i and γ_{i+1}. Such a domain is shown in Figure 3.11(b). By splitting this domain by lines of constant $\lambda = \lambda_j = i/n'$ into small sectors (see Figure 3.11(b)) and applying the result Eq. (3.6.8) to each of the small sectors one can show that for vanishing curvature the parallel transport along the nearby lines γ_i and γ_{i+1} gives the same result. Since it is valid for each of the domains V_i, one concludes that the result of the parallel transport from p to p' along γ and γ' is the same.

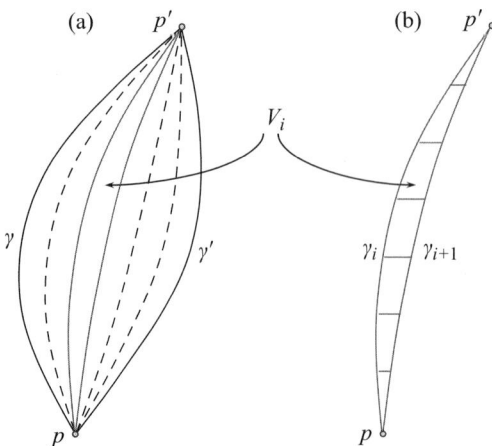

(a)   (b)

Fig. 3.11 Illustration of the proof of the theorem. The left figure (a) shows a one-parameter family of curves connecting two points p and p' in a simply connected region. The right figure (b) shows a surface element between two nearby curves γ_i and γ_{i+1}.

By assumption, the curvature tensor vanishes in a simply connected region V. Let us choose a point $p \in V$ and let $e_{\hat{v}}^{\mu}$ be an orthonormal basis at p. Let p' be another point of V and $\gamma(p, p')$ be a curve connecting p and p'. We define $e_{\hat{v}}^{\mu}(p')$ by parallel transport of $e_{\hat{v}}^{\mu}(p)$ along $\gamma(p, p')$. Since the curvature vanishes in V, $e_{\hat{v}}^{\mu}(p')$ does not depend on the choice of $\gamma(p, p')$. The scalar product of any two vectors is invariant under a parallel transport and, one has

$$g_{\mu\nu}(p') \, e_{\hat{\alpha}}^{\mu}(p') \, e_{\hat{\beta}}^{\nu}(p') = \eta_{\hat{\alpha}\hat{\beta}}. \tag{3.6.9}$$

The equation of the parallel transport implies

$$\frac{\partial e_{\hat{\alpha}}^{\mu}}{\partial x^{\nu}} = -\Gamma^{\mu}{}_{\nu\lambda} \, e_{\hat{\alpha}}^{\lambda}. \tag{3.6.10}$$

Let us define new coordinates $X^{\hat{\alpha}}$ by relations

$$\frac{\partial x^{\mu}}{\partial X^{\hat{\alpha}}} = e_{\hat{\alpha}}^{\mu}, \tag{3.6.11}$$

A first-order partial differential equation (3.6.11) has solutions since the integrability conditions

$$\frac{\partial e_{\hat{\alpha}}^{\mu}}{\partial X^{\hat{\beta}}} = \frac{\partial e_{\hat{\beta}}^{\mu}}{\partial X^{\hat{\alpha}}} \tag{3.6.12}$$

are satisfied. Indeed,

$$\frac{\partial e_{\hat{\alpha}}^{\mu}}{\partial X^{\hat{\beta}}} = e_{\hat{\beta}}^{\nu} \frac{\partial e_{\hat{\alpha}}^{\mu}}{\partial x^{\nu}} = -\Gamma^{\mu}{}_{\nu\lambda} \, e_{\hat{\beta}}^{\nu} \, e_{\hat{\alpha}}^{\lambda} = e_{\hat{\alpha}}^{\nu} \frac{\partial e_{\hat{\beta}}^{\mu}}{\partial x^{\nu}} = \frac{\partial e_{\hat{\beta}}^{\mu}}{\partial X^{\hat{\alpha}}}. \tag{3.6.13}$$

To complete the proof, it is sufficient to find the metric in the 'new' coordinates $X^{\hat{\alpha}}$. One has

$$g_{\hat{\alpha}\hat{\beta}} = \frac{\partial x^{\mu}}{\partial X^{\hat{\alpha}}} \frac{\partial x^{\nu}}{\partial X^{\hat{\beta}}} \, g_{\mu\nu} = g_{\mu\nu} \, e_{\hat{\alpha}}^{\mu} \, e_{\hat{\beta}}^{\nu} = \eta_{\hat{\alpha}\hat{\beta}}. \tag{3.6.14}$$

Problem 3.10: *By direct calculation of the Riemann tensor prove that the Rindler metric*

$$ds^2 = -x^2 dt^2 + dx^2 + \sum_{i=1}^{D-2} (dx_i)^2 \tag{3.6.15}$$

can be transformed to the flat one. Find this transformation.

Problem 3.11: *Calculate the curvature tensor of a round two-dimensional sphere of radius a.*

Conformal flatness

Suppose there exist two metrics $g_{\mu\nu}$ and $\tilde{g}_{\mu\nu}$ on a manifold. They are called *conformally related* if there exists a scalar function Ω such that

$$\tilde{g}_{\mu\nu} = \Omega^2 g_{\mu\nu}. \tag{3.6.16}$$

Since both metrics are smooth non-degenerate tensors, the function Ω must be a smooth non-vanishing (strictly positive) function.

Theorem: *The metric $g_{\mu\nu}$ in a simply connected region V of the Riemannian manifold M^D ($D \geq 4$) is conformally flat (i.e. there exist coordinates X^α in V in which $g_{\mu\nu} = \Omega^2(X)\,\eta_{\mu\nu}$) if and only if the Weyl tensor vanishes in this region.*

In a three-dimensional spacetime a metric $g_{\mu\nu}$ in a simply connected region V is conformally flat if and only if the tensor

$$R_{\mu\nu;\lambda} - R_{\mu\lambda;\nu} - \frac{1}{4}(g_{\mu\nu}R_{;\lambda} - g_{\mu\lambda}R_{;\nu}), \tag{3.6.17}$$

vanishes. Any two-dimensional metric is conformally flat.

> *Each of two metrics, which enter Eq. (3.6.16), can be used to define the Riemannian geometry on the same differential manifold. It is evident that such objects as the covariant derivatives and the curvature tensors for these two geometries are related. Appendix C contains corresponding useful formulas.*

3.6.2 Bianchi identities

Bianchi identities

We demonstrate now that first-order covariant derivatives of the curvature tensor obey the following symmetry relation, called the *Bianchi identity*

$$R^\sigma{}_{\alpha\,[\nu\kappa;\mu]} = 0. \tag{3.6.18}$$

Here, we use the same trick as before: Because this is a tensor relation, it is sufficient to prove it in a specially chosen coordinate system. We use the *Riemann normal coordinates* where

$$R^\sigma{}_{\alpha\nu\kappa;\mu} \doteq \Gamma^\sigma{}_{\alpha\kappa,\nu\mu} - \Gamma^\sigma{}_{\alpha\nu,\kappa\mu}. \tag{3.6.19}$$

The relation Eq. (3.6.18) directly follows from Eq. (3.6.19) after using the symmetry property of the Christoffel symbols. Since the metric commutes with the covariant derivative one can rewrite the Bianchi identity Eq. (3.6.18) in the form

$$R_{\sigma\alpha\,[\nu\kappa;\mu]} = 0. \tag{3.6.20}$$

Contracted Bianchi identities

The Bianchi identity can be written as

$$R^\alpha{}_{\cdot\sigma\beta\gamma;\delta} + R^\alpha{}_{\cdot\sigma\delta\beta;\gamma} + R^\alpha{}_{\cdot\sigma\gamma\delta;\beta} = 0. \tag{3.6.21}$$

By contracting this relation with respect to the indices α and β we obtain

$$R^\alpha{}_{\cdot\sigma\gamma\delta;\alpha} = R_{\sigma\delta;\gamma} - R_{\sigma\gamma;\delta}. \tag{3.6.22}$$

Multiplying this relation by $g^{\sigma\gamma}$ one finds

$$2R^{\alpha}{}_{\delta;\alpha} - R_{;\delta} = 0, \tag{3.6.23}$$

or equivalently

$$G^{\alpha\beta}{}_{;\beta} = 0, \tag{3.6.24}$$

where $G_{\alpha\beta}$ is the Einstein tensor.

Schur's theorem

We say that the curvature of a spacetime is direction independent if the curvature tensor can be presented in the form

$$R_{\mu\nu\alpha\beta} = b \left(g_{\mu\alpha} g_{\nu\beta} - g_{\mu\beta} g_{\nu\alpha} \right). \tag{3.6.25}$$

Simple calculation gives

$$R_{\alpha\beta} = (D-1) b g_{\alpha\beta}, \qquad R = D(D-1) b. \tag{3.6.26}$$

The corresponding Einstein tensor is

$$G_{\alpha\beta} = -\frac{1}{2}(D-1)(D-2) b g_{\alpha\beta}. \tag{3.6.27}$$

For $D \geq 3$ from Eq. (3.6.24) it follows that,

$$b_{,\alpha} = 0. \tag{3.6.28}$$

Thus, b is constant. This result allows the following formulation:

Schur's theorem: *If the Riemann tensor at each point is direction independent, it does not vary from one point to another.*

3.6.3   Geodesic deviation and tidal forces

We have already mentioned that non-vanishing tidal forces acting on a freely falling body in the gravitational field indicate that the gravitational field cannot be 'gauged away' by means of coordinate transformations. This means that the spacetime is curved and the Riemann tensor does not vanish. Let us discuss this point in more detail.

A free particle in a curved spacetime moves along a geodesic. In the *Fermi coordinates* $\Gamma^{\alpha}{}_{\mu\nu} \doteq 0$ along its worldline. This means that if a size l of a 'laboratory' associated with this line is much smaller that the characteristic curvature scale L, then experiments in such a geodesically moving laboratory will give the same results as those conducted in the inertial frame laboratory in a flat spacetime. This conclusion is not correct if l is comparable to L, since the curvature corrections, which are of the order of $(l/L)^2$, become important. This happens, for example, when one considers the motion of an *extended body*. Gravitational forces acting differently on different parts of such a body result in the *tidal forces*, which 'try' to distort the body.

We have already discussed this effect in the Newtonian gravity in Section 1.3.1. Now establish the corresponding relativistic equations that are valid in arbitrary spacetime. For this purpose, let us consider a beam of freely moving particles and find out how a separation between two close geodesics for nearby particles changes with time. The corresponding equation is known as the *geodesic deviation equation*.

Riemannian operator

For any two vectors $\boldsymbol{u}$ and $\boldsymbol{v}$ the *Riemannian operator* $\mathcal{R}(\boldsymbol{u}, \boldsymbol{v})$ acting on a third vector $\boldsymbol{w}$ is defined by the relation

$$\mathcal{R}(\boldsymbol{u}, \boldsymbol{v})\boldsymbol{w} = (\nabla_{\boldsymbol{u}}\nabla_{\boldsymbol{v}} - \nabla_{\boldsymbol{v}}\nabla_{\boldsymbol{u}} - \nabla_{[\boldsymbol{u},\boldsymbol{v}]})\boldsymbol{w}. \tag{3.6.29}$$

Let us show that

$$(\mathcal{R}(\boldsymbol{u}, \boldsymbol{v})\boldsymbol{w})^{\mu} = R^{\mu}{}_{\nu\alpha\beta} w^{\nu} u^{\alpha} v^{\beta}. \tag{3.6.30}$$

The right-hand side of Eq. (3.6.29) is

$$u^{\alpha}(v^{\beta}w^{\mu}{}_{;\beta})_{;\alpha} - v^{\alpha}(u^{\beta}w^{\mu}{}_{;\beta})_{;\alpha} - (u^{\alpha}v^{\beta}{}_{;\alpha} - v^{\alpha}u^{\beta}{}_{;\alpha})w^{\mu}{}_{;\beta}$$
$$= u^{\alpha}v^{\beta}(w^{\mu}{}_{;\beta\alpha} - w^{\mu}{}_{;\alpha\beta}) = R^{\mu}{}_{\nu\alpha\beta} w^{\nu} u^{\alpha} v^{\beta}. \tag{3.6.31}$$

The last equality was obtained by using the expression Eq. (3.5.5) for the commutator of the covariant derivatives. This proves Eq. (3.6.30).

Geodesic deviation equation

Consider a one-parameter family of geodesics $x^{\mu} = x^{\mu}(\lambda, \sigma)$, where λ is an affine (proper time) parameter, and σ 'enumerates' the geodesics. Denote by n^{μ} a vector connecting the points of the same λ on two close geodesics with parameters σ and $\sigma + \delta\sigma$. Since the vector $\boldsymbol{n}$ is 'frozen' into the congruence, it obeys the relation Eq. (3.4.1)

$$\mathcal{L}_{\boldsymbol{u}}\boldsymbol{n} = [\boldsymbol{u}, \boldsymbol{n}] = 0. \tag{3.6.32}$$

Let us use the relation Eq. (3.6.29) taking $\boldsymbol{v} = \boldsymbol{n}$ and $\boldsymbol{w} = \boldsymbol{u}$

$$\mathcal{R}(\boldsymbol{u}, \boldsymbol{n})\boldsymbol{u} = \nabla_{\boldsymbol{u}}\nabla_{\boldsymbol{n}}\boldsymbol{u} - \nabla_{\boldsymbol{n}}\nabla_{\boldsymbol{u}}\boldsymbol{u} = \nabla_{\boldsymbol{u}}\nabla_{\boldsymbol{n}}\boldsymbol{u} = \nabla_{\boldsymbol{u}}\nabla_{\boldsymbol{u}}\boldsymbol{n}. \tag{3.6.33}$$

We used here the geodesic equation $\nabla_{\boldsymbol{u}}\boldsymbol{u} = 0$ and the relation Eq. (3.6.32) that implies that $\nabla_{\boldsymbol{n}}\boldsymbol{u} = \nabla_{\boldsymbol{u}}\boldsymbol{n}$. Using Eq. (3.6.30) one obtains

$$\frac{D^2 n^{\alpha}}{d\lambda^2} \equiv (\nabla_{\boldsymbol{u}}\nabla_{\boldsymbol{u}}\boldsymbol{n})^{\alpha} = -R^{\alpha}{}_{\sigma\mu\nu} u^{\sigma} u^{\nu} n^{\mu}. \tag{3.6.34}$$

The obtained *geodesic deviation equation* shows that two inertially moving close particles have a relative acceleration. This tidal acceleration, induced by the curvature, is proportional to the distance between the particles.

3.7 Spacetime Symmetries

3.7.1 Killing vectors

Let f_t be a one-parameter family of diffeomorphisms of M^D onto itself generated by the vector field ξ

$$x^\mu \to x^\mu(t), \qquad \frac{dx^\mu(t)}{dt} = \xi^\mu(x). \tag{3.7.1}$$

Any vector at an initial point t_0 is dragged along the trajectory $x^\mu(t)$ by this map. Let η and ζ be two such Lie-propagated vectors. We call a map a *symmetry transformation*, if for any choice of these vectors their scalar product remains invariant along this trajectory. This can be written as

$$\mathcal{L}_\xi(\zeta \cdot \eta) = 0. \tag{3.7.2}$$

To have this property the vector ξ must obey the following Killing equation

$$\mathcal{L}_\xi\, g_{\mu\nu} = g_{\alpha\nu}\, \nabla_\mu \xi^\alpha + g_{\mu\alpha}\, \nabla_\nu \xi^\alpha = 0. \tag{3.7.3}$$

Lowering the indices in this relation we obtain the Killing equation in the form that is already familiar to us

$$\xi_{(\mu;\nu)} = 0. \tag{3.7.4}$$

3.7.2 Symmetry algebra

It is evident that a linear combination of two Killing vectors is again a Killing vector, that is, the Killing vectors form a linear space. If ξ^μ and η^μ are two Killing vectors, then their commutator

$$\zeta^\mu = [\xi, \eta]^\mu = \xi^\alpha\, \eta^\mu{}_{,\alpha} - \eta^\alpha\, \xi^\mu{}_{,\alpha}, \tag{3.7.5}$$

is again a Killing vector. To show this, let us use the relations Eq. (3.4.7)

$$\mathcal{L}_\zeta = [\mathcal{L}_\xi, \mathcal{L}_\eta] = \mathcal{L}_\xi\, \mathcal{L}_\eta - \mathcal{L}_\eta\, \mathcal{L}_\xi. \tag{3.7.6}$$

Applying this relation to the metric one obtains

$$\mathcal{L}_\zeta g = \mathcal{L}_\xi\, \mathcal{L}_\eta g - \mathcal{L}_\eta\, \mathcal{L}_\xi g = 0. \tag{3.7.7}$$

The linear space of the Killing vectors with a product operation defined by their commutator forms the *Lie algebra*. Let $\{\xi_A\}$ $(A = 1, \ldots, r)$ be a basis in the space of the Killing vectors, then

$$[\xi_A, \xi_B] = C^C{}_{AB}\, \xi_C. \tag{3.7.8}$$

Here, $C^C{}_{AB}$ are the *structure constants* obeying the relations

$$C^C{}_{AB} = -C^C{}_{BA}, \qquad C^E{}_{[AB}\, C^K{}_{C]E} = 0. \tag{3.7.9}$$

The second relation is called the *Jacobi identity*.

Theorems (Lie):

1. *A Lie group determines uniquely a Lie algebra.*
2. *A Lie algebra determines uniquely a (simply connected)* Lie group.
3. *Any set of constants $C^A{}_{BC}$ obeying Eq. (3.7.9) determines a Lie algebra.*

3.7.3   Stationary and static spacetimes

Let ξ^μ be a Killing vector. Consider a congruence formed by its integral lines

$$\frac{dx^\mu}{dt} = \xi^\mu. \tag{3.7.10}$$

Let (t, y^i) be coordinates such that y^i are fixed for a given integral curve, and the parameter t is defined by Eq. (3.7.10). (Note that there exists a coordinate freedom $t \to t + f(y^i)$.)

In the coordinates (t, y^i) one has $\xi^\mu \doteq \delta_0^\mu$ and the Killing equation Eq. (3.7.3) implies $\Gamma_{\nu\mu 0} + \Gamma_{\mu\nu 0} = 0$. This relation gives

$$g_{\mu\nu,0} \doteq 0. \tag{3.7.11}$$

This means that in the chosen coordinates the metric coefficients do not depend on one of the coordinates, t. If the vector field ξ is time-like in some simply connected domain V, we call a spacetime *stationary* in V. In such a domain one can always find coordinates in which the metric does not depend on time t.

A stationary spacetime is called *static* if the Killing vector ξ satisfies an additional condition

$$\xi_{[\mu} \xi_{\alpha;\beta]} = 0. \tag{3.7.12}$$

The Frobenius theorem implies that the condition Eq. (3.7.12) is equivalent to

$$\xi_\mu = \alpha \, t_{,\mu}. \tag{3.7.13}$$

In other words, ξ^μ is orthogonal to $t = \text{const}$ surfaces. One can use y^i as coordinates at $t = \text{const}$ surfaces. In coordinates (t, y^i) the *static metric* takes the form

$$ds^2 = -e^{2U} \, dt^2 + h_{ij} \, dy^i \, dy^j, \qquad \xi^2 \equiv \xi^\alpha \xi_\alpha = -e^{2U}, \tag{3.7.14}$$

and the metric coefficients U and h_{ij} do not depend on the time t.

3.7.4   Properties of the Killing vectors

The existence of one or more Killing vectors imposes constrains on the metric. If a solution of the Killing equation exists it uniquely determines the Killing vector in a simply connected domain V. The only ambiguity is the choice of the normalization, which can be fixed by normalizing the Killing vector at one chosen point of V.[7] A Killing vector obeys the following equation

---

[7]If the domain (or the spacetime) is not simply connected the problem of the normalization becomes non-trivial. We discuss this problem in detail in Section 10.8.

$$\xi_{\alpha;\beta;\gamma} = R_{\alpha\beta\gamma}{}^{\sigma}\xi_{\sigma}. \tag{3.7.15}$$

In order to prove this relation it is sufficient to add two and subtract one of the following three relations, in which the commutators of the covariant derivatives act on $\boldsymbol{\xi}$,

$$\xi_{\alpha;\beta;\gamma} - \xi_{\alpha;\gamma;\beta} = R_{\gamma\beta\alpha}{}^{\sigma}\xi_{\sigma},$$

$$\xi_{\gamma;\alpha;\beta} - \xi_{\gamma;\beta;\alpha} = R_{\beta\alpha\gamma}{}^{\sigma}\xi_{\sigma},$$

$$\xi_{\beta;\gamma;\alpha} - \xi_{\beta;\alpha;\gamma} = R_{\alpha\gamma\beta}{}^{\sigma}\xi_{\sigma},$$

and use the symmetry properties of the curvature tensor. Equation (3.7.15) plays the role of the integrability condition for the system of the first-order partial differential equations Eq. (3.7.4).

Consider a Killing vector at a point p. It has D components. Its first covariant derivative $\xi_{\mu;\nu}$ is antisymmetric and has $\frac{1}{2}D(D-1)$ independent components. Equation (3.7.15) shows that the second (as well as higher) derivatives of $\boldsymbol{\xi}$ at the point p are uniquely determined by $\boldsymbol{\xi}$ and its first derivatives. This means that the largest possible number of independent Killing vector fields that can exist on the D-dimensional manifold is $\frac{1}{2}D(D+1)$. For the four-dimensional manifold this number is 10. The Minkowski spacetime is an example of the spacetime with the largest possible group of symmetry. This is a 10-parameter Poincaré group.

Contracting Eq. (3.7.15) with respect to the indices β and γ, one obtains

$$\xi_{\alpha;\beta}{}^{;\beta} = -R_{\alpha\beta}\xi^{\beta}. \tag{3.7.16}$$

Denote $F_{\alpha\beta} = \xi_{\beta;\alpha} - \xi_{\alpha;\beta}$, then Eq. (3.7.16) implies that in a vacuum spacetime ($R_{\alpha\beta} = 0$)

$$F_{\alpha}{}^{\beta}{}_{;\beta} = 0. \tag{3.7.17}$$

That is, in vacuum spacetimes the Killing vector can be identified with the electromagnetic field potential A_{μ} in the Lorentz gauge $A^{\mu}{}_{;\mu} = 0$. For such a potential the source-free Maxwell equations are automatically satisfied.

Problem 3.12: *Prove that in a static spacetime Eq. (3.7.14) the Killing vector obeys the relation*

$$\xi_{\mu;\nu} = \xi_{\mu}w_{\nu} - \xi_{\nu}w_{\mu}, \tag{3.7.18}$$

where

$$w_{\mu} = U_{;\mu}, \qquad \xi^{2} \equiv \xi^{\alpha}\xi_{\alpha} = -e^{2U}. \tag{3.7.19}$$

Denote by $u^{\mu} = \xi^{\mu}/|\xi^2|^{1/2}$. Consider an observer with the velocity u^{μ} that is at rest in the static spacetime. Its motion is non-geodesic, since there must exist external non-gravitational forces that support the observer at rest. Prove that w^{μ} is the four-vector of a proper acceleration of the static (Killing) observer.

Problem 3.13: *Prove that in a static spacetime the acceleration vector w_{μ} obeys the relations*

$$w_{\mu;\nu} = -w_{\mu}w_{\nu} + \zeta_{\mu\nu}w_{\lambda}w^{\lambda} - R_{\mu\alpha\nu\beta}\zeta^{\alpha\beta},$$

$$\Box U = w_{\mu}{}^{;\mu} = -R_{\alpha\beta}\zeta^{\alpha\beta}, \tag{3.7.20}$$

where $\zeta^{\alpha\beta} = \xi^{\alpha}\xi^{\beta}/\xi^2$.

Problem 3.14:   *Let ξ^μ be a Killing vector in a static spacetime. Denote $\eta_\mu = \xi_\mu/\xi^2$. Using Eq. (3.7.18) prove that*

$$\eta_{\mu;\nu} = -\frac{\xi_\mu w_\nu + \xi_\nu w_\mu}{\xi^2}. \tag{3.7.21}$$

Equation (3.7.21) shows that the one-form η is closed, $d\eta = 0$, so that in a simply connected region it can be presented as $\eta_\mu dx^\mu = dt$. Surface $t = \mathrm{const}$ is orthogonal to ξ, and hence to the velocity of the Killing observer. In the vicinity of any point, elements of the surface $t = \mathrm{const}$ orthogonal to u form a set of events simultaneous from the point of view of the observer u. The proved relations show that in a simply connected domain of a static spacetime the simultaneous events for the Killing observers are 'in agreement' and form a surface $t = \mathrm{const}$. Namely, this parameter t enters as a time coordinate in the metric Eq. (3.7.14).

Conservation law

In accordance with the Noether theorem, spacetime symmetries imply conservation laws for conservative systems. For a particle and light motion in a spacetime with symmetries the conservation laws take a very simple form.

Let $x^\mu(\lambda)$ be a worldline of a particle or a light ray. We choose the parameter λ to coincide with the proper time τ (for the particle) or with an affine parameter (for the light ray). Denote $u^\mu = dx^\mu/d\lambda$. Suppose ξ^μ is a Killing vector, then for a geodesic motion the following quantity is conserved

$$\varepsilon = u_\mu \xi^\mu. \tag{3.7.22}$$

Indeed,

$$\frac{d\varepsilon}{d\lambda} = u^\nu (u^\mu \xi_\mu)_{;\nu} = u^\nu u^\mu_{;\nu} \xi_\mu + u^\mu u^\nu \xi_{\mu;\nu} = 0. \tag{3.7.23}$$

The last relation takes place because the motion is geodesic, $u^\nu u^\mu_{;\nu} = 0$, and $\xi_{\mu;\nu}$ is an antisymmetric tensor.

In a stationary spacetime with the symmetry generator ξ the corresponding conserved quantity $-\varepsilon$ is proportional to the particle energy. The energy of the particle of mass m is $E = -mu^\mu \xi_\mu$. To obtain the energy of a photon one must choose the affine parameter λ so that $u^\mu = p^\mu$, where p^μ is the momentum of the photon.

Problem 3.15:   *Consider a static spacetime with the Killing vector ξ^μ. Let $\alpha = |\xi^2|^{1/2}$ be a redshift factor. The frequency of a photon with the momentum p^μ measured by a Killing observer u^μ is*

$$\omega = -p_\mu u^\mu. \tag{3.7.24}$$

Show that the redshifted frequency $\alpha\omega$ remains constant along the photon trajectory.

3.7.5   Killing tensor and Killing–Yano tensor

Killing tensor and conservation law

Killing vectors allow generalizations. Two generalizations are considered most often. The first one is a symmetric rank n tensor $K_{\mu_1\ldots\mu_n}$ that obeys the following relation:

$$K_{(\mu_1...\mu_n;\nu)} = 0. \tag{3.7.25}$$

Such a tensor is called a *Killing tensor*.

In the presence of the Killing tensor one also has a conservation law. The corresponding conserved quantity is

$$\varepsilon = K_{\mu_1...\mu_n} u^{\mu_1} \cdots u^{\mu_n}. \tag{3.7.26}$$

▌ **Problem 3.16:** *Prove that* $\varepsilon = K_{\mu_1...\mu_n} u^{\mu_1} \cdots u^{\mu_n}$ *is conserved.*

Another generalization of the Killing vector is an antisymmetric rank n tensor $f_{\mu_1...\mu_n}$ tensor (or n-form) obeying the equation

$$\nabla_{(\nu} f_{\mu_1)...\mu_n} = 0. \tag{3.7.27}$$

Such a tensor is called a *Killing–Yano tensor*.

Since the Killing–Yano tensor is skew-symmetric, one cannot construct a non-trivial object similar to Eq. (3.7.26) by contracting it with the velocity vector. Nevertheless, the Killing–Yano tensor also generates a conservation law. The reason for this is that the Killing–Yano tensor allows one to construct a Killing tensor. Namely, if $f_{\mu_1\mu_2...\mu_n}$ is a Killing–Yano tensor, then

$$K_{\alpha\beta} = f_{\alpha\mu_2...\mu_n} f_\beta^{\ \mu_2...\mu_n} \tag{3.7.28}$$

is a rank-2 Killing tensor, i.e. it obeys Eq. (3.7.25)

$$K_{(\alpha\beta;\lambda)} = 0. \tag{3.7.29}$$

To prove this we first notice that $K_{\alpha\beta}$ is symmetric. We also have

$$K_{\alpha\beta;\lambda} = (\nabla_\lambda f_{\alpha\mu_2...\mu_n}) f_\beta^{\ \mu_2...\mu_n} + f_\alpha^{\ \mu_2...\mu_n} (\nabla_\lambda f_{\beta\mu_2...\mu_n}). \tag{3.7.30}$$

Note that for a symmetrization with respect to the indices λ and α the first term on the right-hand side vanishes. Similarly, the second term vanishes because of the symmetrization with respect to λ and β. Thus, for the symmetrization with respect to all 3 indices α, β, and λ the right-hand side vanishes and, hence, $K_{\alpha\beta}$ is a Killing tensor.

One can say that the square of the Killing–Yano tensor is always a rank-2 Killing tensor. The other way around is not generally true. Namely, not for any rank-2 Killing tensor K does there exist the Killing–Yano tensor that is its 'square root'.

Let us mention another important property of the Killing–Yano tensor. Suppose $f_{\mu\nu}$ is a Killing–Yano tensor, then for a geodesic motion the following vector $n_\mu = f_{\mu\nu} u^\nu$ is parallel propagated. Indeed,

$$u^\lambda n_{\mu;\lambda} = f_{\mu(\nu;\lambda)} u^\nu u^\lambda + f_{\mu\nu} u^\lambda u^\nu_{;\lambda} = 0. \tag{3.7.31}$$

To prove this relation we used the definition of the Killing–Yano tensor and the geodesic equation of motion.

Since the conservation laws for the Killing and Killing–Yano tensors are not related to explicit spacetime symmetries like in the case of the Killing vector, the symmetries connected with Killing and Killing–Yano tensors are called *hidden symmetries*.

3.8 Submanifold

3.8.1 Definition of a submanifold

Definition: *Let M^D be a smooth manifold of dimension D. We call its subset N^n $(n < D)$ a submanifold, if the following properties are satisfied:*

1. *N^n is a smooth manifold of dimension n.*
2. *If $p \subset N^n$, then there exist such coordinates x^α in the region U near p in M^D in which for a subset $(N^n \cap U) = V$ one has the equation*

$$x^{n+1} = x^{n+2} = \ldots = x^D = 0, \qquad (3.8.1)$$

and $x^1, \ldots, x^n$ are coordinates on $N^n \cap U$ near $p \in N^n$.

The difference $D - n$ of the dimensions of M^D and N^n is called a *codimension of a submanifold*. In the physical literature the space M^D, in which N^n is embedded, is often called a *bulk space*.

3.8.2 Induced metric

Equations (3.8.1) determine an embedding of a submanifold N^n in the bulk space M^D. There is another way to characterize this embedding. Namely, let y^i, $i = 1, \ldots, n$, be coordinates on N^n. Then, the relations describing the embedding can be written also in the form:

$$x^\alpha = x^\alpha(y^i), \qquad \alpha = 1, \ldots, D. \qquad (3.8.2)$$

The metric $g_{\mu\nu}$ in the bulk spacetime M^D allows one to determine the intervals between any two close points located in N^n. The restriction of the bulk metric to N^n generates an *induced metric h*. For any two points y^i and $y^i + dy^i$ in N^n one has

$$ds^2 = g_{\alpha\beta} \frac{\partial x^\alpha}{\partial y^i} \frac{\partial x^\beta}{\partial y^j} \, dy^i \, dy^j. \qquad (3.8.3)$$

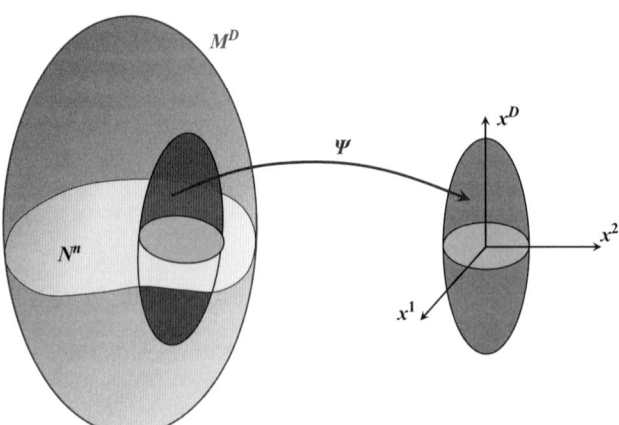

Fig. 3.12 Submanifold N^n of the manifold M^D.

The induced metric is

$$dl^2 = h_{ij} dy^i \, dy^j \,, \qquad h_{ij} = g_{\alpha\beta} \, \frac{\partial x^\alpha}{\partial y^i} \, \frac{\partial x^\beta}{\partial y^j}. \tag{3.8.4}$$

The induced metric h_{ij} allows one to define a covariant derivative and curvature of the subspace. To distinguish the induced covariant derivative from the covariant derivative with respect to the bulk metric we denote the former as follows: $(\ldots)_{|i}$. We denote also by γ^i_{jk} and γ_{ijk} the Christoffel symbols for the metric h_{ij}, and by $\mathcal{R}_{ijkl}$, $\mathcal{R}_{jl} = h^{ik}\mathcal{R}_{ijkl}$, and $\mathcal{R} = h^{jl}\mathcal{R}_{jl}$ the corresponding Riemann and Ricci tensors, and the Ricci scalar.

3.8.3 Extrinsic curvature

The *induced metric* determines internal geometric properties of a submanifold and makes it a Riemannian space. The induced geometry depends on the embedding of the submanifold in the bulk spacetime but it does not specify this embedding uniquely.

> *Here is a simple example that illustrates this fact. Consider a piece of paper. Its local internal geometry remains the same flat 2-dimensional one for different embeddings, e.g., if it is embedded either as a flat piece or it is rolled up into a cylinder.*

The quantity that serves to characterize how the submanifold is embedded is called an *extrinsic curvature*. The extrinsic curvature can be introduced for any number of codimensions. Let us define it first for the simplest case of a submanifold of a codimension 1. Let y^i, $(i = 1, \ldots, D-1)$, be coordinates on submanifold Σ of the dimension $D-1$. Let Σ be determined by relations

$$x^\alpha = x^\alpha(y^i), \qquad \alpha = 0, 1, \ldots, D-1. \tag{3.8.5}$$

We determine a normal to Σ as a vector n^α obeying the conditions

$$n_\alpha \frac{dx^\alpha(y^i)}{dy^i} = 0. \tag{3.8.6}$$

We call Σ space-like, time-like, or null if $n^2 < 0, > 0$, or $= 0$, respectively. We consider here space-like and time-like submanifolds only. By properly normalizing the vector $\boldsymbol{n}$ it can be made a unit one, so that

$$n_\alpha n^\alpha = \varepsilon(n) = \begin{cases} -1, & \text{space-like surface;} \\ +1, & \text{time-like surface.} \end{cases} \tag{3.8.7}$$

Instead of using the explicit equations Eq. (3.8.5), in which the surface Σ is given in a parametric form, one can define it by the following 'constraint' equation

$$F(x^\alpha) = 0. \tag{3.8.8}$$

In this case one has $n_\alpha \sim F_{,\alpha}$.

The *extrinsic curvature* of the submanifold Σ is defined by the relation[8]

$$K_{ij} = -\frac{\partial x^\alpha}{\partial y^i} \frac{\partial x^\beta}{\partial y^j} n_{\beta;\alpha}. \tag{3.8.9}$$

Note that now we have two independent set of coordinates, $\{x^\alpha\}$ and $\{y^i\}$, which can be transformed independently. For this reason objects connected with a submanifold can have different transformation laws with respect to 'external' and 'internal' coordinate transformations.

K_{ij} is a symmetric rank 2 tensor with respect to the 'internal' coordinate transformations in the submanifold. Under the coordinate transformations in the bulk space it behaves as a scalar. The symmetry property of $K_{ij} = K_{(ij)}$ follows from the relations

$$\begin{aligned}
K_{ij} &= -\frac{\partial x^\alpha}{\partial y^i} \frac{\partial x^\beta}{\partial y^j} \left(\frac{\partial n_\beta}{\partial x^\alpha} - \Gamma^\lambda_{\alpha\beta} n_\lambda \right) \\
&= n_\beta \frac{\partial^2 x^\beta}{\partial y^i \partial y^j} + \Gamma^\lambda_{\alpha\beta} n_\lambda \frac{\partial x^\alpha}{\partial y^i} \frac{\partial x^\beta}{\partial y^j} = n_\beta x^\beta_{;ij},
\end{aligned} \tag{3.8.10}$$

where the property $n_\beta \, \partial x^\beta / \partial y^i = 0$ has been used.

The notion of an extrinsic curvature can be easily generalized to the case when the number of codimensions is greater than 1. Let us consider Eq. (3.8.6) again. For the number of codimensions m it has m linearly independent solutions n^A, $(A = 1, \dots, m)$. We choose these vectors to be mutually orthogonal and assume that none of them is null. We denote by $\varepsilon(n^A) = (n^A)^2$. Since the spacetime signature is $(-, +, \dots, +)$ no more than one of $\varepsilon(n^A)$ can be negative. We normalize these vectors by the conditions

$$(n^A, n^B) = \varepsilon(n^A)\delta^{AB}. \tag{3.8.11}$$

The tensors of the extrinsic curvature, corresponding to the vectors n^A, are defined as

$$K^A_{ij} = -\frac{\partial x^\alpha}{\partial y^i} \frac{\partial x^\beta}{\partial y^j} n^A_{\beta;\alpha}. \tag{3.8.12}$$

They are symmetric with respect to the indices i and j.

3.8.4 Gaussian normal coordinates

Consider a submanifold Σ of codimension 1. In its vicinity one can introduce *Gaussian normal coordinates* in which the line element takes the form

---

[8] Our definition of the extrinsic curvature tensor coincides with that of the "Gravitation" by Misner, Thorne and Wheeler (1973), but it may differ by a sign from some other books. Our definition assumes that for a two-sphere of a radius r in a flat three-dimensional space the trace of the extrinsic curvature is negative and equal to $K = -2/r$.

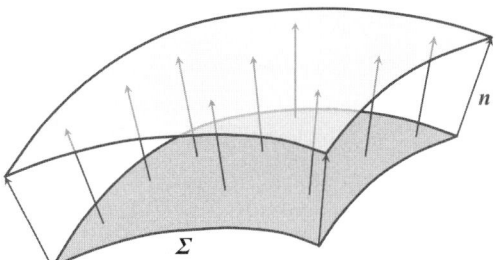

Fig. 3.13 Illustration to the definition of the Gaussian normal coordinates.

$$ds^2 = \varepsilon(n)\, d\tau^2 + h_{ij}(\tau, y)\, dy^i\, dy^j. \tag{3.8.13}$$

To define these coordinates let us consider a set of unit normal vectors n at Σ and consider geodesics $x^\mu(\tau, y^i)$ in M^D that at Σ obey relations

$$\left. \frac{dx^\mu(\tau, y^i)}{d\tau} \right|_\Sigma = n^\mu(y^i). \tag{3.8.14}$$

Denote by τ the proper distance (time) along these geodesics and choose it so that $\tau = 0$ on Σ. There exists a neighborhood of Σ where the geodesics of this family do not intersect each other. We restrict ourselves by writing the metric only in this domain. Let p be a point in this neighborhood. There is only one geodesic passing through it. We parametrize this point by (τ, y^i), where τ is the proper distance from p to Σ in the direction n, and y^i are the corresponding coordinates on Σ, where the geodesic crosses it. In these coordinates $g_{\tau\tau} = \varepsilon(n)$. Let us show that in the neighborhood of Σ the metric coefficient g_{0i} vanishes. For every i the vector $k^\mu_{(i)} = x^\mu(\tau, y)_{,i}$ is 'frozen' into the congruence. Hence,

$$[n, k] = 0. \tag{3.8.15}$$

At Σ one has $(n, k) = 0$. Let us show that this relation remains valid out of Σ. One has

$$n^\mu (n^\nu k_\nu)_{;\mu} = n^\mu n^\nu_{;\mu} k_\nu + n^\mu n^\nu k_{\nu;\mu}. \tag{3.8.16}$$

The first term on the right-hand side vanishes, because the congruence is geodesic. The second term can be transformed using Eq. (3.8.15) as follows

$$n^\mu n^\nu k_{\nu;\mu} = k^\mu n^\nu n_{\nu;\mu} = \frac{1}{2} k^\mu (n^2)_{;\mu} = 0. \tag{3.8.17}$$

Here, we used the normalization condition $n^2 = \varepsilon(n) = \text{const}$. Thus, $(n, k) = 0$ in the neighborhood of Σ, so that $g_{0i} = 0$.

In the Gaussian normal coordinates the *extrinsic curvature* reads

$$K_{ij} = -\frac{1}{2}\,\varepsilon(n)\,\partial_\tau h_{ij}. \tag{3.8.18}$$

3.9   Integration

3.9.1   Unity decomposition

For a finite domain U that is covered by one chart the integral of a scalar function can be easily defined. It is sufficient to use the invariant volume element

$$d^D v = \sqrt{|g|}\, dx^D\,, \qquad g = \det(g_{\mu\nu}). \qquad (3.9.1)$$

This volume element is invariant under the coordinate transformations, and hence the integral of a scalar function φ

$$J = \int_U d^D v\, \varphi \qquad (3.9.2)$$

is well defined.

The problem may arise when the integration is taken over the total manifold that is not covered by a single coordinate chart. In this case the following construction, known as a *unity decomposition*, is useful.

Definition: *Let M^D be a manifold, and $\{U_i, \Psi_i\} = \mathcal{A}$ be an atlas on it. One has a unity decomposition on M^D consistent with $\mathcal{A}$ if there exists a set of smooth scalar functions $E = \{e_i\}$ that possesses the following properties:*

1. *e_i is non-vanishing only in one region U_i, that is, $e_i \big|_{M \setminus U_i} = 0$.*
2. *For each point $p \in M$ there exists only a finite number of functions from E non-vanishing at p.*
3. *$\sum_i e_i = 1$.*

Theorem: *For each atlas $\mathcal{A}$ there exists a unity decomposition consistent with $\mathcal{A}$.*

Let F be a scalar function on M^D, and E be a unity decomposition, then the integral of F over M^D is defined as follows

$$\int_M F = \sum_i \int_{U_i} e_i(x)\, F(x)\, \sqrt{|g(x)|}\, d^D x, \qquad (3.9.3)$$

where x^μ are coordinates on U.

Theorem: *The value $\int_M F$ does not depend on the choice of the unity decomposition.*

3.9.2   Integration over a submanifold

The presence of a metric allows one to introduce an integration operation that is covariantly defined. One can also reformulate in a totally covariant manner such results as the integration by parts and the Stokes' theorems. To demonstrate these relations let us consider the integration over submanifolds.

Let N^n be a submanifold of M^D and locally its embedding is defined by the relations $x^\alpha = x^\alpha(y^i)$. Let us introduce the following object:

$$\frac{\partial(x^{\nu_1}, \ldots, x^{\nu_n})}{\partial(y^1, \ldots, y^n)} = \begin{vmatrix} \dfrac{\partial x^{\nu_1}}{\partial y^1} & \cdots & \dfrac{\partial x^{\nu_n}}{\partial y^1} \\ \vdots & \vdots & \vdots \\ \dfrac{\partial x^{\nu_1}}{\partial y^n} & \cdots & \dfrac{\partial x^{\nu_n}}{\partial y^n} \end{vmatrix}. \tag{3.9.4}$$

Let $p = D - n$ be a number of codimensions. Then, the volume element on N^n is defined as follows:

$$d\sigma_{\mu_1 \ldots \mu_p} = \frac{1}{p!\, n!} e_{\mu_1 \ldots \mu_p \nu_1 \ldots \nu_n} \frac{\partial(x^{\nu_1}, \ldots, x^{\nu_n})}{\partial(y^1, \ldots, y^n)} dy^1 \cdots dy^n. \tag{3.9.5}$$

This volume element has the following properties:

1. It is a skew-symmetric tensor of rank p (p-form) under the coordinate transformations in the bulk space M^D.
2. It is invariant (scalar) under the coordinate transformations in the submanifold N^{D-p}.

The volume element on N^n can also be written as follows:

$$d\sigma_{\mu_1 \ldots \mu_p} = n^{(1)}_{[\mu_1} \ldots n^{(p)}_{\mu_p]} \sqrt{|h|}\, d^n y, \tag{3.9.6}$$

where $n^{(A)}_\mu$ ($A = 1, \ldots, p$) is a complete set of mutually orthogonal unit vectors normal to N^n, and $\sqrt{|h|}\, d^n y$ is an invariant volume scalar on the submanifold. If $f^{\mu_1 \ldots \mu_p}$ is a skew-symmetric tensor of rank p on N^n, its integral over the submanifold is defined as

$$\int_{N^{D-p}} f^{\mu_1 \ldots \mu_p} d\sigma_{\mu_1 \ldots \mu_p} = \int_{N^{D-p}} f \sqrt{|h|}\, d^n y, \tag{3.9.7}$$

where

$$f(y) = (f^{\mu_1 \ldots \mu_p}(x))|_{x^\mu = x^\mu(y^i)}\, n^{(1)}_{[\mu_1} \ldots n^{(p)}_{\mu_p]}. \tag{3.9.8}$$

The invariant δ-function $\delta(N^n)$ with a support on the submanifold N^n is defined as follows

$$\int_{M^D} F(x) \delta(N^n) \sqrt{|g|}\, d^D x = \int_{N^n} F(x)|_{x^\mu = x^\mu(y)} \sqrt{|h|}\, d^n y, \tag{3.9.9}$$

where F is an arbitrary smooth function on the manifold.

3.9.3 Stokes' theorems

Let V be a domain in the manifold M^D and $\Omega_{\mu_1 \ldots \mu_D} = F e_{\mu_1 \ldots \mu_D}$, then the integral $\int_V \Omega$ is defined by the following relation:

$$\int_V \Omega = \int_V F \sqrt{|g|}\, d^D x. \tag{3.9.10}$$

It is possible to prove the following general result known as *Stokes' theorem*

$$\int_V d\omega = \int_{\partial V} \omega, \tag{3.9.11}$$

where ω is a $(D-1)$-form. The following relations follow from Stokes' theorem:

$$\int_V \varphi^\alpha{}_{;\alpha}\, d^D v = \int_{\partial V} \varphi^\alpha\, d\sigma_\alpha\,, \qquad \int_\Sigma \varphi^{\alpha\beta}{}_{;\beta}\, d\sigma_\alpha = \int_{\partial\Sigma} \varphi^{\alpha\beta}\, d\sigma_{\alpha\beta}. \qquad (3.9.12)$$

In the second relation, dim $\Sigma = D - 1$, and $\varphi^{\alpha\beta}$ is a skew-symmetric tensor.

3.9.4 Minimal and geodesic submanifolds

Let Σ be a submanifold of codimension 1 and Eq. (3.8.13) be *Gaussian normal coordinates* associated with it, so that $\tau = 0$ is the equation of Σ. We assume here that τ is the time-like coordinate. Denote

$$\mathcal{A}(\Sigma) = \int_\Sigma \sqrt{h}\, d^{D-1}y. \qquad (3.9.13)$$

Consider a surface that is obtained by a small deformation of Σ, and let the equation of this surface be $\tau = \delta n(y)$. The corresponding variation of the functional $\mathcal{A}$ can be written in the form

$$\delta\mathcal{A}(\Sigma) \doteq \int_\Sigma \frac{\partial\sqrt{h}}{\partial\tau} \delta n(y)\, d^{D-1}y. \qquad (3.9.14)$$

One has

$$\frac{\partial\sqrt{h}}{\partial\tau} \doteq \frac{1}{2}h^{ij}\frac{\partial h_{ij}}{\partial\tau} = -h^{ij}K_{ij} \equiv -\mathrm{tr}(K). \qquad (3.9.15)$$

Thus, for the variation of $\mathcal{A}$ one has the following relation

$$\delta\mathcal{A}(\Sigma) \doteq -\int_\Sigma \mathrm{tr}(K)\delta n(y)\, d^{D-1}y. \qquad (3.9.16)$$

A surface Σ is called a *minimal surface* if it provides an extremum to the functional $\mathcal{A}$. This occurs when $\delta\mathcal{A}(\Sigma)$ vanishes for the arbitrary variation $\delta n(y)$. Thus, the equation of the minimal surface is

$$\mathrm{tr}(K) = 0. \qquad (3.9.17)$$

This definition can be easily generalized to the case when the codimension of Σ is arbitrary. Such a surface is a *minimal submanifold* if

$$\mathrm{tr}(K)^A = h^{ij}K^A_{ij} = 0, \qquad (3.9.18)$$

where K^A_{ij} is defined by Eq. (3.8.12).

In the general case, a geodesic in the bulk space that connects two points on the submanifold N^n ($n < D$) does not belong to N^n. A submanifold for which each geodesic in the bulk space, which connects two points in the submanifold N^n always belongs to N^n is called a *geodesic submanifold*. The necessary and sufficient conditions when this happens are

$$K^A_{ij} = 0\,, \qquad A = 1,\ldots,D-n. \qquad (3.9.19)$$

4

Particle Motion in Curved Spacetime

4.1 Equations of Motion

4.1.1 Lagrangian approach

A particle in a curved spacetime moves along a time-like geodesic. If u is its velocity, the equation of motion is $\nabla_u u = 0$, or equivalently,

$$u^\alpha \nabla_\alpha u^\mu = 0. \tag{4.1.1}$$

This means that the velocity is parallelly propagated along the trajectory. This makes geodesics similar to straight lines in the flat spacetime. Another property of geodesics, namely that they are curves with the longest proper time between given two points, is also the straightforward generalization of the flat spacetime properties. This second property of time-like geodesics means that the action for a particle of mass m

$$S[x^\mu(\tau)] = \int_1^2 d\tau\, L, \qquad L = -m\sqrt{-g_{\mu\nu}\frac{dx^\mu}{d\tau}\frac{dx^\nu}{d\tau}}, \tag{4.1.2}$$

with fixed intitial and final points is extremal for the geodesic.

In order to derive the geodesic equation determined by this extremum we impose the constraint

$$g_{\mu\nu}u^\mu u^\nu = -1, \tag{4.1.3}$$

which fixes the meaning of τ as the proper time. Then, the variation of the action Eq. (4.1.2) leads to the equation of motion in the standard form

$$\frac{d^2x^\mu}{d\tau^2} + \Gamma^\mu_{\nu\lambda}\frac{dx^\nu}{d\tau}\frac{dx^\lambda}{d\tau} = 0. \tag{4.1.4}$$

The parameter τ is the *proper time* along the geodesic.

A general way to work with a system with constraints is to introduce them by using the *Lagrange multipliers*. For our case a convenient form of the action is

$$\tilde{S}[x^\mu(\tau), \eta(\tau)] = \int_1^2 d\tau\, \tilde{L}, \qquad \tilde{L} = \frac{1}{2}\left[\eta^{-1}g_{\mu\nu}\frac{dx^\mu}{d\tau}\frac{dx^\nu}{d\tau} - m^2\eta\right]. \tag{4.1.5}$$

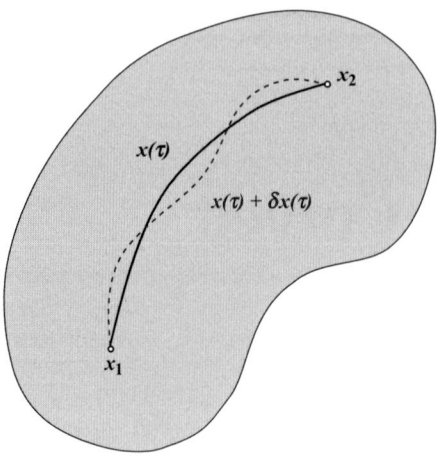

Fig. 4.1 The geodesic line between points x_1^μ and x_2^μ and the variation around it.

Here, $\eta(\tau)$ is a Lagrange multiplier. The action Eq. (4.1.5) is parametrization invariant, provided η transforms as follows:

$$\eta'(\tau')d\tau' = \eta(\tau)d\tau. \tag{4.1.6}$$

This action does not contain radicals and can be used for both time-like and null trajectories.
   The variation of the action $\tilde{S}[x^\mu(\tau), \eta(\tau)]$ with respect to $x^\mu(\tau)$ is

$$\delta\tilde{S} = \left(\frac{\partial\tilde{L}}{\partial\dot{x}^\mu}\delta x^\mu\right)\bigg|_{\tau_1}^{\tau_2} + \int_1^2 d\tau\left[\frac{\partial\tilde{L}}{\partial x^\mu} - \frac{d}{d\tau}\frac{\partial\tilde{L}}{\partial\dot{x}^\mu}\right]\delta x^\mu(\tau). \tag{4.1.7}$$

For the fixed boundary values x_1^μ and x_2^μ the first term on the right-hand side vanishes, so that the required equation of motion is

$$\frac{\partial\tilde{L}}{\partial x^\mu} - \frac{d}{d\tau}\frac{\partial\tilde{L}}{\partial\dot{x}^\mu} = 0. \tag{4.1.8}$$

The variation with respect to the Lagrange multiplier η gives

$$\eta^2 = -\frac{1}{m^2}g_{\mu\nu}\frac{dx^\mu}{d\tau}\frac{dx^\nu}{d\tau}. \tag{4.1.9}$$

Substituting η in the expression for $\tilde{L}$ one obtains $\tilde{L} = L$, where L is defined by Eq. (4.1.2). Hence, Eq. (4.1.8) is equivalent to the geodesic equation Eq. (4.1.4) provided τ is an *affine parameter*.

4.1.2   Hamiltonian form of the equation of motion

The Hamiltonian

$$H = p_\mu\frac{dx^\mu}{d\tau} - L, \qquad p_\mu = \frac{\partial L}{\partial\dot{x}^\mu}, \tag{4.1.10}$$

for the action Eq. (4.1.2) vanishes identically. This is the direct consequence of the reparame-trization invariance of the action.

The proper Hamiltonian can be obtained if one starts with the action Eq. (4.1.5). In fact, one has

$$p_\mu = \frac{\partial \tilde{L}}{\partial \dot{x}^\mu} = \eta^{-1} g_{\mu\nu} \frac{dx^\nu}{d\tau},$$

$$\tilde{H} = p_\mu \frac{dx^\mu}{d\tau} - \tilde{L} = \frac{1}{2}\eta\,[g^{\mu\nu}p_\mu p_\nu + m^2]. \qquad (4.1.11)$$

The corresponding action in the Hamiltonian form is

$$\tilde{S}_H[x^\mu(\tau), p_\mu(\tau), \eta(\tau)] = \int d\tau\,(p_\mu \dot{x}^\mu - \tilde{H}). \qquad (4.1.12)$$

The Hamilton equations of motion are

$$\frac{dx^\mu}{d\tau} = \frac{\partial \tilde{H}}{\partial p_\mu}, \qquad \frac{dp_\mu}{d\tau} = -\frac{\partial \tilde{H}}{\partial x^\mu}. \qquad (4.1.13)$$

The variation of Eq. (4.1.12) with respect to η gives the constraint equation

$$g^{\mu\nu}p_\mu p_\nu + m^2 = 0. \qquad (4.1.14)$$

This is a *mass shell* constraint for a particle.

Let us denote

$$d\sigma = \eta\,d\tau, \qquad d(\ )/d\sigma = (\dot{\ }),$$

$$H = \frac{1}{2}\left(g^{\mu\nu}p_\mu p_\nu + m^2\right). \qquad (4.1.15)$$

Then, the Hamilton equations Eq. (4.1.13) take the form

$$\dot{x}^\mu = \frac{\partial H}{\partial p_\mu}, \qquad \dot{p}_\mu = -\frac{\partial H}{\partial x^\mu}. \qquad (4.1.16)$$

This means that one can use the Hamiltonian Eq. (4.1.15) as a starting point to study the particle motion in a Riemannian manifold. It should be emphasized that this Hamiltonian has a well-defined limit for $m \to 0$, so that it can be used for the description of photons (zero-mass particles) propagation as well.

Problem 4.1: *Denote* $u^\mu = \dot{x}^\mu$. *Prove that the Hamilton equations Eq. (4.1.16) are equivalent to the following second-order equation:*

$$u^\nu u^\mu{}_{;\nu} = 0. \qquad (4.1.17)$$

That is, the motion of particles and light described by the Hamiltonian Eq. (4.1.15) is geodesic, as it should be.

4.1.3   Hamilton–Jacobi equation

In the previous subsections we considered the actions Eq. (4.1.2) and Eq. (4.1.5) as functionals of worldlines $x^\mu(\tau)$ connecting two points, the initial one $x_1^\mu = x^\mu(\tau_1)$ and the final one $x_2^\mu = x^\mu(\tau_2)$. Now we consider the same actions but from a different point of view. We fix the initial point x_1^μ but keep the final point x_2^μ arbitrary. We calculate the value of the action as a function of the final point, assuming that the equations of motion are satisfied. We fix the gauge choice by putting $\eta = 1$, so that τ is a proper time (or affine) parameter. We denote the obtained function by $S(x_2)$. For simplicity, we omit the index 2. Using Eq. (4.1.7) we obtain

$$\delta S = p_\mu \delta x^\mu, \tag{4.1.18}$$

and hence

$$\frac{\partial S}{\partial x^\mu} = p_\mu. \tag{4.1.19}$$

Substituting this relation into Eq. (4.1.14) one obtains

$$g^{\mu\nu} \frac{\partial S}{\partial x^\mu} \frac{\partial S}{\partial x^\nu} + m^2 = 0. \tag{4.1.20}$$

This is the *Hamilton–Jacobi equation* for a relativistic particle. Sometimes, this equation is written in a slightly different form. One introduces an extra parameter λ and considers a function $\bar S = \bar S(x^\mu, \lambda)$ obeying the equation

$$-\frac{\partial \bar S}{\partial \lambda} = \frac{1}{2} g^{\mu\nu} \frac{\partial \bar S}{\partial x^\mu} \frac{\partial \bar S}{\partial x^\nu}. \tag{4.1.21}$$

The previous form Eq. (4.1.20) is easily restored from Eq. (4.1.21) by the following substitution:

$$\bar S = S + \frac{1}{2} m^2 \lambda. \tag{4.1.22}$$

4.1.4   Reduced action formalism

The phase space, i.e. the space consisting of all possible values of position and momentum variables, constructed in the previous subsections is redundant. For example, in a 4-dimensional spacetime it has 8 dimensions. On the other hand, in order to define the motion of a particle it is sufficient to fix at the initial moment of time 6 quantities, 3 for the position and 3 for its velocity or momentum. The constraint equation Eq. (4.1.14) reflects this redundancy. Let us discuss now a reduced version of the action for the relativistic particle where this redundancy is removed.

Consider a *foliation of the spacetime* by a one parameter set of 3-dimensional surfaces Σ_t. We assume that these surfaces are space-like and their equation is $t(x^\mu)$=const. Let y^i ($i = 1, 2, 3$) be coordinates on Σ_t. For a given foliation equations $y^i = y^i(t)$ determine the trajectory of a particle. We denote

$$U = -g_{tt}, \qquad A_i = g_{ti}, \qquad V = g^{ij} A_i A_j + U, \tag{4.1.23}$$

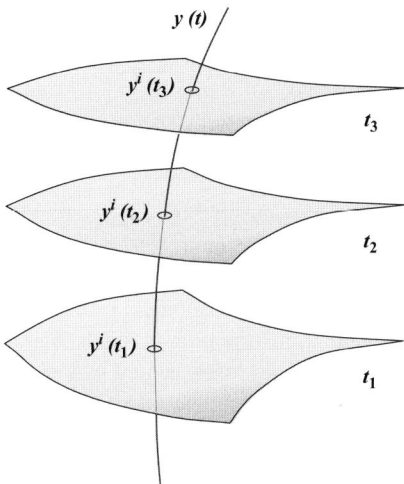

Fig. 4.2 The foliation of the spacetime by a set of 3-dimensional space-like surfaces Σ_t.

where g^{ij} is the inverse matrix to g_{ij}. Then, one can rewrite the Lagrangian Eq. (4.1.2) in the form ($\dot{y}^i = dy^i/dt$)

$$\mathcal{L} = -m\sqrt{U - 2A_i\dot{y}^i - g_{ij}\dot{y}^i\dot{y}^j}. \tag{4.1.24}$$

The momentum p_i conjugated to y^i is

$$p_i = \frac{\partial \mathcal{L}}{\partial \dot{y}^i} = \frac{m(g_{ij}\dot{y}^j + A_i)}{\sqrt{U - 2A_i\dot{y}^i - g_{ij}\dot{y}^i\dot{y}^j}}. \tag{4.1.25}$$

Straightforward calculations give the following expression for the *reduced Hamiltonian* $\mathcal{H} = p_i\dot{y}^i - \mathcal{L}$:

$$\mathcal{H} = \sqrt{V}\sqrt{m^2 + g^{ij}p_ip_j} - p_iA^i. \tag{4.1.26}$$

Similarly, a reduced action for a particle can be obtained starting with the action Eq. (4.1.5). Namely, one has

$$\tilde{S}[y^i(t), \eta(t)] = \int dt\tilde{\mathcal{L}}, \quad \tilde{\mathcal{L}} = \frac{1}{2}[\eta(t)^{-1}(g_{ij}\dot{y}^i\dot{y}^j + 2A_i\dot{y}^i - U) - \eta(t)m^2]. \tag{4.1.27}$$

Denote $p_i = \partial\tilde{\mathcal{L}}/d\dot{y}^i$, then the corresponding *reduced Hamiltonian* is

$$\tilde{\mathcal{H}} = \frac{1}{2}\eta(t)[g^{ij}p_ip_j + m^2] - A^ip_i + \frac{1}{2}\eta(t)^{-1}V. \tag{4.1.28}$$

The action in the Hamiltonian form takes the form

$$\tilde{S}_H[y^i(t), p_i(t), \eta(t)] = \int dt[p_i\dot{q}^i - \tilde{\mathcal{H}}]. \tag{4.1.29}$$

Solving the equation $\partial \tilde{\mathcal{H}}/\partial \eta = 0$, which is obtained by the variation of $\tilde{S}_H$ with respect to η, and substituting the obtained expression for η into Eq. (4.1.28) we reproduce Eq. (4.1.26).

4.2 Phase Space

4.2.1 Hamiltonian mechanics

In general relativity both Lagrangian and Hamiltonian approaches result in the same geodesic equation for the motion of particles and light propagation. The Hamiltonian approach is often very useful for study of general problems of mechanics. The reason is that it is more flexible. It allows canonical transformations that mix coordinates and momenta. The variety of these transformation is larger than the standard coordinate transformations adopted in the Lagrangian approach. Using canonical transformations opens up an additional possibility to simplify the equations of motion, by simplifying the corresponding Hamiltonian.

A Hamiltonian system is defined on the *phase space* **P** with the coordinates (p_i, q_i). If the configuration space, defined by the coordinates q_i, has n dimensions, the corresponding phase space is a $2n$-dimensional one. For the *Hamiltonian $H(p, q, t)$* the equations of motion are

$$\dot{q}_i = \frac{\partial H}{\partial p_i}, \qquad \dot{p}_i = -\frac{\partial H}{\partial q_i}, \qquad \frac{dH}{dt} = \frac{\partial H}{\partial t}. \tag{4.2.1}$$

It is useful to rewrite the first two of these equations in matrix notations

$$\frac{d}{dt}\begin{pmatrix} p_i \\ q_i \end{pmatrix} = \begin{pmatrix} 0 & -\delta_{ij} \\ \delta_{ij} & 0 \end{pmatrix} \begin{pmatrix} \partial H/\partial p_j \\ \partial H/\partial q_j \end{pmatrix}. \tag{4.2.2}$$

Denote by z^A a collective variable, by Ω^{AB} the skew-symmetric matrix, by Ω_{AB} its inverse, and by $I = \delta_{ij}$ the unity matrix. Then one has

$$z^A = \begin{pmatrix} p_i \\ q_i \end{pmatrix}, \qquad \Omega^{AB} = \begin{pmatrix} 0 & -I \\ I & 0 \end{pmatrix}, \qquad \Omega_{AB} = \begin{pmatrix} 0 & I \\ -I & 0 \end{pmatrix}. \tag{4.2.3}$$

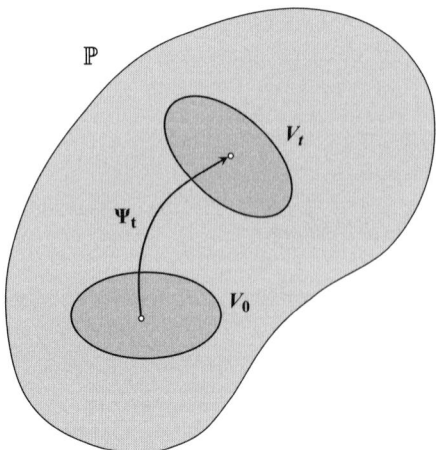

Fig. 4.3 The Hamiltonian map Ψ_t of the domain V_0 of the phase space **P** to V_t.

The variables (p_i, q_i) are canonical coordinates in the phase space **P**. The Hamiltonian equations of motion Eq. (4.2.1) can be rewritten in the form

$$\dot{z}^A = \Omega^{AB} \frac{\partial H}{\partial z^B}, \qquad \frac{dH}{dt} = \frac{\partial H}{\partial t}. \tag{4.2.4}$$

In the canonical coordinates (p_i, q_i), Ω^{AB} is a skew-symmetric non-degenerate $n \times n$ matrix with constant coefficients. It has the inverse matrix Ω_{AB}, which is also constant, antisymmetric and non-degenerate. The corresponding 2-form Ω is

$$\Omega = \sum_{i=1}^{n} dp_i \wedge dq_i. \tag{4.2.5}$$

It is evident that $d\Omega = 0$. Hence, the 2-form Ω is closed and (at least locally in some domain) it can be written in the form

$$\Omega = d\alpha, \qquad \alpha = \sum_{i=1}^{n} p_i dq_i. \tag{4.2.6}$$

4.2.2   Liouville theorem

Consider a domain V_0 in the *phase space* **P** and determine its *phase volume* as follows:

$$\mathcal{V}_0 = \int_{V_0} dq_1 \ldots dq_n dp_1 \ldots dp_n. \tag{4.2.7}$$

Let $\mathbf{z} = \mathbf{z}(t_0)$ be a point in the phase space representing the system at the initial moment of time t_0. As a result of motion Eq. (4.2.4), its position at the moment t is $\mathbf{z}(t)$. The Hamiltonian map $\mathbf{z} \to \mathbf{z}(t)$ determines a transformation Ψ_t of the phase space. Under this transformation the initial domain V_0 transforms into a domain V_t. Let

$$\mathcal{V}_t = \int_{V_t} dq_1 \ldots dq_n dp_1 \ldots dp_n \tag{4.2.8}$$

be its phase volume.

According to the *Liouville theorem*, the phase volume remains the same under the Hamiltonian map, that is

$$\mathcal{V}_t = \mathcal{V}_0. \tag{4.2.9}$$

To prove this theorem, let us make coordinate transformations and write Eq. (4.2.8) in the form

$$\mathcal{V}_t = \int_{V_0} \det \left| \frac{\partial z^A(t)}{\partial z^B} \right| dq_1 \ldots dq_n dp_1 \ldots dp_n. \tag{4.2.10}$$

For small $t - t_0$ one has

$$z^A(t) \simeq z^A(t_0) + (t - t_0) \, \Omega^{AC} \left. \frac{\partial H}{\partial z^C} \right|_{t=t_0}. \tag{4.2.11}$$

Hence,

$$\frac{\partial z^A(t)}{\partial z^B} = \delta_B^A + (t - t_0)\, \Omega^{AC} \left.\frac{\partial^2 H}{\partial z^B \partial z^C}\right|_{t=t_0}. \tag{4.2.12}$$

If I is a unit matrix and ϵ is a small parameter one has

$$\det(I + \epsilon A) = 1 + \epsilon \operatorname{Tr}(A) + O(\epsilon^2). \tag{4.2.13}$$

Taking the trace of the matrix on the right-hand side of Eq. (4.2.12) one can see that the second term, which depends on $t - t_0$, vanishes. It happens because Ω^{AB} is an antisymmetric matrix, while $\frac{\partial^2 H}{\partial z^A \partial z^B}$ is a symmetric one, so that

$$\Omega^{AB} \frac{\partial^2 H}{\partial z^A \partial z^B} = 0. \tag{4.2.14}$$

As a result, one has

$$\dot{\mathcal{V}}_t = 0, \tag{4.2.15}$$

and therefore the phase volume remains constant, $\mathcal{V}_t = \mathcal{V}_0$, under the Hamiltonian map Ψ_t.

4.2.3 Kinetic theory in curved spacetime

For a particle motion in a curved spacetime the natural choice of the canonically conjugated quantities is $\{p_\mu, x^\mu\}$, where x^μ is a position of the particle and p_μ is its momentum. Consider an ensemble of particles that move in a curved spacetime along geodesics without collisions, see Figure 4.4. We choose one of these geodesics, say γ_0, as a fiducial one and consider geodesics nearby it. We call such a set of geodesics a *pencil of geodesics*. Denote by t time along γ_0. We also consider foliation of the spacetime by 3D space-like surfaces and denote by Σ_t a surface of this foliation that crosses the *fiducial geodesic* at time t. For a given Σ_t the pencil of geodesics Π determines the phase volume element $\Delta \mathcal{V}_t$. According to the Liouville theorem, this volume element does not depend on time t.

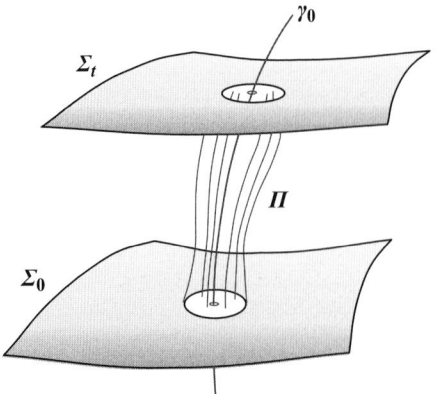

Fig. 4.4 An ensamble of collisionless particles that move in a curved spacetime along geodesics.

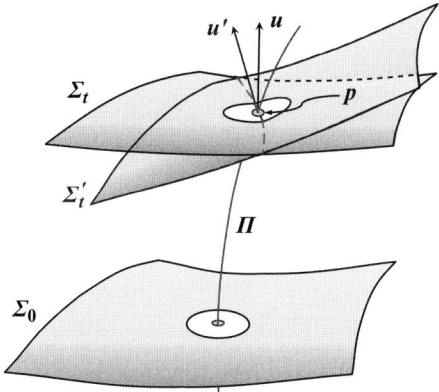

Fig. 4.5 Two foliations Σ_t and Σ'_t of the phase space are shown. At the initial moment $t = 0$ the foliation surfaces coincide. The fiducial geodesic intersects the foliation surfaces Σ_t and Σ'_t at a point p. $\boldsymbol{u}$ and $\boldsymbol{u}'$ are unit time-like vectors at p orthogonal to Σ and Σ', respectively.

Consider now another foliation of the spacetime. Suppose it is formed by a one-dimensional set of 3-dimensional surfaces Σ'_t. We suppose also that at some initial moment of time t_0, the initial surfaces Σ_{t_0} and Σ'_{t_0} for both foliations are the same. Hence, the initial element of the phase volume $\Delta\mathcal{V}_{t_0}$ for a given pencil of geodesics Π is the same as well. As a result of the Liouville theorem for both evolutions, described in the Σ_t and Σ'_t foliations the final elements of the phase volumes $\Delta\mathcal{V}_t$ and $\Delta\mathcal{V}'_t$ are the same. In other words, the element of the phase volume for a given pencil of geodesics Π does not depend on the choice of the foliation.

This result has another interpretation. Consider a point p on a chosen *fiducial geodesic* of the pencil Π. Let Σ_t and Σ'_t be the surfaces of the two foliations passing through p. Let $\boldsymbol{u}$ and $\boldsymbol{u}'$ be unit time-like vectors at p orthogonal to Σ and Σ', respectively (see Figure 4.5). One can interpret these vectors as a 4-velocity of 2 local observers at p, moving with some velocity with respect to each other. The above discussed independence of the element of the phase volume for the pencil Π on the foliation implies its independence on the motion of a local observer at p, that is, its local Lorentz invariance. (For more details, see (Misner *et al.* 1973), p.585.)

Let N be a number of particles (trajectories) in a chosen pencil of geodesics and $\mathcal{V}$ be their phase volume, then the *number density in phase space* or *distribution function* is defined as $N/\mathcal{V}$. For a given swarm of particles this object depends on the choice of a fiducial geodesic, but it does not depend on the choice of the point on it and on the choice of an observer at this point.

The element of the phase volume in a local comoving with the observer frame is

$$d\mathcal{V} = d^3p\, d^3x. \tag{4.2.16}$$

Here, d^3x is the 3D space volume, and d^3p is a 3D volume in the space of momenta. In both cases the quantities are determined with respect to the chosen local frame. Consider two observers at the same point, but moving with some relative velocity to each other. The phase volume elements in these two frames are related according to the Lorentz transformation. It happens that the phase volume element $d\mathcal{V}$ is Lorentz invariant despite the fact that neither d^3x nor d^3p is separately invariant. Indeed, for a particle with the mass m the quantity

$$\int d^3p\, dp_0\, \theta(p_0)\delta(-p_0^2 + p^2 + m^2) = \frac{1}{2E}\int d^3p, \qquad E = \sqrt{p^2 + m^2} = |p_0|, \qquad (4.2.17)$$

is Lorentz invariant by construction. Therefore, the combination d^3p/E is Lorentz invariant. On the other hand, $u_0 d^3x = d^4x/d\tau$, where τ is a proper time, is also Lorentz invariant. Taking into account that for the particle $E = |p_0| = m|u_0|$, we get $d\mathcal{V} = d^3p\, d^3x = (d^3p/E)(mu_0\, d^3x)$ is the product of two Lorentz invariant quantities. According to quantum mechanics the number of states corresponding to this phase volume for the canonical quantization of the system is

$$d\Gamma = \frac{d^3p\, d^3x}{h^3} = \frac{d^3p\, d^3x}{(2\pi\hbar)^3}. \qquad (4.2.18)$$

Massless particles (photons) provide an important application of these results. Let us consider emission and absorption of photons. To be more concrete, let us consider an astronomer observing the radiation with a telescope. We choose a local reference frame in which this telescope is at rest. We assume that an astronomer observes photons in the frequency interval $(\nu, \nu + \Delta\nu)$ arriving from the Z-direction within a small solid angle $\Delta\Omega$. These photons have energy $h\nu$ and momentum $h\nu/c$, where h is the Planck constant. The aperture of the telescope S is placed orthogonally to the Z-axis. The phase volume $\mathcal{V}$ of the photons registered by the telescope during the time interval Δt is a product of the space volume $S\Delta t$ by the momentum volume $h^3\nu^2\Delta\nu\Delta\Omega$

$$\Delta\Gamma = \frac{\Delta\vec{p}\,\Delta\vec{x}}{h^3} = S\Delta t\, \nu^2\Delta\nu\Delta\Omega. \qquad (4.2.19)$$

Hence, the number density of photons in the phase space is

$$\mathcal{N} = \frac{\Delta N}{S\Delta t\nu^2\Delta\nu\Delta\Omega}. \qquad (4.2.20)$$

The standard measure of the intensity of light adopted in optics is the *specific intensity I_ν*, which is defined as 'the energy per unit area per unit time per unit frequency per unit solid angle crossing a surface orthogonal to the beam',

$$I_\nu = \frac{h\nu\Delta N}{S\Delta t\Delta\nu\Delta\Omega}. \qquad (4.2.21)$$

Comparison of Eq. (4.2.20) and Eq. (4.2.21) gives

$$\mathcal{N} = h^{-1}(I_\nu/\nu^3). \qquad (4.2.22)$$

4.2.4   Phase space: General case

In different applications it is often useful to consider a generalization of the above-described notion of the phase space. Let M^{2m} be an even-dimensional differential manifold. A non-degenerate closed 2-form Ω on M^{2m} is called a *symplectic structure*. A pair (M^{2m}, Ω) is called a *symplectic manifold*. According to the general definition of the manifold, there exists a set of open 2m-dimensional domains O_α covering the manifold M^{2m}. For each of these domains one has a one-to-one map $O_\alpha \to U_\alpha$ onto a region of the Euclidean 2m-dimensional space

$\mathbb{R}^{2m}$. This map introduces coordinates z^A, $A = 1, \ldots, 2m$ in the domain O_α of the symplectic manifold. In these coordinates the symplectic form $\mathbf{\Omega}$ reads

$$\mathbf{\Omega} = \Omega_{AB}\, dz^A \wedge dz^B. \tag{4.2.23}$$

Under the transformation from one coordinate chart U_α to another $U_{\alpha'}$, Ω_{AB} transforms as a skew-symmetric tensor. Because the 2-form is non-degenerate, its matrix rank is $2m$ and it has the inverse Ω^{AB},

$$\Omega^{AB}\Omega_{BC} = \delta^A_C, \tag{4.2.24}$$

which is also skew-symmetric and non-degenerate. Similar to the metric, the tensors Ω_{AB} and Ω^{AB} can be used to establish one-to-one correspondence between the objects in M^{2m} with the upper and lower position of indices.

Consider a scalar function H on a symplectic manifold. Its gradient $H_{,A} = \partial H / \partial z^A$ is a one-form, while $\eta^A = \Omega^{AB} H_{,B}$ is a vector field. The function H is called a *Hamiltonian* and the vector field η is called a *Hamiltonian vector field*. The symplectic manifold $(M^{2m}, \mathbf{\Omega})$ with the Hamiltonian H on it is called a *phase space*.

Consider integral lines of the Hamiltonian vector field

$$\frac{dz^A}{dt} = \Omega^{AB} H_{,B}. \tag{4.2.25}$$

This equation defines the map $z^A(0) \rightarrow z^A(t)$. For a fixed value of t this is a diffeomorphism of the symplectic manifold onto itself called a *Hamiltonian flow*. It is possible to show that this flow preserves the symplectic structure

$$\mathcal{L}_\eta \mathbf{\Omega} = 0, \tag{4.2.26}$$

where $\mathcal{L}_\eta$ is the Lie derivative along the vector η. The proof can be found, e.g., in (Arnold 1989).

For any two functions F and G on the symplectic manifold the *Poisson bracket* $\{F, G\}$ is defined by the relation

$$\{F, G\} = -\Omega^{AB} F_{,A} G_{,B}. \tag{4.2.27}$$

The Poisson brackets for any three functions A, B, and C on the symplectic manifold obey the *Jacobi identity*

$$\{\{A, B\}, C\} + \{\{B, C\}, A\} + \{\{C, A\}, B\} = 0. \tag{4.2.28}$$

For general coordinates z^A the form Ω_{AB} is a function of z^A. But it is always possible to find such coordinates in which Ω_{AB} takes a simple form. Namely, according to the *Darboux theorem*, if $\mathbf{\Omega}$ is a non-degenerate closed 2-form in $\mathbb{R}^{2m}$, then in the vicinity of any point z^A of $\mathbb{R}^{2m}$ it is always possible to choose such coordinates $(p_1, \ldots, p_m; q_1, \ldots, q_m)$ in which the symplectic form takes the standard canonical form

$$\mathbf{\Omega} = \sum_{i=1}^{m} dp_i \wedge dq_i. \tag{4.2.29}$$

This result can be used to show that one can always find such an atlas on the symplectic manifold in which Ω has the canonical form Eq. (4.2.29). Such an atlas is called a *canonical symplectic atlas*. This atlas allows one to extend the results of the local nature to the case of the general symplectic manifold. This generalization is applicable to the results that are invariant under the canonical transformations and are proved for the standard phase space with the form Eq. (4.2.29). In particular, the Poisson bracket Eq. (4.2.27) in the canonical coordinates Eq. (4.2.29) takes the form

$$\{F, G\} = \sum_{i=1}^{m} \left(\frac{\partial F}{\partial p_i} \frac{\partial G}{\partial q_i} - \frac{\partial F}{\partial q_i} \frac{\partial G}{\partial p_i} \right). \tag{4.2.30}$$

4.3   Complete Integrability

4.3.1   Liouville theorem on complete integrability

Integrability is an important property of some dynamical systems. In its original sense, integrability means that a system of equations can be solved by quadratures. Integrability is closely linked to the existence of integrals (or constants) of motion. The integrability property depends on:

1. how many constants of motion exist;
2. how precisely they are related;
3. how the phase space is foliated by their level sets.

A system of differential equations is said to be completely integrable by quadratures if its solutions can be found after a finite number of steps involving algebraical operations and integration of given functions. Integrability and chaotic motion are in some sense opposite properties of dynamical systems. But integrability is rare and exceptional, while chaotic nature is generic. There are several important examples of completely integrable mechanical systems. Here are some of them:

1. motion in Euclidean space under a central potential;
2. motion at the two Newtonian centers;
3. geodesics on a surface of a triaxial ellipsoid;
4. rigid-body motion about a fixed point;
5. the Neumann model.[1]

As we shall see later, the geodesic motion of test particles and light in the gravitational field of a stationary axisymmetric black hole (described by the Kerr metric) is completely

---

[1] This model describes the motion of N particles on S^{N-1} under the action of harmonics forces on each of them. The corresponding Lagrangian is

$$L = \frac{1}{2} \left[\sum_{k=1}^{N} (\dot{q}_k^2 - \omega_k^2 q_k^2) + \lambda (\sum_{k=1}^{N} q_k^2 - 1) \right]$$

integrable. Moreover, this result is also valid for the geodesic motion in the higher-dimensional spacetimes for metrics describing stationary vacuum rotating black holes. This means that the study of the geodesic motion in four-and higher-dimensional black hole spacetimes gives a new physically important class of completely integrable dynamical systems.

Let us discuss the complete integrability of dynamical systems in more detail (Babelon 2006). Let H be a Hamiltonian. A scalar function $F(z^A)$ on the phase space is a *first integral of motion* if its *Poisson bracket* with the Hamiltonian vanishes

$$\{F, H\} = 0. \tag{4.3.1}$$

Consider a value of F on the integral lines of the Hamiltonian flow

$$F(\lambda) = F(z^A(t)). \tag{4.3.2}$$

Using Eq. (4.2.25) we get

$$\frac{dF(t)}{dt} = F_{,A} \Omega^{AB} H_{,B} = -\{F, H\} = \{H, F\} = 0. \tag{4.3.3}$$

Thus, along any integral line of the *Hamiltonian flow* the value of $F(t)$ remains constant. This justifies the definition Eq. (4.3.1).

In the general case, in order to solve the system of $2m$ first-order ordinary differential equations Eq. (4.2.25) one needs to know its $2m$ first integrals. The remarkable property of the canonical equations for the *Hamiltonian flow* on a symplectic manifold is that under some conditions, in order to integrate the system, it is sufficient to know only m of its first integrals. In fact, any of these first integrals allows one to decrease the order of the system not by one but by two units.

To describe the corresponding results let us introduce the following notion. Let F and G be two functions on the symplectic manifold. One says that they are in *involution* if their Poisson bracket vanishes

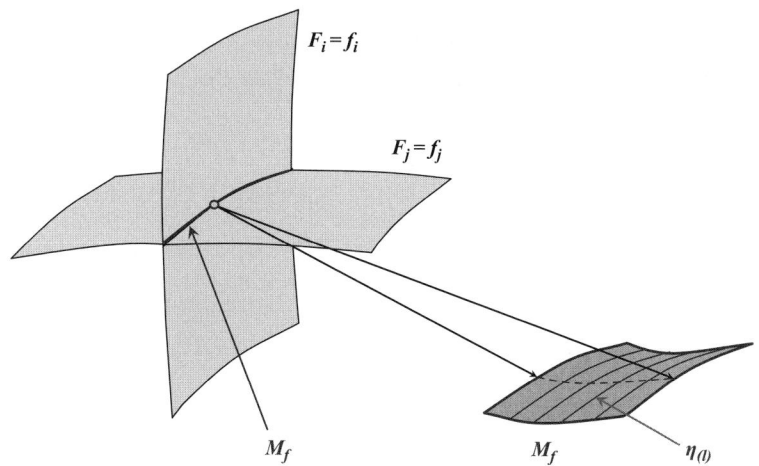

Fig. 4.6 Submanifold M_f and the Hamiltonian flux on it.

$$\{F, G\} = 0. \tag{4.3.4}$$

Liouville (1855) proved the following **theorem**: *If a system with a Hamiltonian H in the 2m phase space has m independent first integrals in involution, $F^{(1)} = H, F^{(2)}, \ldots, F^{(m)}$, then the system can be integrated by quadratures.*

The main idea behind the *Liouville theorem* is that one can use the integrals of motion $F^{(i)}$ as m coordinates on the phase space. The involution condition guarantees that m vector fields $\Omega^{AB} F_{,B}^{(i)}$ are tangent to the level set of $F^{(i)}$, and that they commute with each other. This allows one to find the coordinates on the level set of $F^{(i)}$, in which the Hamilton equations of motion are simply integrated.

Now, we give a sketch of the proof of the *Liouville integrability theorem*. Suppose there exist m first integrals $F^{(i)}$ ($i = 1, \ldots, m$) in involution. We choose $H = F^{(1)}$. Consider a subset M_f of the symplectic manifold M^{2m} determined by the relations

$$F^{(i)}(x) = f_i, \qquad i = 1, \ldots, m. \tag{4.3.5}$$

The independence of the first integrals means that the gradients $F_{,A}^{(i)}$ are linearly independent at each point of M_f. This implies that M_f is a m-dimensional *submanifold* of M^{2m}. The involution conditions give

$$\{F^{(i)}, F^{(j)}\} = -F_{,A}^{(i)} \Omega^{AB} F_{,B}^{(j)} = 0. \tag{4.3.6}$$

Thus, the vectors

$$\eta^{(i)A} = \Omega^{AB} F_{,B}^{(i)}, \tag{4.3.7}$$

are orthogonal to the normal to M_f covectors $F_{,A}^{(i)}$ and hence the vectors $\eta^{(i)}$ are tangent to M_f. Since Ω is non-degenerate, the vectors $\eta^{(i)A}$ are linear independent and span the tangent to M_f spaces. This result also implies that M_f is invariant under the phase flows generated by any of the 'Hamiltonians' $F^{(i)}$.

The vectors $\eta^{(i)}$ commute with each other. To prove this, we consider two vectors

$$\eta^A = \Omega^{AB} F_{,B}, \qquad \zeta^A = \Omega^{AB} G_{,B}, \tag{4.3.8}$$

corresponding to the first two integrals F and G in involution. Calculating the commutator (see Eq. (3.4.1)) one obtains

$$[\eta, \zeta]^A = \mathcal{A}^A + \mathcal{B}^A,$$
$$\mathcal{A}^A = \Omega^{AC} \Omega^{BD} (F_{,D} G_{,CB} - G_{,D} F_{,CB}), \tag{4.3.9}$$
$$\mathcal{B}^A = \Omega^{BD} \Omega^{AC}_{\ \ ,B} (F_{,D} G_{,C} - F_{,C} G_{,D}).$$

Taking the partial derivative of the involution condition for F and G one gets

$$\left(\Omega^{BD} G_{,B} F_{,D}\right)_{,C} = \Omega^{BD} [G_{,CB} F_{,D} - G_{,D} F_{,CB}] + \Omega^{BD}_{\ \ ,C} G_{,B} F_{,D} = 0. \tag{4.3.10}$$

Here, we have used the property that the partial derivatives commute and Ω^{AB} is skew-symmetric. Using Eq. (4.3.10) one can write $\mathcal{A}^A$ in the form

$$\mathcal{A}^A = -\Omega^{AC}\Omega^{BD}{}_{,C}G_{,B}F_{,D}. \tag{4.3.11}$$

Thus, one has

$$[\eta, \zeta]^A = \left(\Omega^{BD}\Omega^{AC}{}_{,B} - \Omega^{BC}\Omega^{AD}{}_{,B} + \Omega^{AB}\Omega^{DC}{}_{,B}\right)F_{,D}G_{,C}. \tag{4.3.12}$$

The following relations are valid

$$\Omega^{KN}{}_{,B} = -\Omega^{MN}\Omega^{KL}\Omega_{LM,B},$$
$$\Omega_{LM,B} + \Omega_{BL,M} + \Omega_{MB,L} = 0. \tag{4.3.13}$$

The second of these relation is simply a condition that the 2-form Ω is closed. (4.3.13) to transform the derivatives of Ω^{AB} to derivative of Ω_{AB}. Then, using the second Eq. (4.3.13), one can finally demonstrate that the right-hand side of Eq. (4.3.12) vanishes. This result implies that all the vectors Eq. (4.3.7) commute.

In the canonical coordinates (p_i, q_i)

$$\Omega = d\alpha = \sum_i dp_i \wedge dq_i, \qquad \alpha = \sum_i p_i dq_i. \tag{4.3.14}$$

We shall construct a canonical transformation $(p_i, q_i) \rightarrow (f_i, \psi_i)$, where f_i are conserved quantities. Thus, we get

$$\Omega = \sum_i df_i \wedge d\psi_i. \tag{4.3.15}$$

In these new variables the dynamical equations Eq. (4.3.3) become trivial

$$\dot{f}_i = \{H, f_i\} = 0,$$
$$\dot{\psi}_i = \{H, \psi_i\} = \frac{\partial H}{\partial f_i} = \omega_i = \text{const.} \tag{4.3.16}$$

A solution of these equations is

$$f_i = \text{const}, \qquad \psi_i(t) = \psi_i(0) + \omega_i t. \tag{4.3.17}$$

Since M_f is the level submanifold defined by the equations $F^{(i)}(p, q) = f_i$, one can solve these equations to obtain $p_i = p_i(f, q)$ on M_f. To construct the required canonical transformation consider a function

$$W(f, q) = \int_{m_0}^m \alpha = \int_{q_0}^q \sum_i p_i(f, q) dq_i. \tag{4.3.18}$$

Here, m_0 and m are two points on M_f. The integral is taken along a path on M_f connecting these points. Equation (4.3.6) implies that being restricted on M_f the symplectic form vanishes

$$d\alpha|_{M_f} = \Omega|_{M_f} = 0. \tag{4.3.19}$$

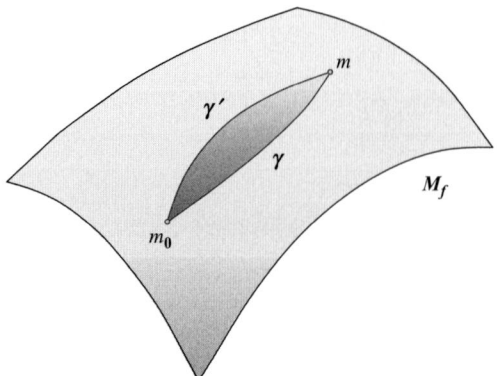

Fig. 4.7 Independence of a function $W(f,q)$ on the choice of the path on M_f.

Since α is closed on M_f, the Stoke's theorem implies that the integral Eq. (4.3.18) does not depend on the choice of the path (see Figure 4.7). We define ψ_i as

$$\psi_i = \frac{\partial W}{\partial f_i}, \tag{4.3.20}$$

so that

$$dW = \sum_i (\psi_i df_i + p_i dq_i). \tag{4.3.21}$$

Since $d^2 W = 0$, one has

$$\Omega = \sum_i dp_i \wedge dq_i = \sum_i df_i \wedge d\psi_i. \tag{4.3.22}$$

To construct the required canonical transformation one should use only an algebraical operation to find $p_i = p_i(f,q)$ and one quadrature to calculate the function W. This finishes the proof of the Liouville theorem.

If the level submanifold M_f is compact and connected, it is diffeomorphic to the m-dimensional torus $\mathbb{T}^m$. In this case, the coordinates ψ_i can be chosen so that each of them varies from 0 to 2π, the end points of this interval being identified. The coordinates (f_i, ψ_i) are called the *action-angle variables* (Arnold 1989).

4.3.2 Complete integrability and Killing tensors

Let us discuss now the question: What are the conditions for the complete integrability of the equations for particle and light motion in a curved spacetime? In this case x^μ are coordinates and $p_\mu = g_{\mu\nu}\dot{x}^\mu$ are the momentum.

Let us assume that the Hamilton equations Eq. (4.1.16) have an integral of motion of the form

$$\mathcal{K} = K^{\mu_1 \dots \mu_s}(x) p_{\mu_1} \dots p_{\mu_s} \tag{4.3.23}$$

that is $\mathcal{K}$ is constant along a geodesic, i.e. $\dot{\mathcal{K}} = 0$. In the Hamilton approach

$$\dot{\mathcal{K}} = \{H, \mathcal{K}\} = \frac{\partial \mathcal{K}}{\partial x^\alpha} \frac{\partial H}{\partial p_\alpha} - \frac{\partial H}{\partial x^\alpha} \frac{\partial \mathcal{K}}{\partial p_\alpha} = \frac{\partial \mathcal{K}}{\partial x^\alpha} \dot{x}^\alpha + \frac{\partial \mathcal{K}}{\partial p_\alpha} \dot{p}_\alpha. \tag{4.3.24}$$

Using Eq. (4.1.15) we have

$$\dot{x}^\mu = \{H, x^\mu\} = p^\mu, \qquad \dot{p}_\mu = \{H, p_\mu\} = \frac{1}{2} g_{\alpha\beta,\mu} p^\alpha p^\beta. \tag{4.3.25}$$

Substitution of these expressions into Eq. (4.3.24) gives

$$\dot{\mathcal{K}} = K^{\mu_1 \dots \mu_s}{}_{,\alpha} p^\alpha p_{\mu_1} \cdots p_{\mu_s} + s \frac{1}{2} g_{\alpha\beta,\mu_s} K^{\mu_1, \dots, \mu_s} p_{\mu_1} \cdots p_{\mu_{s-1}} p^\alpha p^\beta. \tag{4.3.26}$$

This expression can be rewritten in a covariant form, if we use the covariant derivative corresponding to the metric $g_{\mu\nu}$,

$$\dot{\mathcal{K}} = K^{\mu_1 \dots \mu_s;\alpha} p_{\mu_1} \cdots p_{\mu_s} p_\alpha. \tag{4.3.27}$$

If the condition $\dot{\mathcal{K}} = 0$ is valid for arbitrary geodesics then Eq. (4.3.27) must vanish for an arbitrary choice of p^μ. Therefore, the tensor $\mathbf{K}$ is to obey the following equation

$$K^{(\mu_1 \dots \mu_s;\alpha)} = 0. \tag{4.3.28}$$

Thus, $K_{\mu_1, \dots, \mu_s}$ is a *Killing tensor*.

Suppose we have two integrals of motion, $\mathcal{K}_{(s)}$ and $\mathcal{K}_{(t)}$, generated by the Killing tensors $K^{\mu_1 \dots \mu_s}_{(s)}$ and $K^{\nu_1 \dots \nu_t}_{(t)}$, respectively. Using the *Jacobi identity* Eq. (4.2.28) it is easy to check that $\{\mathcal{K}_{(s)}, \mathcal{K}_{(t)}\}$ commutes with the Hamiltonian H and hence is an integral of motion. This commutator is a monomial of the power $s + t - 1$ in the momentum. The corresponding Killing tensor of the rank $s + t - 1$ is

$$[\mathbf{K}_{(s)}, \mathbf{K}_{(t)}] = \mathbf{K}_{(s+t-1)}. \tag{4.3.29}$$

The commutator in the left-hand side of this relation is a *Schouten-Nijenhuis bracket*. Written in the components it is

$$K^{\mu_1 \dots \mu_{s-1} \lambda \nu_1 \dots \nu_{t-1}}_{(s+t-1)} = s K^{\epsilon(\mu_1 \dots \mu_{s-1}}_{(s)} \partial_\epsilon K^{\lambda \nu_1 \dots \nu_{t-1})}_{(t)} - t K^{\epsilon(\nu_1 \dots \nu_{t-1}}_{(t)} \partial_\epsilon K^{\lambda \mu_1 \dots \mu_{s-1})}_{(s)}. \tag{4.3.30}$$

Killing tensors form a Lie subalgebra of a Lie algebra of all totally symmetric contravariant tensor fields on the manifold with respect to the *Schouten-Nijenhuis bracket* (Norris 1997).

> **Problem 4.2:** *Prove that $K^{\mu_1 \dots \mu_{s-1} \lambda \nu_1 \dots \nu_{t-1}}_{(s+t-1)}$ defined by Eq. (4.3.30) is a tensor. Show that it can also be written in the form that includes only covariant derivatives*
>
> $$K^{\mu_1 \dots \mu_{s-1} \lambda \nu_1 \dots \nu_{t-1}}_{(s+t-1)} = s K^{\epsilon(\mu_1 \dots \mu_{s-1}}_{(s)} \nabla_\epsilon K^{\lambda \nu_1 \dots \nu_{t-1})}_{(t)} - t K^{\epsilon(\nu_1 \dots \nu_{t-1}}_{(t)} \nabla_\epsilon K^{\lambda \mu_1 \dots \mu_{s-1})}_{(s)}. \tag{4.3.31}$$

The condition that the integrals of motion $\mathcal{K}_{(s)}$ and $\mathcal{K}_{(t)}$ are in involution implies that

$$[\mathbf{K}_{(s)}, \mathbf{K}_{(t)}] = 0. \tag{4.3.32}$$

If one of the Killing tensors is of rank 1, i.e. it is a Killing vector $\boldsymbol{\xi}$, the *Schouten–Nijenhuis bracket* of it with a Killing tensor of rank k gives again a Killing tensor of rank k. In this case, the Schouten–Nijenhuis brackets reduce to the Lie derivative with respect to the Killing vector,

$$[\boldsymbol{\xi}, \boldsymbol{K}_{(t)}] = \mathcal{L}_{\boldsymbol{\xi}} \boldsymbol{K}_{(t)}. \tag{4.3.33}$$

4.3.3 Separation of variables and integrability

Complete integrability of the Hamiltonian systems is closely related to the separation of variables in the Hamilton–Jacobi equation. For the Hamiltonian $H(P, Q)$, $P = p_1, \ldots, p_m$ and $Q = q_1, \ldots, q_m$, the *Hamilton–Jacobi equation* is

$$H(\partial_Q S, Q) = 0, \qquad \partial_Q S = (\partial_{q_1} S, \ldots, \partial_{q_m} S). \tag{4.3.34}$$

If the variable q_1 and the derivative $\partial_{q_1} S$ enter this equation only as the combination $\Phi_1(\partial_{q_1} S, q_1)$, then one says that the variable q_1 is separated. In such a case, one may try to search for a solution in the form

$$S = S_1(q_1) + \bar{S}(q_2, \ldots, q_m). \tag{4.3.35}$$

Taking

$$\Phi_1(\partial_{q_1} S, q_1) = C_1, \tag{4.3.36}$$

one obtains an equation with less number of variables

$$H_1(\partial_{q_2} \bar{S}, \ldots, \partial_{q_m} \bar{S}, q_2, \ldots, q_m; C_1) = 0. \tag{4.3.37}$$

Let $\bar{S}(q_2, \ldots, q_m; C_1)$ be a solution of this equation depending on the parameter C_1, then Eq. (4.3.35) is a solution of Eq. (4.3.34) if S_1 obeys an ordinary differential equation Eq. (4.3.36), which is easily solvable by quadratures. If the variable q_2 can be separated in the new equation Eq. (4.3.37), one can repeat this procedure again. One says that the *Hamilton–Jacobi equation* Eq. (4.3.34) allows a *complete separation of variables* if after m steps one obtains a solution of the initial equation Eq. (4.3.34) which contains m constants C_i,

$$S = S_1(q_1, C_1) + S_2(q_2; C_1, C_2) + \ldots + S_m(q_m; C_1, \ldots, C_m). \tag{4.3.38}$$

In this case, one obtains the *complete solution of the Hamilton–Jacobi equation* which depends on m parameters, and the corresponding Hamilton equations are integrable by quadratures (*Jacobi theorem*).

For a completely separable Hamilton–Jacobi equation the constants $C_1, \ldots, C_m$ can be considered as functions on the phase space where they are integrals of motion. In the case when these integrals on motion are independent and in involution, the system is completely integrable in the sense of Liouville.

5

Einstein Equations

5.1 Einstein–Hilbert Action

The *equivalence principle* implies that a gravitational field should be described by a metric **g**. We have already demonstrated that physical laws in an external gravitational field can be obtained by using the *covariance principle*, provided these laws are known in the Minkowski spacetime. Now, we demonstrate that the equations of the gravitational field itself can be derived by applying the same covariance principle.

> *In our derivation we shall use an approach based on the* stationary action principle. *This principle is well known and widely used in theoretical physics. There are deep reasons for this. The roots of the stationary action principle are connected to quantum theory. In the Feynman path integral approach an amplitude of a probability of a transition of a quantum system from one quantum state to another is given by the functional integral of* exp(iS/ℏ), *where S is the classical action of the system. In the classical limit, when S ≫ ℏ, the main contribution to this integral is given by the extremum of the action. That is how the classical equations of motion follow from the quantum theory.*

To derive the *Einstein equations* we first obtain the *Einstein–Hilbert action* describing gravity. We establish this action in four-dimensional spacetime. Later, we discuss its generalization to the higher-dimensional gravitational theory (see Section 5.5).

First, we assume that the theory of gravity is local, so that its action can be written in the form

$$S = \int \mathcal{L} \, d^4 x, \tag{5.1.1}$$

where $\mathcal{L}$ is a local Lagrangian density depending only on the metric. By this postulate we exclude a large class of generalizations of the Einstein gravity.[1] Thus, we write

$$\mathcal{L} = \mathcal{L}(\mathbf{g}, \ldots). \tag{5.1.2}$$

Here . . . denote terms that contain the metric derivatives. The covariance principle implies that the action S is a scalar. Hence,

$$\mathcal{L} = \sqrt{-g} \, L(\mathbf{g}, \ldots), \tag{5.1.3}$$

where $L(\mathbf{g}, \ldots)$ is a local scalar constructed from the metric and its derivatives.

We require that the gravity equations of motion are not higher than the second order in derivatives. In the presence of higher derivatives a theory is usually pathological. For

---

[1] For example, in the bimetric theory besides $g_{\mu\nu}$ another (external) metric is present. In the scalar-tensor theories in addition to the metric a scalar field is present.

example, it may have tachionic states, which imply instability of the theory, or it may have negative-energy states. In the latter case and in the presence of a (self)-interaction, it is possible that the creation of particles with positive energy is accompanied by a simultaneous production of negative-energy particles. In this process the total energy does not change but the vacuum becomes unstable. It is definitely not what one would expect from the vacuum and the quantization of such a theory is inconsistent.

A Lagrangian that produces second-order equations contains the metric and its first derivatives. It may also contain second derivatives of the metric provided they enter linearly and the coefficients in these terms depend on the metric but not on its derivatives.

> *For instance, consider a term that has the structure $g^n (\partial g)^m (\partial^2 g)^k$, where n, m and k are integer numbers. If $k > 1$ the variation of the $(\partial^2 g)^k$ term would contain fourth derivatives. If $k = 1$ and $m \geq 1$ a variation of $(\partial g)^m$ after the integration by parts would give terms where an extra derivative is acting on the term with the second derivative, which results in the appearance of a third derivative in the equation of motion. Thus, in order to have the second-order equations either $k = 0$ or $k = 1$ and $m = 0$.*

Thus, the Lagrangian density must be of the form

$$\mathcal{L} = \sqrt{-g}\, L(g, \partial g, \partial^2 g), \qquad (5.1.4)$$

where $\partial^2 g$ may enter only linearly.

Since L is a scalar constructed from the metric and its derivatives (up to the second order), it is a scalar function of curvature. The Ricci scalar is the only scalar made of the curvature that contains the second derivatives of the metric linearly. Thus, the gravitational action can be written in the form

$$S = C \int d^4x \sqrt{-g}\, (R - 2\Lambda). \qquad (5.1.5)$$

The constant C describes the strength of the gravitational interaction and is related to the Newtonian constant, while the other constant Λ corresponds to the term without derivatives and is called the *cosmological constant*. The choice of the sign and the factor 2 in this term is simply a matter of convenience.

Let us discuss the dimensionality of the gravitational action. The interval ds^2 has the dimensionality l^2 (that is, $(length)^2$). We assume that the metric is dimensionless, then $[dx] = l$ and the dimensionality of R is l^{-2}. Λ has the same dimensionality as R. Thus, the dimensionality of the integral in the right-hand side of Eq. (5.1.5) is l^2. The coefficient C must contain the speed of light c (since this is a relativistic theory) and the Newtonian coupling constant G (since this is a theory of gravity). The proper choice of C that gives the correct dimensionality[2] of the action leads to the following form of the gravitational action

$$S[g] = \frac{c^3}{16\pi G} \int d^4x \sqrt{-g}\, (R - 2\Lambda). \qquad (5.1.6)$$

---

[2]The dimensionality of the action in the CGS units is $[Action] = [E \cdot T] = \frac{g \cdot cm^2}{s}$. We also have $[G] = \frac{cm^3}{s^2 \cdot g}$ and $[c] = \frac{cm}{s}$. Thus, $\left[\frac{c^3}{G}\right] = \frac{g}{s} = [Action]/[cm^2]$. Note that in the integration here $d^4x = dx^0 dx^1 dx^2 dx^3$, where the coordinate $x^0 = ct$ is also measured in [cm].

An ambiguity in the dimensionless factor, which we have chosen to be $1/(16\pi)$, is uniquely fixed by the requirement that the gravity theory has a correct Newtonian limit (see Section 5.3). The action Eq. (5.1.6) is called the *Einstein–Hilbert action*.

5.2 Einstein Equations

5.2.1 Variation of the Einstein–Hilbert action

We consider the metric with the lower position of indices as the functional argument of the action Eq. (5.1.6).[3] Let us express the variation of the objects that enter the action in terms of the variation of the metric $\delta g_{\alpha\beta}$. By the variation of the identity

$$g^{\alpha\epsilon} g_{\epsilon\beta} = \delta^\alpha{}_\beta \tag{5.2.1}$$

one obtains $\delta g^{\alpha\epsilon} g_{\epsilon\beta} + g^{\alpha\epsilon} \delta g_{\epsilon\beta} = 0$. Hence,

$$\delta g^{\alpha\beta} = -g^{\alpha\mu} g^{\beta\nu} \delta g_{\mu\nu}. \tag{5.2.2}$$

For the variation of the determinant of the metric $g = \det g_{\mu\nu} \equiv \det g$ one has

$$\delta g = g\, g^{\mu\nu}\, \delta g_{\mu\nu}. \tag{5.2.3}$$

To prove this relation let us consider a square matrix $\mathbf{g}$ with components $g_{\mu\nu}$. Define the (μ, ν) minor of $\mathbf{g}$, denoted $m^{\mu\nu}$, as the determinant of the matrix that results from deleting the row μ and the column ν of $\mathbf{g}$. The inverse matrix is defined by Eq. (5.2.1). Its components are

$$g^{\mu\nu} = \frac{c^{\mu\nu}}{\det g}, \qquad c^{\mu\nu} = (-1)^{\mu+\nu} m^{\nu\mu}. \tag{5.2.4}$$

Thus, one has

$$\det g = \sum_\nu g_{\mu\nu} c^{\nu\mu}, \qquad (\text{no summation over } \mu). \tag{5.2.5}$$

Note that in this decomposition of the determinant a term with given indices μ and ν enters only in the combination with the corresponding cofactor coefficient. Thus, the variation of $\det g$ is $\delta(\det g) = \delta g_{\mu\nu} c^{\nu\mu}$. Using Eq. (5.2.4) and the symmetry of $g_{\mu\nu}$ one obtains Eq. (5.2.3).

The variation formula Eq. (5.2.3) implies

$$\delta\left(\sqrt{-g}\right) = \frac{1}{2}\sqrt{-g}\, g^{\alpha\beta}\, \delta g_{\alpha\beta} = -\frac{1}{2}\sqrt{-g}\, g_{\alpha\beta}\, \delta g^{\alpha\beta}. \tag{5.2.6}$$

As the next step, we find variation of the Christoffel symbols. We shall use the following trick. As we know, the Christoffel symbol is not a tensor, but if we compare two metrics on the same manifold, the difference between their Christoffel symbols is a tensor. In particular, this means that the variation of the Christoffel symbol $\delta\Gamma^\lambda{}_{\mu\nu}$ is a tensor. Let us write this variation as follows

---

[3] Other choice of the functional argument, for example $g^{\mu\nu}$, gives an equivalent set of equations for the gravitational field.

$$\delta\Gamma^\lambda{}_{\mu\nu} = -g^{\lambda\rho}\,\delta g_{\rho\sigma}\,\Gamma^\sigma{}_{\mu\nu} + \frac{1}{2}g^{\lambda\rho}\left[\delta g_{\rho\mu,\nu} + \delta g_{\rho\nu,\mu} - \delta g_{\mu\nu,\rho}\right]$$

$$= \frac{1}{2}g^{\lambda\rho}\left[\delta g_{\rho\mu;\nu} + \delta g_{\rho\nu;\mu} - \delta g_{\mu\nu;\rho}\right]. \tag{5.2.7}$$

The last equality is evident in the Riemann normal coordinates. But since $\delta\Gamma^\lambda{}_{\mu\nu}$ is a tensor, it is valid in arbitrary coordinates. Thus, we proved that

$$\delta\Gamma^\lambda{}_{\mu\nu} = \frac{1}{2}g^{\lambda\rho}\left[\delta g_{\rho\mu;\nu} + \delta g_{\rho\nu;\mu} - \delta g_{\mu\nu;\rho}\right]. \tag{5.2.8}$$

Using the same trick it is easy to check that

$$\delta R^\lambda{}_{\alpha\mu\beta} = \left(\delta\Gamma^\lambda{}_{\alpha\beta}\right)_{;\mu} - \left(\delta\Gamma^\lambda{}_{\alpha\mu}\right)_{;\beta},$$

$$\delta R_{\alpha\beta} = \left(\delta\Gamma^\lambda{}_{\alpha\beta}\right)_{;\lambda} - \left(\delta\Gamma^\lambda{}_{\alpha\lambda}\right)_{;\beta}, \tag{5.2.9}$$

and

$$\delta R_{\alpha\mu\beta\nu} = \frac{1}{2}\left(\delta g_{\alpha\nu;\mu\beta} + \delta g_{\mu\beta;\alpha\nu} - \delta g_{\mu\nu;\alpha\beta} - \delta g_{\alpha\beta;\mu\nu}\right.$$

$$\left. + \delta g_{\alpha\epsilon}R^\epsilon{}_{\mu\beta\nu} - \delta g_{\mu\epsilon}R^\epsilon{}_{\alpha\beta\nu}\right). \tag{5.2.10}$$

We also have

$$\delta\left(\sqrt{-g}\,R\right) = \delta\left(\sqrt{-g}\,g^{\alpha\beta}\,R_{\alpha\beta}\right)$$

$$= \frac{1}{2}\sqrt{-g}\,g^{\alpha\beta}\,R\,\delta g_{\alpha\beta} - \sqrt{-g}\,R^{\alpha\beta}\,\delta g_{\alpha\beta} + \sqrt{-g}\,g^{\alpha\beta}\,\delta R_{\alpha\beta}, \tag{5.2.11}$$

so that

$$\delta\left(\sqrt{-g}\,R\right) = -\sqrt{-g}\,G^{\alpha\beta}\,\delta g_{\alpha\beta} + \sqrt{-g}g^{\alpha\beta}\,\delta R_{\alpha\beta}. \tag{5.2.12}$$

Here,

$$G_{\alpha\beta} = R_{\alpha\beta} - \frac{1}{2}g_{\alpha\beta}\,R \tag{5.2.13}$$

is the *Einstein tensor*. The integral of the second term in the variation Eq. (5.2.12) has the form

$$\int d^4x\sqrt{-g}\,(\delta f^\lambda)_{;\lambda}, \qquad \delta f^\lambda = g^{\alpha\beta}\delta\Gamma^\lambda{}_{\alpha\beta} - g^{\beta\lambda}\delta\Gamma^\alpha{}_{\alpha\beta}. \tag{5.2.14}$$

Using Stokes' theorem, Eq. (3.9.12), the integral of a total derivative in Eq. (5.2.14) can be reduced to the surface integral and, hence, it does not contribute to the equations of motion. Thus, the variation of the *Einstein–Hilbert action* is

$$\frac{1}{\sqrt{-g}}\frac{\delta S}{\delta g_{\alpha\beta}} = -\frac{c^3}{16\pi G}(G^{\alpha\beta} + \Lambda g^{\alpha\beta}). \tag{5.2.15}$$

Strictly speaking, one must be more accurate when one deals with the surface terms, which come from the integration in Eq. (5.2.14). The reason is that the Einstein Lagrangian contains second derivatives of the metric. As a result, to banish boundary terms one has to fix at the boundary not only the metric, but also its normal derivatives. But for an arbitrary choice of these boundary conditions the variation procedure is not well defined. This problem can be solved by adding a special boundary term to the Einstein–Hilbert action. Appendix E contains more detailed discussion of the problem and describes the covariant form of the corresponding boundary term.

5.2.2   Einstein equations

The vacuum *Einstein equations* with the cosmological constant are

$$G_{\alpha\beta} + \Lambda g_{\alpha\beta} = 0. \tag{5.2.16}$$

In the presence of matter the Einstein–Hilbert action must be modified by adding the matter action, which we write in the form

$$S_m[\Phi, g] = \frac{1}{c} \int d^4x \sqrt{-g} \, L_m(\Phi, g). \tag{5.2.17}$$

Here, Φ^A are the dynamical variables of matter fields and A is an index enumerating its components. The Lagrangian depends on Φ^A and its derivatives. It depends also on the metric and its derivatives. If one knows a theory in a flat spacetime, in order to obtain the theory in the curved spacetime, it is sufficient to write the corresponding action in curved coordinates and instead of the flat metric and partial derivatives and integrals to use their covariant versions. After this, one can use the obtained action in an arbitrary metric.

In a more general case the matter action in the curved spacetime can contain terms describing the interaction of the matter with the curvature. Such terms, which contain second (and higher) derivatives of the metric, are not fixed by the corresponding theory in the Minkowski spacetime.

In accordance with the *covariance principle* we require that S_m is invariant under the coordinate transformations. Let us consider a general variation of the action Eq. (5.2.17)

$$\delta S_m[\Phi, g] = \int d^4x \sqrt{-g} \left(\frac{\delta S_m}{\delta \Phi^A} \delta \Phi^A + \frac{\delta S_m}{\delta g_{\alpha\beta}} \delta g_{\alpha\beta} \right). \tag{5.2.18}$$

The variation of this action with respect to Φ^A gives dynamical equations for matter fields

$$\frac{\delta S_m}{\delta \Phi^A} = 0. \tag{5.2.19}$$

We assume that these equations are satisfied, so that the first term of the integrand vanishes. The variation of S_m over the metric gives the symmetric tensor of rank two

$$\frac{1}{c} T^{\alpha\beta} = \frac{2}{\sqrt{-g}} \frac{\delta S_m}{\delta g_{\alpha\beta}}. \tag{5.2.20}$$

The tensor $T_{\alpha\beta}$ is called the (metric) *stress-energy tensor* or *energy-momentum tensor* of the matter. The extremum of the action $S[g] + S_m[\Phi, g]$ with respect to the metric variations gives the *Einstein equations* in the form

$$G_{\alpha\beta} + \Lambda g_{\alpha\beta} = \frac{8\pi G}{c^4} T_{\alpha\beta}. \tag{5.2.21}$$

The terms on the left-hand side depend only on the spacetime geometry, while on the right-hand side we have the stress-energy tensor of the matter fields.[4]

5.2.3 Stress-energy tensor

Let us discuss the properties of the stress-energy tensor. Under the coordinate transformation

$$x^\mu \to x^\mu + \xi^\mu(x), \tag{5.2.22}$$

where $\boldsymbol{\xi}$ is the generator of the transformation, the metric transforms as follows[5]

$$g_{\mu\nu} \to g_{\mu\nu} + \delta_{\boldsymbol{\xi}} g_{\mu\nu}, \qquad \delta_{\boldsymbol{\xi}} g_{\mu\nu} = -2\xi_{(\mu;\nu)}. \tag{5.2.24}$$

Since S_m is a scalar, one has

$$0 = \delta_{\boldsymbol{\xi}} S_m = -\int d^4x \sqrt{-g}\, T^{\mu\nu} \xi_{(\mu;\nu)}. \tag{5.2.25}$$

Integrating by parts and using the fact that $\boldsymbol{\xi}$ is an arbitrary vector one concludes that the stress-energy tensor for any covariant theory obeys the following *covariant conservation law*

$$T^{\mu\nu}{}_{;\nu} = 0. \tag{5.2.26}$$

The Bianchi identity $G^{\mu\nu}{}_{;\nu} = 0$ shows a consistency of the condition Eq. (5.2.26) with the Einstein equations Eq. (5.2.21).

In a spacetime that admits symmetry the covariant conservation law can be used to obtain a conserved quantity connected with this symmetry. Let $\boldsymbol{\xi}$ be a *Killing vector*, then one has

$$(T^{\mu\nu}\xi_\nu)_{;\mu} = T^{\mu\nu}{}_{;\mu}\xi_\nu + T^{\mu\nu}\xi_{(\nu;\mu)} = 0. \tag{5.2.27}$$

Integrating this relation over the four-dimensional volume V with a boundary $\Sigma = \partial V$ and using the Stokes' theorem one gets

$$\int_\Sigma T^{\mu\nu} \xi_\nu d\sigma_\mu = 0. \tag{5.2.28}$$

Let us assume that the spacetime is asymptotically flat and choose two space-like Cauchy surfaces Σ_- and Σ_+ extended to the spatial infinity. We assume that Σ_+ lies in the future of Σ_-. We assume also that at far distances the stress-energy tensor falls rapidly enough. If we apply the relation Eq. (5.2.28) to this case and take into account that the integral over the boundary at infinity vanishes then we obtain

---

[4]From now on we put the speed of light $c = 1$.

[5]The variation of the metric can be written in the form

$$\delta g_{\mu\nu} = -\mathcal{L}_{\boldsymbol{\xi}}\, g_{\mu\nu}. \tag{5.2.23}$$

In a special case of a manifold with a symmetry, when $\boldsymbol{\xi}$ is a Killing vector, $\mathcal{L}_{\boldsymbol{\xi}}\, g_{\mu\nu} = 0$. This equation is simply the definition of the symmetry transformation.

$$\int_{\Sigma_-} T^{\mu\nu}\xi_\nu d\sigma_\mu = \int_{\Sigma_+} T^{\mu\nu}\xi_\nu d\sigma_\mu. \tag{5.2.29}$$

Note that in this relation both elements of volume $d\sigma_\mu$ are chosen future-directed. For Σ_- this means that an additional minus sign with respect the general relation where the direction of $d\sigma_\mu$ is chosen everywhere to be out with respect to the volume V. Relation Eq. (5.2.29) means that

$$P[\xi] = \int_\Sigma T^{\mu\nu}\xi_\nu d\sigma_\mu \tag{5.2.30}$$

does not depend on the choice of the Cauchy surface Σ and, hence, is a conserved quantity.

5.2.4 Examples

Fluid

Here, we give a few examples of the stress-energy tensors. The first example is a fluid with density ρ and pressure p. The *stress-energy tensor* for this system is

$$T_{\mu\nu} = (\rho + p)\, u_\mu u_\nu + p\, g_{\mu\nu}. \tag{5.2.31}$$

If an observer is comoving with the fluid, the projection of this tensor on the vector $\boldsymbol{u}$ is

$$\rho = T_{\mu\nu}\, u^\mu u^\nu. \tag{5.2.32}$$

This has the meaning of the matter density as measured by the observer. The spatial components of this tensor, as measured by a comoving observer, are

$$p\, h_{\mu\nu} = h_{\mu\alpha} h_{\nu\beta}\, T^{\alpha\beta}. \tag{5.2.33}$$

Here, $h_{\mu\nu} = g_{\mu\nu} + u_\mu u_\nu$ is the projection operator (see Eq. (3.2.16)).

Electromagnetic field

Another example is the case of the electromagnetic field. The *action* for the field A_μ is

$$S[A] = -\frac{1}{16\pi}\int_V F_{\mu\nu} F^{\mu\nu}\sqrt{-g}\, d^4x + \int_V A_\mu J^\mu\sqrt{-g}\, d^4x, \tag{5.2.34}$$

$$F_{\mu\nu} = 2A_{[\nu,\mu]} = \partial_\mu A_\nu - \partial_\nu A_\mu.$$

The second term in the action describes the interaction of the electromagnetic field with an electric current J^μ. The Maxwell equations are

$$F^{\mu\nu}{}_{;\nu} = 4\pi J^\mu, \qquad F_{[\mu\nu;\alpha]} = 0. \tag{5.2.35}$$

The first set of Maxwell equations, written in terms of the vector potential in the Lorentz gauge $A_\mu{}^{;\mu} = 0$, reads

$$\Box A_\mu - R_\mu{}^\nu A_\nu = -4\pi J^\mu. \tag{5.2.36}$$

For a free electromagnetic field the current $J^\mu = 0$. Its energy-momentum tensor is

$$T_{\mu\nu} = \frac{1}{4\pi} \left(F_{\mu\alpha} F_\nu{}^\alpha - \frac{1}{4} g_{\mu\nu} F_{\alpha\beta} F^{\alpha\beta} \right). \tag{5.2.37}$$

Problem 5.1: *Prove that the action for the Maxwell field for $J^\mu = 0$ is conformally invariant, i.e. invariant under the transformations*

$$g_{\mu\nu} \to \tilde{g}_{\mu\nu} = \omega^2 g_{\mu\nu}, \qquad A_\mu \to \tilde{A}_\mu = A_\mu. \tag{5.2.38}$$

Here, ω is function of x. Show that the conformal invariance of the action implies that the trace of the corresponding stress-energy tensor vanishes.

Non-minimal scalar field

The action for a massive scalar field, which is non-minimally coupled to the scalar curvature (the term proportional to the coupling constant ξ), is

$$S[\varphi] = -\frac{1}{8\pi} \int_V d^4x \sqrt{-g} \left[g^{\mu\nu} \varphi_{;\mu} \varphi_{;\nu} + m^2 \varphi^2 + \xi R \varphi^2 \right] + \int_V d^4x \sqrt{-g}\, j\varphi. \tag{5.2.39}$$

Here, the non-minimal coupling ξ is a dimensionless parameter. The dynamical equation for the scalar field is

$$\left(\Box - m^2 - \xi R \right) \varphi = -4\pi\, j. \tag{5.2.40}$$

In the case of a free scalar field $j = 0$ and the corresponding *stress-energy tensor* has the form

$$4\pi\, T_{\mu\nu} = \nabla_\mu \varphi \nabla_\nu \varphi - \frac{1}{2} g_{\mu\nu} \nabla^\epsilon \varphi \nabla_\epsilon \varphi - \frac{1}{2} g_{\mu\nu} m^2 \varphi^2$$
$$- \xi \left[\nabla_\mu \nabla_\nu \left(\varphi^2 \right) - R_{\mu\nu}\, \varphi^2 + g_{\mu\nu} \left(-\Box(\varphi^2) + \frac{1}{2} R \varphi^2 \right) \right]. \tag{5.2.41}$$

Note that when $\xi \neq 0$ the action Eq. (5.2.39) contains a term with the second derivatives of the metric (see discussion on page 131). Such a term with a special choice $\xi = 1/6$ makes the corresponding theory for $m = 0$ conformally invariant.

Problem 5.2: *Consider a massless scalar field, $m = 0$, in the absence of a source, $j = 0$. Using Eq. (5.2.40) show that for $\xi = 1/6$ the trace of the stress-energy tensor Eq. (5.2.41) vanishes. Show that the corresponding action Eq. (5.2.39) is conformally invariant. (Use relation Eq. (C.0.7) for the conformal transformation of the curvature scalar.)*

5.2.5 Energy conditions

In order to solve the Einstein equations Eq. (5.2.21) one needs to specify the stress-energy tensor generating a gravitational field. It should be emphasized that the stress-energy tensor contains the metric. For this reason, for consistency we have to start with a complete set of equations for the matter and metric.

A remarkable property of the Einstein equations is that quite often in order to prove some general results concerning the properties of their solutions, it is sufficient to know only some general properties of the stress-energy tensor generating the solution. There is a set of requirements on the stress energy tensor that proves to be helpful. These conditions are known as the *energy conditions*.

The weak energy condition is a requirement that for every future-directed time-like vector field u^μ the stress-energy tensor $T_{\mu\nu}$ obeys the relation

$$\rho \equiv T_{\mu\nu} u^\mu u^\nu \geq 0. \tag{5.2.42}$$

If u^μ is a unit vector, the quantity ρ has the meaning of the energy density of the matter in a local reference frame connected with u^μ. Thus, Eq. (5.2.42) guarantees that the energy density is non-negative in any local reference frame.

The strong energy condition requires that for every future-directed time-like vector field u^μ the stress-energy tensor $T_{\mu\nu}$ obeys the relation

$$\left(T_{\mu\nu} - \frac{1}{2} g_{\mu\nu} g^{\lambda\rho} T_{\lambda\rho} \right) u^\mu u^\nu \geq 0. \tag{5.2.43}$$

The null energy condition requires that for every future-directed null vector field l^μ the stress-energy tensor $T_{\mu\nu}$ obeys the relation

$$T_{\mu\nu} l^\mu l^\nu \geq 0. \tag{5.2.44}$$

The dominant energy condition requires that for every future-directed time-like or null vector field v^μ the vector $T_{\mu\nu} v^\nu$ is also a future-directed time-like or null vector.

Sometimes, it is convenient to use the integrated (average) versions of the corresponding energy condition. For example, consider a null vector field l^μ and let γ be its integral line, that is, a solution of the equation

$$\frac{dx^\mu}{d\lambda} = l^\mu(x). \tag{5.2.45}$$

The averaged null energy condition requires that for any such integral line γ the following relation is valid:

$$\int_\gamma T_{\mu\nu} l^\mu l^\nu d\lambda \geq 0. \tag{5.2.46}$$

5.3 Linearized Gravity

5.3.1 Linearized Einstein equations

One can expect that in the strong-field limit the Einstein theory of gravity is quite different from the Newton gravity. On the other hand, in the weak-field regime the Einstein equations must properly reproduce the results of the latter one. Let us study this limit.

Suppose the metric g is a solution of the Einstein equations. One may consider a solution $g + \delta g$ that only slightly differs from g. Subtracting the unperturbed equation

$$G = 8\pi G T \tag{5.3.1}$$

from the perturbed one

$$G + \delta G = 8\pi G (T + \delta T) \tag{5.3.2}$$

we obtain an equation for the metric perturbation

$$\delta G = 8\pi G \, \delta T. \tag{5.3.3}$$

Note that an explicit expression for the perturbation δG can be easily found by using the expression Eq. (5.2.9). The equation Eq. (5.3.3) can be used to study, for example, gravitational-wave generation and their propagation in the curved background.

Note that strictly speaking, when one compares the metrics g and $g + \delta g$, one must be sure that these metrics are taken in the *same* coordinates. But, in fact, even if the coordinates of the background solution g are fixed, there still remains a possibility to perform the coordinate transformations

$$x^\mu \rightarrow x^\mu + \xi^\mu(x), \tag{5.3.4}$$

which preserve g up to the same order as δg. This means that the quantity δg is not uniquely fixed but it allows a *gauge freedom*

$$\delta g_{\mu\nu} \rightarrow \delta g_{\mu\nu} - 2\xi_{(\mu;\nu)}. \tag{5.3.5}$$

To study the weak-field limit of the Einstein equations we use Eq. (5.3.3). In this case, one must choose the unperturbed metric to be flat and write

$$g_{\mu\nu} = \eta_{\mu\nu} + h_{\mu\nu}, \tag{5.3.6}$$

where $h = \delta g$. We shall use the Cartesian coordinates and the flat $\eta = \text{diag}(-1, 1, \ldots, 1)$. Under the gauge transformations Eq. (5.3.5) the perturbations transform as follows

$$h_{\hat\mu\hat\nu} \rightarrow h_{\hat\mu\hat\nu} - 2\xi_{(\hat\mu,\hat\nu)}. \tag{5.3.7}$$

This gauge freedom will be used to simplify the form of the linearized equations.

To linearize the Einstein equations on the flat background it is convenient to start directly with the expression for the curvature tensor Eq. (3.5.7) and use the fact that in the Cartesian coordinates the Christoffel symbols vanish. Then, one obtains

$$\delta R_{\hat\alpha\hat\mu\hat\beta\hat\nu} \doteq \frac{1}{2}\left(h_{\hat\alpha\hat\nu,\hat\mu\hat\beta} + h_{\hat\mu\hat\beta,\hat\alpha\hat\nu} - h_{\hat\mu\hat\nu,\hat\alpha\hat\beta} - h_{\hat\alpha\hat\beta,\hat\mu\hat\nu}\right). \tag{5.3.8}$$

Let us define the following object

$$\bar h_{\hat\mu\hat\nu} = h_{\hat\mu\hat\nu} - \frac{1}{2}\eta_{\hat\mu\hat\nu}\, h^{\hat\alpha}{}_{\hat\alpha}. \tag{5.3.9}$$

In order to fix the gauge it is convenient to impose the following conditions

$$\bar h_{\hat\nu}{}^{\hat\mu}{}_{,\hat\mu} = 0. \tag{5.3.10}$$

In this gauge the linearized Einstein equations take the form

$$\Box \bar h_{\hat\mu\hat\nu} \equiv \bar h_{\hat\mu\hat\nu,\hat\alpha}{}^{,\hat\alpha} = -16\pi G\, T_{\hat\mu\hat\nu}. \tag{5.3.11}$$

For known $\bar{h}_{\hat{\mu}\hat{\nu}}$ the metric perturbation $h_{\hat{\mu}\hat{\nu}}$ can be easily found

$$h_{\hat{\mu}\hat{\nu}} = \bar{h}_{\hat{\mu}\hat{\nu}} - \frac{1}{2}\eta_{\hat{\mu}\hat{\nu}}\,\bar{h}^{\hat{\alpha}}_{\hat{\alpha}}. \tag{5.3.12}$$

5.3.2 Newtonian limit

There exists a special case of the weak-field approximation known as the *Newtonian limit* of the Einstein gravity. Consider a non-relativistic particle or a compact distributed matter source for which the pressure is negligible in comparison with the energy density. In both cases there exists a reference frame in which T_{00} dominates and one can neglect the other components of the stress-energy tensor. In the leading order one has

$$T_{\hat{\mu}\hat{\nu}} = \rho\,\delta^0_{\hat{\mu}}\delta^0_{\hat{\nu}}. \tag{5.3.13}$$

For such matter, the gravitational field $\bar{h}_{\hat{\mu}\hat{\nu}}$ has only one non-vanishing component, $\bar{h}_{00}$, which obeys the equation

$$\triangle\bar{h}_{00} = -16\pi G\rho, \tag{5.3.14}$$

where $\triangle$ is the Laplace operator in the flat coordinates. This equation coincides with the equation for the Newtonian potential φ, if we take $\bar{h}_{00} = -4\varphi$ (cf. Eq. (1.3.1)). Using this result one can write

$$h_{00} = -2\varphi, \qquad h_{ij} = -2\varphi\,\delta_{ij}, \tag{5.3.15}$$

where

$$\triangle\varphi = 4\pi G\rho. \tag{5.3.16}$$

Hence, in the weak-field approximation, the metric describing the gravitational field of a static distribution of matter is[6]

$$ds^2 = -(1 + 2\varphi)(dX^0)^2 + (1 - 2\varphi)d\vec{X}^2,$$
$$d\vec{X}^2 = (dX^1)^2 + (dX^2)^2 + (dX^3)^2. \tag{5.3.17}$$

Let us show that a particle in the background Eq. (5.3.17) behaves as one expects from the Newtonian theory of gravity. Consider a geodesic motion of a particle in the metric Eq. (5.3.17). We assume that the velocity of this particle is much smaller than the speed of light, $v \ll 1$. Calculating the Christoffel symbols for the metric Eq. (5.3.17) one obtains

$$\Gamma_{ktt} = -\Gamma_{tkt} = \varphi_{,k}, \qquad \Gamma_{kij} = -\frac{1}{2}(\varphi_{,j}\delta_{ik} + \varphi_{,i}\delta_{jk} - \varphi_{,k}\delta_{ij}). \tag{5.3.18}$$

---

[6]Here and in what follows, we denote by $\vec{X}$ a 3-dimensional vector in a subspace X^0 =const. Similarly, in a spacetime with D dimensions $\vec{X}$ is a $(D-1)$-dimensional vector.

The equation of geodesic motion is

$$\frac{d^2 X^{\hat{\mu}}}{d\tau^2} = -\Gamma^{\hat{\mu}}_{\hat{\alpha}\hat{\beta}} \frac{dX^{\hat{\alpha}}}{d\tau} \frac{dX^{\hat{\beta}}}{d\tau}.$$ (5.3.19)

Let us assume that the particle is initially at rest, $dX^{\hat{\mu}}/d\tau = (1,0,0,0)$. Then, in the leading order one has

$$\frac{d^2 X^0}{d\tau^2}\bigg|_{\tau=0} = 0, \qquad \frac{d^2 X^i}{d\tau^2}\bigg|_{\tau=0} = -\Gamma^i{}_{00} = -\varphi^{,i}.$$ (5.3.20)

Thus, the acceleration of a particle is

$$a^i = \frac{d^2 X^i}{dX^{02}} = -\varphi^{,i}.$$ (5.3.21)

This is simply the standard Newton law.[7]

5.3.3   Komar mass

For black holes the gravitational field in their vicinity is not weak. Nevertheless, the Newtonian limit can be used for the interpretation of some of their global parameters, such as the mass. The reason is that far away from an isolated compact object the gravitational field becomes weak. The spacetime in this region is almost flat. In such an *asymptotically flat spacetime* the weak-field expression Eq. (5.3.17) can be used to approximate the metric in the asymptotic domain. At far distance from the static gravitating body

$$\varphi = -\frac{GM}{r},$$ (5.3.22)

and the free-fall acceleration is

$$a^i = -\frac{GM}{r^2} n^i.$$ (5.3.23)

Here, n^i is a unit vector of the external normal to a 2D sphere σ of the radius r. One can relate the mass of the object to the following integral

$$GM = -\frac{1}{4\pi} \int_\sigma a^i n_i \, d^2\sigma.$$ (5.3.24)

The expression Eq. (5.3.24) can be written in a covariant form. The gravitational field Eq. (5.3.17) is static and its Killing vector $\boldsymbol{\xi}$ has components

$$\xi^\mu = \delta^\mu_0, \qquad \xi_\mu = -(1 + 2\varphi)\delta^0_\mu.$$ (5.3.25)

---

[7]Let us recall that we reproduced the Newtonian law equations Eq. (5.3.16) and Eq. (5.3.21) in their standard form because of the proper choice of the dimensionless coefficient, $(16\pi)^{-1}$, in the Einstein–Hilbert action Eq. (5.1.6).

One also has

$$\xi_{\mu;\nu} = -2\varphi_{,i}\,\delta^0_{[\mu}\delta^i_{\nu]}. \tag{5.3.26}$$

An observer that is at rest at far distance has the velocity $u^\mu \approx \delta^\mu_0$. Thus, Eq. (5.3.24) can be written in the form[8]

$$GM = \frac{1}{4\pi}\int_\sigma \xi^{\mu;\nu}d\sigma_{\mu\nu}, \qquad d\sigma_{\mu\nu} = n_{[\mu}u_{\nu]}d^2\sigma. \tag{5.3.27}$$

This definition of the mass of a gravitating system is called the *Komar mass*. The relation Eq. (5.3.27) is an evident analog of the electric-charge definition that uses the Gauss law.

We obtained the expression for M using a special surface, a sphere of radius r taken at the fixed moment of time. Let us show that in fact this definition is much more general. Consider a static spacetime, and let ξ be its Killing vector field. Then, using the relation Eq. (3.7.16)

$$\xi^{\mu;\nu}{}_{;\nu} = -R^\mu{}_\nu\,\xi^\nu \tag{5.3.28}$$

and Stokes' theorem Eq. (3.9.12) one has

$$\int_{\partial\Sigma}\xi^{\mu;\nu}d\sigma_{\mu\nu} = -\int_\Sigma R^\mu{}_\nu\,\xi^\nu\,d\sigma_\mu. \tag{5.3.29}$$

Here, Σ is a 3-dimensional surface with the boundary $\partial\Sigma$. Suppose we deform the boundary surface. Denote by σ_0 and σ_1 its initial and final position, respectively. If between these surfaces $R_{\mu\nu} = 0$ one has

$$I[\sigma_0] = I[\sigma_1], \qquad I[\sigma] = \int_\sigma \xi^{\mu;\nu}d\sigma_{\mu\nu}. \tag{5.3.30}$$

This means that in the Ricci flat spacetime, that is for the vacuum solutions of the Einstein equations, the quantity $I[\sigma]$ does not depend on the choice of a 2-dimensional closed surface σ. In particular, the mass M defined by Eq. (5.3.27) does not depend on the particular choice of a 2-dimensional surface that surrounds a static matter distribution.

In the next section we study the asymptotic form of the metric for rotating objects. The gravitational field in this case is not static but stationary, since it is not invariant under the time reflection. In this case there also exists a Killing vector that is time-like at infinity and has the asymptotic form Eq. (5.3.25) there. The expression Eq. (5.3.27) defines the mass in this case as well. Let us emphasize, that Eq. (5.3.27) can be used to calculate the mass of a compact object performing the integration in the domain where the gravitational field is not weak. For example, for a static (or stationary) vacuum black hole its mass can be found by using Eq. (5.3.27) in the region close to its horizon.

---

[8]Note that we use the definition Eq. (3.9.5) for the surface element $d\sigma_{\mu\nu}$, which agrees with that employed by (Bardeen *et al.* 1973; Carter 1973), but is half that in (Carter 1979; Lightman *et al.* 1975).

5.3.4   Gravitational field of rotating sources

Multipole expansion

A stationary (not necessarily static) weak gravitational field generated by matter distribution obeys the equations

$$\Delta \bar{h}_{\hat{\mu}\hat{\nu}} = -16\pi\,G\,T_{\hat{\mu}\hat{\nu}}. \tag{5.3.31}$$

We focus our attention to the far-distance behavior of solutions and use the standard *multipole expansion* technique. The *Green function* of the flat Laplacian

$$\Delta \mathcal{G}(\vec{X}, \vec{Y}) = -\delta^3(\vec{X} - \vec{Y}) \tag{5.3.32}$$

is

$$\mathcal{G}(\vec{X}, \vec{Y}) = \frac{1}{4\pi |\vec{X} - \vec{Y}|}. \tag{5.3.33}$$

Using this Green function we can write a solution of Eq. (5.3.31) in the form

$$\bar{h}_{\hat{\mu}\hat{\nu}}(\vec{X}) = 4G \int \frac{T_{\hat{\mu}\hat{\nu}}(\vec{Y})}{|\vec{X} - \vec{Y}|}\, d^3 Y. \tag{5.3.34}$$

We assume that the matter is located within a sphere of the finite radius L, so that the integration is restricted to the domain where $|\vec{Y}| \leq L$. For the point of 'observation', $\vec{X}$, located at the far distance $r = |\vec{X}| \to \infty$, one has

$$|\vec{X} - \vec{Y}|^2 = r^2 - 2(\vec{X}, \vec{Y}) + \cdots = r^2 \left(1 - \frac{2(\vec{X}, \vec{Y})}{r^2} + \cdots\right),$$

$$\frac{1}{|\vec{X} - \vec{Y}|} = \frac{1}{r}\left(1 + \frac{(\vec{X}, \vec{Y})}{r^2} + \cdots\right). \tag{5.3.35}$$

These relations allow us to write the following asymptotic expansion for $\bar{h}_{\hat{\mu}\hat{\nu}}$

$$\bar{h}_{\hat{\mu}\hat{\nu}}(\vec{X}) = \frac{4G}{r}\int T_{\hat{\mu}\hat{\nu}}(\vec{Y})\, d^3 Y + \frac{4\,GX^k}{r^3}\int Y_k\, T_{\hat{\mu}\hat{\nu}}(\vec{Y})\, d^3 Y + \cdots. \tag{5.3.36}$$

Mass and angular momentum

The expansion Eq. (5.3.36) can be written in a simpler and more convenient form that contains only global characteristics of the matter sources. We assume that the source of the field is matter dominated, that is $|T_{ij}| \ll |T_{00}|$. Since the background geometry is flat, it possesses ten Killing vectors–generators of the Poincaré transformations. Let $\boldsymbol{\xi}_{(\hat{\mu})}$ be four vectors generating translations and $\boldsymbol{\zeta}_{(\hat{\mu})(\hat{\nu})}$ be six vectors generating boosts and rotations

$$\boldsymbol{\xi}_{(\hat{\mu})} = \xi^\alpha_{(\hat{\mu})}\partial_\alpha = \partial_{\hat{\mu}}, \qquad \boldsymbol{\zeta}_{(\hat{\mu})(\hat{\nu})} = X_{\hat{\mu}}\partial_{\hat{\nu}} - X_{\hat{\nu}}\partial_{\hat{\mu}}. \tag{5.3.37}$$

We denote the corresponding conserved quantities as

$$P_{\hat{\mu}} = P[\boldsymbol{\xi}_{(\hat{\mu})}], \qquad J_{\hat{\mu}\hat{\nu}} = P[\boldsymbol{\zeta}_{(\hat{\mu})(\hat{\nu})}]. \tag{5.3.38}$$

Thus, the energy-momentum vector of the matter $P^{\hat\mu}$ and its angular momentum tensor $J^{\hat\mu\hat\nu}$ are defined as follows

$$P^{\hat\mu} = \int T^{0\hat\mu} d^3X, \qquad J^{\hat\mu\hat\nu} = \int \left(X^{\hat\mu} T^{\hat\nu 0} - X^{\hat\nu} T^{\hat\mu 0}\right) d^3X. \qquad (5.3.39)$$

In the frame where the system as a whole is at rest one has

$$\int T_{0i} d^3X = 0, \qquad M = P^0 = \int T_{00} d^3X. \qquad (5.3.40)$$

We choose the center-of-mass at the origin of the coordinates

$$\int X^k T_{00} d^3X = 0. \qquad (5.3.41)$$

The angular momentum tensor is skew-symmetric and it has only spatial components J^{ij}. The expression for J^{ij} can be slightly simplified. The conservation of the energy-momentum $T^{\hat\mu i}{}_{,i} = 0$ implies

$$\left(X^m X^k T^{\hat\mu l}\right)_{,l} = X^k T^{\hat\mu m} + X^m T^{\hat\mu k}. \qquad (5.3.42)$$

Integrating this equality over a compact region that contains the matter, and using the fact that $T_{\hat\mu\hat\nu}$ vanishes outside this region, one obtains

$$\int X^m T^{k\hat\mu} d^3X = - \int X^k T^{m\hat\mu} d^3X. \qquad (5.3.43)$$

Hence,

$$J^{kl} = 2 \int X^k T^{l0} d^3X. \qquad (5.3.44)$$

Asymptotic form of the metric

By using Eq. (5.3.36) one recovers $h_{\hat\mu\hat\nu}$ and obtains

$$h_{00} \approx \frac{2\,GM}{r}, \qquad h_{ij} \approx \frac{2\,GM}{r} \delta_{ij}, \qquad h_{0i} \approx \frac{2\,GJ_{ik}X^k}{r^3}. \qquad (5.3.45)$$

J_{ik} is a 3×3 antisymmetric matrix. By means of 3-dimensional rotations, any such matrix can be transformed into the following canonical form:

$$J = \begin{pmatrix} 0 & J & 0 \\ -J & 0 & 0 \\ 0 & 0 & 0 \end{pmatrix}. \qquad (5.3.46)$$

The first two expressions in Eq. (5.3.45) are invariant under the spatial rotations. In the new coordinates the expression Eq. (5.3.45) is still valid and J_{ij} has the form of Eq. (5.3.46). Let us denote $t = X^0$ and introduce spherical coordinates (r, θ, ϕ)

$$X^1 = r \sin\theta \cos\phi, \qquad X^2 = r \sin\theta \sin\phi, \qquad X^3 = r \cos\theta. \qquad (5.3.47)$$

In these coordinates the asymptotic form of the metric Eq. (5.3.45) is

$$ds^2 = -\left(1 - \frac{2GM}{r}\right)dt^2 - \frac{4GJ\sin^2\theta}{r}\,dt d\phi + \left(1 + \frac{2GM}{r}\right)(dr^2 + r^2\,d\omega^2), \quad (5.3.48)$$

where $d\omega^2 = d\theta^2 + \sin^2\theta d\phi^2$ is a metric on a unit round sphere S^2.

■ **Problem 5.3:**   *Prove this.*

Problem 5.4:   *Metric Eq. (5.3.48) is axisymmetric and $\xi_{(\phi)} = \partial_\phi$ is a Killing vector. Denote*

$$J = -\frac{1}{8\pi G}\int_\sigma \xi_{(\phi)}^{\mu;\nu}\,d\sigma_{\mu\nu}, \qquad (5.3.49)$$

where the integral is taken over any two-dimensional surface σ (see footnote on page 139) surrounding the matter. Prove that J defined by Eq. (5.3.49) is identical to J that enters Eq. (5.3.48).

5.4   Gravitational radiation

5.4.1   Gravitational waves

When a source of the gravitational field moves, the curvature of the spacetime changes in time to reflect the change of the position of the source. This motion can generate *gravitational waves*, that is ripples in the spacetime curvature propagating freely through space. For the gravitational waves generated by a compact source the curvature at far distance r (in the *wave zone*) falls as $1/r$, that is much slower than the curvature ($\sim 1/r^3$) generated by a static source. The gravitational waves in many aspects are similar to the electromagnetic ones. In particular, they carry energy and momentum.

The emission of the gravitational waves by close binary systems, containing neutron stars and black holes, plays the crucial role in their evolution.[9] Roughly speaking, there exist two phases:

- long stage, during which due to the radiation of the gravitational ways the distance between the components slowly decreases;
- fast final stage of coalessence and the formation of the final black hole.

During the first longer stage the gravitational field is relatively weak and the motion of the components of the binary is non-relativistic. The *weak-field approximation* is a good starting point for the study of this stage. In this section we discuss the gravitational-waves generation and propagation in the weak-field approximation.

Study of this second 'fast' stage is a real challenge. The problems that arise in attempts at an analytical study are evident. The gravitational field in such a problem is strong, time dependent, and asymmetric. Thus, one needs to use general relativity in all its complexity. Only recently did it became possible to model such events in computer simulations. We postpone the discussion of the obtained results of these computations to Section 10.3.

---

[9]In Section 10.3 we discuss this problem in detail.

5.4.2   Propagation of gravitational waves

A *weak gravitational wave* is a propagating solution of the linearized Einstein equations Eq. (5.3.11)

$$\Box \bar{h}_{\hat{\mu}\hat{\nu}} = -16\pi G\, T_{\hat{\mu}\hat{\nu}}. \tag{5.4.1}$$

The conservation law

$$T^{\hat{\mu}\hat{\nu}}{}_{,\hat{\nu}} = 0 \tag{5.4.2}$$

is consistent with the gauge-fixing condition Eq. (5.3.10), which was imposed to obtain this equation. Let us recall that we are working in the Cartesian coordinates and use the flat background metric $\eta_{\hat{\mu}\hat{\nu}}$ to operate with indices.

Outside the matter $\bar{h}_{\hat{\mu}\hat{\nu}}$ obeys the homogeneous equation

$$\Box \bar{h}_{\hat{\mu}\hat{\nu}} = 0. \tag{5.4.3}$$

The condition Eq. (5.3.10) does not fix $\bar{h}_{\hat{\mu}\hat{\nu}}$ completely. There remains freedom in the gauge transformations Eq. (5.3.7), provided the vector field ξ^{μ}, generating these transformations, obeys the equations

$$\Box \xi^{\hat{\mu}} = 0. \tag{5.4.4}$$

These equations are similar to Eq. (5.4.3). It is possible to show (see, e.g., (Wald 1984)) that the remaining freedom ξ^{μ} can be used to impose the following additional conditions

$$\bar{h}^{\hat{\mu}}_{\hat{\mu}} = 0, \qquad \bar{h}_{\hat{0}\hat{i}} = 0, \qquad \hat{i} = 1, 2, 3. \tag{5.4.5}$$

This is a so-called *transverse traceless gauge*. This specialization of the gauge has a simple physical meaning: Under the perturbation $\bar{h}_{\hat{\mu}\hat{\nu}}$ satisfying Eq. (5.4.5) free particles at rest remain at the fixed values of the coordinates even, when the proper distance between them changes. To indicate that the amplitude $\bar{h}_{\hat{\mu}\hat{\nu}}$ is written in this gauge one usually writes it as follows $\bar{h}^{\mathrm{TT}}_{\hat{\mu}\hat{\nu}}$. The first of the conditions Eq. (5.4.5) implies that $\bar{h}^{\mathrm{TT}}_{\hat{\mu}\hat{\nu}} = h^{\mathrm{TT}}_{\hat{\mu}\hat{\nu}}$.

Four conditions, Eq. (5.4.5) together with four conditions of Eq. (5.3.10) leave free only two of ten components of $\bar{h}_{\hat{\mu}\hat{\nu}}$. Outside, the sources these two functions can be found as solutions of the wave equation. To specify these solutions one has to fix at some initial moment of time four functions of three variables (their initial values and the initial values of their time derivatives). This means that a free gravitational field has two degrees of freedom, the same number as the free electromagnetic field. These states correspond to two states of polarization.

5.4.3   Generation of gravitational waves

Solutions of the homogeneous equation (5.4.3) (with $T_{\hat{\mu}\hat{\nu}} = 0$) describe freely propagating waves. Let us impose a condition that originally there were no incoming free gravitational waves, so that all the waves are generated by the motion of the matter. The corresponding solution of the inhomogeneous equation (5.4.1) can be found by using a *retarded Green function* of the box-operator

$$G^{\text{ret}}(T,\vec{X};T',\vec{Y}) = \frac{1}{4\pi} \frac{\theta(T-T')\,\delta(T-T'-|\vec{X}-\vec{Y}|)}{|\vec{X}-\vec{Y}|}. \tag{5.4.6}$$

The result is

$$\bar{h}_{\hat{\mu}\hat{\nu}}(T,\vec{X}) = 4G \int d^3Y \, \frac{T_{\hat{\mu}\hat{\nu}}(T-|\vec{X}-\vec{Y}|,\vec{Y})}{|\vec{X}-\vec{Y}|}. \tag{5.4.7}$$

The integration is performed over the intersection of the past-directed null cone with the top at the point $(T,\vec{X})$ and the region occupied by the matter source.

We assume that the source is compact and is located near the origin of the coordinate system. Consider a point $(T,\vec{X})$ located far from the source origin. Denote $r = |\vec{X}|$. We assume that the size L of the source is much smaller than r, $L \ll r$. In this approximation $|\vec{X}-\vec{Y}| \approx r \equiv |\vec{X}|$. In the slow-motion approximation one can change the integration region in Eq. (5.4.7) by a plane $T' = T - r = $const. The result is

$$\bar{h}^{\hat{\mu}\hat{\nu}}(T,\vec{X}) \approx \frac{4G}{r} \int d^3Y \, T^{\hat{\mu}\hat{\nu}}(T-r,\vec{Y}). \tag{5.4.8}$$

For a stationary source this relation reduces to Eq. (5.3.34). We already know that after fixing the position of the center-of-mass of the source, the terms $\bar{h}_{\hat{0}\hat{0}}$ and $\bar{h}_{\hat{0}\hat{i}}$ calculated in leading order are determined by conserved quantities. They have asymptotic behavior $\sim r^{-1}$, but the contribution of them to the curvature falls as r^{-3}. For this reason, to describe the gravitational waves generated by slowly moving matter, it is sufficient to focus on the spatial components of $\bar{h}_{\hat{\mu}\hat{\nu}}$.

The conservation law Eq. (5.4.2) implies

$$\partial_T T^{\hat{i}\hat{0}} = -\partial_j T^{\hat{i}\hat{j}}, \qquad \partial_T T^{\hat{0}\hat{0}} = -\partial_j T^{\hat{0}\hat{j}}. \tag{5.4.9}$$

For $T_{\hat{\mu}\hat{\nu}}$ vanishing outside a compact spatial region, by using these relations, integration by parts and *Stokes' theorem* one obtains

$$\int d^3Y \, T^{\hat{i}\hat{j}} = \frac{1}{2}\partial_T \int d^3Y \,(T^{\hat{i}\hat{0}}Y^{\hat{j}} + T^{\hat{j}\hat{0}}Y^{\hat{i}}) = \frac{1}{2}\partial_T^2 \int d^3Y \, T^{\hat{0}\hat{0}}Y^{\hat{i}}Y^{\hat{j}}. \tag{5.4.10}$$

Let $\rho = T^{\hat{0}\hat{0}}$ be the mass density, and denote by $\bar{Q}^{\hat{i}\hat{j}}$ the *quadrupole moment of mass distribution*,

$$\bar{Q}^{\hat{i}\hat{j}} = \int d^3Y \rho Y^{\hat{i}}Y^{\hat{j}}. \tag{5.4.11}$$

Then, one obtains

$$\bar{h}^{\hat{i}\hat{j}}(T,\vec{X}) = \frac{2G}{r} \frac{d^2\bar{Q}^{\hat{i}\hat{j}}}{dT^2}\bigg|_{T'=T-r}. \tag{5.4.12}$$

This is a famous *quadrupole formula* for the amplitude of the *gravitational radiation* in the linear and slow-motion approximation.[10]

5.4.4 Outgoing radiation in *TT*-gauge

In the general case, the asymptotic of the gravitational wave Eq. (5.4.12) does not obey *TT*-gauge condition. To satisfy this condition it is necessary to perform an additional gauge transformation.

Consider a point at large distance r from the source and introduce an orthonormal tetrad (u, n, e_1, e_2) at this point:

- $u^{\hat{\mu}} = (1, \vec{0})$ is a four-velocity vector of an observer at rest at the point of observation;
- $n^{\hat{\mu}} = (0, \vec{n})$ with $\vec{n} = \vec{X}/r$ is a unit vector in the radial direction;
- two other mutually orthonormal vectors e_1 and e_2 are chosen in the plane orthogonal to u and n.

We also denote $l = u + n$ and $k = u - n$. The vectors l and k are null. The asymptote of the gravitational wave at large r is of the form

$$\bar{h}^{\hat{\mu}\hat{\nu}} \sim \frac{\bar{H}^{\hat{\mu}\hat{\nu}}(u, \theta, \phi)}{r}, \tag{5.4.13}$$

where $u = T - r$ and θ and ϕ are usual spherical coordinates. One has $l_{\hat{\mu}} = u_{,\hat{\mu}}$. Derivatives of $\bar{h}^{\hat{\mu}\hat{\nu}}$ in the directions n and $e_{1,2}$ are of the order of r^{-2}. Hence, in the leading order in r one gets

$$\bar{h}^{\hat{\mu}\hat{\nu}}_{,\hat{\lambda}} \sim \frac{\dot{\bar{H}}^{\hat{\mu}\hat{\nu}} l_{\hat{\lambda}}}{r}. \tag{5.4.14}$$

We denote by a dot the derivative ∂_u with respect to the *retarded time u*. Thus, the gauge condition Eq. (5.3.10) implies

$$\dot{\bar{H}}^{\hat{\mu}\hat{\nu}} l_{\hat{\nu}} = 0. \tag{5.4.15}$$

Since $l_{\hat{\nu}}$ is time independent, one can choose

$$\bar{H}^{\hat{\mu}\hat{\nu}} l_{\hat{\nu}} = 0. \tag{5.4.16}$$

This means that $\bar{H}^{\hat{\mu}\hat{\nu}}$ can be written in the form

$$\bar{H}^{\hat{\mu}\hat{\nu}} = a_0 l^{\hat{\mu}} l^{\hat{\nu}} + a_1 l^{(\hat{\mu}} e_1^{\hat{\nu})} + + a_2 l^{(\hat{\mu}} e_2^{\hat{\nu})} + b_1 e_1^{\hat{\mu}} e_1^{\hat{\nu}} + b_2 e_2^{\hat{\mu}} e_2^{\hat{\nu}} + b_3 e_1^{(\hat{\mu}} e_2^{\hat{\nu})}. \tag{5.4.17}$$

The vector k is not present in the decomposition Eq. (5.4.17) since it would violate the condition Eq. (5.4.15).

The remaining gauge transformation is

$$\bar{h}_{\hat{\mu}\hat{\nu}} \rightarrow \bar{h}_{\hat{\mu}\hat{\nu}} - 2\xi_{(\hat{\mu},\hat{\nu})} + \eta_{\hat{\mu}\hat{\nu}} \xi_{\hat{\lambda}}^{,\hat{\lambda}}. \tag{5.4.18}$$

---

[10]For discussion of the general multipole decomposition of the gravitational perturbations see, e.g., the review (Thorne 1980).

Since outside the sources ξ obeys the same equation as $\bar{h}^{\hat{\mu}\hat{v}}$, at large r it can be written in a form similar to Eq. (5.4.13)

$$\xi_{\hat{\mu}} \sim \frac{\zeta_{\hat{\mu}}(u,\theta,\phi)}{r}. \tag{5.4.19}$$

Under the transformation Eq. (5.4.18) $\bar{H}^{\hat{\mu}\hat{v}}$ changes as follows

$$\bar{H}_{\hat{\mu}\hat{v}} \to \bar{H}_{\hat{\mu}\hat{v}} - 2\dot{\zeta}_{(\hat{\mu}}l_{\hat{v})} + \eta_{\hat{\mu}\hat{v}}\dot{\zeta}_{\hat{\lambda}}l^{\hat{\lambda}}. \tag{5.4.20}$$

One also has

$$\bar{H}^{\hat{\mu}}_{\hat{\mu}} \to \bar{H}^{\hat{\mu}}_{\hat{\mu}} + 2\dot{\zeta}_{\hat{\mu}}l^{\hat{\mu}}. \tag{5.4.21}$$

Let us write

$$\dot{\zeta} = c_k k + c_l l + c_1 e_1 + c_2 e_2. \tag{5.4.22}$$

By the proper choice of the coefficient c_k one can make $\bar{H}^{\hat{\mu}}_{\hat{\mu}} = 0$. As a consequence, for the decomposition Eq. (5.4.17) one has

$$b_2 = -b_1. \tag{5.4.23}$$

In a similar way one can choose the coefficients c_l and c_2 in Eq. (5.4.22) to banish a_0, a_1, and a_2 in Eq. (5.4.17). Denote

$$e_+^{\hat{\mu}\hat{v}} = e_1^{\hat{\mu}}e_1^{\hat{v}} - e_2^{\hat{\mu}}e_2^{\hat{v}}, \qquad e_\times^{\hat{\mu}\hat{v}} = e_1^{\hat{\mu}}e_2^{\hat{v}} + e_2^{\hat{\mu}}e_1^{\hat{v}}. \tag{5.4.24}$$

As a result of the performed gauge transformations one obtains

$$\bar{h} = h \sim (h_+ e_+ + h_\times e_\times)/r. \tag{5.4.25}$$

Components h_+ and $h_\times$ describe two linearly polarized states of the gravitational radiation. It should be emphasized that the transformations Eq. (5.4.22) that preserve the condition Eq. (5.4.23) do not change the value of $b_2 - b_1$ and b_3. This means that if the property $\bar{H}^{\hat{\mu}}_{\hat{\mu}} = 0$ is already satisfied, then to obtain the coefficients h_+ and $h_\times$ it is sufficient to neglect in the matrix $\bar{H}_{\hat{\mu}\hat{v}}$ all the components except the coefficients $b_1 - b_2$ and b_3. It is evident that Eq. (5.4.25) gives the asymptotic form of the gravitational amplitude in the TT-gauge.

The traceless part of h can be obtained if instead of $\bar{Q}^{\tilde{ij}}$ in the quadrupole expansion Eq. (5.4.12) to use its traceless part

$$Q^{\tilde{ij}} = \bar{Q}^{\tilde{ij}} - \frac{1}{3}\delta^{\tilde{ij}}\bar{Q}^k_k = \int d^3Y \rho \left(Y^{\hat{i}}Y^{\hat{j}} - \frac{1}{3}\delta^{\tilde{ij}}|\vec{Y}|^2 \right). \tag{5.4.26}$$

Problem 5.5: *Two compact objects with the masses M_1 and M_2 separated by a distance R revolve around the center-of-mass of the system along the circular trajectories. Find the corresponding traceless tensor of the quadrupole moment $Q_{\tilde{ij}}$ for this system.*

Solution: *Let us choose the origin of the coordinates in the center-of-mass of the system. We get*

$$\vec{R}_1 = \frac{M_2}{M_1 + M_2}\vec{R}, \qquad \vec{R}_2 = -\frac{M_1}{M_1 + M_2}\vec{R}, \qquad \vec{R} = \vec{R}_1 - \vec{R}_2. \tag{5.4.27}$$

For the circular orbits $R = |\vec{R}| = \text{const}$. Let us choose the orbit in the $(X\text{-}Y)$-plane and denote by ψ a polar angle

$$X = R \cos \psi, \qquad Y = R \sin \psi. \tag{5.4.28}$$

For the circular Keplerian orbits

$$\psi = \Omega T, \qquad \Omega = R^{-3/2} \sqrt{G(M_1 + M_2)}. \tag{5.4.29}$$

The traceless quadrupole moment is

$$Q_{\hat{i}\hat{j}} = M_1 \vec{R}_{1\hat{i}} \vec{R}_{1\hat{j}} + M_2 \vec{R}_{2\hat{i}} \vec{R}_{2\hat{j}} - \frac{1}{3} \left(M_1 |\vec{R}_1|^2 + M_2 |\vec{R}_2|^2 \right) \delta_{\hat{i}\hat{j}}. \tag{5.4.30}$$

This gives us the following non-vanishing components of $Q_{\hat{i}\hat{j}}$

$$Q_{xx} = \mu R^2 (\cos^2 \psi - 1/3), \qquad Q_{yy} = \mu R^2 (\sin^2 \psi - 1/3),$$
$$Q_{xy} = \mu R^2 \sin \psi \cos \psi, \qquad Q_{zz} = -\frac{1}{3} \mu R^2. \tag{5.4.31}$$

Here, $\mu = M_1 M_2 / (M_1 + M_2)$.

Let us estimate an amplitude of the gravitational waves emitted by a binary system. Using relations Eqs. (5.4.29) and (5.4.31) one has

$$\ddot{Q} \sim 2\mu \Omega^2 R^2 = \frac{2G^2 \mu M}{R}. \tag{5.4.32}$$

Using Eq. (5.4.25) one obtains

$$h \sim \frac{2G^2 \mu M}{rR}. \tag{5.4.33}$$

Here, $M = M_1 + M_2$ is the total mass of the system and $\mu = M_1 M_2 / M$ is its *reduced mass*.

5.4.5 Energy of the gravitational waves

An effective stress-energy tensor of the gravitational waves is

$$T_{\hat{\mu}\hat{\nu}}^{\text{gw}} = \frac{1}{32\pi G} \langle h_{\hat{i}\hat{j},\hat{\mu}} h^{\hat{i}\hat{j}}{}_{,\hat{\nu}} \rangle, \tag{5.4.34}$$

where $h_{\hat{i}\hat{j}}$ is the amplitude of the gravitational wave in the TT-gauge. The angle brackets mean that the average is taken over several wavelengths. The derivation of this relation and the detailed discussion of the averaging procedure can be found in (Misner *et al.* 1973). In the quadrupole approximation at far distance from the source one has

$$T_{\hat{\mu}\hat{\nu}}^{\text{gw}} = \frac{l_{\hat{\mu}} l_{\hat{\nu}}}{32\pi G r^2} \langle \dot{h}_+^2 + \dot{h}_\times^2 \rangle. \tag{5.4.35}$$

The power of the gravitational radiation emitted by the source can be obtained by integrating Eq. (5.4.35) over a sphere of radius r. The result is

$$-\frac{dE}{dT} = \frac{G}{5}\dddot{\tilde{Q}}_{ij}\dddot{\tilde{Q}}^{ij}.$$ (5.4.36)

The negative sign means that due to radiation the system loses its energy.

Problem 5.6:   *Two compact objects with masses M_1 and M_2 separated by distance R revolve around the center-of-mass of the system along the circular trajectories. Show that the power of the gravitational waves emitted by this system is*

$$-\frac{dE}{dT} = \frac{32G^4 M_1^2 M_2^2 (M_1 + M_2)}{5R^5}.$$ (5.4.37)

A solution can be found in (Landau and Lifshitz 1980; Lightman et al. 1975).

Problem 5.7:   *As a result of the gravitational radiation a binary system loses the energy and its radius shrinks. Consider a circular motion of two compact objects of masses M_1 and M_2, which at the initial moment of time are separated by distance R_0. Calculate the lifetime of this system.*

Solution: *The energy of the system is*

$$E = -GM_1 M_2 / R.$$ (5.4.38)

Using Eq. (5.4.37) one gets

$$\frac{dR}{dT} = \frac{2R^2}{GM_1 M_2}\frac{dE}{dT} = -\frac{A}{R^3}, \qquad A = \frac{64}{5}G^3 M_1 M_2 (M_1 + M_2).$$ (5.4.39)

This equation can be easily integrated and gives

$$R_0^4 - R^4 = 4AT.$$ (5.4.40)

The time required to reach $R = 0$ is

$$T_0 = \frac{R_0^4}{4A}.$$ (5.4.41)

It should be stressed that the orbit velocity of the compact objects grows with time. Formally, when the radius between them becomes small enough, this velocity reaches the speed of light. This means that the slow-motion approximation breaks down and the quadrupole formula for the gravitational radiation does not work. However, if the initial radius R_0 is large enough and initially the motion was non-relativistic, the expression Eq. (5.4.41) can be used for the estimation of the lifetime of the binary system. The reason is that, as it follows from Eq. (5.4.40), the main contribution to the lifetime T_0 comes from the largest values of R. The evolution of the system becomes faster as R decreases. For example, after the radius halves the remaing lifetime becomes 16 times shorter than the original one.

For a binary system where the components have equal mass M and are separated by distance R_0 the estimation Eq. (5.4.41) takes a very simple form

$$T_0 = \frac{5}{64}\frac{R_0^4}{r_S^3},$$ (5.4.42)

where $r_S = 2GM$ is the gravitational radius of the total mass M.

> *When similar compact bodies move along the elliptic Kepler orbits the expression for the power of the emitted gravitational waves is more complicated. For such a motion one has*
>
> $$a(1 - e^2)R^{-1} = 1 + e\cos\psi, \quad d\psi/dT = R^{-2}\left[G(M_1 + M_2)a(1 - e^2)\right]^{1/2}. \tag{5.4.43}$$
>
> *Here, e is an eccentricity and a is the semi-major axis of the orbit. The calculations give the following result*
>
> $$-\frac{dE}{dT} = \frac{8G^4 M_1^2 M_2^2 (M_1 + M_2)}{15a^5 (1 - e^2)^5}(1 + e\cos\psi)^4\left[12(1 + e\cos\psi)^2 + e^2\sin^2\psi\right]. \tag{5.4.44}$$
>
> *After averaging over the period one obtains the following expression for the average power (Landau and Lifshitz 1980)*
>
> $$-\frac{d\bar{E}}{dT} = \frac{32G^4 M_1^2 M_2^2 (M_1 + M_2)}{5a^5 (1 - e^2)^{7/2}}\left(1 + \frac{73}{24}e^2 + \frac{37}{96}e^4\right). \tag{5.4.45}$$

5.5 Gravity in Higher-Dimensions

5.5.1 Higher-dimensional Einstein equations

Our derivation of the Einstein–Hilbert action, presented in Section 5.1, can be repeated practically without changes in a spacetime with the number of dimensions higher than 4. In should be emphasized that the dimensionality of the action is the same for any number of dimensions,[11] while the Newton coupling constant has the dimensionality

$$[G^{(D)}] = \frac{cm^{D-1}}{s^2 g}. \tag{5.5.1}$$

The Einstein–Hilbert action in D dimensions is

$$S[g] = \frac{c^3}{16\pi G^{(D)}}\int (R - 2\Lambda)\sqrt{-g}\, d^D x. \tag{5.5.2}$$

We keep the same coefficient $1/16\pi$ as before. This is a matter of agreement and convenience since there are no generally adopted coefficients in the higher-dimensional Newton law. Repeating the calculations similar to the 4D case one obtains the *higher-dimensional Einstein equations*

$$G_{\alpha\beta} + \Lambda g_{\alpha\beta} = 8\pi G^{(D)} T_{\alpha\beta}, \qquad G_{\alpha\beta} = R_{\alpha\beta} - \frac{1}{2}g_{\alpha\beta}R. \tag{5.5.3}$$

If the spacetime is not empty the stress-energy tensor in the right-hand side of this equation can be found by varying the corresponding matter action.

---

[11] It must be so if the Planck constant has this property and the classical stationary action principle is a consequence of the quantum-mechanical path-integral approach.

5.5.2 Linearized gravity in higher dimensions

Linearization of the higher-dimensional Einstein equations over a flat background can be done similarly to the 4D case. Namely, we write

$$g_{\hat{\mu}\hat{\nu}} = \eta_{\hat{\mu}\hat{\nu}} + h_{\hat{\mu}\hat{\nu}}, \tag{5.5.4}$$

and introduce new variables for the metric perturbations

$$\bar{h}_{\hat{\mu}\hat{\nu}} = h_{\hat{\mu}\hat{\nu}} - \frac{1}{2}\eta_{\hat{\mu}\hat{\nu}} h^{\hat{\alpha}}_{\ \hat{\alpha}}, \qquad h_{\hat{\mu}\hat{\nu}} = \bar{h}_{\hat{\mu}\hat{\nu}} - \frac{1}{D-2}\eta_{\hat{\mu}\hat{\nu}} \bar{h}^{\hat{\alpha}}_{\ \hat{\alpha}}. \tag{5.5.5}$$

To fix the gauge freedom $h_{\hat{\mu}\hat{\nu}} \to h_{\hat{\mu}\hat{\nu}} - 2\xi_{(\hat{\mu},\hat{\nu})}$, we impose the conditions

$$\bar{h}_{\hat{\nu}}^{\ \hat{\mu}}{}_{,\hat{\mu}} = 0. \tag{5.5.6}$$

In this gauge the linearized Einstein equations take the form

$$\Box \bar{h}_{\hat{\mu}\hat{\nu}} \equiv \bar{h}_{\hat{\mu}\hat{\nu},\hat{\alpha}}^{\ \ \ \hat{\alpha}} = -16\pi G^{(D)} T_{\hat{\mu}\hat{\nu}}. \tag{5.5.7}$$

For known $\bar{h}_{\hat{\mu}\hat{\nu}}$ the metric perturbation $h_{\hat{\mu}\hat{\nu}}$ can be easily found (see Eq. (5.5.5)).
For a stationary source one needs to solve the equations

$$\Delta \bar{h}_{\hat{\mu}\hat{\nu}} = -16\pi G^{(D)} T_{\hat{\mu}\hat{\nu}}. \tag{5.5.8}$$

The Green function of the higher-dimensional Laplace equation

$$\Delta G(\vec{X}, \vec{Y}) = -\delta^{D-1}(\vec{X} - \vec{Y}), \tag{5.5.9}$$

is

$$G(\vec{X}, \vec{Y}) = \frac{\Gamma((D-3)/2)}{4\pi^{(D-1)/2}|\vec{X} - \vec{Y}|^{D-3}}. \tag{5.5.10}$$

This allows one to write

$$\bar{h}_{\mu\nu}(X) = \beta_D G^{(D)} \int \frac{T_{\mu\nu}(\vec{Y})}{|\vec{X} - \vec{Y}|^{D-3}} d^{D-1}Y, \qquad \beta_D = \frac{4\Gamma((D-3)/2)}{\pi^{(D-3)/2}}. \tag{5.5.11}$$

In the four-dimensional spacetime $\beta_4 = 4$, and the expression Eq. (5.5.11) coincides with Eq. (5.3.34).
To obtain the asymptotic form of $\bar{h}_{\mu\nu}$ at large distance r we, as before, begin with the expression Eq. (5.3.35), which gives

$$\frac{1}{|\vec{X} - \vec{Y}|^{D-3}} = \frac{1}{r^{D-3}}\left(1 + (D-3)\frac{(\vec{X}, \vec{Y})}{r^2} + \cdots\right). \tag{5.5.12}$$

Thus, we have

$$\bar{h}_{\mu\nu}(\vec{X}) \simeq \frac{\beta_D G^{(D)}}{r^{D-3}}\left[\int T_{\mu\nu}(\vec{Y}) d^{D-1}Y + \frac{(D-3)X^k}{r^2}\int Y_k T_{\mu\nu}(\vec{Y}) d^{D-1}Y\right]. \tag{5.5.13}$$

Using the frame in which our system is at rest and the origin coincides with the center-of-mass one obtains

$$h_{00} \approx \frac{(D-3)\beta_D}{(D-2)} \frac{G^{(D)}M}{r^{D-3}}, \qquad h_{ij} \approx \frac{\beta_D}{(D-2)} \frac{G^{(D)}M}{r^{D-3}} \delta_{ij},$$

$$h_{0i} \approx \frac{(D-3)\beta_D}{2} \frac{G^{(D)}J_{ik}X^k}{r^{D-1}}, \tag{5.5.14}$$

where

$$M = \int T_{00}\, d^{D-1}X, \qquad J^{kl} = 2\int X^k\, T^{l0}\, d^{D-1}X. \tag{5.5.15}$$

In higher dimensions, by suitable rigid rotations of the spatial $(D-1)$ coordinates, the $(D-1) \times (D-1)$ matrix J_{ij} can be transformed into the following canonical form:

$$J = \begin{pmatrix} 0 & J_1 & 0 & 0 & \cdots \\ -J_1 & 0 & 0 & 0 & \cdots \\ 0 & 0 & 0 & J_2 & \cdots \\ 0 & 0 & -J_2 & 0 & \cdots \\ \cdots & \cdots & \cdots & \cdots & \cdots \end{pmatrix}. \tag{5.5.16}$$

It is convenient to write the number of dimensions D of the spacetime in the form

$$D = 2n + \varepsilon, \tag{5.5.17}$$

where $\varepsilon = 0$ for even and $\varepsilon = 1$ for odd number of dimensions. This often helps to write the formulas for these two cases in a unique way.

For even D the last row and column of the matrix Eq. (5.5.16) vanish. Thus, in higher dimensions there are n independent components of the angular momentum. The parameters J_k $(k = 1, 2, \ldots n)$ are the components of the angular momentum tensor in the planes $(X^{2k-1}-X^{2k})$, i.e. in the planes of rotation of the system. The subscript k enumerates these two planes. In a stationary axially symmetric spacetime the parameters M (the *Komar mass*) and J_k can be written in the following covariant form:

$$G^{(D)}M = \frac{D-2}{D-3} \frac{1}{8\pi} \int \nabla^\alpha \xi^\beta_{(t)}\, d\sigma_{\alpha\beta}, \tag{5.5.18}$$

$$G^{(D)}J_k = -\frac{1}{8\pi} \int \nabla^\alpha \xi^\beta_{(k)}\, d\sigma_{\alpha\beta}, \tag{5.5.19}$$

where $d\sigma_{\alpha\beta}$ is defined by Eq. (3.9.5) (see also footnote on page 139), and the integration is taken over a $(D-2)$ sphere surrounding the matter distribution. In four dimensions these expressions coincide with the relations Eq. (5.3.27) and Eq. (5.3.49).

In an asymptotically flat D-dimensional spacetime there is a natural higher-dimensional general-ization of a stationary axially symmetric spacetime. Namely, this is a spacetime in which besides a time-like (at infinity) Killing vector $\xi_{(0)}$ there exist n space-like (at infinity) Killing vectors ξ_k that mutually commute and that commute with $\xi_{(0)}$. The vectors ξ_k generate symmetry transformations (*rotations*) *in two-dimensional mutually orthogonal planes. In a flat spacetime this set of n commuting Killing vectors generates the* Cartan subgroup $U(1)^n$ *of the rotation group* $SO(D-1)$.

5.5.3   Newtonian limit in higher dimensions

In a special case of a non-relativistic matter source the stress-energy tensor in the Cartesian coordinates is

$$T_{\hat{\mu}\hat{\nu}} = \rho\, \delta^0_{\hat{\mu}} \delta^0_{\hat{\nu}}. \tag{5.5.20}$$

The Newtonian potential φ is defined as a solution of the equation

$$\Delta\varphi = \frac{D-3}{D-2}\, 8\pi G^{(D)}\rho. \tag{5.5.21}$$

The solution of this equation is

$$\varphi = -\frac{4\Gamma((D-1)/2)}{(D-2)\,\pi^{(D-3)/2}} \cdot \frac{G^{(D)}M}{r^{D-3}}, \tag{5.5.22}$$

where

$$M = \int \rho\, d^{D-1}x \tag{5.5.23}$$

is the total mass of the system Eq. (5.5.15).

The metric in the weak-field approximation reads

$$ds^2 = -(1+2\varphi)\, dt^2 + \left(1 - \frac{2}{D-3}\varphi\right) \delta_{ij}\, dX^i dX^j. \tag{5.5.24}$$

The Christoffel symbols for this metric are

$$\Gamma_{ktt} = -\Gamma_{tkt} = \varphi_{,k}, \qquad \Gamma_{kij} = -\frac{1}{D-3}(\varphi_{,j}\delta_{ik} + \varphi_{,i}\delta_{jk} - \varphi_{,k}\delta_{ij}). \tag{5.5.25}$$

Using the geodesic equation, one finds that a particle that is initially at rest experiences an acceleration

$$a^i = \frac{d^2X^i}{dX^{0^2}} = -\varphi^{,i}. \tag{5.5.26}$$

Relations Equations (5.5.22) and (5.5.26) give the D-dimensional version of the Newton laws.

To summarize this section let us make a few general remarks about the properties of the higher-dimensional gravity. If we compare the gravitational field of a compact object in four and higher dimensions we see that:

- In higher dimensions the force of the gravitational attraction at large distance, $F \sim r^{-(D-2)}$, is weaker than in four dimensions, where $F \sim r^{-2}$.
- At small distance the situation is opposite.

- In the higher dimensions there exist n independent components of the angular momentum instead of one. As a result, the asymptotic metric has n off-diagonal terms.

- The existence of stable Keplerian orbits in 4D is a result of the delicate competition of the attractive Newtonian potential, $\sim r^{-1}$, and the repulsive centrifugal one, $\sim r^{-2}$. In the higher dimensions the Newtonian potention is $\sim r^{-(D-3)}$, while the centrifugal one has the same form. This makes impossible the existence of stable bounded orbits in $D \geq 6$. Stability analysis is more delicate in $D = 5$ (see Section 7.11 for more discussion).

5.5.4   The gravitational field of ultrarelativistic objects

Weak-field approach

There is another case when the weak-field approximation is useful and allows one to obtain interesting results. This is the problem of finding of the gravitational field of objects moving with a velocity close to the speed of light.

> *An example of such a problem is the following: What is the gravitational interaction between two narrow beams of light? It had been shown long ago that for parallel beams moving in the same direction the force vanishes. (The proof of this result can be found in the book by Tolman, which was published in 1934. The last edition of this book is (Tolman 2010)). This result has quite a simple explanation. One can always use for the calculations a reference frame moving fast enough in the same direction as photons. In such a frame the frequency of photons (and hence their energy) is redshifted and can be made arbitrarily small. For this reason the resulting gravitational force between the beams vanishes. One can arrive at the same conclusion considering two photons. In the lowest order the amplitude of the gravitational scattering of such photons must depend on the invariants, constructed from their momenta $\boldsymbol{p}_1$ and $\boldsymbol{p}_2$. But for parallel photons moving in the same direction all such invariants vanish.*

The case of two parallel beams with opposite velocities is more interesting. For this system the energy in the center-of-mass is non-vanishing invariant and it determines, for example, a non-vanishing gravitational interaction between the beams. This problem is important for the discussion of the black hole formation in the collision of two ultrarelativistic particles. In the models with large extra dimensions a possibility of micro-black hole creation in the collider experiments (see Section 10.5) is widely discussed. For this purpose we consider in this section the gravitational field of ultrarelativistic particles in four and higher dimensions.

A standard way to obtain such a solution is to consider a black hole of the mass m, to boost it up to the speed v, and then to take the limit $v \to 1$. It should be emphasized that in this limit the energy of the black hole, $m\gamma$, grows as the Lorentz factor $\gamma = (1 - v^2)^{-1/2}$, and the gravitational field becomes singular in this limit. In order to obtain a well-defined result one has to fix the energy $m\gamma$ rather than the mass, so that $m \sim \gamma^{-1} \to 0$. This procedure is called the *Penrose limit*. Note that because of the Lorentz contraction the scales in the direction of motion shrink as γ^{-1}, while the scale in the transverse direction remains unchanged. As a result, in the Penrose limit the gravitational field is contracted in one direction and becomes located mainly at the null plane 'comoving' with the black hole. The metric obtained in such a manner for a boosted black hole is known as the *Aichelburg–Sexl metric*.

We demonstrate now that the same metric can be obtained in a much simpler way by using the *weak-field approximation*. Non-linear effects of the strong gravity are important at the distance from the black hole of the order of its gravitational radius $r_S = 2m$. In the Penrose limit the bare mass of the black hole m as well as its gravitational radius become small. This means that at a finite distance from the source the gravitational field in the Penrose limit practically coinsides with the limit of the corresponding linear approximation.

Let us consider the Penrose limit of a weak-field solution. We perform calculations in an arbitrary number of dimensions $D \geq 4$. We also assume that the source of the field has a non-vanishing angular momentum. This will allows us, for example, to obtain a gravitational field of circularly polarized light beams of ultrarelativistic particles with spin. Our starting point is the metric for a massive spinning point-like object in the linearized gravity. We write the linearized gravitational field in the form $g_{\mu\nu} = \eta_{\mu\nu} + h_{\mu\nu}$, where $\eta_{\mu\nu}$ is the flat metric. We denote by D the total number of spacetime dimensions and by $(\bar{t}, \vec{X}) = (\bar{t}, \bar{\xi}, x^a)$ the Cartesian coordinates in the Minkowski spacetime. In what follows we shall consider the motion of the point-like spinning object with a constant velocity. We start with a zero-velocity solution, and boost it in the $\bar{\xi}$ direction. We denote the other spatial coordinates in the direction transverse to the motion by $\mathbf{x} = (x^a)$.

The linearized Einstein equations have the form

$$\Box h_{\mu\nu} = -16\pi G^{(D)} \left(T_{\mu\nu} - \frac{1}{D-2} \eta_{\mu\nu} \eta^{\alpha\beta} T_{\alpha\beta} \right), \tag{5.5.27}$$

where $G^{(D)}$ is the D-dimensional gravitational coupling constant. If a compact (point-like) rotating object is considered in the center-of-mass rest frame then the stress-energy components are

$$T_{\bar{t}\bar{t}} = M\delta^{D-1}(\vec{X}), \qquad T_{\bar{t}a} = \frac{1}{2} J_{ab} \partial_b \delta^{D-1}(\vec{X}), \qquad T_{\bar{t}\bar{\xi}} = 0, \tag{5.5.28}$$

where M is the mass and J_{ab} is the antisymmetric angular momentum tensor of the body. For simplicity, we assume that the angular momentum is orthogonal to $\bar{\xi}$ axis. In higher dimensions the number of different planes orthogonal to the chosen axis and the number of independent components of the angular momenta is equal to the integer part $[(D-2)/2]$. It is non-zero only for $D \geq 4$.

By solving Eq. (5.5.27), one obtains the linearized metric in the form

$$ds^2 = -d\bar{t}^2 + d\bar{\xi}^2 + d\mathbf{x}^2 + 2\bar{A}_a dx^a d\bar{t} + \bar{\Phi} \left[d\bar{t}^2 + \frac{1}{D-3}(d\bar{\xi}^2 + d\mathbf{x}^2) \right], \tag{5.5.29}$$

$$\bar{\Phi} = \frac{8\Gamma\left(\frac{D-1}{2}\right)}{(D-2)\pi^{\frac{D-3}{2}}} \frac{G^{(D)}M}{\bar{r}^{D-3}}, \qquad \bar{A}_a = \frac{4\Gamma\left(\frac{D-1}{2}\right)}{\pi^{\frac{D-3}{2}}} \frac{G^{(D)} J_{ab} x^b}{\bar{r}^{D-1}}. \tag{5.5.30}$$

Here,

$$\bar{r}^2 = \bar{\xi}^2 + r^2, \qquad r^2 = \mathbf{x}^2. \tag{5.5.31}$$

Because of the linearity, a solution for an extended object can be written in the same form (5.5.29) with the following functions $\bar{\Phi}$ and $\bar{A}_{ab}$

$$\bar{\Phi} = \frac{8\Gamma\left(\frac{D-1}{2}\right)}{(D-2)\pi^{\frac{D-3}{2}}} G^{(D)} \int dx' d\bar{\xi}' \frac{\bar{\varepsilon}(\bar{\xi}', x')}{[(\bar{\xi}' - \bar{\xi})^2 + (x' - \mathbf{x})^2]^{(D-3)/2}}, \tag{5.5.32}$$

$$\bar{A}_a = \frac{4\Gamma\left(\frac{D-1}{2}\right)}{\pi^{\frac{D-3}{2}}} G^{(D)} \int dx' d\bar{\xi}' \frac{\bar{j}_{ab}(\bar{\xi}', x')(x'^b - x^b)}{[(\bar{\xi}' - \bar{\xi})^2 + (\mathbf{x}' - \mathbf{x})^2]^{(D-1)/2}}. \tag{5.5.33}$$

Here, $\bar{\varepsilon}(\bar{\xi}, \mathbf{x})$ and $\bar{j}_{ab}(\bar{\xi}, \mathbf{x})$ are the mass and angular momentum densities, respectively.

Boosting the source

To obtain a metric for a spinning source moving with the speed of light we boost the obtained solution. We denote by t and ξ the coordinates in the frame that is moving along the $\bar{\xi}$-axis with the velocity β (in the negative direction)

$$\bar{\xi} = \gamma(\xi - \beta t) = \frac{\gamma}{\sqrt{2}}[(1 - \beta)v + (1 + \beta)u],$$

$$\bar{t} = \gamma(t - \beta\xi) = \frac{\gamma}{\sqrt{2}}[(1 - \beta)v - (1 + \beta)u]. \tag{5.5.34}$$

Here, $\gamma = (1 - \beta^2)^{-1/2}$, $u = (t - \xi)/\sqrt{2}$ and $v = (t + \xi)/\sqrt{2}$ are null coordinates in the flat background spacetime. As a result of the boost the length r in the direction orthogonal to the motion does not change, while the length in the $\bar{\xi}$-direction becomes $\gamma^{-1}l$. In order to obtain an object moving with the speed of light and having *finite duration* (length), we assume that transition to the boosted frame is accompanied by the scaling of the initial length of the object with the factor γ. We also assume that the energy, $E = \gamma M$, and the angular momentum, J_{ab}, remain fixed.

In the Penrose limit, that is when $\beta \to 1$, one has

$$\bar{\xi} \sim \sqrt{2}\gamma u, \qquad \bar{t} \sim -\sqrt{2}\gamma u. \tag{5.5.35}$$

Denote

$$\varepsilon(u) = \frac{1}{\sqrt{2}} \int d\mathbf{x} \, T_{uu} = \sqrt{2} \int d\mathbf{x} \, T_{tt},$$

$$j_{ab}(u) = 2 \int d\mathbf{x} \, T_{ua}x_b = 2\sqrt{2} \int d\mathbf{x} \, T_{ta}x_b. \tag{5.5.36}$$

Here, the integrals are taken along the slice $t = $const over the transverse space x^a. The energy and the angular momentum of the boosted object in the Penrose limit take the form

$$E = \gamma\bar{M} = \gamma \int dx d\bar{\xi} \, \bar{\varepsilon}(\bar{\xi}, x^a) = \int du \, \varepsilon(u), \tag{5.5.37}$$

$$J_{ab} = \bar{J}_{ab} = \int dx d\bar{\xi} \, \bar{j}_{ab}(\bar{\xi}, x^a) = \int du \, j_{ab}(u), \tag{5.5.38}$$

where

$$\bar{\varepsilon}(\bar{\xi}, x^a) = \frac{1}{\sqrt{2}\gamma^2} \varepsilon(u)\delta(\mathbf{x}), \qquad \bar{j}_{ab}(\bar{\xi}, x^a) = \frac{1}{\sqrt{2}\gamma} j_{ab}(u)\delta(\mathbf{x}). \tag{5.5.39}$$

To obtain the metric in the Penrose limit, we use the following relation:

$$\lim_{\gamma \to \infty} \frac{\gamma}{(\gamma^2 y^2 + r^2)^{m/2}} = \frac{\sqrt{\pi}\,\Gamma((m-1)/2)}{\Gamma(m/2)}\,\frac{\delta(y)}{r^{m-1}}\,, \qquad m \geq 2. \qquad (5.5.40)$$

After simple calculations we obtain

$$ds^2 = -2\,du\,dv + \delta_{ab}\,dx^a dx^b + 2A_a\,dx^a du + \Phi\,du^2. \qquad (5.5.41)$$

For $D > 4$

$$\Phi = \frac{4\sqrt{2}\,\Gamma\left(\frac{D-4}{2}\right)}{\pi^{\frac{D-4}{2}}}\,\frac{G^{(D)}\varepsilon(u)}{r^{D-4}}\,, \qquad A_a = \frac{4\,\Gamma\left(\frac{D-2}{2}\right)}{\pi^{\frac{D-4}{2}}}\,\frac{G^{(D)}j_{ab}(u)x^b}{r^{D-2}}\,, \qquad (5.5.42)$$

and for $D = 4$

$$\Phi = -8\sqrt{2}\,G\varepsilon(u)\ln(r)\,, \qquad A_a = 4\,\frac{Gj_{ab}(u)x^b}{r^2}. \qquad (5.5.43)$$

The energy and the angular momentum densities ϵ and j_{ab} can be arbitrary functions of the null coordinate u. We also use the standard notation G for the 4D coupling constant. In the special four-dimensional case for an impulsive wave distribution $\varepsilon = E\delta(u)$ and $j_{ab} = 0$ these formulas reproduce the well-known *Aichelburg–Sexl metric* .

The obtained expressions for Φ and A_a formally coincide with potentials for a stationary electromagnetic field of a point-like source in $(D-2)$-dimensional Euclidean space. The corresponding electric-and magnetic-field strengths are $E = \Phi_{,a}$ and $F_{ab} = 2A_{[b,a]}$, respectively. For $D = 4$ the 'magnetic' field strength outside the source vanishes. This means that locally one can banish A_a by the gauge transformation $A_a \to A_a + \partial_a\lambda$, generated by the coordinate transformations (Bonnor 1969). However, one can not put A_a to zero globally in the general case, when the 'magnetic flux'

$$\mathcal{J} = \oint_C A_a dx^a \qquad (5.5.44)$$

calculated for a contour surrounding the source (which does not depend on C) does not vanish. This 'flux' reflects the presence of the spin of the ultrarelativistic particle.
A remarkable fact is that in the 4D case the metric Eq. (5.5.41), Eq. (5.5.43) is an exact solution of the Einstein equations. In the higher-dimensional case the exact solution for the gravitational field of the spinning ultrarelativistic object also has the form Eq. (5.5.41), but the expressions in Eq. (5.5.42) must be slightly modified. Namely, while A_a is unchanged, one has to include in Φ an additional term containing an integral of $F_{ab}F^{ab}$. The corresponding solution is called a gyraton. For discussion of gyratons and their properties see (Frolov and Fursaev 2005; Frolov et al. 2005).

5.5.5   Gravity in a spacetime with compact extra dimensions

The problem of extra dimensions is widely discussed in the literature. In these models it is usually assumed that all matter and fields, except the gravitational one, are confined in the 4D brane representing our usual world. However, the gravity can propagate in the higher-dimensional bulk space as well. An intriguing feature of these models is a prediction of micro-black hole creation in the collider experiments at the already achieved energies or at the energies that will be available in the near future. In Section 10.5 we shall discuss this subject in more detail. Now we use the weak-field approximation to illustrate some important properties of the gravitational field in some of the models with large extra dimensions.

Before going further, we need to clarify an important question. From our everyday experience we know that we are living in the four-dimensional world. How can one combine these observations with an assumption that at small distances the gravity is effectively higher dimensional? One can resolve this paradox by assuming that the extra dimensions are compact. To illustrate this mechanism let us study properties of a static solution of the gravitational equations in the weak-field approximation in such a spacetime with compact extra dimensions.

Let us discuss how compactification modifies the gravitational field. Consider a D-dimensional spacetime with coordinates $(X^0, X^1, \ldots, X^{D-2}, z)$. We impose periodicity condition along the z-axis and identify the points z and $z + L$. Denote

$$\rho^2 = (X^1)^2 + \ldots + (X^{D-2})^2. \tag{5.5.45}$$

Before the compactification the gravitational field of a point-like source has the form (see Eq. (5.5.14))

$$\bar{h}_{00} = \beta_D G^{(D)} M \psi, \quad \bar{h}_{0i} = -\frac{\beta_D}{2} G^{(D)} J_{ik} \psi_{,k}, \quad \psi = (\rho^2 + z^2)^{-(D-3)/2}. \tag{5.5.46}$$

Here, β_D is given in Eq. (5.5.11). For simplicity we assume that the source does not move in the z-direction, so that $T_{tz} = 0$. When the space along the direction z is made periodic with the period L the solution ψ of the Laplace equation is modified. It can be found by using the method of images. The result is

$$\psi_L = \sum_{k=-\infty}^{\infty} \frac{1}{[\rho^2 + (z + kL)^2]^{(D-3)/2}}. \tag{5.5.47}$$

Problem 5.8: *Show that the leading asymptote of ψ_L at $\rho/L \to \infty$ is*

$$\psi \sim \frac{B_D}{L \rho^{D-4}}, \tag{5.5.48}$$

where

$$B_D = \frac{\sqrt{\pi}\, \Gamma(D/2 - 2)}{\Gamma((D - 3)/2)}. \tag{5.5.49}$$

Solution: *Denote*

$$x_k = z/\rho + kL/\rho, \qquad \Delta x_k = x_{k+1} - x_k = L/\rho, \tag{5.5.50}$$

then Eq. (5.5.47) can be written in the form

$$\psi_L = \frac{1}{L\rho^{D-4}} \sum_{k=-\infty}^{\infty} \frac{\Delta x_k}{(1 + x_k^2)^{(D-3)/2}}. \tag{5.5.51}$$

For small L/ρ the series that enters the right-hand side of this relation can be approximated by the following integral

$$B_D = \int_{-\infty}^{\infty} \frac{dx}{(1 + x^2)^{(D-3)/2}}. \tag{5.5.52}$$

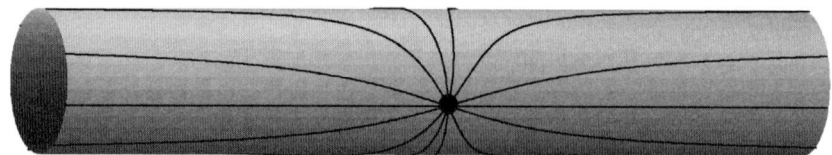

Fig. 5.1 Gravitational field of a point-like mass in a spacetime with compact extra dimension.

▮ *The calculation of the integral gives Eq. (5.5.49). As a result, we obtain the relation Eq. (5.5.48).*

Note that

$$B_D = \beta_{D-1}/\beta_D, \tag{5.5.53}$$

where β_D is given by Eq. (5.5.11). Thus, for $\rho \gg L$ one has

$$\psi_L \approx \frac{\beta_{D-1}}{\beta_D} \frac{1}{L\rho^{D-4}}. \tag{5.5.54}$$

In this asymptotic regime

$$\bar{h}_{00} = \beta_{D-1} G^{(D)} M \tilde{\psi}, \quad \bar{h}_{0i} = -\frac{\beta_{D-1}}{2} G^{(D)} J_{ik} \tilde{\psi}_{,k}, \quad \tilde{\psi} = \rho^{-(D-4)}. \tag{5.5.55}$$

This gives the correct gravitational field in the $(D-1)$-dimensional spacetime provided we identify the gravitational coupling constant $G^{(D-1)}$ in this spacetime with

$$G^{(D-1)} = G^{(D)}/L. \tag{5.5.56}$$

In the case when k coordinates are compactified and the spacetime is $\mathbb{R}^{D-k} \times \mathbb{T}^k$, where $\mathbb{T}^k$ denotes a flat k-torus, then one has a similar result

$$G^{(D-k)} = G^{(D)}/V^{(k)}, \tag{5.5.57}$$

where $V^{(k)}$ is the volume of the k-torus.

When $D - 4$ of D dimensions are compactified, the gravitational field of a point-like source in the $\rho \gg L$ limit coincides with the Newtonian 4D result. The deflection from the Newtonian law at $\rho \sim L$ can be used to test the presence of the extra dimensions.

Relation Eq. (5.5.57) shows that though the 4D gravitational coupling constant is very small, the bulk constant $G^{(4+k)}$ can be large, provided the size of the extra dimensions is big enough. This observation opens up an interesting possibility to 'explain' the *hierarchy problem*, namely: *Why the gravity in our world is so much weaker than the electromagnetic field and other interactions?* It is this property, that was one of the main initial motivations for the study of models with large extra dimensions.

Let us consider two protons at the distance $r \ll L$. In the brane world models the electromagnetic field is confined on the 4D brane, so that the presence of extra dimensions does not modify Coulomb's law. The electromagnetic repulsion force between two protons is $F_e = e^2/r^2$. Since the gravity can propagate in the bulk D-dimensional space the gravitational attraction of the protons is $F_G^{(D)} \sim G^{(D)} m_p^2/r^{D-2}$. In the four-dimensional case one has

$$\beta = \left| \frac{F_g^{(4)}}{F_e} \right| = \frac{Gm_p^2}{e^2} \approx 8 \times 10^{-37}. \tag{5.5.58}$$

This ratio does not depend on a scale. In the higher-dimensional case there always exists a scale at which the gravitational and electromagnetic interactions become of the same strength. The corresponding size $r_\star$ is

$$r_\star \equiv [G^{(D)} m_p^2/e^2]^{1/k} = \beta^{1/k} L, \tag{5.5.59}$$

where $k = D - 4$ is the number of extra dimensions. For example, for $L \sim 10^{-2}$cm and $k = 2$ one has $r_\star \sim 10^{-20}$ cm.

Let us illustrate the above general relations by a special example of the $5D$ flat spacetime with one compact dimension. In this particular case the summation in the series Eq. (5.5.47) for the potential ψ_L can be done explicitly. The result is

$$\psi_L = \sum_{k=-\infty}^{\infty} \frac{1}{\rho^2 + (z + kL)^2} = \frac{\pi}{L\rho} \frac{\cosh(\pi\rho/L)\sinh(\pi\rho/L)}{\cosh^2(\pi\rho/L) - \cos^2(\pi z/L)}. \tag{5.5.60}$$

The asymptote of this function at large ρ is

$$\psi_L \sim \pi/L\rho. \tag{5.5.61}$$

This results agrees with Eq. (5.5.48) since $B_5 = \pi$. Near the source, $\rho \to 0$ and $|z| \to 0$, one has

$$\psi_L \sim \frac{1}{r^2} = \frac{1}{\rho^2 + z^2}. \tag{5.5.62}$$

The critical scale when the function ψ_L changes its regimes is L. The obtained results show that in a spacetime with compactified spatial extra dimensions one correctly reproduces the four-dimensional weak-field behavior at distances greater than the compactification radius. A deviation from the four-dimensional law will be of the order of $\sim \exp(-\pi\rho/L)$. We postpone further discussion of other interesting consequences of the hypothesis of large extra dimensions to Section 10.5.

Problem 5.9: *Consider a $(2n + 1)$-dimensional spacetime with one compactified spatial dimension of size L. It has the topology $\mathbb{R}^{2n} \times \mathbb{T}^1$. Let Γ_c be an equipotential surface $\psi_L = c$. Find a critical value $c = c_\star$ at which the topology of equipotential surfaces Γ_c changes from S^{2n-1} to $S^{2(n-1)} \times S^1$. Show that for this value the critical surface Γ_c has a self-intersection at $(\rho = 0, z = \pm L/2)$ and its form is a two-folded cone. Calculate the angle θ_n of this cone.*

Solution: *The gravitational potential Eq. (5.5.47) is*

$$\psi_L(\rho, z) = \sum_{k=-\infty}^{\infty} \frac{1}{[\rho^2 + (z + kL)^2]^{n-1}}. \tag{5.5.63}$$

It is easy to check that $\partial_\rho \psi_L = \partial_z \psi_L = 0$ at $(\rho = 0, z = \pm L/2)$ and hence the function ψ_L has a critical point there. Near this point ψ_L has the following expansion

$$\psi_L = c_\star + \alpha \, \rho^2 + \beta \, \tilde{z}^2 + \gamma \, \tilde{z} \rho + \ldots, \quad \tilde{z} = L/2 - z, \tag{5.5.64}$$

where α, β and γ are determined by the second partial derivatives of ψ_L at the critical point. Simple calculations give

$$c_* = 2(2^{2n-2} - 1)\zeta(2n - 2)\frac{1}{L^{2n-2}}, \qquad \gamma = 0,$$

$$\alpha = -(n-1)\frac{A}{L^{2n}}, \qquad \beta = (2n-1)(n-1)\frac{A}{L^{2n}}. \tag{5.5.65}$$

Here, $A = \sum_{k=-\infty}^{\infty}(k + 1/2)^{-2n} = 2(2^{2n} - 1)\zeta(2n)$, and $\zeta(z)$ is the Riemann zeta-function, $\zeta(s) = \sum_{k=1}^{\infty} k^{-s}$. Thus, near the critical point one has

$$\psi = c_* + \frac{(n-1)A}{L^{2n}}[(2n-1)\bar{z}^2 - \rho^2] + \dots. \tag{5.5.66}$$

The condition $\psi = c_*$ leads to $(2n-1)z^2 \simeq \rho^2$. The angle of the cone of the critical surface is

$$\theta_n = \arctan(d\rho/dz)_{\rho_0} = \arctan(\sqrt{2n-1}). \tag{5.5.67}$$

The obtained result concerning the behavior of the gravitational potential near the critical point allows a for-reaching generalization. Consider the potential ψ in the N-dimensional space. Outside the sources it is a solution of the corresponding Laplace equation $\triangle_N \psi = 0$. Suppose ψ has a critical point p, where $\psi_{,a} = 0$, and $\det \psi_{,ab} \neq 0$. Since two quadratic forms, flat metric and $\psi_{ab} = \psi_{,ab}$, can be diagonalized simultaneously at p, there exist such flat coordinates in which near p one has

$$dl^2 = \sum_{a=1}^{N} dX_a^2, \qquad \psi_{ab} = diag(\psi_1, \dots, \psi_N). \tag{5.5.68}$$

Let us consider a case when

$$\psi = \psi_c + \alpha\rho^2 + \beta z^2 + \dots, \tag{5.5.69}$$

where $\rho^2 = X_1^2 + \dots + X_p^2$ and $z^2 = X_{p+1}^2 + \dots + X_{p+q}^2$, and $p + q = N$. The validity of the vacuum equation $\triangle_N \psi = 0$ near the critical point implies $p\alpha + q\beta = 0$. For the equipotential surface passing through the critical point p one has

$$\frac{d\rho}{dz} = \frac{\rho}{z} = \sqrt{\frac{-\beta}{\alpha}} = \sqrt{\frac{p}{q}}. \tag{5.5.70}$$

This relation immediately gives the result Eq. (5.5.67) for $N = D - 1 = 2n$, $p = N - 1 = 2n - 1$, and $q = 1$.

5.5.6 The gravitational field of a codimension 2 brane

As one more application of the higher-dimensional linearized gravitational equations let us consider the problem of the gravitational field of a brane. Let Σ be a time-like submanifold of codimension q and $p_{\mu\nu}$ be a projector on Σ. If n^A, $A = 1, \dots, q$ are unit mutually orthogonal vectors normal to Σ, then

$$p_{\mu\nu} = g_{\mu\nu} - \sum_{A=1}^{q} n_\mu^A n_\nu^A, \qquad p_\mu^\mu = D - q. \tag{5.5.71}$$

The distribution of the matter for the brane is

$$T_{\mu\nu} = t_{\mu\nu}\,\delta(\Sigma), \qquad\qquad t_{\mu\nu} = \mu\,p_{\mu\nu}. \tag{5.5.72}$$

In what follows we assume that the *brane tension* μ is small. We assume also that the brane is static and it has a plane configuration. We choose the Cartesian coordinates so that the brane equations are

$$X^{D-q} = \ldots = X^{D-1} = 0, \tag{5.5.73}$$

and use the additional coordinates $(X^0, \ldots, X^{D-q-1})$ as the intrinsic coordinates on the brane.

We focus on a special case when the number of codimensions is $q = 2$, and show that the gravitational field of such a brane is very simple. Namely, the corresponding spacetime is locally flat outside the brane, and it has a *conical singularity* on the brane surface. Denote $x = X^{D-2}$ and $y = X^{D-1}$. For the static brane of the codimension 2 the linearized equations Eq. (5.5.8) written for $h_{\mu\nu}$ give

$$\Delta h_{ab} = 0, \quad \text{for} \qquad 0 \le a,b \le D - 3,$$

$$(\partial_x^2 + \partial_y^2)h_{ij} = 16\pi\,G^{(D)}\mu\,\delta_{ij}\delta(x)\delta(y), \quad \text{for} \quad D - 2 \le i,j \le D - 1. \tag{5.5.74}$$

Using the symmetry of the problem and Eq. (5.5.74) one can write a solution of these equations in the form

$$h_{ab} = 0, \qquad h_{ij} = \frac{1}{2}h\,\delta_{ij}, \qquad h = 16G^{(D)}\mu\,\ln(\tilde\rho/\tilde\rho_0). \tag{5.5.75}$$

Here, $\tilde\rho^2 = x^2 + y^2$ and $\tilde\rho_0$ is a constant. Thus, the metric has the form

$$ds^2 = -(dX^0)^2 + (dX^1)^2 + \ldots + (dX^{D-2})^2 + (1 - h/2)(d\tilde\rho^2 + \tilde\rho^2 d\tilde\theta^2). \tag{5.5.76}$$

Here, $(\tilde\rho, \tilde\theta)$ are polar coordinates in the (x, y)-plane. Let us introduce new coordinates

$$\rho^2 = (1 - 8G^{(D)}\mu)^{-1}(1 - 8G^{(D)}\mu\,\ln(\tilde\rho/\tilde\rho_0))\tilde\rho^2, \qquad \theta = (1 - 4G\mu)\tilde\theta. \tag{5.5.77}$$

In the linear approximation corresponding to the small parameter μ the metric becomes

$$ds^2 = -(dX^0)^2 + (dX^1)^2 + \ldots + (dX^{D-2})^2 + d\rho^2 + \rho^2 d\theta^2. \tag{5.5.78}$$

The spacetime is a direct sum of $(D-2)$-dimensional Minkowski spacetime $\mathbb{R}^{D-2}$ and a two-dimensional cone $\mathbb{C}^2$. Its metric differs from the D-dimensional flat metric only by the property that the period of θ is not 2π but

$$0 \le \theta \le 2\pi(1 - 4G^{(D)}\mu). \tag{5.5.79}$$

The quantity $\Delta\theta = 8\pi\,G^{(D)}\mu$ is an *angle deficit*. To summarize, outside the brane of codimension 2 the spacetime is locally flat. Globally, the metric of such a spacetime is a direct sum of the $D - 2$ Minkowski metric and a metric of a 2-plane with the angle deficit. A special case of such branes, when $D = 4$ and $q = 2$, is a static straight *cosmic string* in a four-dimensional spacetime.

6

Spherically Symmetric Black Holes

6.1 Spherically Symmetric Gravitational Field

6.1.1 Spherically symmetric spacetime

A spacetime is called *spherically symmetric* if there exist coordinates in which its metric takes the form

$$ds^2 = \gamma_{AB}\, dx^A\, dx^B + r^2\, d\omega^2, \tag{6.1.1}$$

where $\gamma_{AB} = \gamma_{AB}(x)$, $r = r(x)$ $(A, B = 0, 1)$, and $d\omega^2$ is a metric on a unit 2-dimensional sphere S^2

$$d\omega^2 = \omega_{XY}\, d\zeta^X d\zeta^Y \doteq d\theta^2 + \sin^2\theta\, d\phi^2. \tag{6.1.2}$$

This metric admits 3 Killing vectors

$$\xi_1 = -\cos\phi\, \partial_\theta + \cot\theta \sin\phi\, \partial_\phi,$$
$$\xi_2 = \sin\phi\, \partial_\theta + \cot\theta \cos\phi\, \partial_\phi, \tag{6.1.3}$$
$$\xi_3 = \partial_\phi.$$

> **Problem 6.1:** *Prove that the commutators of these vector fields are of the form*
> $$[\xi_i, \xi_j] = \epsilon_{ijk}\, \xi_k, \tag{6.1.4}$$
> *where ϵ_{ijk} is a 3D Levi-Civita symbol.*

In the flat spacetime the vectors ξ_i are the generators of the rotation group, that is usual operators L_X, L_Y and L_Z of the angular momentum. Under the action of the symmetry transformations a point remains on a surface of constant radius r. Any two points on this 2D surface can be obtained from one another by the action of the symmetry transformation. Such a surface is called a *transitivity surface* of the symmetry group.

Besides the continuous isometries generated by the Killing vectors, the metric Eq. (6.1.1) possesses discrete symmetries:

$$\phi \rightarrow -\phi, \qquad \theta \rightarrow \pi - \theta. \tag{6.1.5}$$

Problem 6.2: Let a, a^X, and a^X_Y be scalar, vector, and tensor functions on S^2 that are invariant under all the isometries of a two-sphere. Prove that

$$a = const, \qquad a^X = 0, \qquad a^X_Y = const\, \delta^X_Y. \tag{6.1.6}$$

In a *spherically symmetric spacetime* a tensor Q^μ_ν that obeys this symmetry satisfies the conditions

$$\mathcal{L}_{\xi_i} Q^\mu_\nu = 0, \qquad i = 1, 2, 3. \tag{6.1.7}$$

In the coordinates (x^A, ζ^X) it has the following form

$$Q^\mu_\nu = \begin{pmatrix} Q^0_0 & Q^0_1 & 0 & 0 \\ Q^0_1 & Q^1_1 & 0 & 0 \\ 0 & 0 & \hat{Q} & 0 \\ 0 & 0 & 0 & \hat{Q} \end{pmatrix}, \qquad \partial Q^\mu_\nu / \partial \zeta^X = 0. \tag{6.1.8}$$

In particular, this is valid for the Einstein tensor G^μ_ν. The conservation law $G^\mu_{\nu;\mu} = 0$, implies

$$G^A_{B:A} + \frac{2r_{:A}}{r} G^A_B - \frac{2r_{:B}}{r} \hat{G} = 0. \tag{6.1.9}$$

Here, $(\ldots)_{:A}$ is the covariant derivative in the 2D metric γ_{AB}.

6.1.2 Dimensional reduction of the action

In order to study the spherically symmetric spacetimes it is sufficient to substitute the metric ansatz Eq. (6.1.1) into the Einstein equations. But there exists an alternative way. Namely, one can substitute the ansatz Eq. (6.1.1) into the *Einstein–Hilbert action* Eq. (5.1.6). Since neither R nor Λ depends on the angular variables, one can integrate over them. As a result, the Einstein–Hilbert action reduces to the action S_{sph}, which is a functional of the two-dimensional metric γ_{AB} and the scalar field r, $S_{\text{sph}} = S_{\text{sph}}[\gamma, r]$. Both of these field variables depend only on the x^A coordinates, so that the original 4D problem is reduced to the 2D one. Let us note that in the general case, substitution of an ansatz chosen for the metric into the action does not guarantee that after the variation of the reduced action the obtained equations reproduce correctly the reduced Einstein equations.

In a spherically symmetric case the variation of S_{sph} with respect to γ_{AB} and r gives a set of equations that is equivalent to the Einstein equations. Indeed, the variation of the Einstein–Hilbert action Eq. (5.1.6) is[1]

$$\delta S[g] = -\frac{1}{16\pi} \int_V d^4x \sqrt{-g}\, \mathcal{E}^{\mu\nu}\, \delta g_{\mu\nu}, \qquad \mathcal{E}^{\mu\nu} = G^{\mu\nu} + \Lambda g^{\mu\nu}. \tag{6.1.10}$$

---

[1] From now on by default we use units in which $G = c = 1$. In special cases, when it is important, we will write these constants explicitly.

Similarly, the variation of the matter action $S_m[\Phi, g]$ reads

$$\delta S_m[\Phi, g] = \frac{1}{2} \int_V d^4x \sqrt{-g}\, T^{\mu\nu}\, \delta g_{\mu\nu}, \qquad T^{\mu\nu} = \frac{2}{\sqrt{-g}} \frac{\delta S_m}{\delta g_{\mu\nu}}. \tag{6.1.11}$$

It is expressed in terms of the integal of the *stress-energy tensor* Eq. (5.2.20).

In a spherically symmetric case both tensors $\mathcal{E}_\mu{}^\nu$ and $T_\mu{}^\nu$ have the structure of Eq. (6.1.8). Their components do not depend on the angular variables. To obtain the reduced equations it is sufficient to impose these constraints on the quantities that enter the integrals in Eq. (6.1.10) and Eq. (6.1.11). Denote

$$\delta\gamma_{AB} = \frac{1}{4\pi} \int d^2\varsigma \sqrt{\omega}\,\delta g_{AB}, \qquad \delta(r^2) = \frac{1}{4\pi} \int d^2\varsigma \sqrt{\omega}\,\delta g_{\theta\theta}. \tag{6.1.12}$$

Then, Eq. (6.1.10) and Eq. (6.1.11) take the form

$$\delta S = -\frac{1}{4} \int d^2x \sqrt{-\gamma} \left[r^2 \mathcal{E}^{AB}\,\delta\gamma_{AB} + 4\, r\, \hat{\mathcal{E}}\,\delta r \right],$$

$$\delta S_m = -\pi \int d^2x \sqrt{-\gamma} \left[r^2 T^{AB}\,\delta\gamma_{AB} + 4\, r\, \hat{T}\,\delta r \right]. \tag{6.1.13}$$

The variations in the right-hand side of these equations are identical to the variations of the reduced actions, that this, the actions in which the spherical symmetry ansatz is substituted to from the very beginning. In other words, the reduced Einstein equations,

$$\mathcal{E}_A{}^B = 8\pi T_A{}^B, \qquad \hat{\mathcal{E}} = 8\pi \hat{T} \tag{6.1.14}$$

can be obtained directly by using the relations

$$\mathcal{E}_A{}^B = -\frac{4}{r^2\sqrt{-\gamma}} \frac{\delta S_{\text{sph}}}{\delta\gamma_{AB}}, \qquad \hat{\mathcal{E}} = -\frac{1}{r\sqrt{-\gamma}} \frac{\delta S_{\text{sph}}}{\delta r},$$

$$T_A{}^B = -\frac{4}{r^2\sqrt{-\gamma}} \frac{\delta S_{m,\text{sph}}}{\delta\gamma_{AB}}, \qquad \hat{T} = -\frac{1}{r\sqrt{-\gamma}} \frac{\delta S_{m,\text{sph}}}{\delta r}. \tag{6.1.15}$$

Here, S_{sph} and $S_{m,\text{sph}}$ are the reduced gravitational and matter actions.

One can write the reduced action S_{sph} in a 2D covariant form in terms of the 2D geometry γ_{AB} and the scalar function r. The four-dimensional Ricci scalar and $\sqrt{-g}$ can be written as follows

$$R = \mathcal{R} + \frac{2}{r^2} - 4\frac{r^{:A}_{:A}}{r} - 2\frac{r^{:A} r_{:A}}{r^2}, \qquad \sqrt{-g} = r^2\sqrt{-\gamma}\sqrt{\omega}. \tag{6.1.16}$$

Here, $\mathcal{R}$ is the Ricci scalar of the 2D metric γ_{AB}, the covariant derivative $(\ldots)_{:A}$ is defined with respect to the metric γ_{AB}, and all indixes are raised using γ^{AB}. Substituting this expression into the reduced Einstein–Hilbert action, integrating by parts the term containing $r^{:A}_{:A}$, and integrating over the angular variables, one gets

$$S_{\text{sph}}[\gamma, r] \equiv \frac{1}{4} \int_{M^2} d^2x \sqrt{-\gamma} \left[\mathcal{R}\, r^2 + 2\, r^{:A} r_{:A} + 2 - 2\Lambda r^2 \right]. \tag{6.1.17}$$

One may consider $\varphi \sim \ln r$ as a new scalar 2D field, which together with γ determine 2D gravity. Thus, the action Eq. (6.1.17) is a special version of an action for a *dilaton 2D gravity*.

The variation of S_{sph} over γ_{AB} and r gives the reduced Einstein equation Eq. (6.1.14), where

$$\mathcal{E}_{AB} \equiv -2\frac{r_{:AB}}{r} + \gamma_{AB}\left(-\frac{1}{r^2} + \frac{r^{:C}r_{:C}}{r^2} + 2\frac{r_{:C}^{:C}}{r} + \Lambda\right),$$

$$\hat{\mathcal{E}} \equiv \frac{r_{:C}^{:C}}{r} - \frac{1}{2}\mathcal{R} + \Lambda. \tag{6.1.18}$$

Here, all the operations are defined with respect to the metric γ_{AB}.

▌**Problem 6.3:** *Using relations (6.1.18) with $\Lambda = 0$ prove the validity of the relation Eq. (6.1.9).*

6.1.3 Generalized Birkhoff's theorem

We denote by e_{AB} and e^{AB} the two-dimensional antisymmetric tensors

$$e_{AB} = \sqrt{-\gamma}\,\epsilon_{AB}, \qquad e^{AB} = -\frac{1}{\sqrt{-\gamma}}\epsilon_{AB}, \tag{6.1.19}$$

where ϵ_{AB} is the antisymmetric *Levi-Civita symbol* ($\epsilon_{01} = 1$).

Lemma 1: *Consider 2D manifold with the metric γ_{AB}. Let f be a function on it obeying the relation*

$$f_{:AB} = F\,\gamma_{AB} \tag{6.1.20}$$

for some function F. Then, $\xi^A \equiv e^{AB}f_{:B}$ is a Killing vector.

Proof: Indeed,

$$\xi_{A:B} = e_A{}^C f_{:CB} = F e_A{}^C \gamma_{CB} = F e_{AB}. \tag{6.1.21}$$

Hence, ξ^A satisfies the Killing equation $\xi_{(A:B)} = 0$.

Lemma 2: *Let ξ be a Killing vector on a two-dimensional manifold with the metric γ_{AB}. Then, the equation*

$$f_{:A} = e_{AB}\,\xi^B \tag{6.1.22}$$

has a solution f that obeys the relation

$$f_{:AB} = F\,\gamma_{AB}, \qquad F = \frac{1}{2}f_{:C}^{:C}. \tag{6.1.23}$$

Proof: Denote $\tilde{f}_A = e_{AB}\xi^B$. As any 2D antisymmetric tensor, $\xi_{A:B}$ can be written as $\xi_{A:B} = F e_{AB}$. Thus, one has

$$\tilde{f}_{A:C} = e_{AB}\,e^B{}_C F = F\gamma_{AC}. \tag{6.1.24}$$

This relation shows that the integrability condition $\tilde{f}_{[A,C]} = 0$ of Eq. (6.1.22) is satisfied. It also proves the relation Eq. (6.1.23).

Generalized Birkhoff's theorem: *If the stress-energy tensor generating a spherically symmetric gravitational field obeys the condition*

$$T_{AB} = \frac{1}{2} T_C^C \gamma_{AB}, \tag{6.1.25}$$

then the corresponding solution of the Einstein equations possesses an additional Killing vector field.

Proof: If Eq. (6.1.25) is satisfied, then Eq. (6.1.18) implies that $r_{:AB} \sim \gamma_{AB}$, and hence, $\xi^A = e^{AB} r_{:B}$ is a Killing vector field for the two-dimensional metric γ_{AB}. Moreover, $\xi^A \partial_A r = 0$. Thus, $\xi^\mu = \delta_A^\mu \xi^A$ is a Killing vector field for the four-dimensional metric Eq. (6.1.1), which is evidently linearly independent of the Killing vectors Eq. (6.1.3).

6.2 Schwarzschild–de Sitter Metric

6.2.1 Spherically symmetric vacuum solutions

Theorem: *A spherically symmetric vacuum solution of Einstein equations with a cosmological constant is determined by one essential constant (mass M), and in the regions where $g^{\mu\nu} \nabla_\mu r \nabla_\nu r \neq 0$ it can be written in the form*

$$ds^2 = -g\, dt^2 + \frac{dr^2}{g} + r^2\, d\omega^2, \quad g = 1 - \frac{2M}{r} - \frac{\Lambda}{3} r^2. \tag{6.2.1}$$

For $\Lambda > 0$ this solution is known as the Schwarzschild–de Sitter metric. *For $\Lambda < 0$ it is the Schwarzschild–anti-de Sitter metric.*

Proof: We start the proof with the following remark. The *Frobenius theorem* implies that any vector obeying the condition $\xi_{[\alpha} \xi_{\beta;\gamma]} = 0$ can be presented in the form $\xi_\alpha = \beta t_{,\alpha}$ (see Chapter 3, Eqs. (3.7.12) and (3.7.13)). Since any two-dimensional vector obeys the condition

$$\xi_{[A} \xi_{B:C]} = 0, \tag{6.2.2}$$

this theorem shows that the Killing vector $\xi_A = e_{AB} r^{:B}$ (which exists according to the Birkhoff's theorem) can be written in the form

$$\xi_A = -\beta\, t_{,A}, \tag{6.2.3}$$

where $t = t(x^A)$. By construction $\gamma^{AB} t_{,A} r_{,B} = 0$, and hence, in (t, r) coordinates

$$d\gamma^2 = \gamma_{tt}\, dt^2 + \gamma_{rr}\, dr^2.$$

Using the relation $e_{AC}\, e^{BC} = -\delta_A{}^B$ one obtains

$$\beta^2 \gamma^{tt} = \xi_A \xi^A = -r_{:A} r^{:A} = -\gamma^{rr}. \tag{6.2.4}$$

We denote $\gamma^{rr} = g$, then in (t, r) coordinates the metric and the Killing vector have the form

$$d\gamma^2 = -\frac{\beta^2}{g} dt^2 + \frac{dr^2}{g}, \qquad \xi^A = \frac{g}{\beta} \delta^A{}_t. \tag{6.2.5}$$

It is easy to see that g does not depend on t ($\partial_t g = 0$). Indeed,

$$\xi^A g_{:A} = 2\xi^A r_{:AB} r^B \sim \xi^A r^B \gamma_{AB} = 0. \tag{6.2.6}$$

The Killing equations $\xi_{(A:B)} = 0$ imply

$$-\beta_{(:B} \delta^t_{A)} + \Gamma^t_{AB} \beta = 0. \tag{6.2.7}$$

The non-vanishing components of the 2D Christoffel symbols Γ^t_{AB} are

$$\Gamma^t_{tt} = \frac{\dot{\beta}}{\beta}, \quad \Gamma^t_{tr} = \frac{\beta'}{\beta} - \frac{g'}{2g}, \quad (\ldots)\dot{} \equiv \partial_t(\ldots), \quad (\ldots)' \equiv \partial_r(\ldots). \tag{6.2.8}$$

The only non-trivial relation Eq. (6.2.7) (for $A = t$ and $B = r$) gives

$$\frac{1}{2} \beta' - \frac{g'}{2g} \beta = 0. \tag{6.2.9}$$

A solution of this equation is $\beta = \beta_0(t) g$. Using the transformation $t \to t(t')$ one can make $\beta_0 = 1$. Thus,

$$d\gamma^2 = -g \, dt^2 + \frac{dr^2}{g}. \tag{6.2.10}$$

To summarize, we showed that any spherically symmetric solution of the Einstein equations satisfying the condition Eq. (6.1.25) of the generalized Birkhoff's theorem in the region where ξ is a time-like vector can be written in the form

$$ds^2 = -g \, dt^2 + \frac{dr^2}{g} + r^2 \, d\omega^2, \qquad g = g(r). \tag{6.2.11}$$

We consider now a vacuum solution with a cosmological constant. To find the function $g(r)$, it is sufficient to use the trace of Eq. (6.1.18)

$$\mathcal{E}^{AB} \gamma_{AB} = 0, \tag{6.2.12}$$

which has the form

$$r r^{:A}_{:A} + r_{:A} r^{:A} - 1 + \Lambda r^2 = 0. \tag{6.2.13}$$

One has

$$\sqrt{-\gamma} = 1, \qquad r_{:A} r^{:A} = g, \qquad r^{:A}_{:A} = g', \tag{6.2.14}$$

and Eq. (6.2.13) takes the form

$$r g' + g - 1 + \Lambda r^2 = 0. \tag{6.2.15}$$

The general solution to this equation is

$$g = 1 - \frac{r_S}{r} - \frac{\Lambda}{3} r^2. \tag{6.2.16}$$

For $\Lambda > 0$ the solution Eqs. (6.2.11) and (6.2.16) is called a Schwarzschild–de Sitter metric, and the *Schwarzschild–anti-de Sitter metric* for $\Lambda < 0$. When $\Lambda = 0$ this is the *Schwarzschild spacetime*.

6.2.2 Schwarzschild metric

Schwarzschild radius

The cosmological constant at the present epoch is very small, so that for the discussion of astrophysical black holes one can take $\Lambda = 0$. The corresponding *Schwarzschild metric*, is asymptotically flat. Near infinity this metric can be approximated by the linearized gravity solution with the Newtonian potential $\varphi = -M/r$. This allows one to conclude that M is the mass of the gravitating object. The constant M also coincides with the *Komar mass* for this system. The radius $r_S = 2M$ is known as the *gravitational radius* or the *Schwarzschild radius*. In physical units, after restoring G and c constants, it has the value

$$r_S = \frac{2GM}{c^2}. \tag{6.2.17}$$

The *Schwarzschild metric*

$$ds^2 = -\left(1 - \frac{r_S}{r}\right) dt^2 + \left(1 - \frac{r_S}{r}\right)^{-1} dr^2 + r^2 d\omega^2 \tag{6.2.18}$$

describes the gravitational field in vacuum, outside a spherical distribution of matter. This matter may be either static or have radial motion preserving the spherical symmetry. According to *Birkhoff's theorem*, the external metric does not depend on such motion. In the absence of matter, the metric Eq. (6.2.18) describes an exterior of a spherically symmetric static black hole. In this case, r_S is the radius of its *event horizon*.

In the Newtonian theory a vacuum solution outside a point-like source is regular for $r > 0$ and it has a singularity at the location of the source. The Schwarzschild solution is quite different. It is also singular at $r = 0$, but it contains an additional apparent 'singularity' at the gravitational radius $r = r_S$, where $g_{tt} = (g^{rr})^{-1} = 0$. Using the relation Eq. (6.2.4)

$$\xi_A \xi^A = -r_{;A} r^{;A} = -(1 - r_S/r) \tag{6.2.19}$$

one can conclude that at the surface $r = r_S$ the Killing vector ξ becomes null (or vanishes). Outside $r = r_S$ the coordinate t is the time-like coordinate and r is the spatial one. Inside the Schwarzschild radius the gradient of r is time-like, and hence, r can be used as a time coordinate, while the coordinate t orthogonal to it becomes a spatial one.

The null surface $r = r_S$ separating the black hole exterior and interior is, in fact, a regular surface of the spacetime manifold. This can be tested by calculating curvature invariants. For example, the so-called *Kretschman invariant* for the Schwarzschild metric

$$\mathcal{R}^2 \equiv R_{\mu\nu\lambda\rho}R^{\mu\nu\lambda\rho} = \frac{12r_S^2}{r^6} \tag{6.2.20}$$

is finite at the gravitational radius. The apparent degeneracy of the metric on the Schwarzschild radius is the result of the choice of the coordinates. In the other coordinate systems the horizon is perfectly regular. At the center of the black hole at $r = 0$ the curvature is infinite. Near this sigularity the tidal forces grow infinitely. This is a physical singularity. It cannot be removed by coordinate transformation.

Schwarzschild metric in the vicinity of the horizon

To understand better what is 'wrong' with the Schwarzschild coordinates (t, r) near the gravitational radius it is instructive to analyze the asymptotic form of the metric Eq. (6.2.18) near r_S. Consider the $(t$-$r)$-sector of the Schwarzschild metric

$$d\gamma^2 = -g\,dt^2 + \frac{dr^2}{g} \tag{6.2.21}$$

in the vicinity of the horizon: $r = r_S(1 + y)$, $y \ll 1$. In this region $g \simeq y$. We denote by ρ the proper length distance from the horizon

$$\rho = \int_{r_S}^{r} \frac{dr}{\sqrt{g}}. \tag{6.2.22}$$

For small y one has

$$\rho \simeq 2r_S\sqrt{y}, \tag{6.2.23}$$

and, hence, the metric takes the form

$$d\gamma^2 = -\kappa^2\,\rho^2\,dt^2 + d\rho^2, \tag{6.2.24}$$

where $\kappa = 1/(2r_S) = 1/(4M)$ is the *surface gravity* of the black hole.

The near-horizon metric Eq. (6.2.24) is simply the *2D* Rindler metric. The complete metric of the Schwarzschild spacetime near the horizon has the asymptotic form

$$ds^2 \approx -\kappa^2\,\rho^2\,dt^2 + d\rho^2 + r_S^2\,d\omega^2. \tag{6.2.25}$$

In the transverse to the $(t - r)$ directions this is a geometry of a sphere of the radius r_S. If we consider a region that in the transverse direction has a size much smaller than r_S, then this sphere can be approximated by a 2D plane. In such a near-horizon region this approximation gives the *4D* Rindler metric.

Main facts of near-horizon physics

Many properties of the black hole in the near-horizon region directly follow from the analysis in the Rindler space (see Chapter 2).

1. For a particle (light ray) falling into a black hole it takes a finite proper time (affine parameter) to reach the event horizon.

2. The time t measured by an external observer for the same process is infinitely large.

3. The redshift factor for the light emitted by an object freely falling into the black hole as measured by a distant observer, is $\sim e^{-\kappa t}$ (see, e.g., Eq. (2.4.24)), where κ is the surface gravity of the black hole.

4. Infinite redshift surface ($\xi_{(t)}^2 = 0$) coincides with the event horizon.

The spacetime described in the Schwarzschild coordinates is *geodesically incomplete*. The Rindler approach allows one to conclude that beyond the Schwarzschild horizon there exists a continuation of the geometry. In particular, one can expect that there must exist regions where the Killing vector becomes space-like.

6.2.3   Embedding of the Schwarzschild metric in a flat space

Before we consider the global properties of spherically symmetric black holes, we make a few remarks on the problem of embedding for the Schwarzschild solution in a flat spacetime. Let $\mathbb{R}^6$ be a 6D space with the metric

$$dS^2 = -(dz^1)^2 - (dz^2)^2 + (dz^3)^2 + (dz^4)^2 + (dz^5)^2 + (dz^6)^2. \tag{6.2.26}$$

Consider a 4D subspace M^4 in $\mathbb{R}^6$ defined by the relations $(r > r_S)$

$$z^1 = r_S\sqrt{\frac{r - r_S}{r}} \cos\left(\frac{t}{r_S}\right), \qquad z^2 = r_S\sqrt{\frac{r - r_S}{r}} \sin\left(\frac{t}{r_S}\right), \qquad z^3 = F(r),$$

$$z^4 = r\sin\theta\cos\phi, \qquad\qquad z^5 = r\sin\theta\sin\phi, \qquad\qquad z^6 = r\cos\theta. \tag{6.2.27}$$

Here, $F(r)$ is a solution of the equation

$$\left(\frac{dF}{dr}\right)^2 = \frac{1}{r - r_S}\left(\frac{r_S^4}{4r^3} + r\right). \tag{6.2.28}$$

Problem 6.4:   *Show that the Schwarzschild metric Eq. (6.2.18) in its exterior domain $r > r_S$ coincides with the metric on M^4 induced by the embedding Eq. (6.2.27) (Eisenhart 1966).*

Notes on the embedding problem. *The above result is an example of a general embedding problem. Suppose we have manifold M^n with a metric $g_{\mu\nu}$. Consider an N-dimensional linear space $\mathbb{R}^N$ with the metric*

$$dS^2 = \sum_{a=1}^{N} c_a(dz^a)^2, \tag{6.2.29}$$

where c_a are constants that take values ± 1. Consider its n-dimensional subspace $z^a = z^a(x^\mu)$. The space M^n is isometrically embedded into $\mathbb{R}^N$ if the induced metric on it coincides with $g_{\mu\nu}$

$$\sum_{a=1}^{N} c_a \frac{\partial z^a}{\partial x^\mu}\frac{\partial z^a}{\partial x^\nu} = g_{\mu\nu}. \tag{6.2.30}$$

This is a system of first-order partial differential equations. The system contains $n(n + 1)/2$ equations for N functions $z^a(x^\mu)$. In a generic case such a system has a solution if $N = n(n + 1)/2$. Thus, the general M^n can be immersed in a flat space $\mathbb{R}^N$ when this condition is satisfied (Eisenhart 1966).
    By a coordinate transformation the metric $g_{\mu\nu}$ at any point $p \in M^n$ can be put into the form $g_{\mu\nu} \doteq \eta_{\mu\nu}$. At this point n of the functions $z^a(x)$ can be identified with x^μ. Hence, Eq. (6.2.30) must

have at least as many positive and as many negative coefficients c_a as there are positive and negative terms in the induced metric at this point *(Eisenhart 1966)*.

For the Riemannian geometry with a positive definite metric the following Cartan–Janet theorem *(Cartan 1927; Janet 1926) is valid: Every real analytic Riemannian manifold of dimension n can be locally real analytically isometrically embedded into* $\mathbb{E}^N$ *with* $N = n(n+1)/2$. *The so-called* fundamental theorem of Riemannian geometry *(Nash 1956; Berger 2003) states that every smooth Riemannian manifold of dimension n can be smoothly isometrically embedded in a Euclidean space* $\mathbb{E}^N$ *with* $N = (n+2)(n+3)/2$.

6.2.4 Euclidean black hole

In the 'physical' spacetime the signature of the metric is $(-,+,+,+)$. The Schwarzschild metric is static and it allows an analytical continuation to the Euclidean one. This continuation can be obtained by making the Wick's rotation $t = it_E$. The corresponding space, called a *Euclidean black hole*,[2] has interesting mathematical properties and has important physical applications. We demonstrate now that by a proper choice of the period of the Euclidean time t_E the metric can be made regular.

The two-dimensional part of the Euclidean metric in the (t_E, r)-sector is

$$d\gamma_E^2 = g\, dt_E^2 + \frac{dr^2}{g}. \tag{6.2.32}$$

Near the *Euclidean horizon* $r = r_S$ is has the form

$$d\gamma_E^2 \approx \kappa^2 \rho^2\, dt_E^2 + d\rho^2, \tag{6.2.33}$$

where $\kappa = 1/(2r_S) = 1/(4M)$ is the *surface gravity*. In the general case this metric has a *conical singularity*. This singularity vanishes if t_E is a periodic coordinate with the period

$$t_E \in \left(0, \frac{2\pi}{\kappa}\right). \tag{6.2.34}$$

The metric

$$ds_E^2 = d\gamma_E^2 + r^2\, d\omega^2 \tag{6.2.35}$$

with the periodicity property Eq. (6.2.34) describes a regular Euclidean space that has the topology $\mathbb{R}^2 \times S^2$. This regular four-dimensional Euclidean space (see Figure 6.1) is called the *Euclidean black hole* or the *Gibbons–Hawking instanton*. The following quantity

$$\Theta_H = \frac{\kappa}{2\pi} \tag{6.2.36}$$

is called the *Hawking temperature* of the black hole.

---

[2] The Euclidean black hole is a solution of the Euclidean version of the Einstein equations. These equations can be obtained from the Einstein action (see Appendix E)

$$S_E[g] = -\frac{c^3}{16\pi G} \int (R - 2\Lambda)\, \sqrt{g}\, d^D x. \tag{6.2.31}$$

Note that this action has a different total sign in comparison with Eq. (5.1.6).

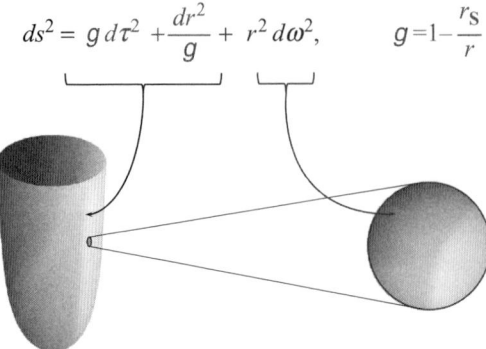

$$ds^2 = g \, d\tau^2 + \frac{dr^2}{g} + r^2 \, d\omega^2, \qquad g = 1 - \frac{r_S}{r}$$

Fig. 6.1 Gibbons–Hawking instanton.

We shall see in Charter 9 that the Hawking temperature plays an important role in black hole physics. Leaving this discussion for later we just give some hints here why the quantity Θ_H is called the temperature. It is well known that in quantum field theory the thermal Green functions for the state of the field with the temperature Θ can be obtained by the analytic continuation of the Euclidean Green functions calculated on a space with a periodic Euclidean time, provided this period is Θ^{-1}. We shall demonstrate in Chapter 9 that in the case of the black hole there exists an equilibrium state of a quantum field in the black hole geometry, provided this state is thermal and its temperature is Eq. (6.2.36). In the absence of the surrounding thermal bath, the black hole is a source of thermal radiation with the Hawking temperature.

6.3   Global Structure of the Schwarzschild Spacetime

6.3.1   Kruskal metric

The Schwarzschild metric does not cover a complete spacetime. In this sense, the Schwarzschild coordinates (t, r) are similar to the *Rindler coordinates* (τ, ρ) acting in the R_+ domain. In this section we describe an analog of null coordinates $U = (X^0 - X^1)/\sqrt{2}$ and $V = (X^0 + X^1)/\sqrt{2}$ that cover the complete manifold of the Schwarzschild black hole. They are called *Kruskal coordinates*.

In order to obtain a solution that covers the complete vacuum spherically symmetric spacetime, we use the coordinate-fixing conditions $g_{00} = g_{11} = 0$ and write the metric as

$$ds^2 = 2 B \, dU \, dV + r^2 \, d\omega^2, \qquad B = B(U, V), \qquad r = r(U, V). \tag{6.3.1}$$

It is convenient to use the dimensionless form of this metric:

$$d\tilde{s}^2 = r_S^{-2} \, ds^2 = 2 b \, dU \, dV + \tilde{r}^2 \, d\omega^2, \qquad r_S = 2M, \qquad b = \frac{B}{r_S^2}, \qquad \tilde{r} = \frac{r}{r_S}. \tag{6.3.2}$$

The non-vanishing Christoffel symbols calculated for the metric Eq. (6.3.2) are

$$\Gamma^U_{UU} = \frac{b_{,U}}{b}, \qquad \Gamma^V_{VV} = \frac{b_{,V}}{b}, \tag{6.3.3}$$

and the vacuum Einstein equations Eq. (6.1.18) written in dimensionless form are

$$\tilde{r}_{;UU} = \tilde{r}_{;VV} = 0, \tag{6.3.4}$$

$$\tilde{r}\,\tilde{r}_{;A}^{:A} + \tilde{r}_{;A}\tilde{r}^{:A} - 1 = 0. \tag{6.3.5}$$

Equations (6.3.4), written in explicit form

$$\partial_U^2 \tilde{r} - \frac{b_{,U}}{b}\,\partial_U \tilde{r} = 0, \qquad \partial_V^2 \tilde{r} - \frac{b_{,V}}{b}\,\partial_V \tilde{r} = 0, \tag{6.3.6}$$

have first integrals

$$\partial_U \tilde{r} = \alpha(V)\,b, \qquad \partial_V \tilde{r} = \beta(U)\,b. \tag{6.3.7}$$

Using the coordinate freedom $U \to \tilde{U} = F(U)$ and $V \to \tilde{V} = G(V)$, one can always put $\alpha = \frac{1}{2}V$ and $\beta = \frac{1}{2}U$. Then, Eqs. (6.37) imply

$$\dot{\tilde{r}} = \frac{1}{2}\,b, \qquad \tilde{r} = \tilde{r}(z), \qquad b = b(z), \tag{6.3.8}$$

where $z = UV$, and the dot means the derivative with respect to z. Equation (6.3.5) now reads

$$\tilde{r}\left(1 + \frac{\tilde{r}\,\ddot{\tilde{r}}}{\dot{\tilde{r}}}\right) + z\,\dot{\tilde{r}} - 1 = 0. \tag{6.3.9}$$

Let us define $Y = (z/z') - 1$, where $(\ )' = d(\ )/d\tilde{r}$, then Eq. (6.3.9) can be written in the form

$$\tilde{r}\,Y' + Y = 0. \tag{6.3.10}$$

We have $Y = c/\tilde{r}$. For $c = 0$, the metric Eq. (6.3.2) is flat. If $c \neq 0$, one can use a scaling transformation $\tilde{r} \to C\tilde{r}$ to put $c = -1$. This gives

$$\frac{z'}{z} = \frac{\tilde{r}}{\tilde{r} - 1}. \tag{6.3.11}$$

The sign of z changes under the reflection $U \to -U$. We choose the sign of z so that the solution of the above equation is

$$z = -(\tilde{r} - 1)\exp(\tilde{r} - 1). \tag{6.3.12}$$

Finally, we get

$$-UV = (\tilde{r} - 1)\exp(\tilde{r} - 1), \qquad b = -\frac{2}{\tilde{r}}\exp[-(\tilde{r} - 1)]. \tag{6.3.13}$$

Thus, the spacetime metric, known as the *Kruskal metric*, is

$$ds^2 = 2B\,dU\,dV + r^2\,d\omega^2, \tag{6.3.14}$$

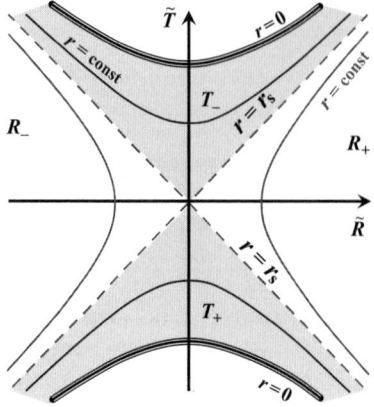

Fig. 6.2 Spherically symmetric vacuum spacetime in Kruskal coordinates.

where

$$-UV = \left(\frac{r}{2M} - 1\right) \exp\left(\frac{r}{2M} - 1\right),$$

$$B = -\frac{16M^3}{r} \exp\left[-\left(\frac{r}{2M} - 1\right)\right].$$
(6.3.15)

Light trajectories for the radial motion are defined by the equations $V = $ const (for an incoming ray) and $U = $ const (for an outgoing one).

Instead of the null coordinates U and V it is possible to introduce time-like and space-like coordinates $\tilde{T} = (V + U)/\sqrt{2}$ and $\tilde{R} = (V - U)/\sqrt{2}$. In these coordinates the metric takes the form the

$$ds^2 = \frac{2r_S^3}{r} e^{-(r/r_S - 1)} (-d\tilde{T}^2 + d\tilde{R}^2) + r^2(d\theta^2 + \sin^2\theta \, d\phi^2),$$
(6.3.16)

where r is a function of $\tilde{T}$ and $\tilde{R}$:

$$2\left(\frac{r}{r_S} - 1\right) e^{(r/r_S - 1)} = \tilde{R}^2 - \tilde{T}^2.$$
(6.3.17)

The Kruskal metric Eq. (6.3.14) is invariant under the following discrete symmetries

$$\mathbf{I}: U \to -U, \qquad V \to -V,$$

$$\mathbf{T}: U \to -V, \qquad V \to -U,$$
(6.3.18)

$$\mathbf{P}: U \to V, \qquad V \to U.$$

The global structure of the *Kruskal spacetime* is illustrated in Figure 6.2. Each point of this diagram is a 2D sphere. $\mathbf{I}$ is a reflection with respect to the point $U = V = 0$. It maps $R_+ \leftrightarrow R_-$ and $T_+ \leftrightarrow T_-$.

6.3.2 Carter–Penrose diagram

Figure 6.2 is very useful for the discussion of the causal properties of the global solution. Straight lines $U = $ const and $V = $ const on this diagram are null. The future-directed local null cones on this 2D diagram are defined by relations

$$dU = 0, \quad dV \geq 0, \qquad \text{and} \qquad dV = 0, \quad dU \geq 0. \qquad (6.3.19)$$

Any radial future-directed time-like or null vector lies either inside the local null cone Eq. (6.3.19) or on its boundary.

There is a special modification of the spacetime diagrams, similar to the Kruskal one, which makes the global causal structure of the spacetime, including its properties at infinity, more profound. Let us denote

$$\mathcal{U} = \arctan U, \qquad \mathcal{V} = \arctan V. \qquad (6.3.20)$$

This coordinate transformation brings 'infinities' $U = \pm\infty$ and $V = \pm\infty$ to the finite values in the coordinates space. Since for constant values of U and V the new coordinates are again constant, they are also null. The local null cones in the new coordinates are again given by straight lines at the angle $\pm\pi/4$ with respect to a horizontal line. The Kruskal diagram in the coordinates $(\mathcal{U}, \mathcal{V})$ is shown in Figure 6.3. Such spacetime diagrams where infinity is brought to finite coordinate distance are known as *Carter–Penrose conformal diagrams*.

Relations Eq. (6.3.20) imply that

$$\mathcal{U} = \tan \mathcal{U}, \qquad dU = \frac{d\mathcal{U}}{\cos^2 \mathcal{U}},$$
$$\qquad (6.3.21)$$
$$V = \tan \mathcal{V}, \qquad dV = \frac{d\mathcal{V}}{\cos^2 \mathcal{V}}.$$

The Kruskal metric in the new coordinates $(\mathcal{U}, \mathcal{V})$ has the form

$$ds^2 = \frac{2 B \, d\mathcal{U} \, d\mathcal{V}}{\cos^2 \mathcal{U} \, \cos^2 \mathcal{V}} + r^2 \, d\omega^2. \qquad (6.3.22)$$

The metric component $g_{\mathcal{U}\mathcal{V}}$, as well as r^2, becomes infinitely large at the 'boundary' $\mathcal{U} = \pm\pi/2$ and $\mathcal{V} = \pm\pi/2$. This is the 'price' one has to pay for bringing the 'infinities' to a finite coordinate distance.

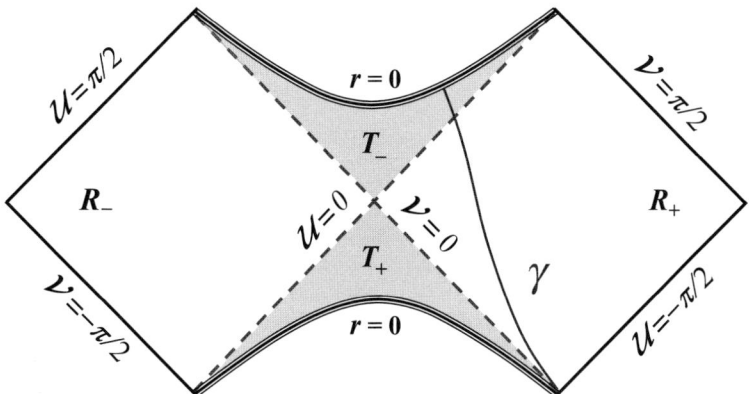

Fig. 6.3 Carter–Penrose diagram for the vacuum spacetime.

Penrose demonstrated that for a wide class of so called asymptotically flat spacetimes such asymptotic behavior of the metric is universal, so that one can write the metric ds^2 in the form

$$ds^2 = \Omega^2 d\bar{s}^2, \tag{6.3.23}$$

where the conformal factor Ω diverges at the points of infinity, while the coordinates in $d\bar{s}^2$ and the metric itself remain finite. We discuss asymptotically flat spacetimes and the properties of their conformal infinity in Section 10.1.

6.3.3   R and T domains in Kruskal spacetime

The Killing vector $\boldsymbol{\xi}_{(t)}$ in Kruskal coordinates is

$$\xi_{(t)}^{\mu} \partial_{\mu} = (1/4M)(V \partial_V - U \partial_U). \tag{6.3.24}$$

It is null at the null 3D surfaces (horizons) $V = 0$ and $U = 0$. At the intersection of the horizons, that is when $U = V = 0$, the Killing vector vanishes. This is a 2D sphere of the radius r_S is called a *horizon bifurcation surface*.

Problem 6.5:   *Show that the extrinsic curvature tensor (see Eq. (3.8.12)) of the horizon bifurcation surface vanishes, and hence, it is a* geodesic submanifold.

$\boldsymbol{\xi}_{(t)}$ is time-like in the regions $R_\pm$ and space-like in $T_\pm$. It is future-directed in R_+ and past-directed in R_-. The coordinate r is spatial in $R_\pm$. In R_+ it changes from r_S (at the horizon) to infinity. Inside $T_\pm$ domains a surface $r = \text{const}$ is space-like, and hence $r_{,\mu}$ is a time-like vector. r grows monotonically from 0 to r_S in T_+ and decreases monotonically from r_S to 0 in T_-.

The spacetime has the singularity at $r = 0$. Motion of particles and light in the region T_- always results in a decrease of r. It continues until the particle reaches the singularity. By studying local null cones one can show that a future-directed causal line that begins in the T_- region can never reach the R_+ domain. This means that an external observer 'living' in R_+ region cannot get any information from the T_- region. This region is the *black hole interior*.

In different regions of the spacetime, Kruskal coordinates (U, V) are related to Schwarzschild coordinates (t, r) as follows: In the wedges $R_\pm \equiv R_\epsilon$: $\epsilon U < 0$, $\epsilon V > 0$, the coordinate transformations are

$$\begin{cases} U = -\epsilon \left(\dfrac{r}{2M} - 1\right)^{\frac{1}{2}} \exp\left[\dfrac{1}{2}\left(\dfrac{r}{2M} - 1\right) - \dfrac{t}{4M}\right], \\[3mm] V = \epsilon \left(\dfrac{r}{2M} - 1\right)^{\frac{1}{2}} \exp\left[\dfrac{1}{2}\left(\dfrac{r}{2M} - 1\right) + \dfrac{t}{4M}\right]. \end{cases} \tag{6.3.25}$$

In the wedges $T_\pm \equiv T_\epsilon$: $\epsilon U < 0$, $\epsilon V < 0$, the coordinate transformations are

$$\begin{cases} U = -\epsilon \left(1 - \dfrac{r}{2M}\right)^{\frac{1}{2}} \exp\left[\dfrac{1}{2}\left(\dfrac{r}{2M} - 1\right) - \dfrac{t}{4M}\right], \\[3mm] V = -\epsilon \left(1 - \dfrac{r}{2M}\right)^{\frac{1}{2}} \exp\left[\dfrac{1}{2}\left(\dfrac{r}{2M} - 1\right) + \dfrac{t}{4M}\right]. \end{cases} \tag{6.3.26}$$

In the regions $R_\pm$ the Schwarzschild metric is

$$ds^2 = -g\,dt^2 + \frac{dr^2}{g} + r^2\,d\omega^2, \qquad g = 1 - \frac{r_S}{r}, \qquad r_S = 2M. \tag{6.3.27}$$

In the regions $T_\pm$ it takes the form

$$ds^2 = -\frac{dr^2}{\tilde{g}} + \tilde{g}\,dt^2 + r^2\,d\omega^2, \qquad \tilde{g} = \frac{r_S}{r} - 1. \tag{6.3.28}$$

We wrote this metric in the form that makes it clear that in $T_\pm$, where $r < r_S$, the Killing 'time' coordinate is in fact space-like, while the radial coordinate r is time-like. We return to the properties of the metric in $T_\pm$ domains later.

> *The Kruskal metric describes an empty spacetime. Sometimes it is called an eternal black hole. In fact, besides the black hole interior T_- and the exterior R_+, it contains the region T_+, called a white hole, and R_-. The physical meaning of these regions can be clarified by analyzing concrete models of the matter source, generating the spherically symmetric solution. For example, consider a static spherically symmetric star of radius $r_0 > r_S$, and suppose that at some moment of time $t = 0$ it loses its stability and shrinks. Suppose that as a result of a gravitational collapse, its surface radius decreases and crosses the Schwarzschild radius. Further evolution of the star is an inevitable continuous contraction, until the curvature becomes infinitely large and a singularity is formed.*
>
> *The vacuum metric is valid only outside the surface of the collapsing body. This surface is represented on the Kruskal diagram (see Figure 6.3) by a time-like curve γ. Before $t = 0$ this line coincides with $r = r_0$, while for $t > 0$ γ enters T_- region and reaches the singularity. The Kruskal diagram must be cut along this line and the Kruskal metric is valid only in the external region located to the right of the curve γ. To determine the complete metric one needs to solve the Einstein equations inside the matter (to the left of γ) and to glue this solution with the Kruskal metric along the line γ. As a result of this procedure, the regions T_+ and R_- do not appear in the complete solution.*
>
> *It is clear that one may also consider a matter, e.g., a dust ball, which begins its expansion in the T_+ domain, crosses the Schwarzschild surface, reaches the maximal expansion in R_+, and then contracts until it enters the T_- region and collapses to the singularity. In order to stitch this solution to the Kruskal metric, one uses parts of the domains, $T_\pm$ and R_+. A part of the ball history, when it moves in the T_+ domain corresponds to the white hole, while further collapse into the T_- region describes the formation of the black hole. There exist, at least formally, solutions for a similar dust ball, which after expansion in the T_+ domain, enters the R_- region. For such a configuration, known as a semi-closed world, one uses all four domains of the Kruskal diagram. It should be emphasized that at the stage of the white hole classical and quantum instabilities are present. Whether such objects as white holes and semi-closed worlds exist in our universe is an open question. For more detail, see, e.g., (Zel'dovich and Novikov 1971a, 1971b; Frolov and Novikov 1998).*

6.3.4 Einstein–Rosen bridge

The equation $U = -V$ determines a three-dimensional space-like slice of the Kruskal spacetime. This slice passes through the bifurcation surface of the horizons. It has two branches $\Sigma_\pm$ located in the $R_\pm$ domains, respectively. This subspace $\Sigma = \Sigma_+ \cup \Sigma_-$ is called the *Einstein–Rosen bridge* (see Figure 6.4). It has the topology $\mathbb{R} \times S^2$. The internal geometry of Σ is

$$dl^2 = \frac{dr^2}{1 - 2M/r} + r^2\,d\omega^2 = \Omega^4\,dl_0^2, \tag{6.3.29}$$

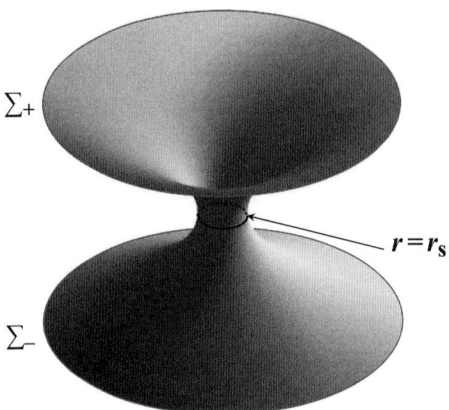

Σ_+

$r = r_s$

Σ_-

Fig. 6.4 The embedding diagram for the *Einstein–Rosen bridge*.

$$\Omega = 1 + \frac{M}{2\rho}, \qquad dl_0^2 = d\rho^2 + \rho^2(d\theta^2 + \sin^2\theta\, d\phi^2). \tag{6.3.30}$$

Here, r is defined as a function of ρ by the relation

$$r = \rho\,(1 + M/2\rho)^2. \tag{6.3.31}$$

r has the minimum $r = 2M$ at $\rho = M/2$, corresponding to the *throat of the Einstein–Rosen bridge*, which is the junction surface of the branches Σ_+ and Σ_-. The region $(0 < \rho < M/2)$ corresponds to Σ_-, while the region $(M/2 < \rho < \infty)$ corresponds to Σ_+. r grows infinitely when either $\rho \to 0$ or $\rho \to \infty$.

The geometry of a two-dimensional section $\theta = \pi/2$ of the metric Eq. (6.3.29) can be embedded in a flat three-dimensional space as a revolution surface

$$dL^2 = dz^2 + dr^2 + r^2\, d\phi^2 = dr^2(1 + z'^2) + r^2\, d\phi^2, \tag{6.3.32}$$

where $z = \pm 2\sqrt{2M(r - 2M)}$. The corresponding *embedding diagram* is shown in Figure 6.4.

6.4   Black Hole Interior

A black hole interior is a mysterious region of the Schwarzschild spacetime. Matter that has entered the interior region T_- cannot escape from it and come back to R_+. No exotic cataclysmic events can change this conclusion, unless the basic principle of the causality is violated. If we believe that the speed of light is the highest possible speed of propagation of information-carrying signals, this conclusion is absolute. From a 'practical' point of view, what happens inside a black hole is not important, since it does not affect observations performed in the black hole exterior. This conclusion is correct at least for classical phenomena. For the discussion of quantum effects in black holes, and especially of some fundamental problems, such as the information-loss 'paradox', the physics in the inner region of the black hole might become important. Let us look at some properties of the *black hole interior*.

The Schwarzschild metric in the region $r < r_S = 2M$ is

$$ds^2 = -\frac{dr^2}{(r_S/r) - 1} + \left[(r_S/r) - 1\right]dz^2 + r^2\,d\omega^2. \qquad (6.4.1)$$

In order to make it more transparent that now the coordinate t is space-like we change the notation t to $z \in (-\infty, \infty)$. One can introduce the proper time coordinate τ

$$\tau = -\int_{r_S}^{r} \frac{\sqrt{r}\,dr}{\sqrt{r_S - r}}. \qquad (6.4.2)$$

Performing the integration one gets

$$\tau = \frac{\pi}{4}r_S - \frac{r_S}{2}\arctan\frac{2r - r_S}{2\sqrt{r(r_S - r)}} + \sqrt{r(r_S - r)}. \qquad (6.4.3)$$

We choose the integration constant so that $\tau = 0$ at $r = r_S$. At $r = 0$ the proper time $\tau = \tau_0 \equiv r_S\pi/2$. At a given τ the three-dimensional space slice has the topology of $R^1 \times S^2$. Formally, the metric of the black hole interior is similar to a solution for a homogeneous anisotropic contracting universe.

> *If outside a black hole there is a matter distribution or a source of some physical field, the Schwarzschild metric is distorted. For static sources this distortion is also static in the black hole exterior. If the distribution violates the spherical symmetry, the symmetry of the metric is also broken. As a result, the horizon geometry will differ from the geometry of a round sphere. The deformation of the metric in the black hole exterior also modifies the internal geometry. For static sources this deformation is invariant under displacement along z, but it will depend on r, i.e. on the time inside the black hole. In this sense, the problem of the study of the black hole interior is a dynamical problem. (For more details see, e.g., (Chandrasekhar 1983; Frolov and Shoom 2007).)*

Since r is a monotonically decreasing function of τ, any motion in the black hole interior is the motion to the 'center', $r = 0$. Along any such trajectory the Kretschmann

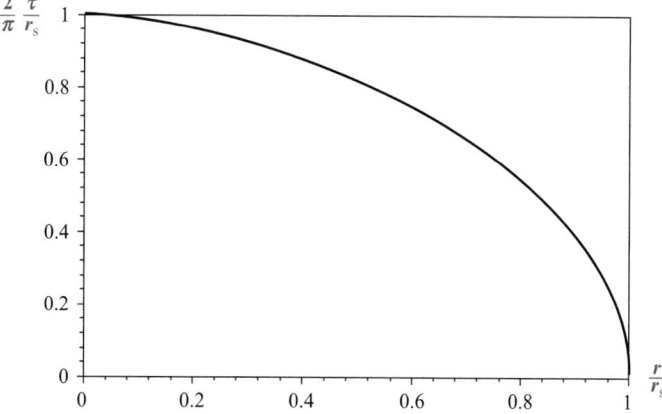

Fig. 6.5 Proper time τ inside the black hole as a function of r/r_S.

curvature-invariant $\mathcal{R}^2 = R_{\alpha\beta\gamma\delta} R^{\alpha\beta\gamma\delta} = 12\, r_S^2/r^6$ grows. Thus, all the matter, originally collapsed into the black hole, becomes more and more compressed.

At $r \ll r_S$ the metric can be approximated as

$$ds^2 \approx -d\tau^2 + \left(\frac{3(\tau_0 - \tau)}{4M}\right)^{-2/3} dz^2 + \left(\frac{3(\tau_0 - \tau)}{4M}\right)^{4/3} (2M)^2\, d\omega^2. \tag{6.4.4}$$

As proper time τ grows, the radius $r \approx (\tau_0 - \tau)^{-1/3}$ decreases.

The spacetime inside the black hole can be described as an evolution of anisotropic homogeneous three-dimensional space. The metric has Kasner-type behavior near the singularity: the contraction of space in two directions, (θ, ϕ), is accompanied by expansion in the third direction, z.

> *Such behavior is a consequence of the classical equations that are valid until the space-time curvature becomes comparable to the Planck one. Particle creation and vacuum polarization may change this regime. At any rate, the existence of the region with infinitely growing curvature indicates that general relativity is not sufficient to give a complete picture and it must be upgraded to describe properly the spacetime near the singularity. Quantum gravity (for example, in the form of the string theory) is required.*

6.5   Painlevè–Gullstrand Metric

The Schwarzschild spacetime can be represented in many different coordinate systems, each being convenient for different applications. In the acoustic analog models of gravity (see Section 10.4) the *Painlevè–Gullstrand metric* appears naturally. The Schwarzschild geometry in the *Painlevè–Gullstrand coordinates* is

$$ds^2 = -g\, d\tilde{t}^2 \pm 2\sqrt{1-g}\, d\tilde{t}dr + dr^2 + r^2\, d\omega^2, \tag{6.5.1}$$

where for the Schwarzschild black hole

$$g = 1 - \frac{r_S}{r}, \qquad r_S = 2M \tag{6.5.2}$$

and

$$g = 1 - \frac{r_S}{r} - \frac{\Lambda}{3} r^2, \tag{6.5.3}$$

in the case of the Schwarzschild–(anti) de Sitter spacetime. The Painlevè–Gullstrand coordinate $\tilde{t}$ is related to the Schwarzschild coordinates by

$$\tilde{t} = t \pm \int \frac{\sqrt{1-g(r)}}{g(r)}\, dr. \tag{6.5.4}$$

Here, the sign $'+'$ corresponds to the coordinate patch covering the usual Schwarzschild domain R_+ and the domain T_- of the maximally extended Schwarzschild spacetime, while the sign $'-'$ corresponds to the domains R_+ and T_+ (see Figure 6.2).

The Painlevè–Gullstrand line element has a very compelling property that all constant-time $\tilde{t}$ slices are flat, i.e. the 3D curvature tensor of every slice vanishes. This property remains valid for an arbitrary function $g(r)$.

6.6 Eddington–Finkelstein Coordinates

Another useful representation of the Schwarzschild black hole is related to coordinates associated with the free-moving radial photons. The geodesics of radially moving photons are described by the equation

$$t = \text{const} \pm \int \frac{dr}{g}, \tag{6.6.1}$$

where the sign $-$ corresponds to ingoing photons and $+$ to outgoing photons. It is convenient to introduce a *tortoise coordinate*

$$r_* \equiv \int \frac{dr}{g} = \int \frac{dr}{1 - \frac{r_S}{r}} = r + r_S \ln \left| \frac{r}{r_S} - 1 \right| \tag{6.6.2}$$

and a new null coordinate

$$v = t + r_* \tag{6.6.3}$$

which is called an *advanced time*. If we use v coordinate instead of time t, the Schwarzschild metric Eq. (6.2.18) takes the form

$$ds^2 = - \left(1 - \frac{r_S}{r}\right) dv^2 + 2\, dv dr + r^2\, d\omega^2, \qquad d\omega^2 = d\theta^2 + \sin^2 \theta\, d\phi^2. \tag{6.6.4}$$

Despite the fact that the g_{vv} component of the metric vanishes on the horizon, this metric is regular because the g_{rv} is finite and makes the metric non-degenerate. The coordinates (v, r, θ, ϕ) are called the *Eddington–Finkelstein coordinates*. These coordinates cover the domain $R_+ \cup T_-$ of the Kruskal spacetime.

Sometimes, it is convenient to use the *retarded-time* coordinate

$$u = t - r_* \tag{6.6.5}$$

instead of the Schwarzschild time t, then the metric reads

$$ds^2 = - \left(1 - \frac{r_S}{r}\right) du^2 - 2\, du dr + r^2\, d\omega^2. \tag{6.6.6}$$

These coordinates are called the outgoing Eddington–Finkelstein coordinates. They cover the domain $R_+ \cup T_+$ of the Kruskal spacetime.

6.7 Charged Black Holes

If an electrically charged particle falls into the Schwarzschild black hole it becomes charged. To describe such a charged black hole one has to solve the Einstein–Maxwell equations and take into account the stress-energy tensor of the electromagnetic field. The spherically symmetric solution of the problem can be found in a similar way as the Schwarzschild solution.

It is easy to check that for the spherically symmetric electric field the condition Eq. (6.1.25) is satisfied, so the generalized Birkhoff's theorem is valid and the metric is of the form Eq. (6.2.11)

$$ds^2 = -g\, dt^2 + g^{-1}\, dr^2 + r^2\, d\omega^2. \tag{6.7.1}$$

Solving the remaining non-trivial Einstein and Maxwell equations one finds

$$g = 1 - \frac{2M}{r} + \frac{Q^2}{r^2} - \frac{\Lambda}{3}r^2, \qquad A_\mu = -\delta_\mu^t \frac{Q}{r}. \tag{6.7.2}$$

For $\Lambda = 0$ this solutions is known as the *Reissner–Nordström spacetime*. Besides the cosmological constant the charged black hole is characterized by two parameters: the mass M and the electric charge Q.

The radius of the horizon of the charged black hole is

$$r_+ = M + \sqrt{M^2 - Q^2}. \tag{6.7.3}$$

This is an analog of the Schwarzschild radius,[3] and for $Q = 0$, $r_+ = r_S$.

For the charged black hole the function g has two roots

$$r_\pm = M \pm \sqrt{M^2 - Q^2}. \tag{6.7.4}$$

The larger one, r_+, is the event horizon, while the smaller one, r_+, is an inner or Cauchy horizon, located inside the black hole.

Note that in astrophysical applications the electric charge is usually negligibly small. This is because the electromagnetic coupling constant is many orders of magnitude stronger than the gravitational one. For two electrons the electromagnetic interaction is proportional to $\alpha = e^2/\hbar c = 1/137.036$, while for their gravitational attraction $Gm_e^2/\hbar c = 1.75 \times 10^{-45}$. Because of this huge disparity electrically charged black holes in the interstellar medium will attract charges of the opposite sign and repel charges of the same sign. Eventually they become almost neutral. The charge Q of such a black hole of mass M obeys the inequality

$$\frac{Q}{M} \leq \frac{m_e}{e} \sim 0.5 \times 10^{-21}. \tag{6.7.5}$$

If there were magnetic monopoles in Nature then black holes could acquire a magnetic charge. The metric of the magnetically charged black hole coincides with Eq. (6.7.1) but the parameter Q is to be understood as the magnetic charge. The vector potential in this case is $A_\mu = -\delta_\mu^\phi Q \cos\theta$.

6.8 Higher-Dimensional Spherical Black Holes

Higher-dimensional vacuum Einstein equations also have solutions describing spherically symmetric black holes. The properties of higher-dimensional black holes are in many aspects similar to those of their 4D 'cousins'. A D-dimensional spherically symmetric spacetime is defined as a space where there exist coordinates in which the metric takes the form

$$ds^2 = \gamma_{AB}dx^A dx^B + r^2 d\omega_{D-2}^2, \qquad (A, B = 0, 1). \tag{6.8.1}$$

---

[3] We denote the gravitational radius of an uncharged static black hole (the Schwarzschild black hole) as r_S. In the case of rotating or charged black holes there are two gravitational horizons: an external horizon at $r = r_+$, which is an analog of the Schwarzschild horizon, and an internal horizon at $r = r_- \leq r_+$, which does not exist in the Schwarzschild case. In the case of the Schwarzschild–de Sitter metric ($\Lambda > 0$) there is also an additional (de Sitter) horizon outside the black hole.

Here, $d\omega_{D-2}^2$ is the metric on a unit $(D - 2)$-dimensional sphere S^{D-2}. Substituting this ansatz into the D-dimensional *Einstein–Hilbert action*, Eq. (5.5.2), and integrating over the angular variables one obtains the reduced action. This action is a functional of the 2D metric γ_{AB} and the scalar function r defined on the 2D manifold. By varying the reduced action with respect to γ_{AB} and r, one reproduces the Einstein equations for the D-dimensional spherically symmetric spacetime. If the matter distribution has the property $T_{AB} \sim \gamma_{AB}$, one can prove the generalized Birkhoff's theorem is valid and the spacetime has an additional Killing vector. The metric Eq. (6.8.1) in this case has the form

$$ds^2 = -g\, dt^2 + \frac{dr^2}{g} + r^2\, d\omega_{D-2}^2, \tag{6.8.2}$$

where g is a function of r. The form of this function is determined by the Einstein equations. For the vacuum case with the cosmological constant Λ it is

$$g = 1 - \left(\frac{r_S}{r}\right)^{D-3} - \frac{2\Lambda}{(D-1)(D-2)}\, r^2. \tag{6.8.3}$$

For $\Lambda = 0$

$$g = 1 - \left(\frac{r_S}{r}\right)^{D-3}. \tag{6.8.4}$$

This vacuum spherically symmetric solution is known as the *Tangherlini metric*. It describes a higher-dimensional spherically symmetric black hole in an asymptotically flat spacetime.

The *Tangherlini spacetime* is asymptotically flat. At far distance the gravitational field is weak. Writing $g = 1 + 2\varphi$ and comparing φ with the Newtonian potential Eq. (5.5.22), one finds

$$r_S^{D-3} = \frac{8\,\Gamma\left(\frac{D-1}{2}\right) G^{(D)} M}{(D-2)\,\pi^{(D-3)/2}}, \tag{6.8.5}$$

where M is the mass of the black hole and $G^{(D)}$ is the D-dimensional gravitational constant.

The metric Eq. (6.8.2) with the function g given by Eq. (6.8.4) has a coordinate singularity at the *Schwarzschild radius* r_S. The global structure of the complete space-time for this metric is similar to the Kruskal 4D solution. It also contains four regions $R_\pm$ and $T_\pm$. These regions are separated by null surfaces (horizons) where $r = r_S$. The *horizon bifurcation surface* is a $(D - 2)$-dimensional sphere of radius r_S. It is a geodesic submanifold.

After the *Wick's rotation* $t = it_E$ this metric becomes

$$ds_E^2 = g\, dt_E^2 + \frac{dr^2}{g} + r^2\, d\omega_{D-2}^2. \tag{6.8.6}$$

It describes the *Euclidean black hole* that is regular at the *Euclidean horizon* $r = r_S$ if the Euclidean time is periodic, $t_E \in (0, 2\pi/\kappa)$ with

$$\kappa = \frac{D - 3}{2\, r_S}. \tag{6.8.7}$$

The parameter κ is the *surface gravity* of the black hole. The corresponding *Hawking temperature* is

$$\Theta_H = \frac{\kappa}{2\pi}.$$ (6.8.8)

In the case of the Schwarzschild–(anti) de Sitter metric with g given by Eq. (6.8.3) the surface gravity should be calculated at the horizon $r = r_+$, defined by the condition $g(r_+) = 0$. Then, the surface gravity is

$$\kappa = \frac{1}{2}\left|\frac{\partial g}{\partial r}\right|_{r=r_+} = \left|\frac{D-3}{2r_+} - \frac{\Lambda r_+}{D-2}\right|.$$ (6.8.9)

7

Particles and Light Motion in Schwarzschild Spacetime

7.1 Equations of Motion

7.1.1 Orbits are planar

A region inside a black hole cannot be seen by an external observer. This follows from the definition of a black hole. This means that all the information concerning the black hole one obtains by studying the matter behavior in its exterior. At large distances the gravitational field of any compact object is weak and is determined by the mass of the object. From far away a black hole and a neutron star of the same mass look very much alike. Thus, for identification of a black hole, besides measuring its mass, it is very important to trace such characteristics of a particle motion and a field propagation that have a 'personal impact' of the strong gravity near the black hole. In this chapter we study particle motion and light propagation in the vicinity of black holes. In this study we focus mainly on those features that are characteristic of black holes only.

We consider motion of particles and light in the Schwarzschild spacetime. In the general case to obtain a particle trajectory one needs to integrate geodesic equations. Symmetries of the Schwarzschild spacetime imply the existence of integrals of motion. We shall demonstrate that they are sufficient to make a problem *completely integrable*.

The particle worldline is $x^\mu = x^\mu(\tau)$, where τ is the proper time. Its four-velocity is $u^\mu = dx^\mu/d\tau$. Let a particle at the initial moment of time be at a point p and have a velocity u_0^μ. The Schwarzschild metric is invariant under rigid rotations of the sphere. Using this freedom we can always choose the initial point p at the equatorial plane, so that it has the coordinates $(t_0, r_0, \theta_0 = \pi/2, \phi_0 = 0)$. There still remains a possibility to perform a solid rotation that preserves the position of the point p. We use this freedom to put $u^\theta(\tau_0) = 0$. In other words, the particle initially is in the equatorial plane and its velocity is tangent to this plane. Let us show that the equation of motion implies $\ddot\theta_p = 0$, that is this particle remains in the equatorial plane. This follows from the geodesic equation

$$\frac{d^2\theta}{d\tau^2} = -\Gamma^\theta_{\mu\nu} \frac{dx^\mu}{d\tau} \frac{dx^\nu}{d\tau}. \tag{7.1.1}$$

Consider this equation at the initial point. For our choice of the initial data the terms on the right-hand side, which contain $\Gamma^\theta_{\mu\nu}$ with either μ or ν equal to θ, vanish. If both μ and ν are different from θ, $\Gamma^\theta_{\mu\nu} = 0$ because of the $\theta \to \pi - \theta$ symmetry. Thus, we prove that *all the orbits are planar.* This property of a motion in a spherically symmetric potential is well known in the non-relativistic theory. This property is also valid for the motion of light. The only difference in its proof is that one has to use the affine parameter λ instead of the proper time τ.

7.1.2 Integrals of motion

The above remarks allow one to reduce the problem of particle and light motion to study the geodesics in the following 3-dimensional spacetime

$$ds^2 = -g\,dt^2 + \frac{dr^2}{g} + r^2\,d\phi^2, \qquad g = 1 - \frac{r_S}{r}. \tag{7.1.2}$$

This is a geometry of the equatorial section $\theta = \pi/2$ of the Schwarzschild metric.[1]

To deal with particles and light simultaneously, it is convenient to use the 4-momentum p^μ instead of the 4-velocity u^μ. In both cases we can write

$$p^\mu = \dot{x}^\mu \equiv dx^\mu/d\lambda. \tag{7.1.3}$$

In the case of a massless particle, λ is a properly chosen affine parameter. For a particle of mass m the affine parameter $\lambda = \tau/m$, where τ is the proper time.

The normalization of the momentum $p^2 = -m^2$ gives

$$-g\,\dot{t}^2 + g^{-1}\,\dot{r}^2 + r^2\,\dot{\phi}^2 = -m^2. \tag{7.1.4}$$

The metric Eq. (7.1.2) has two Killing vectors $\xi_{(t)} = \partial_t$ and $\xi_{(\phi)} = \partial_\phi$ and, correspondingly, two conserved quantities: the energy E and the angular momentum L

$$E \equiv -\xi^\mu_{(t)} p_\mu = g\,\dot{t},$$
$$L \equiv \xi^\mu_{(\phi)} p_\mu = r^2\,\dot{\phi}. \tag{7.1.5}$$

Substituting these equations into Eq. (7.1.4) one gets

$$\dot{r}^2 = E^2 - g\,(m^2 + L^2/r^2),$$
$$\dot{\phi} = L/r^2, \tag{7.1.6}$$
$$\dot{t} = g^{-1}E.$$

In particular, these formulas give the following equation for the time dependence of r

$$\frac{1}{g^2}\left(\frac{dr}{dt}\right)^2 = 1 - \frac{g}{E^2}\left(m^2 + \frac{L^2}{r^2}\right). \tag{7.1.7}$$

In the Newtonian limit one has

$$\dot{t} = m, \qquad E = m + E_{\text{non-rel}}, \qquad g = 1 + 2\varphi, \tag{7.1.8}$$

---

[1] A similar expression is valid for the 'equatorial' metric for a D-dimensional spherically symmetric black hole. The only difference is that in this case $g = 1 - (r_S/r)^{D-3}$ (see Section 7.11).

where φ is the Newtonian gravitational potential and $E_{\text{non-rel}}$ is the non-relativistic energy. Then, Eq. (7.1.6) gives

$$E_{\text{non-rel}} = \frac{1}{2}m\left(\frac{dr}{dt}\right)^2 + \left(m\varphi + \frac{L^2}{2mr^2}\right). \tag{7.1.9}$$

This is the standard equation for the non-relativistic particle motion in the spherically symmetric gravitational field. The first term on the right-hand side of Eq. (7.1.9) is kinetic energy, while the second one (in the brackets) is effective potential, which includes the gravitational potential and the angular momentum contribution.

7.1.3 First-order equations for particle motion

Equations (7.1.6) are written in the form that allows one to consider both cases, massive particle and light motion, simultaneously. For a particle motion, which we consider first, the affine parameter $\lambda = \tau/m$, used in the equations, is proportional to the proper time τ. The corresponding equations are

$$\left(\frac{dr}{d\tau}\right)^2 = \mathcal{E}^2 - U, \qquad U = g\left(1 + \frac{\mathcal{L}^2}{r^2}\right),$$

$$\frac{d\phi}{d\tau} = \frac{\mathcal{L}}{r^2}, \tag{7.1.10}$$

$$\frac{dt}{d\tau} = \frac{\mathcal{E}}{g}.$$

The quantities $\mathcal{E} = E/m$ and $\mathcal{L} = L/m$ are the *specific energy* and the *specific angular momentum*, that is, the relativistic energy and the angular momentum per unit mass. We shall call U an *effective potential*.

The form of the equations of motion is simplified if one uses instead of the radial coordinate r another variable proportional to r^{-1}. We use the gravitational radius r_S to introduce dimensionless quantities, since it is the only dimensionful parameter that determines the scales for our problem. Denote

$$\zeta = r_S/r, \qquad \ell = \mathcal{L}/r_S, \qquad \sigma = m\lambda/r_S = \tau/r_S, \qquad \tilde{t} = t/r_S. \tag{7.1.11}$$

We assume that $\mathcal{L} \geq 0$, and hence, $\ell \geq 0$. Then, Eqs. (7.1.10) take the form

$$\zeta^{-4}\zeta'^2 = \mathcal{E}^2 - U, \qquad U = g(1 + \ell^2\zeta^2),$$

$$\phi' = \ell\zeta^2, \tag{7.1.12}$$

$$\tilde{t}' = \mathcal{E}/g.$$

Here $(\ldots)' = d(\ldots)/d\sigma$. The quantity ℓ is the dimensionless *impact parameter*. In the literature the potential U sometimes is denoted as V^2, since the motion of the particle takes place only in the region where $\mathcal{E} \geq V = \sqrt{U}$. A point ζ_{turn} where $\mathcal{E} = V$ is a *radial turning point*. In our analysis we shall use the effective potential U, and refer to V as a *potential function*.

7.1.4 First-order equations for light motion

In the case of massless particles it is convenient to use the other dimensionless variables

$$\zeta = r_S/r, \qquad \ell = L/(Er_S), \qquad \sigma = E\lambda/r_S. \qquad (7.1.13)$$

The complete set of equations of motion takes the form

$$\zeta' = -\epsilon\sqrt{1-\mathcal{U}}, \qquad \mathcal{U} = g\,\ell^2\zeta^2,$$
$$\zeta^{-2}\tilde{t}' = 1/g\ , \qquad \phi' = \ell\,\zeta^2. \qquad (7.1.14)$$

The quantity ℓr_S is an *impact parameter*. We call $\mathcal{U}$ an *effective potential for the light motion*. The parameter ϵ depends on the direction of radial motion. For the outgoing rays, when r increases along the trajectory, $\epsilon = +1$, while for incoming rays $\epsilon = -1$.

By excluding the affine parameter σ one can rewrite these equations in a different form

$$\frac{d\tilde{t}}{d\zeta} = -\epsilon\,\frac{1}{g\,\zeta^2\sqrt{1-g\,\ell^2\zeta^2}}, \qquad \frac{d\phi}{d\zeta} = -\epsilon\,\frac{\ell}{\sqrt{1-g\,\ell^2\zeta^2}}. \qquad (7.1.15)$$

For some further applications it is also useful to know that these two differential equations satisfy the identity

$$\frac{\partial}{\partial\ell}\left(\frac{d\tilde{t}}{d\zeta}\right) = \ell\,\frac{\partial}{\partial\ell}\left(\frac{d\phi}{d\zeta}\right), \qquad (7.1.16)$$

which can be checked by differentiating Eq. (7.1.15) with respect to the parameter ℓ.

7.2 Particle Trajectories

7.2.1 Properties of effective potential

So far we have considered the generic metric Eq. (7.1.2) with an arbitrary function g. From now on we study the case of the Schwarzschild geometry where $g = 1 - \zeta$. The argument ζ takes the values in the interval $(0, 1)$. The point $\zeta = 0$ is the spatial infinity, and $\zeta = 1$ is the black hole horizon. The effective potential for massive particles is

$$U = (1 - \zeta)(1 + \ell^2\zeta^2). \qquad (7.2.1)$$

To describe different types of particle trajectories we need to analyze the properties of the effective potential U. It is convenient to consider this potential as a function of two variables, ζ and ℓ (see Figure 7.1). The effective potential is a cubic polynomial in ζ and quadratic in ℓ. At the horizon ($\zeta = 1$) the effective potential U vanishes, while at infinity ($\zeta = 0$) the potential $U = 1$. For a fixed value of ℓ the potential has extrema at ζ, where

$$U_{,\zeta} = -1 - 3\ell^2\zeta^2 + 2\ell^2\zeta = 0. \qquad (7.2.2)$$

A solution of this equation gives two curves $\zeta = \zeta_\pm(\ell)$ in the (ζ, ℓ)-plane, where

$$\zeta_\pm(\ell) = \frac{1 \pm \sqrt{1 - 3\ell^{-2}}}{3}. \qquad (7.2.3)$$

Both curves begin at $\zeta_0 = 1/3$ and $\ell = \sqrt{3}$ and continue for all $\ell \geq \sqrt{3}$. One has

$$\zeta_- \leq \zeta_0 \leq \zeta_+. \tag{7.2.4}$$

The second derivative of U with respect to ζ is

$$U_{,\zeta\zeta} = 2\ell^2(1 - 3\zeta). \tag{7.2.5}$$

Hence, at $\zeta = \zeta_-$ the function U has the minimum U_-, while at $\zeta = \zeta_+$ it has the maximum U_+

$$U_\pm = \frac{2[\ell(\ell^2 + 9) \pm (\ell^2 - 3)^{3/2}]}{27\ell}. \tag{7.2.6}$$

At $\zeta = \zeta_0$ one has $U_+ = U_- = 8/9$.

In the (ζ, ℓ)-plane the curves $U = U_\pm$ are smooth (see Figure 7.1). Figure 7.2 is obtained by projecting the U-surface along the ζ direction. The singular point at $\ell = \sqrt{3}$, where U_- and U_+ meet one another, is the result of this projection. Near this point

$$U_+ - U_- \simeq \frac{8\sqrt{2}}{9} \left(\frac{\ell}{\sqrt{3}} - 1 \right)^{3/2}. \tag{7.2.7}$$

Additional information concerning properties of the effective potential can be obtained from the following relations:

$$U_{,\ell} = 2(1 - \zeta)\ell\zeta^2, \quad U_{,\ell\ell} = 2(1 - \zeta)\zeta^2, \quad U_{,\zeta\ell} = 4\zeta\ell(1 - 3\zeta/2). \tag{7.2.8}$$

For example, the first of these relations shows that for a fixed ζ, U is a monotonically growing function of ℓ.

Consider the equation

$$U \equiv (1 - \zeta)(1 + \ell^2\zeta^2) = 1. \tag{7.2.9}$$

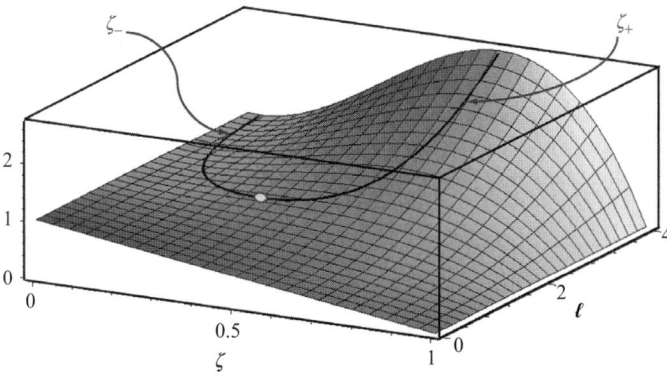

Fig. 7.1 Effective potential $U(\zeta, \ell)$.

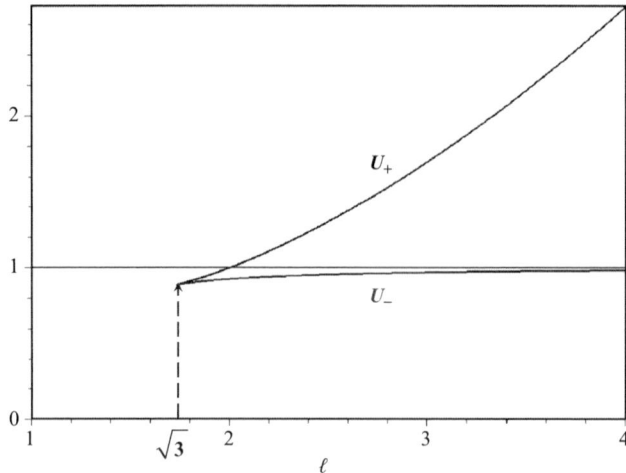

Fig. 7.2 Maximum U_+ and minimum U_- values of the potential U as functions of ℓ. For $U_- < \mathcal{E}^2 <$ min$(U_+, 1)$ particle trajectories are bounded orbits.

Its solution determines two curves $\zeta = \tilde{\zeta}_\pm$ in the (ζ, ℓ)-plane

$$\tilde{\zeta}_\pm = \frac{1 \pm \sqrt{1 - 4\ell^{-2}}}{2}. \qquad (7.2.10)$$

These curves cross one another at the point $\ell = 2$. For $\ell < 2$ the effective potential $U < 1$ for all ζ.

7.2.2 Types of trajectories

Figure 7.3 gives slices of the effective potential for different values of ℓ. This demonstrates that for $0 < \ell < \sqrt{3}$ the effective potential monotonically decreases from 1 at infinity ($\zeta = 0$) to 0 at the horizon ($\zeta = 1$). Since it does not have a minimum, for these values of ℓ bounded motion is not possible. For $\ell > \sqrt{3}$ in the inner part of the interval $\zeta \in (0, 1)$ the effective potential has both a minimum at $\zeta = \zeta_-(\ell)$, and a maximum at $\zeta = \zeta_+(\ell)$. For these values of ℓ the bounded motion is possible. For $\ell = 2$ the maximum reaches the value 1 and it is greater than 1 for $\ell > 2$.

Let us consider motion corresponding to $\ell > 2$. An example of the effective potential U for such ℓ is shown in Figure 7.4. Horizontal lines represent a motion with a given (fixed) value of the specific energy $\mathcal{E}$. The following qualitatively different types of motion are possible:

1. gravitational capture, $\mathcal{E}^2 > U_+$;
2. hyperbolic motion (scattering), $\mathcal{E}^2 \in (1, U_+)$ and $\zeta < \zeta_+$;
3. bounded orbits, $\mathcal{E}^2 \in (U_-, \min(U_+, 1))$ and $\zeta < \zeta_+$;
4. stable circular orbits, $\mathcal{E}^2 = U_-$ and $\zeta = \zeta_-$;
5. unstable circular orbits, $\mathcal{E}^2 = U_+$ and $\zeta = \zeta_+$;

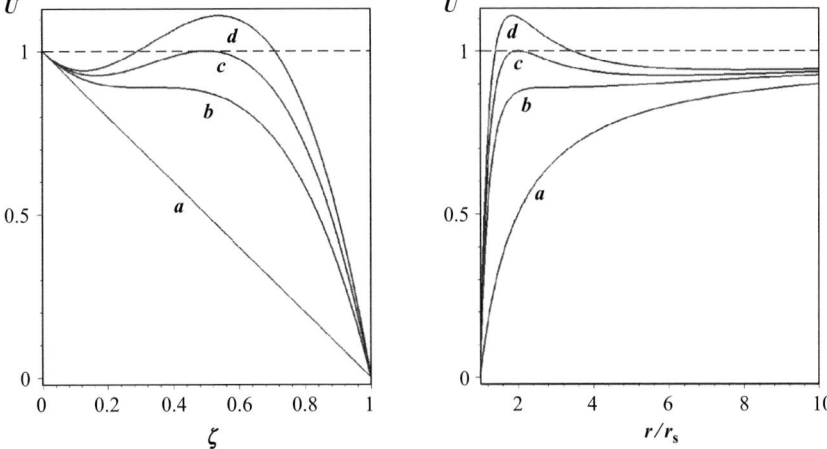

Fig. 7.3 Effective potential. Left figure shows the effective potential U as a function of ζ for a fixed value of ℓ. The lines in this figure correspond to four different values of ℓ, $\ell = 0$ – line a, $\ell = \sqrt{3}$ – line b, $\ell = 2$ – line c, and $\ell = 2.2$ – line d. The right figure shows the same effective potential as a function of the radius r.

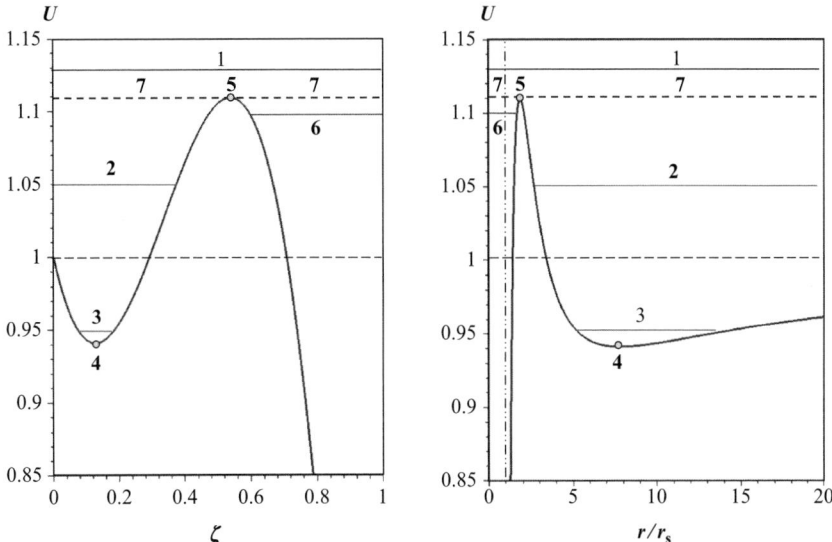

Fig. 7.4 Effective potential U for $\ell = 2.2$. The left figure shows the plot of U as a function of $\zeta = r_S/r$. The same plot for U as a function of r/r_S is shown in the right figure. Horizontal lines show different types of trajectories.

6. near-horizon trapped motion, $\mathcal{E}^2 < U_+$ and $\zeta > \zeta_+$;
7. marginal outer ($\zeta < \zeta_+$) and inner ($\zeta > \zeta_+$) orbits, $\mathcal{E}^2 = U_+$.

The orbits of types 1, 5, 6, and 7 do not exist in the Newtonian gravity. Different types of orbits of the particle in the field of the black hole are shown in Figure 7.5.

7.2.3   Gravitational capture

For a given ℓ the gravitational capture occurs when $\mathcal{E}^2 > U_+$. Let us calculate the capture cross-section

$$\sigma(v) = \pi b^2(v) \tag{7.2.11}$$

for a particle that has the velocity v at infinity. In this relation b is an impact parameter. The momentum of such particle is $p = mv/\sqrt{1 - v^2}$, and the angular momentum is

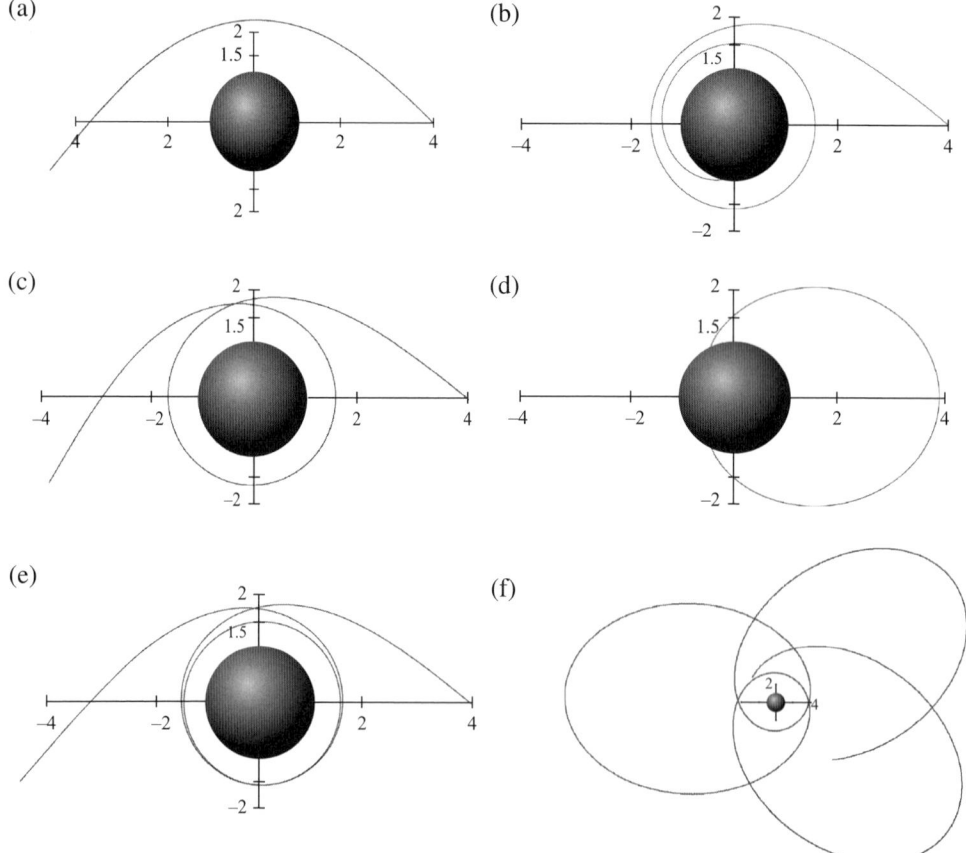

Fig. 7.5 Different types of trajectories of a particle in the gravitational field of a spherically symmetric black hole.

$$L = mr_S\ell = bp. \tag{7.2.12}$$

The relation $E^2 = p^2 + m^2$ and the capture threshold condition $\mathcal{E}^2 = U_+$ imply

$$p = m\sqrt{U_+ - 1}. \tag{7.2.13}$$

Since $p = mv/\sqrt{1 - v^2}$, we have

$$U_+(\ell) \equiv \frac{2[\ell(\ell^2 + 9) + (\ell^2 - 3)^{3/2}]}{27\,\ell} = \frac{1}{1 - v^2}. \tag{7.2.14}$$

From Eq. (7.2.12) and Eq. (7.2.13) we get

$$b = r_S \frac{\ell}{\sqrt{U_+(\ell) - 1}} = r_S\ell(v)\frac{\sqrt{1 - v^2}}{v}. \tag{7.2.15}$$

To summarize, Eq. (7.2.14) determines $\ell(v)$ for a given v, and then Eq. (7.2.15) gives the corresponding impact parameter $b(v)$. Substituting it into Eq. (7.2.11) we obtain the cross-section

$$\sigma(v) = \pi r_S^2 \frac{\ell^2(v)(1 - v^2)}{v^2}. \tag{7.2.16}$$

Consider two limiting cases. For a non-relativistic motion, $v \ll 1$, one has $U_+ \approx 1$ and $\ell \approx 2$. Thus,

$$b \approx \frac{2r_S}{v}, \qquad \sigma_{\text{non}-\text{rel}}(v) \approx \frac{4\pi r_S^2}{v^2}. \tag{7.2.17}$$

For the ultrarelativistic motion $v \to 1$, $\ell \to \infty$, and $U_+ \approx 4\ell^2/27$. The cross-section in this limit is

$$b \approx \frac{r_S\sqrt{27}}{2}, \qquad \sigma_{\text{ultra}-\text{rel}}(v) \approx \frac{27}{4}\pi r_S^2. \tag{7.2.18}$$

7.2.4 Circular and marginal orbits

Circular motion around a black hole is an important special case of motion of a particle at constant radius, that is, when $\zeta = \text{const}$. This motion is represented in Figure 7.4 by the point at the extremum of the effective potential curve. The point at the minimum corresponds to stable motion; and that at the maximum to unstable motion. The latter motion has no analog in Newtonian theory. It is specific to black holes. Generally speaking, motion along an unstable trajectory is not physically feasible. However, if the motion of a particle is represented by a horizontal line $\mathcal{E}^2 = \text{const}$ very close to U_+, then the particle makes many turns around the black hole at radii close to $r = r_S/\zeta_+$ before it moves far away from this value of r. This follows from the analysis of Eq. (7.1.15) for $\mathcal{E} \approx \sqrt{U_+}$.

The maximum and minimum appear on the U curves when $\ell > \sqrt{3}$. If $\ell < \sqrt{3}$, the U curve is monotonic. Hence, motion on circular orbits is possible only if $\ell > \sqrt{3}$. The minima of the curves then lie at $\zeta < \min \zeta_- = 1/3$. The corresponding radius $r > 3r_S$. Thus, stable circular orbits exist only for $r > 3r_S$. At smaller distances there are only unstable circular orbits corresponding to the maxima of the U_+ curves. If $\ell \to \infty$, the coordinates of the maxima on the U_+ curves decrease to $r = 1.5\,r_S$ ($\zeta = 2/3$). For r less than $1.5\,r_S$ even unstable geodesic circular motion becomes impossible.[2]

The critical circular orbit that separates stable trajectories from unstable ones corresponds to $r = 3\,r_S$. This orbit is called an *innermost stable* or *marginally stable orbit*. Particles move along it at the velocity $v = 1/2$, the specific energy of a particle being

$$\mathcal{E}_{\min} = \sqrt{U_-}\Big|_{\ell=\sqrt{3}} = \sqrt{8/9} \approx 0.943. \tag{7.2.19}$$

This is the motion with the maximum possible *binding energy*

$$\Delta\mathcal{E} = (1 - \mathcal{E}_{\min})mc^2 \approx 0.057\,mc^2. \tag{7.2.20}$$

The velocity of motion on (unstable) orbits with $r \le 3r_S$ increases from $c/2$ at $r = 3\,r_S$ to c on the last circular orbit with $r = 1.5\,r_S$. When $r = 2\,r_S$, the particle's energy is $\mathcal{E} = 1$; that is, its circular velocity is equal to the escape velocity. For $r < 2\,r_S$ the escape velocity is smaller than the circular velocity. There is no paradox in this since the circular motion here is unstable and even the slightest perturbation may kick out the particle to infinity; that is, to an orbit corresponding to hyperbolic motion.

In astrophysical black holes matter falling onto a black hole usually forms an *accretion disk*. Particles are moving approximately along circular (Keplerian) orbits. As a result of loss of energy and angular momentum the radius of their orbits slowly decreases. They move closer and closer to the black hole. A particle may lose up to $(1 - E)mc^2 \approx 0.057\,mc^2$ of its proper rest-mass energy mc^2 until it reaches the *innermost stable circular orbit* (ISCO). After this, it falls into the black hole. This scenario gives the estimation Eq. (7.2.20) for the efficiency of a static black hole. Namely, this part of the total energy can be extracted from the accreting matter.

Problem 7.1: *Two particles of masses m_1 and m_2 move in the opposite directions along the same stable circular orbit of the radius $r = r_0$ around a Schwarzschild black hole of the mass M. They collide and form a new particle. What is the mass m of the formed particle? What must be the ratio $\mu = m_2/m_1$ for which this new particle will be captured by the black hole? Neglect energy loss during the particle collision.*

Solution: *We choose the angular momentum of the particle 1 to be positive. The energy and angular momentum of the particles before the collision are*

$$E_1 = m_1\mathcal{E}_0, \quad E_2 = m_2\mathcal{E}_0, \quad L_1 = m_1\ell_0 r_S, \quad L_2 = -m_2\ell_0 r_S. \tag{7.2.21}$$

The condition that the initial orbits are circular implies

$$\ell_0^2\zeta_0(3\zeta_0 - 2) + 1 = 0, \tag{7.2.22}$$

where $\zeta_0 = r_S/r_0$.
Let m be the mass of the formed particle and $E = m\mathcal{E}$ and $L = m\ell r_S$ be its energy and angular momentum. Since during the particle collision their total 4-momentum is conserved one has

$$\mathcal{E} = \frac{m_1 + m_2}{m}\mathcal{E}_0, \qquad \ell = \frac{m_1 - m_2}{m}\ell_0. \tag{7.2.23}$$

---

[2] If there is an external non-gravitational force acting on a particle, then it could move along a circular orbit inside $r = 1.5\,r_S$.

Denote

$$W_\ell(\zeta) = \mathcal{E}^2 - (1 + \ell^2 \zeta^2)(1 - \zeta). \tag{7.2.24}$$

At the moment of collision $\dot{r} = 0$, so that $W_\ell(\zeta_0) = 0$. This condition together with Eq. (7.2.23) allows one to find the mass of the formed particle

$$m^2 = (m_1 + m_2)^2 + 4 m_1 m_2 \ell_0^2 \zeta_0^2, \tag{7.2.25}$$

where ℓ_0 is a function of ζ_0 determined by Eq. (7.2.22). This relation shows that

$$m \geq m_1 + m_2. \tag{7.2.26}$$

Using Eq. (7.2.24) we also obtain

$$|\ell| \leq \frac{|m_1 - m_2|}{m_1 + m_2} \ell_0 \leq \ell_0. \tag{7.2.27}$$

Since at the moment of the collision

$$W_\ell|_{\zeta_0} = 0, \qquad W'_\ell|_{\zeta_0} = 1 - \frac{\ell^2}{\ell_0^2} > 0, \tag{7.2.28}$$

the formed particle always starts its radial motion towards the black hole. It will be captured if between ζ_0 and the horizon, $\zeta = 1$, the function $W_\ell(\zeta)$ remains positive. The maximal value of this function is (see Eq. (7.2.6))

$$W_+(\ell) \equiv (1 + \ell^2 \zeta_0^2)(1 - \zeta_0) - \frac{2[\ell(\ell^2 + 9) + (\ell^2 - 3)^{3/2}]}{27\,\ell}. \tag{7.2.29}$$

The critical value of the specific angular momentum ℓ separating the capture and non-capture regimes is determined by the equation $W_+(\ell) = 0$. This condition implies

$$[4 - (1 - \zeta_0)(1 + 3\zeta_0)\ell^2][1 + \ell^2 \zeta_0(3\zeta_0 - 2)]^2 = 0. \tag{7.2.30}$$

Since $\ell \neq \ell_0$ the second factor (in the square bracket) does not vanish (see Eq. (7.2.22)). Thus, one has

$$\ell^2 = \frac{4}{(1 - \zeta_0)(1 + 3\zeta_0)}. \tag{7.2.31}$$

Using Eq. (7.2.24) one can write the condition $W_\ell|_{\zeta_0} = 0$ in the form

$$\ell^2 = \frac{(1 - \mu)^2}{(1 + \mu)^2 \zeta_0(2 - 3\zeta_0) + 4\mu\zeta_0^2}, \tag{7.2.32}$$

where $\mu = m_2/m_1$. Excluding ℓ^2 from the system Eq. (7.2.31) and Eq. (7.2.32), and solving the obtained equation for μ one finds

$$\mu_\pm = \frac{1 + 10\zeta_0 - 7\zeta_0^2 \pm 4(1 + \zeta_0)\sqrt{2\zeta_0(1 - \zeta_0)}}{(1 - 3\zeta_0)^2}. \tag{7.2.33}$$

It is easy to check that $W_+(\ell)$ is positive when the following inequality is valid

$$\mu_- < \mu < \mu_+. \tag{7.2.34}$$

The obtained inequality shows that the formed particle is captured by the black hole if the mass ratio μ is between μ_- and μ_+. Note that $\mu_- = 1/\mu_+$, as it must be, because of the symmetry $m_1 \leftrightarrow m_2$.

7.3 Kepler's Law

7.3.1 Kepler's law for circular orbits

Let us derive the period for a Keplerian motion along a stable circular orbit of radius R. Stable orbits correspond to the minimum of the effective potential U. The radial position of the stable circular orbit in the dimensionless units $\zeta = r_S/R$ is given by Eq. (7.2.3)

$$\zeta = \zeta_- = \frac{1}{3}(1 - \sqrt{1 - 3\ell^{-2}}). \tag{7.3.1}$$

The specific energy and angular momentum of the particle at this orbit are

$$\mathcal{E} = \sqrt{U}\big|_{\zeta=\zeta_-} = \frac{1 - \zeta_-}{\sqrt{1 - \frac{3}{2}\zeta_-}}, \tag{7.3.2}$$

$$\ell = 1/\sqrt{\zeta_-(2 - 3\zeta_-)},$$

and their ratio is

$$\frac{\ell}{\mathcal{E}} = \frac{1}{\sqrt{2\zeta_-}(1 - \zeta_-)}. \tag{7.3.3}$$

The dimensionless frequency can be derived from Eq. (7.1.12)

$$\tilde{\Omega} = \frac{d\phi}{d\tilde{t}} = \frac{\phi'}{\tilde{t}'} = \ell\,\zeta_-^2(1 - \zeta_-)\,\mathcal{E}^{-1}. \tag{7.3.4}$$

Using Eq. (7.3.3) we can rewrite this expression in the form

$$\tilde{\Omega} = \zeta_-^{3/2}/\sqrt{2}, \quad \zeta_- = r_S/R. \tag{7.3.5}$$

The dimensionful expressions for the frequency Ω and its period T (measured in the coordinate time t) are

$$\Omega = r_S^{-1}\tilde{\Omega} = \left(\frac{r_S}{2}\right)^{1/2} R^{-3/2}, \qquad T = \frac{2\pi}{\Omega}. \tag{7.3.6}$$

So, for arbitrary circular orbits the period satisfies the relation

$$T = 2\pi \left(\frac{R^3}{GM}\right)^{1/2}. \tag{7.3.7}$$

In other words for a circular orbit the square of the period of rotation is proportional to the cube of the radius of the orbit. This is simply the *Kepler's third law*. It is curious that, though this law was derived at first in the Newtonian gravity, for circular orbits it happens to be exact in the Einstein theory too.

The equation of motion of a particle in (t, r, θ, ϕ) coordinates is $(t, R, \pi/2, \Omega t)$ and hence its 4-velocity u^{μ} is

$$u^{\mu} = \frac{dt}{d\tau}(1, 0, 0, \Omega). \tag{7.3.8}$$

Denote by $e_{(t)}$ and $e_{(\phi)}$ unit vectors in t and ϕ directions, then one has

$$u = \gamma(e_{(t)} + v e_{(\phi)}). \tag{7.3.9}$$

The normalization condition gives $\gamma = 1/\sqrt{1 - v^2}$. By comparing Eq. (7.3.9) and Eq. (7.3.8) one gets

$$v = \frac{\Omega R}{\sqrt{g}} = \frac{\Omega R}{\sqrt{1 - r_S/R}}, \tag{7.3.10}$$

or, substituting Eq. (7.3.6) we find for the *Keplerian velocity* v the following expression:

$$v = \frac{\sqrt{r_S}}{\sqrt{2(R - r_S)}}. \tag{7.3.11}$$

7.3.2 Periapsis shift

Let us discuss non-circular bounded motion of particles. For such a motion there exist two radial turning points, $r_2 \leq r_1$. From the analysis of the effective potential we know that in the region $r < 3r_S$ there is no inner turning point. Thus, $3r_S \leq r_2 \leq r_1$. In the Einstein gravity a generic bounded orbit near the spherically symmetric gravitating body is not a closed line. This property is in contrast to the Newtonian gravity, where the orbits are closed ellipses. Let us calculate the angle $\Delta\phi$ that spans the particle while moving from the closest point of the orbit to the black hole (*periapsis*) r_2 to the most distant one r_1 and again to the next periapsis r_2. In the Newtonian gravity this angle is exactly equal to 2π.

We denote

$$\zeta_1 = r_S/r_1, \qquad \zeta_2 = r_S/r_2, \qquad \zeta_1 \leq \zeta_2 \leq 1/3. \tag{7.3.12}$$

The parameters ζ_1 and ζ_2 are uniquely determined by the energy $\mathcal{E}$ and the angular momentum ℓ. They can be found from the following relations:

$$\ell^2 \equiv \frac{1}{q^2} = \frac{1}{\zeta_1 + \zeta_2 - \zeta_1^2 - \zeta_2^2 - \zeta_1\zeta_2}, \qquad \mathcal{E}^2 = \frac{(1 - \zeta_1)(1 - \zeta_2)(\zeta_1 + \zeta_2)}{\zeta_1 + \zeta_2 - \zeta_1^2 - \zeta_2^2 - \zeta_1\zeta_2}. \tag{7.3.13}$$

To obtain ζ_1 and ζ_2 as functions of the energy and the angular momentum one needs to solve these equations. For practical purposes it is much more convenient to specify solutions by the parameters ζ_1 and ζ_2 rather than by the energy and the angular momentum. For a circular motion $\zeta_1 = \zeta_2$. The limit $\zeta_1 = 0$ corresponds to a *hyperbolic motion*.

The following equation for the particle orbit $r = r(\phi)$ is obtained from Eq. (7.1.12)

$$\frac{d\zeta}{d\phi} = -\epsilon \sqrt{F}, \quad F = \zeta^2(\zeta - 1) + q^2(\zeta - 1 + \mathcal{E}^2), \quad q = \ell^{-1}. \tag{7.3.14}$$

Here, $\epsilon = +1$ for the motion with increasing radius, while $\epsilon = -1$ for descending motion. F is a cubic polynomial of ζ that vanishes at the radial turning points, ζ_1 and ζ_2. Thus, it can be written as

$$F = (\zeta - \zeta_1)(\zeta - \zeta_2)(\zeta - \zeta_3). \tag{7.3.15}$$

From the expression for F it follows that

$$\zeta_1 + \zeta_2 + \zeta_3 = 1. \tag{7.3.16}$$

Hence, $\zeta_3 \geq 1/3 \geq \zeta_2 \geq \zeta_1$. The function F is non-negative for $\zeta_1 \leq \zeta \leq \zeta_2$.
Using Eq. (7.3.14) we obtain the following expression for $\Delta\phi$:

$$\Delta\phi = 2 \int_{\zeta_1}^{\zeta_2} \frac{d\zeta}{\sqrt{F}}. \tag{7.3.17}$$

Changing the integration variable $\zeta = \zeta_1 + y(\zeta_2 - \zeta_1)$ and using the relation $\zeta_3 = 1 - \zeta_1 - \zeta_2$, one can express $\Delta\phi$ in terms of ζ_1 and ζ_2 only

$$\Delta\phi = 2 \int_0^1 dy \frac{1}{\sqrt{y(1-y)[(1 - 2\zeta_1 - \zeta_2) - (\zeta_2 - \zeta_1)y]}}. \tag{7.3.18}$$

Calculating the integral one obtains the expression for $\Delta\phi$ in terms of the complete elliptic integral of the first kind

$$\Delta\phi = \frac{4}{\sqrt{1 - 2\zeta_1 - \zeta_2}} \mathbf{K}\left(\sqrt{\frac{\zeta_2 - \zeta_1}{1 - 2\zeta_1 - \zeta_2}}\right). \tag{7.3.19}$$

In most cases, the integrals that enter relations for the particle and light motion in the Schwarzschild geometry, can be calculated explicitly in terms of the elliptic integrals. Note that the arguments of the incomplete elliptic integrals are sometimes defined differently in the literature. We use the representations of the elliptic integrals that are defined by the relations

$$\mathbf{F}(\varphi, k) \equiv \int_0^\varphi (1 - k^2 \sin^2 t)^{-1/2} dt,$$

$$\mathbf{E}(\varphi, k) \equiv \int_0^\varphi (1 - k^2 \sin^2 t)^{1/2} dt, \tag{7.3.20}$$

$$\Pi(\varphi, v, k) \equiv \int_0^\varphi (1 + v \sin^2 t)^{-1}(1 - k^2 \sin^2 t)^{-1/2} dt.$$

The complete elliptic integrals are

$$\mathbf{K}(k) = \mathbf{F}(\pi/2, k), \quad \mathbf{E}(k) = \mathbf{E}(\pi/2, k), \quad \Pi(v, k) = \Pi(\pi/2, v, k). \tag{7.3.21}$$

When $\zeta_2 - \zeta_1 \ll 1$, i.e. for nearly circular orbits at arbitrary radii, or for distant orbits when $\zeta_1 \ll 1$ and $\zeta_2 \ll 1$ but with arbitrary eccentricity $e = (\zeta_2 - \zeta_1)/(\zeta_2 + \zeta_1)$, one can expand $\Delta\phi$ in powers of the small parameter $\zeta_2 - \zeta_1$. Up to the fourth order in this parameter in both cases the result reads

$$\Delta\phi = \frac{2\pi}{\sqrt{1 - \frac{3}{2}(\zeta_1 + \zeta_2)}} + \frac{3\pi}{32}\frac{(\zeta_2 - \zeta_1)^2}{\left(1 - \frac{3}{2}(\zeta_1 + \zeta_2)\right)^{\frac{5}{2}}} + O\left((\zeta_2 - \zeta_1)^4\right). \qquad (7.3.22)$$

7.3.3   Kepler's third law: General case

In a similar manner we can calculate the time required for a planet to move from a periapsis to the next one. In order to do this we use Eq. (7.1.12) and write

$$\frac{dt}{d\zeta} = -\epsilon \frac{r_S \, \mathcal{E}q}{\zeta^2(1 - \zeta)} \frac{1}{\sqrt{F}}, \qquad (7.3.23)$$

where F is given by Eq. (7.3.14). The time interval Δt between two subsequent closest approaches of a particle to the black hole is

$$\Delta t = 2r_S \, \mathcal{E}q \int_{\zeta_1}^{\zeta_2} d\zeta \frac{1}{\zeta^2(1 - \zeta)} \frac{1}{\sqrt{F}}. \qquad (7.3.24)$$

After the change of the integration variable $\zeta = \zeta_1 + x(\zeta_2 - \zeta_1)$ the integration gives the exact formula in terms of the complete elliptic integrals $\mathbf{K}$, $\mathbf{E}$ and Π Eq. (7.3.21)

$$\begin{aligned}
\Delta t = 2r_S \; & \frac{\sqrt{\zeta_1 + \zeta_2}\sqrt{1 - \zeta_1}\sqrt{1 - \zeta_2}}{\zeta_1\zeta_2\sqrt{1 - 2\zeta_1 - \zeta_2}}\left[\mathbf{K}(y) - \frac{1 - 2\zeta_1 - \zeta_2}{1 - \zeta_1 - \zeta_2}\mathbf{E}(y)\right. \\
& -\frac{\zeta_1 + \zeta_2 - \zeta_1^2 - \zeta_2^2 - 2\zeta_1^2\zeta_2 - 2\zeta_1\zeta_2^2}{\zeta_1(1 - \zeta_1 - \zeta_2)}\Pi\left(-\frac{\zeta_2 - \zeta_1}{\zeta_1}, y\right) \\
& \left. +2\frac{\zeta_1\zeta_2}{1 - \zeta_1}\Pi\left(\frac{\zeta_2 - \zeta_1}{1 - \zeta_1}, y\right)\right],
\end{aligned} \qquad (7.3.25)$$

where

$$y = \sqrt{\frac{\zeta_2 - \zeta_1}{1 - 2\zeta_1 - \zeta_2}}.$$

Expansion of the result in powers of the small parameter $\zeta_2 - \zeta_1$ leads to

$$\Delta t = \frac{8\pi r_S}{(\zeta_1 + \zeta_2)^{3/2}\sqrt{1 - \frac{3}{2}(\zeta_1 + \zeta_2)}} + O\left((\zeta_2 - \zeta_1)^2\right). \qquad (7.3.26)$$

This is the third Kepler law with the Einstein gravity corrections.

One can also calculate the average angular velocity of a planet during one period of rotation around the star

$$\begin{aligned}
\frac{\Delta\phi}{\Delta t} &= \frac{(\zeta_1 + \zeta_2)^{3/2}}{4r_S} + O\left((\zeta_2 - \zeta_1)^2\right) \\
&= \left(\frac{r_S}{2}\right)^{1/2}\left(\frac{r_1 + r_2}{2}\right)^{-3/2} + O\left((r_2 - r_1)^2\right).
\end{aligned} \qquad (7.3.27)$$

Note that because of the periapsis shift the time Δt between two subsequent points of closest approach to the central mass is not the same as the period T required to span the angle 2π even for almost circular orbits. Nevertheless, the average angular velocity reproduces that of the circular orbit. This formula is another version of the third Kepler law generalized to the Einstein gravity. Evidently, for circular orbits it is in complete agreement with Eq. (7.3.6).

7.4   Light Propagation

7.4.1   Null-ray orbits

There is a variety of effects connected with action of the gravitational field of a black hole on the light propagation in its vicinity:

1. Light rays are bent so that images of bright objects are distorted.
2. A point-like source may have many images.
3. A cross-section of a beam of light rays after its passing near the black hole may shrink.
4. The beam's shape is distorted.
5. Visible brightness of images depends on their position.
6. Time of a light signal arrival is delayed when it passes near the black hole.
7. Registered frequency of radiation, emitted by an object moving near the black hole, is shifted.

In this section we discuss some of these effects that are important, in particular, for the interpretation of the astrophysical data.

For the motion of null rays there is only one essential parameter, ℓ, which determines the type of the trajectory. The turning points for the radial motion can be found from Eq. (7.1.14), taking into account that $\zeta' = 0$ at these points, i.e.

$$1 - \ell^2 \zeta^2 (1 - \zeta) = 0. \tag{7.4.1}$$

Solving this equation for ℓ one obtains

$$\ell = \frac{1}{\zeta \sqrt{1 - \zeta}}. \tag{7.4.2}$$

Figure 7.6 shows the curve $\ell(\zeta)$. It is easy to check that the minimum of $l(\zeta)$ takes place at $\zeta = 2/3$, and at this point $\ell_{min} = 3\sqrt{3}/2$.

A null-ray orbit is given by the equation $r = r(\phi)$. Instead of r we use the dimensionless parameter $\zeta = r_S/r$. Excluding the affine parameter in Eq. (7.1.14) we obtain the following equation for the null-ray trajectory

$$\frac{d\zeta}{d\phi} = -\epsilon \sqrt{F}, \qquad F = q^2 - \zeta^2(1 - \zeta). \tag{7.4.3}$$

ℓ_{min} is a critical impact parameter ℓ that separates two regimes. For $\ell < \ell_{min}$ a null ray propagates from infinity to the horizon. Such a null ray is captured by the black hole. Thus, the capture cross-section for photons is

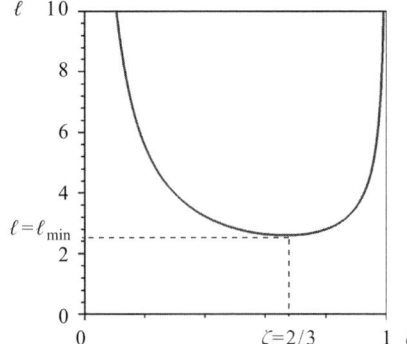

Fig. 7.6 Function $\ell(\zeta)$.

$$\sigma_{\text{photons}} = \pi \ell_{\text{min}}^2 r_S^2 = \frac{27}{4} \pi r_S^2.$$
(7.4.4)

This value, as expected, coincides with Eq. (7.2.18).

For $\ell > \ell_{\text{min}}$ a null-ray has a radial turning point. Let us study null-ray orbits for this case in more detail. Let us parametrize the inverse impact parameter $q = \ell^{-1}$ as follows:

$$q = \frac{2}{3\sqrt{3}} \sin \psi.$$
(7.4.5)

Then,

$$F = (\zeta - \zeta_1)(\zeta - \zeta_2)(\zeta - \zeta_3),$$
(7.4.6)

where

$$\zeta_1 = \frac{1}{3}\left[1 + 2\cos(2\psi/3)\right],$$

$$\zeta_2 = \frac{1}{3}\left[1 - \cos(2\psi/3) + \sqrt{3}\sin(2\psi/3)\right],$$
(7.4.7)

$$\zeta_3 = \frac{1}{3}\left[1 - \cos(2\psi/3) - \sqrt{3}\sin(2\psi/3)\right].$$

In the expression Eq. (7.4.7) we explicitly took into account that in the region of interest one has $\zeta_3 \leq 0 < \zeta \leq \zeta_2 < \zeta_1$. The roots of F, corresponding to the radial turning points, parametrized by ψ are shown in Figure 7.7.

Motion with the radial turning point is possible in the domains of ζ where $F \geq 0$. These domains are:

1. $0 < \zeta < \zeta_2$: This is a motion from infinity to a turning point at ζ_2 and back to infinity.

2. $\zeta_1 < \zeta < 1$: For this motion a null ray starts at the horizon, reaches a maximal radius at ζ_1, and after this falls back onto the black hole.

Any light trajectory has no more than one radial turning point. The first case is the most interesting for applications. Let us study it.

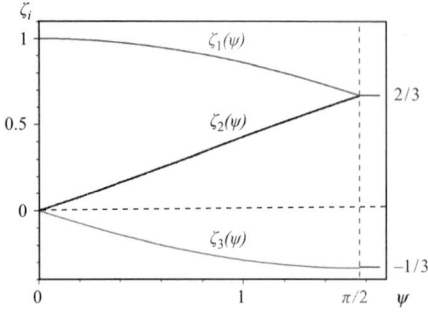

Fig. 7.7 Functions $\zeta_i(\psi)$.

7.4.2   Bending angle

Consider the bending angle $\Delta\psi$ between two points on the light orbit. For fixed radii of these points there is a one-parameter family of orbits connecting them. An orbit can be specified by an inverse impact parameter $q = \ell^{-1}$. However, it is much more convenient to parametrize the null-ray orbits by the parameter ζ_2 of the corresponding radial turning point. The relation between q and ζ_2 is

$$q^2 = \zeta_2^2(1 - \zeta_2).$$

$$(7.4.8)$$

Equations (7.4.5) and (7.4.7) give the required solution of Eq. (7.4.8) in the parametric form.
  We define the following function

$$\Psi(\zeta, \zeta_2) = \int_{\zeta}^{\zeta_2} \frac{d\zeta}{\sqrt{q^2 - \zeta^2(1 - \zeta)}}.$$

$$(7.4.9)$$

This function gives the bending angle between the points with inverse radial coordinates ζ_2 and ζ (see Figure 7.8). Using this function it is easy to write an expression for the bending angle between any two points of the null-ray orbit. The integral in Eq. (7.4.9) can be calculated exactly

$$\Psi(\zeta, \zeta_2) = \frac{2}{\sqrt{\chi}} F(\psi, k).$$

$$(7.4.10)$$

Here

$$k = \frac{1}{\sqrt{2}}\sqrt{1 - \frac{1 - 3\zeta_2}{\chi}}, \quad \chi = \sqrt{(1 - \zeta_2)(1 + 3\zeta_2)},$$

$$(7.4.11)$$

and

$$\psi = \arcsin\left(2\sqrt{\frac{(\zeta_2 - \zeta)\chi}{(3\zeta_2 - 1 + \chi)(1 - \zeta_2 + \chi - 2\zeta)}}\right).$$

$$(7.4.12)$$

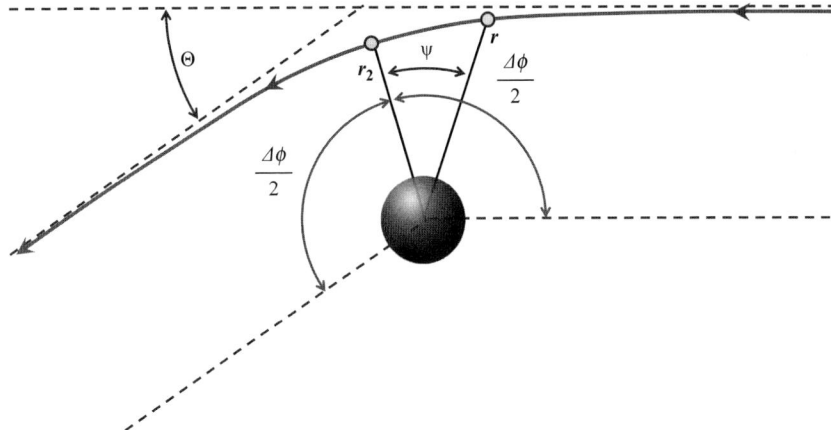

Fig. 7.8 The light deflection by the gravitational field of a black hole. The minimal distance of the null ray to the black hole is $r_2 = r_S/\zeta_2$.

Another equivalent representation for $\Psi(\zeta, \zeta_2)$ is

$$\Psi(\zeta, \zeta_2) = \frac{2\sqrt{2}}{u} F(A, B), \tag{7.4.13}$$

where

$$u = \sqrt{1 - 3\zeta_2 + \sqrt{1 + 2\zeta_2 - 3\zeta_2^2}}, \quad v = \sqrt{3\zeta_2 - 1 + \sqrt{1 + 2\zeta_2 - 3\zeta_2^2}}$$

$$A = \arcsin\left(\frac{\sqrt{2(\zeta_2 - \zeta)}}{v}\right), \qquad B = i\frac{v}{u}. \tag{7.4.14}$$

The parameters u and v are real and positive for the considered range of the parameter $\zeta_2 \in [0, 2/3]$. The function Ψ given by Eq. (7.4.13) is also real in spite of the appearance of i in the argument of the elliptic function. It's plot is depicted in Fig. 7.9.

7.4.3 Light deflection

As a first application let us consider light deflection by a black hole. Namely, consider a null ray that comes from infinity and after reaching the radial turning point returns to infinity. For such rays $\ell > \ell_{min} = 3\sqrt{3}/2$. So we assume that this condition is satisfied. The deflection angle Θ for such a null orbit is (see Figure 7.8)

$$\Theta = \Delta\phi - \pi, \quad \Delta\phi = 2\Psi(0, \zeta_2). \tag{7.4.15}$$

In this relation ζ_2 is the inverse radius parameter for the radial turning point. Using Eq. (7.4.10) and the properties of the elliptic integrals one obtains

$$\Psi(0, \zeta_2) = \frac{2}{\sqrt{\chi}} F(\alpha, k) = \frac{2}{\sqrt{\chi}} [K(k) - F(\varphi, k)] \tag{7.4.16}$$

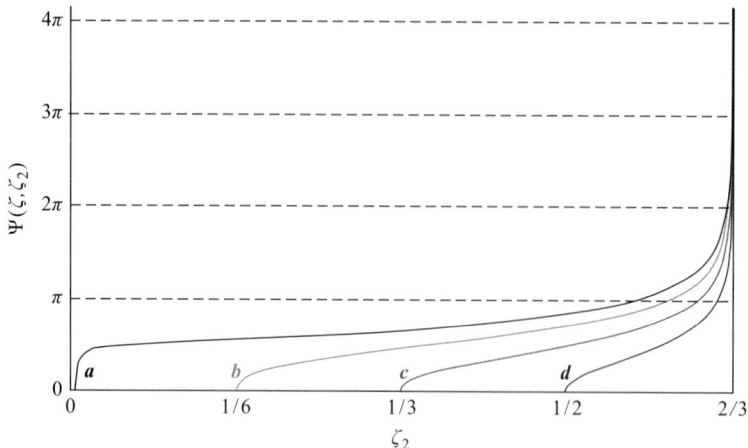

Fig. 7.9 The function $\Psi(\zeta, \zeta_2)$ as a function of ζ_2. The plots (*a*), (*b*),(*c*), and (*d*), correspond to $\zeta = 1/200$, $\zeta = 1/6$, $\zeta = 1/3$, and $\zeta = 1/2$, respectively.

$$\Theta = -\pi + \frac{4}{\sqrt{\chi}} F(\alpha, k) = -\pi + \frac{4}{\sqrt{\chi}} [\mathbf{K}(k) - F(\varphi, k)]. \tag{7.4.17}$$

Here, $\mathbf{K}(k)$ is the complete elliptic integral of the first kind, Eq. (7.3.21), F is the incomplete elliptic integral of the first kind, Eq. (7.3.20), k and χ are given by Eq. (7.4.11), and

$$\varphi = \arcsin\left(\sqrt{\frac{1 - \zeta_2 - \chi}{1 - 3\zeta_2 - \chi}}\right), \quad \alpha \equiv \psi|_{\zeta=0} = \arcsin\left(\sqrt{\frac{3\zeta_2 + 1 - \chi}{3\zeta_2 - 1 + \chi}}\right). \tag{7.4.18}$$

If the impact radius ℓr_S is much bigger than the gravitational radius, i.e. $\ell \gg 1$, then parameter $\zeta_2 \ll 1$. The obtained relations for the deflection angle Θ can be used in order to find its expansion with respect to the small parameter q. Using Eq. (7.4.8) one obtains

$$\zeta_2 = q + \frac{1}{2}q^2 + \frac{5}{8}q^3 + q^4 + \frac{231}{128}q^5 + O(q^6). \tag{7.4.19}$$

Substituting this expansion into Eq. (7.3.20) one finds expansions for the α and k parameters. Using these expansions and series expansion of the elliptic integral $F(\alpha, k)$ one obtains

$$\Theta = 2q + \frac{15\pi}{16}q^2 + \frac{16}{3}q^3 + \frac{3465\pi}{1024}q^4 + \frac{112}{5}q^5 + O(q^6). \tag{7.4.20}$$

The leading term in this expression reproduces the *Einstein formula for light deflection* in the weak gravitational field of static massive objects. The other terms provide higher-order corrections.

This result can also be obtained directly from the integral expression for the bending angle Eq. (7.4.9). After changing the integration variable $\zeta = \zeta_2 y$ in the integral for $\Psi(0, \zeta_2)$ one obtains the following expression for the deflection angle

$$\Theta = -\pi + 2 \int_0^1 \frac{dy}{\sqrt{1 - y^2 - \zeta_2(1 - y^3)}}. \tag{7.4.21}$$

For the small parameter ζ_2 one can rewrite this expression as follows

$$\Theta = \zeta_2 \int_0^1 dy \frac{1-y^3}{(1-y^2)^{3/2}} + \frac{3}{4}\zeta_2^2 \int_0^1 dy \frac{(1-y^3)^2}{(1-y^2)^{5/2}} + O(\zeta_2^3)$$

$$= 2\zeta_2 + \left(\frac{15\pi}{16} - 1\right)\zeta_2^2 + O(\zeta_2^3). \tag{7.4.22}$$

For small ζ_2 one has

$$\zeta_2 = q + \frac{1}{2}q^2 + O(q^3). \tag{7.4.23}$$

Combining these two relations one gets

$$\Theta = 2q + \frac{15\pi}{16}q^2 + O(q^3) = 2\frac{Er_S}{L} + \frac{15\pi}{16}\left(\frac{Er_S}{L}\right)^2 + \dots. \tag{7.4.24}$$

Problem 7.2:   *Calculate the higher order in q terms in the expansion Eq. (7.4.22). Use the obtained result and Eq. (7.4.19) to derive the relation Eq. (7.4.20).*

7.4.4   Geometry of null rays

A light ray emitted at the point p is uniquely specified either by the impact parameter ℓ or by its radial turning point r_2. Sometimes, it is more convenient to use another parametrization. Namely, to specify a ray it is sufficient to give the angle η between the direction of the ray at p and the direction to the black hole. We establish now a relation between the angle η and the other two parameters, ℓ and r_2.

We choose the spherical coordinates so that the ray propagates in the equatorial plane $\theta = \pi/2$. Let $\boldsymbol{u}$ be the four-velocity of an observer that is at rest at the point p

$$u^\mu = (g^{-1/2}, 0, 0, 0). \tag{7.4.25}$$

We choose the other 3 spatial unit vectors orthogonal to $\boldsymbol{u}$ as follows

$$e_{\hat{r}}^\mu = (0, g^{1/2}, 0, 0), \quad e_{\hat{\theta}}^\mu = (0, 0, r^{-1}, 0), \quad e_{\hat{\phi}}^\mu = (0, 0, 0, r^{-1}). \tag{7.4.26}$$

The null-ray orbit $r = r(\phi)$ satisfies the equation

$$\frac{dr}{d\phi} = -r^2\sqrt{F}, \quad F = q^2 - \zeta^2(1-\zeta) \equiv q^2 - \left(\frac{r_S}{r}\right)^2\left(1 - \frac{r_S}{r}\right). \tag{7.4.27}$$

We choose the minus sign in this relation. For this choice the ray emitted at r at the beginning moves to smaller values of the radius, then reaches the turning point at $r_2 < r$ (see Figure 7.10). A unit spatial vector in the direction of the emitted ray is

$$e = \frac{1}{\sqrt{r^2 + g^{-1}(dr/d\phi)^2}}(0, dr/d\phi, 0, 1). \tag{7.4.28}$$

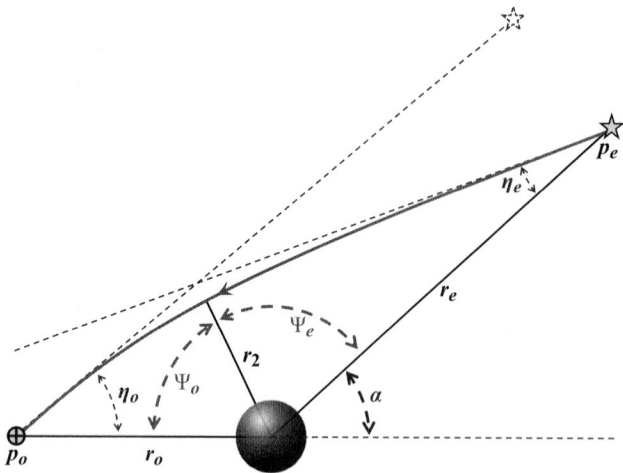

Fig. 7.10 An observer at point p_e emits a null ray at the angle η_e with respect to the direction to the black hole. Another observer at point p_o receives this ray at the angle η_o with respect to the direction to the black hole.

The angle η between the ray and the direction to the black hole is determined by the relation

$$\cos \eta = -(e, e_{\hat{r}}) = -\frac{dr/d\phi}{\sqrt{gr^2 + (dr/d\phi)^2}}. \tag{7.4.29}$$

Denote $\zeta = r_S/r$, then simple calculations give

$$\cos \eta = q^{-1}\sqrt{F} \equiv q^{-1}\sqrt{q^2 - \zeta^2(1 - \zeta)}. \tag{7.4.30}$$

Since $\cos \eta \geq 0$, one has $0 < \eta \leq \pi/2$. For the observer receiving the signal the ray is directed outwards the black hole and the signs in Eqs. (7.4.27) and (7.4.30) are opposite and $\cos \eta' \leq 0$. This angle corresponds to $\eta' = \pi - \eta_o$ as depicted in Figure 7.10. It is convenient to use the symmetry between the emitter and the receiver in the setup of the problem and interchange the observers. Then, one can use the same equations in both cases, but one should just keep in mind the proper direction of the light propagation.

Equation (7.4.30) establishes the relation between the angle η and the impact parameter $\ell = q^{-1}$. One also gets the following relation between η and $\zeta_2 = r_S/r_2$

$$\sin \eta = \sqrt{\frac{\zeta^2 - \zeta^3}{\zeta_2^2 - \zeta_2^3}}. \tag{7.4.31}$$

The latter relation allows one to express ζ_2 in terms of ζ and η

$$\zeta_2 = \frac{P}{\cos\left[\frac{1}{3} \arccos(-3P)\right]}, \quad P = \frac{\sqrt{3}\zeta\sqrt{1 - \zeta}}{2 \sin \eta}. \tag{7.4.32}$$

This relation, written in the radial coordinates, gives

$$r_2 = \frac{2}{\sqrt{3}} \frac{r \sin \eta}{\sqrt{1 - \dfrac{r_S}{r}}} \cos\left[\frac{1}{3} \arccos\left(-\frac{3\sqrt{3}}{2} \frac{r_S}{r \sin \eta}\sqrt{1 - \frac{r_S}{r}}\right)\right]. \tag{7.4.33}$$

Figure 7.10 illustrates a problem that can be solved by using obtained results. Namely, one observer at the point p_e emits a photon at the angle η_e with respect to the direction to the black hole. After passing the radial turning point at r_2 it reaches the second observer at point p_o, which registers its angle η_o with respect to the direction to the black hole. The radii of the observers are r_e and r_o, respectively. In order to find a relation between the angles η_e and η_o one has to find the radius r_2 of the turning point by using Eq. (7.4.33) for the initial data $r = r_e$ and $\eta = \eta_e$. Then, one can use Eq. (7.4.31) for $\zeta = r_S/r_o$ and $\zeta_2 = r_S/r_2$ and find η_o.

7.4.5   Time delay

Another effect of gravity on the propagation of light is the *time delay* of signals in the gravitational field. A simplified version of an experiment that measures this time delay is the following. Consider a spherically symmetric object. As a special case it can be a black hole. The gravitational field outside it is described by the Schwarzschild metric. Consider a beam of light emitted by an observer located at the point p_E, $(r = r_E, \phi = 0)$, which reaches a reflector placed at the point p_R, (r_R, ϕ_0), and returns to the original point (see Figure 7.11). The observer measures the time Δt between the emission of the signal and its return. For a signal passing close to the gravitating body the time interval Δt is larger than in a flat spacetime. This effect is known as the *Shapiro time delay*.

We assume that the observer is at rest, so that this time is $\Delta t = 2\Delta t_{ER}$, where Δt_{ER} is the time of the one-way trip. We assume that the corresponding null-ray orbit between emission and reflection has a radial turning point at $r = r_2$. In order to reach the point p_R the ray must have a special value of the specific angular momentum ℓ, which uniquely specifies r_2. Our aim is to express Δt in terms of observables that can be measured by a person performing this experiment. We solve this problem in two steps:

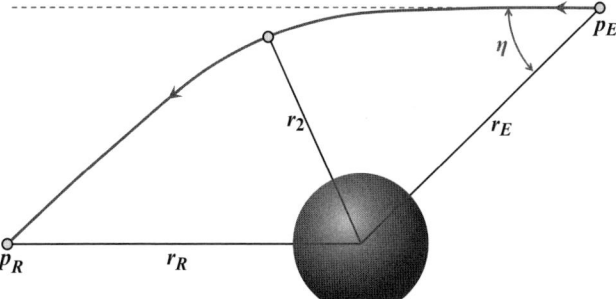

Fig. 7.11 An observer located at the point p_E sends a light pulse to the reflector at the point p_R. On its way to the reflector, the null ray passes at the minimum distance r_2 from the center. After reflection at p_R, the ray returns to the initial point p_E. The observer measures the time between emission of the signal and its return.

1. We first find Δt_{ER} as a function of three parameters r_E, r_R, and r_2.
2. We establish a relation between r_2 and the angle η between the visible position of the reflector and the direction to the center of the gravitating object (black hole).

We start with the first problem. From the equations of motion Eq. (7.1.14) we have

$$
\frac{d\tilde{t}}{d\zeta} = \frac{1}{g\zeta^2\sqrt{1 - \ell^2 g\zeta^2}} = \frac{q}{(1-\zeta)\zeta^2\sqrt{q^2 - \zeta^2 + \zeta^3}}, \qquad g = 1 - \zeta. \tag{7.4.34}
$$

Taking into account the relation $q^2 = 1/\ell^2 = \zeta_2^2 - \zeta_2^3$ one gets for the time of light propagation from the observer located at radius r to the point of closest approach to the black hole at r_2

$$
\tilde{t} = \zeta_2\sqrt{1 - \zeta_2} \int_\zeta^{\zeta_2} d\zeta \, \frac{1}{(1-\zeta)\zeta^2\sqrt{\zeta_2^2 - \zeta_2^3 - \zeta^2 + \zeta^3}}. \tag{7.4.35}
$$

This integral can be calculated exactly and expressed in terms of the elliptic integrals.

$$
\tilde{t} = \frac{u}{\sqrt{2}\,\zeta_2\sqrt{1 - \zeta_2}} \left[-E(A,B) + \left(1 + \frac{2\zeta_2}{u^2}\right) F(A,B) \right]
$$

$$
+ \frac{2\sqrt{2}\,\zeta_2\sqrt{1 - \zeta_2}}{u} \left[\frac{1}{\zeta_2}\Pi\left(A, \frac{C}{\zeta_2}, B\right) + \frac{1}{1 - \zeta_2}\Pi\left(A, -\frac{C}{1 - \zeta_2}, B\right) \right]. \tag{7.4.36}
$$

Here

$$
u = \sqrt{1 - 3\zeta_2 + \sqrt{1 + 2\zeta_2 - 3\zeta_2^2}}, \qquad v = \sqrt{3\zeta_2 - 1 + \sqrt{1 + 2\zeta_2 - 3\zeta_2^2}}
$$

$$
A = \arcsin\left(\frac{\sqrt{2}\sqrt{\zeta_2 - \zeta}}{v}\right), \qquad B = i\frac{v}{u}, \qquad C = \frac{v^2}{2}. \tag{7.4.37}
$$

The parameters u, v, A, and C are real. Equation (7.4.36) allows one to solve the problem exactly, but the corresponding calculations are quite involved. They can be easily done numerically.

In what follows we assume that the null ray passes the black hole at the distance well above the gravitational radius, then $\zeta_2 \ll 1$. We solve the problem perturbatively by expanding the integrand in powers of the small parameter. Substitution of $\zeta = \zeta_2 \sin\beta$ in Eq. (7.4.35) leads to

$$
\tilde{t} = \frac{1}{\zeta_2} \int_{\arcsin(\zeta/\zeta_2)}^{\pi/2} d\beta \, \frac{1}{\sin^2\beta(1 - \zeta_2\sin\beta)\sqrt{1 - \dfrac{\zeta_2}{1 - \zeta_2}\dfrac{\sin^2\beta}{1 + \sin\beta}}}. \tag{7.4.38}
$$

Then, for $\zeta_2 \ll 1$ we have

$$
\tilde{t} = \frac{1}{\zeta_2} \int_{\arcsin(\zeta/\zeta_2)}^{\pi/2} \frac{d\beta}{\sin^2\beta} + \int_{\arcsin(\zeta/\zeta_2)}^{\pi/2} d\beta \left(\frac{1}{\sin\beta} + \frac{1}{2(1 + \sin\beta)} \right) + O(\zeta_2). \tag{7.4.39}
$$

We denote $\zeta/\zeta_2 = \sin\beta$ and recall that $t = r_S\tilde{t}$, then the result reads

$$t = \frac{r_S}{\zeta_2}\frac{1}{\tan\beta} + r_S\left[-\ln\tan\left(\frac{\beta}{2}\right) + \frac{1}{2}\tan\left(\frac{\pi}{4} - \frac{\beta}{2}\right)\right] + O(\zeta_2). \tag{7.4.40}$$

Denote by t_E the time t required for a null ray emitted at r_E to reach the turning point r_2. Restoring the original radial coordinates we can write the result as

$$t_E = \sqrt{r_R^2 - r_S^2} + r_S\ln\left(\frac{r_E + \frac{1}{2}\sqrt{r_E^2 - r_2^2}}{r_2}\right) + \frac{1}{2}\sqrt{\frac{r_E - r_2}{r_E + r_2}}. \tag{7.4.41}$$

A similar formula is valid for the propagation time from the turning point r_2 to the reflector at r_R. The round-trip time Δt is the double of their sum, $\Delta t = 2(t_E + t_R)$. Since the observer is in the gravitational field, the time measured by his/her clock differs from Δt by a redshift factor. Thus, we obtain

$$\Delta\tau = 2\sqrt{1 - \zeta_E}(t_E + t_R),$$

$$\Delta\tau = \sqrt{1 - \frac{r_S}{r_E}}\left\{2\sqrt{r_E^2 - r_S^2} + 2\sqrt{r_R^2 - r_S^2}\right.$$

$$+ r_S\left[2\ln\left(\frac{r_E + \sqrt{r_E^2 - r_2^2}}{r_2}\right) + 2\ln\left(\frac{r_R + \sqrt{r_R^2 - r_2^2}}{r_2}\right)\right. \tag{7.4.42}$$

$$\left.\left. + \sqrt{\frac{r_E - r_2}{r_E + r_2}} + \sqrt{\frac{r_R - r_2}{r_R + r_2}}\right]\right\}.$$

To solve the second part of the problem it is sufficient to use the results of Section 7.4.4. Namely, Eq. (7.4.33) can be used to express r_2 in terms of r_E and η

$$r_2 = \frac{2}{\sqrt{3}}\frac{r_E\sin\eta}{\sqrt{1 - \frac{r_S}{r_E}}}\cos\left[\frac{1}{3}\arccos\left(-\frac{3\sqrt{3}}{2}\frac{r_S}{r_E\sin\eta}\sqrt{1 - \frac{r_S}{r_E}}\right)\right]. \tag{7.4.43}$$

Let us summarize. By measuring the angle η the observer can find r_2 for the coming null ray. It is sufficient to use Eq. (7.4.33). Substituting this quantity in Eq. (7.4.42) one finds the expected value of the time delay $\Delta\tau$. Let us note that the obtained relation contains only observable parameters. The result does not depend on what kind of gravitating object is used. In the real experiment the time delay was observed in the Solar System. Certainly, the described gedanken experiment is simplified. In real observations the emitter and reflector are located at bodies that are not at rest, but move along Keplerian orbits. It should be emphasized that the presented derivation used the approximation $r \gg r_S$. However, the exact formula Eq. (7.4.36), combined with exact formula for the bending angle Eq. (7.4.10) allow one to perform similar calculations beyond this approximation. This might be important for future observations of the light propagation near black holes.

The time-delay effect was first detected in 1964, by Irwin I. Shapiro for reflection of the radar beams off the surface of Venus and Mercury. These experiments have been repeated many times since then, with increasing accuracy. This is one of the important tests of general relativity in the weak-field regime. As an interesting application of the Shapiro delay let us mention recent observations that allowed one to establish the mass of the heaviest known pulsar. The obtained mass of the neutron star PSR J1614-2230 is 1.97M⊙. This system is binary, which besides a neutron star contains a white dwarf. When the white dwarf is in front of the pulsar, the radio waves emitted by the pulsar propagate near the white dwarf and are affected by its gravitational field. Because of the Shapiro delay the time of the arrival of the signal at the Earth depends on the position of the pulsar. This dependence was observed. This gives the mass of the white dwarf, and this allows one to estimate the neutron star mass.

7.4.6   Darwin's formula for bending angle

As we mentioned, a null ray with the impact parameter $\ell = \ell_{min} = 3\sqrt{3}/2$ is a critical one. It comes from infinity and makes an infinite number of turns around the black hole approaching the circle $\zeta = 2/3$. There exists another *critical null ray* approaching the same circle from the inner region ($\zeta > 2/3$). We study now the behavior of the *near-critical rays*.

We use the second of the relations Eq. (7.4.17) for the deflection angle

$$\Theta = -\pi + \frac{4}{\sqrt{\chi}}\left[\mathbf{K}(k) - \mathbf{F}(\varphi, k)\right], \tag{7.4.44}$$

where

$$k = \frac{1}{\sqrt{2}}\sqrt{1 - \frac{1 - 3\zeta_2}{\chi}}, \qquad \chi = \sqrt{(1 - \zeta_2)(1 + 3\zeta_2)}, \tag{7.4.45}$$

$$\varphi = \arcsin\left(\sqrt{\frac{1 - \zeta_2 - \chi}{1 - 3\zeta_2 - \chi}}\right).$$

For the critical ray one has

$$\zeta_2^0 = 2/3, \quad k_0 = 1, \quad \sin\varphi_0 = 1/\sqrt{3}. \tag{7.4.46}$$

The function $\mathbf{F}(\varphi, k)$ is regular in this limit. Using the relation

$$\mathbf{F}(\varphi, 1) = \ln\left(\frac{1 + \sin\varphi}{\cos\varphi}\right), \tag{7.4.47}$$

one obtains

$$\mathbf{F}(\varphi_0, 1) = \frac{1}{2}\ln(2 + \sqrt{3}). \tag{7.4.48}$$

The complete elliptic integral $\mathbf{K}(k)$ has a logarithmic divergence near $k = 1$ and one has

$$\mathbf{K}(k) = \ln(4/k') + O(k'^2 \ln k'), \qquad k' = \sqrt{1 - k^2}. \tag{7.4.49}$$

In order to estimate the contribution of this term to the deflection angle for the near-critical ray we write the angle ψ that enters relations (7.4.5) and (7.4.7) in the form

$$\psi = \pi/2 - y. \tag{7.4.50}$$

Keeping the leading terms in the expansion over small parameter y one obtains

$$\zeta_2 = \frac{2}{3} - \frac{2\sqrt{3}}{9}y, \quad \chi = 1 + \frac{2\sqrt{3}}{9}y,$$

$$k = 1 - \frac{2}{3\sqrt{3}}y, \quad k' = \frac{2}{3^{3/4}}y^{1/2}, \tag{7.4.51}$$

$$\mathbf{K}(k) = \ln 2 + \frac{3}{4}\ln 3 - \frac{1}{2}\ln y.$$

The inverse impact parameter q, defined as $q^2 = \zeta_2^2(1 - \zeta_2)$, and the impact parameter ℓ are

$$q^2 = \frac{4}{27}(1 - y^2), \quad \ell = \ell_{\min} + \frac{3\sqrt{3}}{4}y^2. \tag{7.4.52}$$

Denote $\Delta\ell = \ell - \ell_{\min}$. Using Eq. (7.4.52) we express y in terms of $\Delta\ell$. Using the relations Eq. (7.4.21), Eq. (7.4.51) and Eq. (7.4.52) one obtains

$$\Theta = -\ln \Delta\ell - \pi + \ln \left[\frac{4 \times 3^{9/2}}{(2 + \sqrt{3})^2} \right]. \tag{7.4.53}$$

In this relation Θ can take values greater that 2π. If we restrict the values of the deflection angle θ by 2π we shall write $\Theta = 2m\pi + \theta$, where m is the number of turns that have been made by the ray before its returning to infinity. One has

$$\Delta\ell = A \exp[-(\theta + (2m + 1)\pi)],$$

$$A = \frac{4 \times 3^{9/2}}{(2 + \sqrt{3})^2} = 324(7\sqrt{3} - 12). \tag{7.4.54}$$

This is the so-called Darwin's formula.

7.5 Ray-Tracing in Schwarzschild Spacetime

7.5.1 Scattering data for null rays

It would be interesting to describe how the picture of the sky is deformed by the gravitational field of the black hole. This effect is known as *gravitational lensing*. For an observer on the Earth the sky is described by angular positions of stars, planets, galaxies, etc. This is equivalent to a projection of all their images onto a sphere, like in a planetarium. Suppose an observer, that is located at rest at far distance r_o from the spherical black hole of mass M, is making an observation of a sky behind the black hole. We assume that all the stars are also located far away from the black hole. The presence of the black hole deforms the image of the sky and affects the time of arrival of signals.

We choose spherical coordinates so that a chosen null ray is in the equatorial plane $\theta = \pi/2$ and the spherical coordinates of the observer are $(\theta_+ = \pi/2, \phi_+ = 0)$. A null ray registered by the distant observer is specified by the impact parameter ℓ and the time t_+ of its arrival. A condition that the deflection angle coincides with $-\phi_-$ for the incoming ray is

$$\phi_- = -2\Psi(0, \zeta_2). \tag{7.5.1}$$

Here, ζ_2 is the inverse radius of the turning point for the impact parameter ℓ. In what follows we always assume that $\ell > \ell_{min}$, so that the rays are not captured by the black hole. For such rays $0 < \zeta_2 < 2/3$. The relation between ζ_2 and ℓ is given by Eqs. (7.4.5) and (7.4.7).

For fixed t_+ the time t_-, when the ray was sent, depends on the radius at which we measure this time. Moreover, $|t_\pm| \to \infty$, when the radii of emission and observation becomes infinitely large. In order to define variables in which both initial and final scattering parameters are finite, we proceed as follows. Let us introduce the parameter

$$u_\epsilon = t - \epsilon r_*, \qquad r_* = \int \frac{dr}{g}, \qquad \epsilon = (\pm). \qquad (7.5.2)$$

For $\epsilon = +$ this parameter u_+ is a *retarded time*, while for $\epsilon = -$ u_- is an *advanced time*. Null surfaces $u_+ =$const and $u_- =$const are formed by radial outgoing and incoming null rays, respectively. The angle ϕ is constant along these null generators.

For both incoming and outgoing (not necessary radial) rays $u_\pm$ remain finite in $r \to \infty$ limit. We shall also use dimensionless versions of the coordinates u and r_*

$$U_\pm = \frac{u_\pm}{r_S}, \qquad \tilde{r}_* = \frac{r_*}{r_S}. \qquad (7.5.3)$$

The definition of r_* contains an ambiguity in the choice of the integration constant. To fix this ambiguity we denote by ζ_m the point where the function $\zeta^2(1 - \zeta)$ reaches its maximum

$$\zeta_m = 2/3, \qquad \zeta_m^2(1 - \zeta_m) = 4/27. \qquad (7.5.4)$$

The corresponding radius is $r_m = r_S/\zeta_m = 3r_S/2$. We define

$$r_* = \int_{r_m}^{r} \frac{dr}{g(r)}. \qquad (7.5.5)$$

For the dimensionless version of r_* one has

$$\tilde{r}_* = \frac{1}{\zeta} - \ln \frac{\zeta}{1 - \zeta} - 3/2 + \ln 2. \qquad (7.5.6)$$

Using the equations of motion of the null rays Eq. (7.1.14) we obtain the following relations

$$\frac{dU_\epsilon}{d\zeta} = -\epsilon \frac{1}{\sqrt{F}(q + \sqrt{F})}, \qquad \frac{d\phi}{d\zeta} = -\epsilon \frac{1}{\sqrt{F}}. \qquad (7.5.7)$$

Here, $q = \ell^{-1}$, $F = q^2 - \zeta^2 g$, and ϵ is $+$ for outgoing rays, and $-$ for incoming ones.
At the radial turning point r_2

$$u_\epsilon(r_2) = t_2 - \epsilon r_{2*}. \qquad (7.5.8)$$

Thus,

$$u_+(r_2) - u_-(r_2) = -2r_{2*}. \qquad (7.5.9)$$

Using this relation one obtains

$$u_+|_{\mathcal{J}^+} = u_-|_{\mathcal{J}^-} - 2r_{2*} + \Delta U, \qquad (7.5.10)$$

where U is a smooth function at the turning point, which obeys Eq. (7.5.7), and ΔU is the difference between the values of this function on $\mathcal{J}^+$ and $\mathcal{J}^-$. Integrating Eq. (7.5.7) over the trajectory one gets

$$\Delta U = 2 \int_0^{\zeta_2} \frac{d\zeta}{\sqrt{F}(q + \sqrt{F})},$$

$$\Delta \phi = \phi_+ - \phi_- = 2 \int_0^{\zeta_2} \frac{d\zeta}{\sqrt{F}} = 2\Psi(0, \zeta_2). \tag{7.5.11}$$

Both expressions can be written in terms of the elliptic integrals. For $\Delta\phi$ we have already obtained the explicit answer (see Eq. (7.4.16)).

These results allow the following interpretation. Let us call $\{u_\pm, \theta_\pm, \phi_\pm, \ell_\pm\}$ initial (for $-$) and final (for $+$) scattering data for a null ray. For any ray the impact parameter is constant along the trajectory, thus $\ell_+ = \ell_-$. In the chosen coordinates $\theta_+ = \theta_- = \pi/2$, while the relations Eq. (7.5.10) and Eq. (7.5.11) give the relation between $u_\pm$ and $\phi_\pm$.

To obtain the relations between the initial and final scattering data out of the equatorial plane it is sufficient to perform a rigid rotation. Let us embed the unit two-sphere in $\mathbb{E}^3$ with Cartesian coordinates (X, Y, Z). Denote

$$\boldsymbol{n}_{0,+} = (1, 0, 0), \quad \boldsymbol{n}_+ = (\sin\theta_+ \cos\phi_+, \sin\theta_+ \sin\phi_+, \cos\theta_+). \tag{7.5.12}$$

A unit vector $\boldsymbol{e}$ orthogonal to both of these vectors is

$$\boldsymbol{e} = (0, e_Y, e_Z), \quad e_Y = \frac{1}{\sqrt{1 + \tan^2\theta_+ \sin^2\phi_+}}, \quad e_Z = -\tan\theta_+ \sin\phi_+ e_Y. \tag{7.5.13}$$

The required transformation is a rotation $\boldsymbol{O}$ around $\boldsymbol{e}$ at the angle α, $\cos\alpha = \sin\theta_+ \cos\phi_+$, in the direction from $\boldsymbol{n}_{0,+}$ to $\boldsymbol{n}_+$. The null ray outgoing in the direction $\boldsymbol{n}_{0,+}$ corresponds to the following direction of the incoming ray

$$\boldsymbol{n}_{0,-} = (\cos\phi_-, \sin\phi_-, 0). \tag{7.5.14}$$

A similar relation for an arbitrary $\boldsymbol{n}_+$ is

$$\boldsymbol{n}_- = \boldsymbol{O}\boldsymbol{n}_{0,-}. \tag{7.5.15}$$

It is convenient to consider 3 quantities $(u_\pm, \phi_\pm, \theta_\pm)$ as coordinates on a 3-dimensional manifold $\mathcal{J}^\pm$. Since θ and ϕ are coordinates on a unit sphere, the manifold $\mathcal{J}^\pm$ has the topology $\mathbb{R}^1 \times S^2$. For $\ell > \ell_{\min}$ a null ray starts its motion at $\mathcal{J}^-$ and after scattering by the black hole reaches $\mathcal{J}^+$. (In Section 10.1 it will be demonstrated that $\mathcal{J}^\pm$ has a well-defined geometrical meaning.) For a fixed value of ℓ the null rays establish a map between the *past null infinity* $\mathcal{J}^-$ and the *future null infinity* $\mathcal{J}^+$. The problem of the light scattering by the black hole can be formulated as a problem of finding such a map.

The established relations allow another interpretation, which is more directly connected with observations. Let us fix the position of the observer and choose a 2-plane orthogonal to the direction to the black hole. Choose the center at this plane corresponding to a radial outgoing ray with $\ell = 0$. Points of the plane are determined by a 2-vector $\vec{\ell}$, with a norm $|\vec{\ell}| = \ell$ and direction tangent to the plane of the orbit of the corresponding null ray.

For $|\vec{\ell}| < \ell_{\min} = 3\sqrt{3}/2$ the rays being traced back reach the black hole. This means that no light comes to an observer in this domain of the impact parameters. The region $|\vec{\ell}| \le 3\sqrt{3}/2$ is called a *black hole shadow*. It is a disk of area $27\pi/4$, which in the dimensional units coincides with the capture cross-section of the black hole for light rays. We shall see in Section 8.6.3 that black hole rotation changes the size and deforms the shape of the black hole shadow. In particular, this effect can be used to distinguish between rotating and non-rotating black holes. The images of point-like sources are discussed in Section 7.8. Here, we just mention that if the brightness of the source changes in time the brightness of the image changes as well, but for different ℓ the corresponding time delay, given by Eq. (7.5.10), is different.

7.5.2   Darwin's formula for time delay

The integral for ΔU in Eq. (7.5.11) can be taken exactly, but the result is quite messy. Instead of this sometimes it is convenient to use the following trick. We use the identity Eq. (7.1.16)

$$\frac{\partial}{\partial \ell}\left(\frac{d\tilde{\imath}}{d\varsigma}\right) = \ell\frac{\partial}{\partial \ell}\left(\frac{d\phi}{d\varsigma}\right), \tag{7.5.16}$$

valid locally at any part of the null trajectory. Using Eq. (7.5.2) and the definition of function U, we have

$$\frac{\partial}{\partial \ell}\left(\frac{dU}{d\varsigma}\right) = \frac{\partial}{\partial \ell}\left(\frac{d\tilde{\imath}}{d\varsigma}\right) \tag{7.5.17}$$

Integrating both parts of Eq. (7.5.16) along the complete null trajectory one obtains

$$\partial_\ell \Delta U = 2\ell\partial_\ell \Psi(0, \varsigma_2). \tag{7.5.18}$$

Using the relations between q, ℓ, and ς_2 one gets

$$\ell\partial_\ell\varsigma_2 = -q\partial_q\varsigma_2 = -\frac{2\varsigma_2(1 - \varsigma_2)}{2 - 3\varsigma_2}. \tag{7.5.19}$$

Thus, one has

$$\partial_\ell \Delta U = -\frac{2\varsigma_2(1 - \varsigma_2)}{2 - 3\varsigma_2}\frac{d}{d\varsigma_2}\Psi(0, \varsigma_2). \tag{7.5.20}$$

Equation (7.5.18) can be also written in the form

$$\partial_\ell \Delta U = \ell\partial_\ell \Theta, \tag{7.5.21}$$

where Θ is the deflection angle. This relation allows one to quite easily calculate the time delay for null rays with the impact parameter close to the critical one. Using Eq. (7.4.53) for the deflection angle for near-critical rays one obtains

$$\partial_\ell \Delta U \sim -\frac{\ell}{\ell - \ell_{\min}}. \tag{7.5.22}$$

Integrating this relation one gets

$$\Delta U \sim -\ell_{min} \ln(\ell - \ell_{min}), \quad \ell_{min} = \frac{3\sqrt{3}}{2}. \tag{7.5.23}$$

This is a time-delay analog of the Darwin's formula for the near-critical scattering.

7.6 Black Hole as a Gravitational Lens

7.6.1 Focusing effect of gravity

Gravity attracts everything, including light. Any distribution of matter that obeys the *null-energy condition* produces a focusing effect on the light propagation. In this sense, gravitating masses are similar to an optical lens with distortion.

Consider a narrow beam of light rays and let l be null vectors tangent to the light rays in the affine parametrization: $l^\nu l^\mu{}_{;\nu} = 0$. Let us place an opaque object in the path of the beam and put a screen behind the object at some distance. The theorem (Jordan *et al.* 1961; Sachs 1961) states that: *All parts of the shadow reach the screen simultaneously. The size, shape, and orientation of the shadow depend only on the position of the screen and are independent of the velocity of the observer. If the screen is at a short distance δr from the object, the expansion and distortion of the shadow are $\theta \delta r$ and $\sigma \delta r$, where*

$$\theta = \frac{1}{2} l^\mu{}_{;\mu}, \qquad \sigma = \left(\frac{1}{2} l_{\mu;\nu} l^{\mu;\nu} - \theta^2 \right)^{1/2}. \tag{7.6.1}$$

The area $\mathcal{A}$ of the beam obeys the equations

$$\frac{d\mathcal{A}^{1/2}}{dr} = \theta \mathcal{A}^{1/2},$$

$$\frac{d^2 \mathcal{A}^{1/2}}{dr^2} = -(\sigma^2 + T_{\mu\nu} l^\mu l^\nu) \mathcal{A}^{1/2}. \tag{7.6.2}$$

If the matter obeys the null-energy condition, the right-hand side of the latter relation is non-positive. This shows that a gravity always has a focusing effect.

7.6.2 Image of a point-like source

A black hole as any other gravitating body acts on light as an optical lens with distortion. But a black hole is a very special gravitating object. Its main feature is a strong gravitational field. We focus now on those properties of the gravitational lenses that are specific to black holes.

Looking through the *black hole lens* on the bright sources one can observe that their position is distorted. The brightness of an image depends on the position of the source. Moreover, some of the sources have many images. Let us discuss these effects now.

Consider a light source at point p_e with radius r_e in the gravitational field of a black hole of mass M. An observer registering the radiation is located at the radius r_o. The angle between the directions to the emitter and the observer is Φ. We choose coordinates (t, r, θ, ϕ) so that:

- Both points p_e and p_o are located on the equatorial plane $\theta = \pi/2$.
- Azimuthal angle $\phi = 0$ for p_e and $\phi = \Phi$ for p_o.
- $0 < \Phi < \pi$.

The last condition can always be met by the choice of proper orientation of the $\theta = 0$ axis.

The null rays from r_e that reach the observer lie in the equatorial plane and their orbits are $\phi = \phi(r)$. One has $\phi(r_e) = 0$. There exist two families of null rays, which differ by the sign of the impact parameter. For $\ell > 0$ a condition that a ray meets p_o is

$$\phi_n(r_e) = \Phi + 2\pi n, \quad n \geq 0. \tag{7.6.3}$$

The non-negative parameter n shows how many times the ray turns around the black hole in the anti-clockwise direction before it reaches p_o. We call the ray with $n = 0$, satisfying the condition

$$\phi_0(r_e) = \Phi, \tag{7.6.4}$$

a *primary ray*. For rays with negative ℓ we have

$$\phi_n(r_e) = \Phi - 2\pi n, \quad n > 0. \tag{7.6.5}$$

Here, n is the number of turns of the ray in the clockwise direction. In order to consider both families simultaneously it is convenient to write Eq. (7.6.5) in a different form. Note that null-ray orbits are invariant under simultaneous change $\ell \to -\ell$ and $\phi \to -\phi$. Thus, we can assume that $\ell \geq 0$ always, provided the second condition Eq. (7.6.5) is written in the form

$$|\phi_n(r_e)| = 2\pi n - \Phi, \quad n > 0. \tag{7.6.6}$$

Let us denote

$$\Phi_n^{\pm} = 2\pi n \pm \Phi, \quad n > 0. \tag{7.6.7}$$

We also choose $\Phi_0^{\pm} = \Phi$. To summarize, the condition that the bending angle coincides with $\Phi_n^{\pm}$ determines the impact parameter of the ray for the corresponding image. We shall enumerate the images by $(n_{\pm})$, where $|n|$ is the winding number. In Figure 7.12 the images

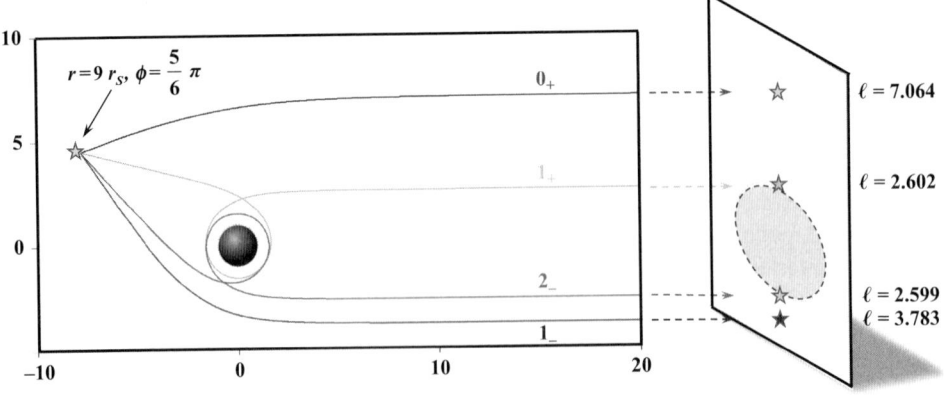

Fig. 7.12 Null rays for the first four images of the object located at $r = 9r_S$ and angle $\phi = 5\pi/6$. Images $1_+, 2_-$ and all images with higher winding numbers concentrate near the critical impact parameter $\ell = \ell_{min} = 3\sqrt{3}/2$.

marked with the sign (+) are above the black hole shadow, and images with sign (−) are below it.

Using Eq. (7.4.10) for the bending angle one can write a condition that determines the inverse impact parameters for the images as follows

$$\Phi_n^{\pm} = \pm\Psi(\zeta_e, \zeta_2) + \Psi(\zeta_o, \zeta_2). \tag{7.6.8}$$

A plus sign is for the case when the radial turning point is between p_e and p_o, while the sign is minus in the opposite case. For the observer located far from the black hole $r_o \gg r_S$, so that $\zeta_o \ll 1$. In the leading-order approximation we put $\zeta_o = 0$ in Eq. (7.6.8)

$$\Phi_n^{\pm} = \Psi_{\pm}(\zeta_e, \zeta_2), \quad \Psi_{\pm}(\zeta, \zeta_2) = \pm\Psi(\zeta_e, \zeta_2) + \Psi(0, \zeta_2). \tag{7.6.9}$$

Figure 7.13 shows the plot of $\Psi_{\pm}(\zeta, \zeta_2)$ as a function of ζ_2 for a fixed value of ζ.

To find the parameter ζ_2 that specifies the null ray, one needs to solve Eq. (7.6.8). It is easy to see that for all the rays, except the primary one, the branch Ψ_{-} does not give solutions, so that for $n > 0$ the equation is

$$2\pi n \pm \Phi = \Psi_{+}(\zeta, \zeta_2), \quad n > 0. \tag{7.6.10}$$

We denote its solutions by $\zeta_2^{n,\pm}$. The images for the sign (+) are seen on the same side of the black hole as the primary image, while the images corresponding to the sign (−) appear on the other side of the black hole. The Figure 7.12 schematically illustrates these images. The images with large winding number n correspond to the impact parameters ℓ close to the critical one, ℓ_{min}. They are located close to the black hole shadow (see Section 7.5.1).

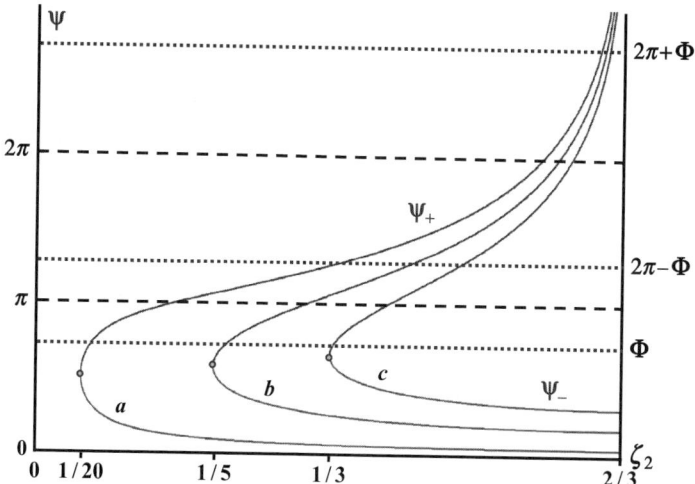

Fig. 7.13 $\Psi_{\pm}(\zeta, \zeta_2)$ as a function of ζ_2 for fixed value of ζ. Curves a, b, and c correspond to the values of $\zeta = 1/20$, $\zeta = 1/5$ and $\zeta = 1/3$, respectively.

7.6.3 Brightness of images

One can expect that the brightness of images is different. We show now that it is really so and only a few first images are bright enough to be observed.

Let us consider again a system that includes an emitter p_e and an observer p_o, placed in the gravitational field of the black hole. But now we choose the coordinates (t, r, θ, ϕ) differently. Namely, we choose the symmetry axis $\theta = 0$ to pass through p_e, and the plane $\phi = 0$ to pass through p_o. For this choice of the spatial coordinates (r, θ, ϕ) one has

$$p_e = (r_e, 0, 0), \quad p_o = (r_o, \Phi, 0). \tag{7.6.11}$$

A null ray connecting these points lies in the $\phi = 0$ plane. In order to reach p_o a ray at p_e must be emitted at the angle η with the direction to the black hole (see Figure 7.14). The condition that the ray passes through both chosen points determines its impact parameter, which, in turn, determines the angle η. This means that for a given configuration there is a relation between Φ and η, which we write in the form

$$\eta = \eta_{n_\pm}(\Phi), \tag{7.6.12}$$

where $(n_\pm)$ is the index enumerating the rays.

Consider a solid angle with the tip at the black hole and restricted by spherical coordinates

$$\theta \in (\Phi - \Delta\Phi/2, \Phi + \Delta\Phi/2), \qquad \phi \in (-\Delta\phi/2, \Delta\phi/2). \tag{7.6.13}$$

The aperture $\mathcal{A}$ of a telescope at p_o can be expressed in terms of the solid angle $\Delta\Omega_o$

$$\mathcal{A} = r_o^2 \Delta\Omega_o, \qquad \Delta\Omega_o = |\sin\Phi|\,\Delta\Phi\Delta\phi. \tag{7.6.14}$$

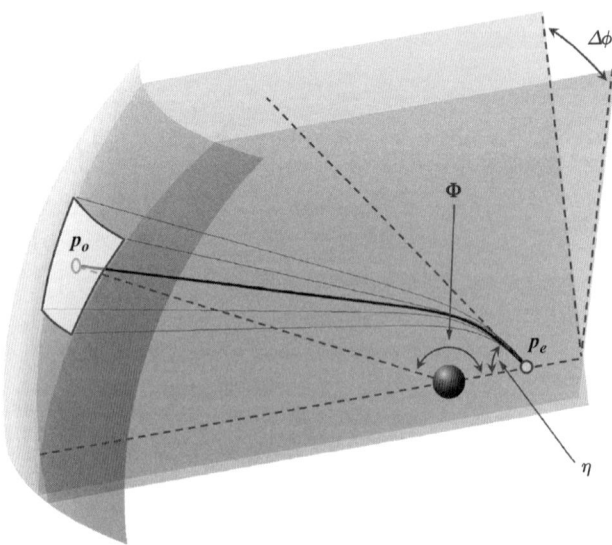

Fig. 7.14 Illustration of the calculation of the brightness of images of an object located near the black hole at the point p_e.

To determine the brightness of the image we consider a beam of null rays emitted from p_e that reach the telescope. For any direction of emission the photons propagate along the planes of the constant value of ϕ. Thus, the solid angle of the beam of photons emitted at p_e, which are collected by the telescope, is

$$d\Omega_e = |\sin\eta|\,\Delta\eta\Delta\phi. \tag{7.6.15}$$

On the other hand, the solution for these null geodesics relates the angle η of the emitted ray and the angular coordinate of the telescope Φ. This gives the condition on the range of angles of the emitted rays that reach the telescope: $\eta = \eta(\Phi) \pm \Delta\eta/2$, where

$$|\Delta\eta| = \left|\frac{d\eta(\Phi)}{d\Phi}\right|\Delta\Phi. \tag{7.6.16}$$

The brightness of a particular spectral line is proportional to the specific flux $\mathcal{F}_\nu$ of photons of the given frequency emitted by the source that reach the observer.

$$\Delta\mathcal{F}_\nu = \int_{\Delta\Omega} d\Omega\,\mathcal{I}_\nu = \frac{\mathcal{F}_\nu}{4\pi}\Delta\Omega_e. \tag{7.6.17}$$

Here, we assumed that in its rest frame the source emits light isotropically in all directions. A similar expression is valid for the total brightness that is obtained by integrating of $\Delta\mathcal{F}_\nu$ over the frequencies.

Combining these results one gets

$$\Delta\mathcal{F}_\nu = \frac{\mathcal{F}_\nu}{4\pi}\frac{\mathcal{A}}{r_o^2}\frac{|\sin\eta|}{|\sin\Phi|}\frac{d\eta}{d\Phi}. \tag{7.6.18}$$

Here, η is given by Eq. (7.6.12). Thus, in order to find the brightness of the images, it is sufficient to know η as a function of Φ. But we have already made the corresponding calculations in Section 7.4.4. Using Eq. (7.4.30) we get

$$|\cos\eta| = \sqrt{1 - \ell^2(\zeta^2 - \zeta^3)}. \tag{7.6.19}$$

To summarize, the steps required for finding the brightness of the image are the following:

- Use Eq. (7.6.9) to find the inverse radius parameter ζ_2 as a function of Φ for a chosen image.
- Find the corresponding impact parameter using the relation

$$\ell = \frac{1}{\zeta_2\sqrt{1 - \zeta_2}}. \tag{7.6.20}$$

- Use Eq. (7.6.19) to establish the relation between η and Φ.
- Use Eq. (7.6.18) to calculate the brightness of the image.

Eq. (7.6.18) can also be written in the form

$$\Delta\mathcal{F}_\nu = \frac{\mathcal{F}_\nu\mathcal{A}}{4\pi r_o^2}\frac{\zeta^2 - \zeta^3}{\sqrt{1 - \ell^2(\zeta^2 - \zeta^3)}}\left|\frac{d(\cos\Phi)}{d\ell}\right|^{-1}. \tag{7.6.21}$$

The appearance of $\sin \Phi$ in the denominator of this expression leads to divergency of the image brightness when $\Phi = n\pi$. This corresponds to the case when the observer is positioned at the focus of the black hole concave lens. Of course, if one takes into account the finite size of the light source or the wave-mechanics corrections to the geometrical-optics approximation this divergency is naturally regularized.

Let us compare the brightness $B_{(n_\pm)}$ of the images of the order $(n_\pm)$. For example, if the order of the images is higher than 2 one can apply Darwin's formula. Note that $\Phi = \pi \pm (\Theta - 2\pi n_\pm - \eta_{n_\pm})$ and, hence,

$$\left| \frac{d\Phi}{d\ell} \right| = \left| \frac{d\Theta}{d\ell} - \frac{d\eta_{n_\pm}}{d\ell} \right| = \frac{1}{\ell - \ell_{\min}} + O(1). \tag{7.6.22}$$

The brightness B is proportional to $\Delta \mathcal{F}_\nu$. All higher-order images have the impact parameters near the $\ell_{\min}$ and, hence, the chief dependence on the impact parameter comes from $d\Phi/d\ell$. Therefore, applying Darwin's formula again we get

$$B \sim \ell - \ell_{\min} = A \exp(-\Theta). \tag{7.6.23}$$

For any two images for which the deflection angle differs by 2π, i. e. when images are on the same side of the black hole shadow and their numbers differ by 1, we get

$$B_{(n_\pm + 1)} = \exp(-2\pi) B_{(n_\pm)} \simeq 0.001867 B_{(n_\pm)}. \tag{7.6.24}$$

Evidently, in practice only the first two images really matter in observations.

7.6.4 Images of circular orbits

Let us consider a visible shape of the circular orbit near the black hole as seen by a distant observer. If the orbit is very far away from the horizon and we look at it from some finite inclination angle to the orbit plane then, evidently, the orbit will look like an ellipse because null rays emitted from the objects on the orbit and observed by us never enter the region of the strong gravitational field. Though, if we look in the direction of the black hole we also find many images of this orbit, concentrated at angles close to the angle corresponding to the critical impact parameter $\ell_{\min} = 3\sqrt{3}/2$. The picture is much more interesting when the orbit is located closer to the black hole and the observer is looking from inclination angles $\iota \sim \pi/2$ (see Figures 7.20 and 7.21). The typical apparent shapes of the first two images of the circular orbit are presented in Figure 7.15. The visible images of the circular orbits, corresponding to a set of different radii and for different inclination angles are depicted in Figures 7.16–7.21. We denote by

$$\alpha = \pi/2 - \iota \tag{7.6.25}$$

the angle complementary to the inclination angle ι.

In order to determine the visible position of the object orbiting a black hole we can use Eq. (7.6.7) and Eq. (7.6.8) that express the angle Φ in terms of the inverse impact parameter q. On the other hand,

$$\cos \Phi = \sin \iota \cos \varphi, \tag{7.6.26}$$

and, hence, we can find the position φ of the object on the orbit as a function of the impact parameter of the ray received by the observer from this object. Now, one has to calculate the

Fig. 7.15 The visible shape of the circular orbit at radius $r = 9r_S$ as seen from the angle $\alpha \equiv \pi/2 - \iota = 5^o$ to the orbit. The main image of the orbit looks here like a pirate hat. The second image is closer to the critical radius $r = (3\sqrt{3}/2)r_S$. It is also depicted on the figure with the arrow showing the direction of rotation. There is also an infinite number of other images. All of them are concentrated near the critical radius and are not shown in this figure. The small dashed circle on the image of the black hole itself is drawn here to provide a scale and corresponds to the impact parameter equal to the gravitational radius r_S.

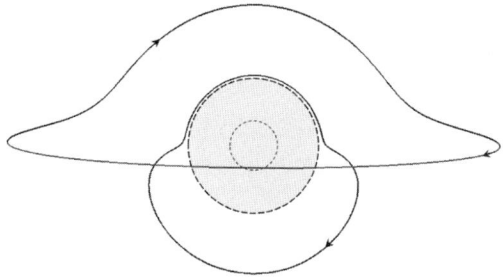

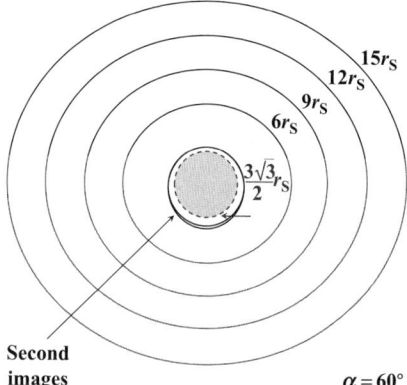

Fig. 7.16 Images of the four circular orbits for the angles $\alpha = 60°$.

Second
images

$\alpha = 60°$

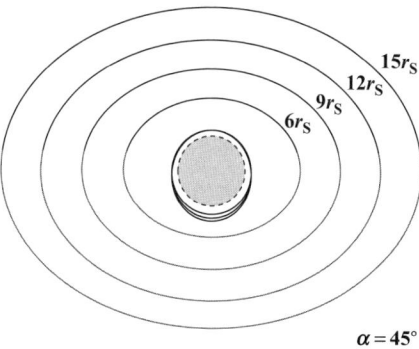

Fig. 7.17 Images of the four circular orbits for the angles $\alpha = 45°$.

$\alpha = 45°$

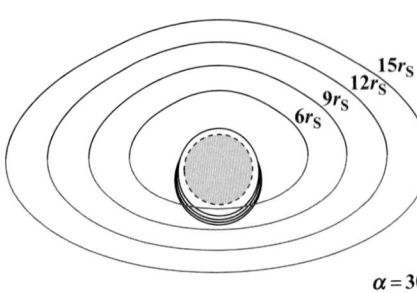

$\alpha = 30°$

Fig. 7.18 Images of the four circular orbits for the angles $\alpha = 30°$.

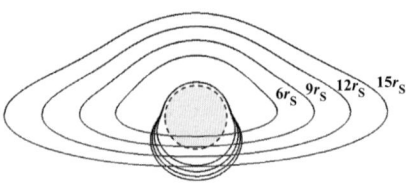

$\alpha = 15°$

Fig. 7.19 Images of the four circular orbits for the angles $\alpha = 15°$.

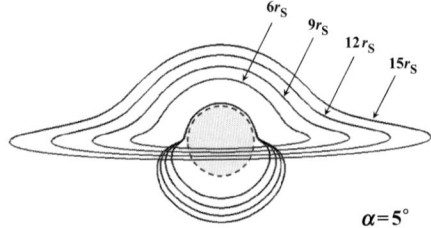

$\alpha = 5°$

Fig. 7.20 Images of the four circular orbits for the angles $\alpha = 5°$.

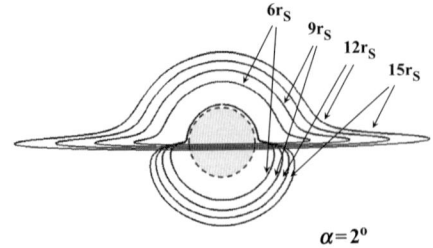

$\alpha = 2°$

Fig. 7.21 Images of the four circular orbits for the angles $\alpha = 2°$.

angle coordinates $\eta_o, \tilde{\varphi}$ of this object on the sky of the observer. The coordinate η_o is given by Eq. (7.4.31), that in the limit $r_o \gg r_S$ reduces to $\eta_o = q^{-1}r_S/r_o$

The angle $\tilde{\varphi}$ may be expressed in terms of angles ι and φ

$$\cos\tilde{\varphi} = \frac{\cos\varphi\cos\iota}{\sqrt{1 - \cos^2\varphi\sin^2\iota}}. \tag{7.6.27}$$

Combining these formulas together with Eq. (7.6.7) and Eq. (7.6.8) we obtain the system of parametric equations

$$\eta_o = q^{-1}r_S/r_o,$$

$$\cos\tilde{\varphi} = \frac{\cos\Phi(q)\cos\iota}{\sin\Phi(q)\sin\iota}. \tag{7.6.28}$$

Note that the black hole gravitational lens creates an infinite number of images of the same object, because the null ray can make many windings around the black hole before it reaches the observer. The first image is usually at a considerable angular distance from the black hole, except the parts of the orbit that is exactly between the observer and the black hole. The second image is a closed line around a black hole that also may deviate considerably from the angular position of the black hole. Other images lie close to the image of the critical light trajectory with $\ell_{\min} = 3\sqrt{3}/2$. If the sky at infinity were to have a non-zero brightness then the black hole would screen the light coming from the sky with the impact radii less than the minimal impact parameter $r_{\min} = \ell_{\min}r_S = (3\sqrt{3}/2)\,r_S \simeq 2.598\,r_S$. This 'black spot' on the sky is symbolically depicted in Figures 7.15–7.21 as a gray disk. The light with less than critical impact parameter emitted by an object either falling into the black hole or moving in front of it can reach the observer. In such a case one can see this object inside the region of the black hole shadow.

Figures 7.16-7.21 represent the images of orbits for different inclination angles $\alpha \equiv \pi/2 - \iota$ and various circular orbit radii $r = 6r_S$, $r = 9r_S$, $r = 12r_S$, and $r = 15r_S$. The orbit $r = 3r_S$ is the closest stable circular orbit for massive objects rotating around the Schwarzschild black hole. Astrophysical observation of the accretions disks show that we see spectral lines from the parts of the disk as close as $r \sim 9r_S$.

7.7 Radiation from an Object Moving Around the Black Hole

7.7.1 Frequency shift of the radiation from a moving particle

Let us consider the following problem. A particle moving on a circular orbit r around a Schwarzschild black hole of mass M emits 3 photons of the same frequency ω_0 in 3 different directions: (1) along the direction of the motion; (2) in the direction opposite to the direction of the motion; (3) in the direction orthogonal to the plane of motion. What are the frequencies of these 3 photons as measured by an observer at rest at infinity?

Let us introduce a local tetrad comoving with the particle. Denote by $e_{\hat{t}}^\mu$, $e_{\hat{r}}^\mu$, $e_{\hat{\theta}}^\mu$, and $e_{\hat{\phi}}^\mu$ unit vectors in the direction of t, r, θ, and ϕ, respectively. A unit vector of 4-velocity of the particle revolving in a circular orbit in the equatorial plane around the black hole is

$$u^\mu = \gamma(e_{\hat{t}}^\mu + ve_{\hat{\phi}}^\mu), \qquad \gamma = \frac{1}{\sqrt{1 - v^2}}. \tag{7.7.1}$$

Here, v is given by Eq. (7.3.11). A unit vector in the (t, ϕ) plane orthogonal to $\boldsymbol{u}$ is $n^\mu = \gamma(v e_{\hat{t}}^\mu + e_{\hat{\phi}}^\mu)$. The vectors $\boldsymbol{p}_\sigma = \omega_0(\boldsymbol{u} + \sigma \boldsymbol{n})$ are null vectors of momentum of a photon of frequency ω_0 (as measured by a moving observer) emitted in the direction of motion, $\sigma = +1$, and in the opposite direction, $\sigma = -1$. For the photon emitted in the direction perpendicular to the plane of motion the momentum is $\boldsymbol{p}_0 = \omega_0(\boldsymbol{u} + \boldsymbol{e}_{\hat{\theta}})$.

For the Schwarzschild metric

$$ds^2 = -g\, dt^2 + g^{-1}\, dr^2 + r^2\, d\omega^2 , \qquad g = 1 - r_S/r \qquad (7.7.2)$$

one has $(\boldsymbol{u}, \boldsymbol{\xi}_{(t)}) = -\gamma\sqrt{g}$ and $(\boldsymbol{n}, \boldsymbol{\xi}_{(t)}) = -\gamma v\sqrt{g}$. Thus,

$$\omega_\sigma^\infty = \omega_0\gamma(1 + \sigma v)\sqrt{g}, \qquad \omega_0^\infty = \omega_0\gamma\sqrt{g}. \qquad (7.7.3)$$

Here, ω^∞ is the frequency measured by an observer located at infinity. Note that the first expression in Eq. (7.7.3) in fact 'works' for the emission orthogonal to the orbit plane if we put $\sigma = 0$ for this case.

What remains to find is how the velocity v depends on the radius r of the circular motion of the particle of mass m. To do this we write

$$mu^\mu = \left(\frac{m\gamma}{\sqrt{g}}, 0, 0, \frac{m v\gamma}{r}\right) = \left(\frac{E}{g}, 0, 0, \frac{L}{r^2}\right). \qquad (7.7.4)$$

Using the expression $\zeta = \frac{1}{3}(1 - \sqrt{1 - 3\ell^{-2}})$ for stable circular orbits (see Eq. (7.2.3)) one gets $\ell = (2\zeta - 3\zeta^2)^{-1/2}$. Recall that $\zeta = r_S/r$ and $\ell = L/(mr_S)$. From $v\gamma = \ell\zeta$ one obtains

$$v = \frac{\sqrt{\zeta}}{\sqrt{2(1 - \zeta)}}, \qquad \gamma = \frac{\sqrt{2(1 - \zeta)}}{\sqrt{2 - 3\zeta}}. \qquad (7.7.5)$$

Substitution of these expressions into Eq. (7.7.3) finally gives

$$\omega_\sigma^\infty = \omega_0\frac{\sqrt{1 - \zeta}\,[\sqrt{2(1 - \zeta)} + \sigma\sqrt{\zeta}]}{\sqrt{2 - 3\zeta}}. \qquad (7.7.6)$$

For the *innermost stable circular orbit* $\zeta = 1/3$ and one has

$$\omega_{+1}^\infty = \sqrt{2}\,\omega_0, \qquad \omega_{-1}^\infty = \frac{\sqrt{2}}{3}\,\omega_0, \qquad \omega_0^\infty = \frac{2\sqrt{2}}{3}\,\omega_0. \qquad (7.7.7)$$

7.7.2 Time dependence of the frequency

We suppose that a point-like object moves around a black hole. If it emits radiation a distant observer can register it. Since the position of the object and its velocity change with time, the parameters of the registered radiation will also be time dependent. What can the observer tell about the characteristics of the motion of the emitting particle by study its radiation? A special example of this problem, which has an important astrophysical interests is the case of the X-ray emission by ions of Fe. Under normal conditions the corresponding Fe α line is extremely sharp, but for ions revolving around a black hole the combined action of the gravitational field and Doppler effects results in the effective broadening of the spectrum after averaging over the period of the rotation. In this section we analyze a slightly simplified version of this problem.

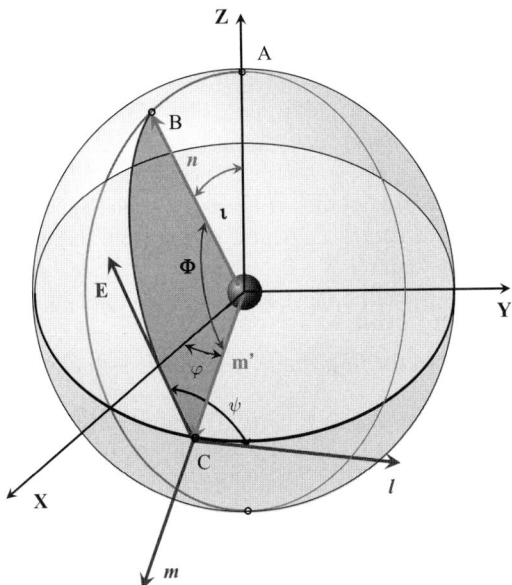

Fig. 7.22 Radiation from a source moving along a circular Keplerian orbit around a Schwarzschild black hole. A unit sphere on the figure corresponds to the 2D section (θ, ϕ) of the spacetime. It also shows different angles that are used in this problem.

We suppose that a point-like object revolves around a spherical black hole of mass M along the circular Keplerian orbit of radius R. Its period and frequency can be found by using Eq. (7.3.6). Suppose that the object emits monochromatic radiation of frequency ω_0, as measured in its own comoving frame. We denote by ι the *inclination angle*, that is the angle between the direction to the distant observer and the vector orthogonal to the plane of the Keplerian orbit. Figure 7.22 schematically shows the main features of the problem. What we would like to find is the frequency of this radiation registered by the distant observer as a function of time t.

For $\iota = 0$ this problem is quite similar to one discussed in the previous subsection. Additional problems that arise for $\iota \neq 0$ are of two kinds:

1. more involved relations between the direction of emitted null rays that reach the observer;

2. time dependence of the position of the point of emission.

Many relations involving angles are greatly simplified by the following trick. To represent the motion of the body we draw a sphere of a unit radius with the internal geometry

$$d\omega^2 = d\theta^2 + \sin^2\theta d\phi^2. \tag{7.7.8}$$

A 'real' sphere on which the orbit is located has the radius R. We include in our consideration this constant factor later. To simplify finding relations between different angles we *embed* this unit sphere into the 3-dimensional flat space $\mathbb{E}^3$ (see Figure 7.22).

We choose the flat coordinates (X, Y, Z) in $\mathbb{E}^3$ so that the plane of the orbit coincides with the (X, Y)-plane. A unit vector $\boldsymbol{k}$ along the axis of rotation is directed along Z. We choose the coordinates so that the direction to the distant observer is in the (X, Z)-plane, and denote a unit vector in this direction by $\boldsymbol{n}$. The end point of this vector on the unit sphere is at the point B. We define the inclination angle ι by the relation

$$\cos \iota = (\boldsymbol{n}, \boldsymbol{k}). \tag{7.7.9}$$

The body is moving along a large circle on the equatorial (X, Y)-plane. Its position at the moment t is represented by the point C. A unit vector from the center O to the point C is $\boldsymbol{m}$. Vector $\boldsymbol{m}$ has the angle ϕ with the X-axis. This angle changes with time. We include this dependence later. Let unit vector $\boldsymbol{l}$ be tangent to the large circle at the point C and directed along ϕ.

The vectors $\boldsymbol{n}$, $\boldsymbol{m}$ and $\boldsymbol{l}$ have the following Cartesian coordinates

$$\boldsymbol{n} = (\sin \iota, 0, \cos \iota), \quad \boldsymbol{m} = (\cos \varphi, \sin \varphi, 0), \quad \boldsymbol{l} = (-\sin \varphi, \cos \varphi, 0). \tag{7.7.10}$$

We denote by Φ the angle between the position of the object, C, and the direction to the distant observer, B,

$$\cos \Phi = (\boldsymbol{n}, \boldsymbol{m}) = \sin \iota \cos \varphi. \tag{7.7.11}$$

The angle Φ changes in the interval $(\pi/2 - \iota, \pi/2 + \iota)$.

Let us connect points C and B by a large circle on the unit sphere and denote by $\boldsymbol{E}$ a unit tangent vector to this circle at the point C, directed from C to B. This vector can be written as a linear combination of the vectors $\boldsymbol{n}$ and $\boldsymbol{m}$. Since $\boldsymbol{E}$ is orthogonal to $\boldsymbol{m}$ and using Eq. (7.7.11) one obtains

$$\boldsymbol{E} = \frac{1}{\sin \Phi} (\boldsymbol{n} - \boldsymbol{m} \cos \Phi). \tag{7.7.12}$$

For this choice of the direction of $\boldsymbol{E}$ one has

$$(\boldsymbol{E}, \boldsymbol{n}) = \sin \Phi \geq 0. \tag{7.7.13}$$

Hence, one has

$$(\boldsymbol{n}, \boldsymbol{l}) = -\sin \iota \sin \varphi,$$

$$(\boldsymbol{E}, \boldsymbol{l}) = \cos \psi = -\frac{\sin \iota \sin \varphi}{\sin \Phi}. \tag{7.7.14}$$

Let us consider now the propagation of a light ray emitted by an object and registered by the observer. Its orbit is planar. This plane is schematically shown at Figure 7.23. The corresponding plane intersects the sphere of a constant radius along the large circle. The corresponding points on the unit sphere representing this plane is a segment of a big circle between the points of the emission C, and observation, B.

The ray can be parametrized by the radius of emission, R, and the radius of its turning point, r_2. Denote $\zeta = r_S/R$ and $\zeta_2 = r_S/r_2$. The bending angle for such a ray is

$$\Psi_\pm(\zeta, \zeta_2) = \pm \Psi(\zeta, \zeta_2) + \Psi(0, \zeta_2). \tag{7.7.15}$$

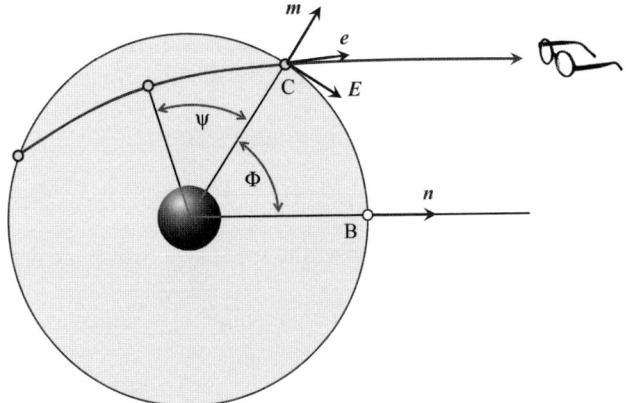

Fig. 7.23 Null-ray orbit in the 2D plane (r, Φ).

Here, $(+)$ corresponds to the rays that include the radial turning point, while $(-)$ corresponds to the rays that do not reach the turning point.

In order to reach the observer the corresponding bending angle for the ray must be equal to Φ or $\Phi \pm n\pi$ for higher-order images. Thus, we have the following equation

$$\Phi \pm \pi n = \Psi_\pm(\zeta, \zeta_2). \tag{7.7.16}$$

For the given radius of the orbit the parameter ζ is fixed. For the given angle Φ and the winding number n this equation determines the position of the turning point ζ_2 and, hence, the impact parameter of the null ray.

We make the following simplifying assumptions:

1. We restrict ourselves by considering the primary images, so that $n = 0$.

2. We assume that Eq. (7.7.16) for $n = 0$ and $\Phi \in (\pi/2 - \iota, \pi/2 + \iota)$ has a solution.

The second condition means that the inclination angle is not too close to $\pi/2$. If the angle $\pi/2 - \iota$ is small there exist primary rays with the impact parameter $\ell < \ell_{min}$. The rays from this part of the orbit arrive in the *black hole shadow* domain. This case requires separate consideration with small modifications.

We denote the solution ζ_2 of the equation

$$\Phi = \Psi_\pm(\zeta, \zeta_2) \tag{7.7.17}$$

by $\zeta_2(\Phi)$. By our assumptions this solution is unique. We also define the corresponding inverse impact parameter $q = \zeta_2\sqrt{1 - \zeta_2}$ by $q(\Phi)$.

Let us return to our problem and calculate the frequency of the received radiation. We assume that at the initial time $t = 0$ the body crosses the X-axis. Thus, its position in the equatorial plane is given by the coordinates

$$R = r_S/\zeta, \qquad \varphi = \Omega t,$$
$$\cos \Phi(t) = \sin \iota \cos(\Omega t) = \sin \iota \cos(\tilde{\Omega} \tilde{t}). \tag{7.7.18}$$

For every point C on the orbit one can calculate the null geodesic connecting it to the observer at infinity. The corresponding inverse impact parameter is $q(\Phi(t))$.

From now on we shall consider vectors in the 4D spacetime, Denote

$$e_{\hat{t}} = (g^{-1/2}, 0, 0, 0), \qquad e_{\hat{r}} = (0, g^{1/2}, 0, 0),$$

$$e_{\hat{\phi}} = (0, 0, 0, R^{-1}) = (0, 0, R^{-1}l), \qquad k = (0, 0, R^{-1}E). \tag{7.7.19}$$

In these relations l and E are 2D vectors tangent to the sphere, which are determined by Eqs. (7.7.10) and (7.7.12), respectively.

We denote by u the 4-velocity of the Keplerian motion in the equatorial plane

$$u = \gamma(e_{\hat{t}} + v e_{\hat{\phi}}) = (\gamma g^{-1/2}, 0, 0, R^{-1} v \gamma). \tag{7.7.20}$$

The quantities v and γ were calculated in the previous subsection, Eq. (7.7.5), and they are

$$v = \frac{\sqrt{\zeta}}{\sqrt{2(1 - \zeta)}}, \qquad \gamma = \frac{\sqrt{2(1 - \zeta)}}{\sqrt{2 - 3\zeta}}, \qquad \zeta = R^{-1}. \tag{7.7.21}$$

The momentum p of the emitted null ray at the point of emission one has

$$p = \lambda(e_{\hat{t}} + \alpha e_{\hat{r}} + \beta k). \tag{7.7.22}$$

The Killing vector $\xi_{(\Phi)} = \partial_\Phi$ generating rotations in the plane of the null-ray propagation calculated at the point of emission is

$$\xi_{(\Phi)}|_e = Rk. \tag{7.7.23}$$

The integrals of motion, redshifted energy ω_∞ and the angular momentum L are

$$\omega_\infty = -(p, \xi_{(t)}) = \lambda\sqrt{g}, \qquad g = 1 - r_S/R = 1 - \zeta,$$

$$L = (p, \xi_{(\Phi)}) = \lambda\beta R. \tag{7.7.24}$$

Using the definition of the impact parameter $b = \ell r_S$, where

$$\ell = L/(r_S \omega_\infty) = \frac{\beta R}{r_S \sqrt{g}} = \frac{\beta}{\zeta \sqrt{1 - \zeta}}, \tag{7.7.25}$$

we find the coefficient $\beta = \ell \zeta \sqrt{1 - \zeta}$. For the emitted frequency ω_0, measured in a frame comoving with the body, one has

$$\omega_0 = -(u, p) = \lambda\gamma(1 - \beta v \cos\psi). \tag{7.7.26}$$

Here, the angle ψ is defined by Eq. (7.7.14). Comparison of Eq. (7.7.24) and Eq. (7.7.26) gives

$$\omega_\infty = \omega_0 \frac{\sqrt{g}}{\gamma(1 - \beta v \cos\psi)}. \tag{7.7.27}$$

Combining the obtained equations we get

$$\omega_\infty = \omega_0 \frac{q\sqrt{2-3\zeta}\,B(\phi)}{q\sqrt{2}\,B(\phi)+\zeta^{3/2}\sin\iota\sin\phi}, \qquad B(\phi)=\sqrt{\sin^2\phi+\cos^2\phi\cos^2\iota}. \qquad (7.7.28)$$

To obtain the required dependence of ω of time t one needs to make the following steps:

1. Substitute ϕ by $\phi(t)=\Omega t=\zeta^{3/2}t/(\sqrt{2}r_S)$.
2. Substitute Φ by $\Phi(t)=\arccos[\sin\iota\cos(\Omega t)]$.
3. Solve equation $\Phi(t)=\Psi_\pm(\zeta,\zeta_2)$ and find $\zeta_2(t)$.
4. Substitute q by $q(t)=\zeta_2(t)\sqrt{1-\zeta_2(t)}$.

After performing these steps Eq. (7.7.28) gives the expression for the time dependence of frequency $\omega_\infty(t)$ measured by the distant observer. Note that only step 3 requires numerical calculations.

It is instructive to derive Eq. (7.728) in a different way, that makes clear its relation to the time-delay effect. We consider the same process as before. Denote by $t_\infty(t)$ the time of arrival of a null ray emitted from the point of the circular orbit at time t. One has $\omega_0 d\tau=\omega_\infty dt_\infty$. Here, τ is a proper time along the object trajectory. Since $dt/d\tau=u^t=\gamma/\sqrt{g}$ one has

$$\omega_\infty = \omega_0\frac{\sqrt{g}}{\gamma\,(dt_\infty/dt)}. \qquad (7.7.29)$$

We write $t_\infty=t+\Delta t(t)$, where Δt is a time delay. Integrating Eq. (7.1.16) along the ray trajectory to the point of observation we obtain

$$\frac{dt_\infty}{dt}=1+\frac{d\Delta t}{d\ell}\frac{d\ell}{dt}=1+r_S\ell\frac{d\Phi}{d\ell}\frac{d\ell}{dt} \qquad (7.7.30)$$

This gives

$$\frac{dt_\infty}{dt}=1+r_S\ell\frac{d\Phi(t)}{dt}. \qquad (7.7.31)$$

To derive this result one has to use the relation

$$\Phi(t)=\Psi_\pm(\zeta,\zeta_2), \qquad (7.7.32)$$

which determines the dependence of ζ_2 on t. Using the property $\ell\partial_\ell=-q\partial_q$ and Eq. (7.7.11) for $\Phi(t)$ we obtain

$$\frac{\omega_0}{\omega_\infty}=\frac{\gamma}{\sqrt{g}}\left(1+\frac{\Omega r_S}{q\sin\Phi}\sin\iota\sin\Omega t\right). \qquad (7.7.33)$$

It is easy to check that this relation is equivalent to Eq. (7.7.28).

The distant observer measures the frequency of photons emitted at the coordinate time t and arrived at the retarded time $t_\infty=t+\Delta t$. So, in order to derive the spectrum of the observed quanta one should first substitute $t(t_\infty)$ into Eq. (7.7.28) and then apply the Fourier transform with respect to t_∞. In this way the spectrum of an originally monochromatic signal gets a distinct shape.

7.8 Equations of Motion in 'Tilted' Spherical Coordinates

To simplify the equations of motion of a test particle we used a special choice of coordinates, namely we oriented the z-axis (that is the direction of $\theta = 0, \pi$) to be orthogonal to the plane of the motion. Let us now check how the equations of motion are modified if the z-axis is tilted and not orthogonal to the orbit plane. This exercise is instructive for comparison with particle motion in the Kerr geometry, where there exists a preferred direction determined by the axis of rotation. For simplicity, we consider particle motion. The equations for null geodesics can be easily obtained in a similar manner.

Denote by τ the proper time. The expression for the *specific energy* $\mathcal{E}$ remains the same

$$\mathcal{E} = g \frac{dt}{d\tau}. \tag{7.8.1}$$

For the Schwarzschild metric $g = 1 - r_S/r$. The *specific azimuthal angular momentum*, which we denote now $\mathcal{L}_z$, is

$$\mathcal{L}_z = r^2 \sin^2 \theta \frac{d\phi}{d\tau}. \tag{7.8.2}$$

One also needs the expression for the conserved *specific total angular momentum*, $\mathcal{L}$,

$$\mathcal{L}^2 = r^4 \left[\left(\frac{d\theta}{d\tau} \right)^2 + \sin^2 \theta \left(\frac{d\phi}{d\tau} \right)^2 \right]. \tag{7.8.3}$$

Using these relations and the normalization condition $g_{\mu\nu}(dx^\mu/d\tau)(dx^\nu/d\tau) = -1$ one obtains the following set of equations

$$\frac{dr}{d\tau} = \pm\sqrt{\mathcal{E}^2 - U}, \qquad U = g\,(1 + \mathcal{L}^2/r^2),$$

$$r^2 \frac{d\theta}{d\tau} = \pm(\mathcal{L}^2 - \mathcal{L}_z^2/\sin^2 \theta),$$

$$r^2 \frac{d\phi}{d\tau} = \mathcal{L}_z/\sin^2 \theta, \tag{7.8.4}$$

$$\frac{dt}{d\tau} = \mathcal{E}g^{-1}.$$

Here, the second equation implies that $\mathcal{L}^2 \geq \mathcal{L}_z^2$. For $\mathcal{L}^2 = \mathcal{L}_z^2$ this equation has a solution $\theta = \pi/2$, and one recovers the equation of motion in the equatorial plane. In the general case, $\theta(\tau)$ changes between θ_0 and $\pi - \theta_0$, where $\sin \theta_0 = \mathcal{L}_z/\mathcal{L}$. This means that the angle between the normal to the trajectory plane and z-axis is $\pi/2 - \theta_0$. One can easily obtain a general solution of the system Eq. (7.8.4) by performing a rigid rotation of the equatorial plane solution.

7.9 Magnetized Schwarzschild Black Hole

7.9.1 Test-field approximation

There exist both theoretical and experimental indications that a magnetic field must be present in the vicinity of black holes. A regular magnetic field can exist near a black hole surrounded by conducting matter (plasma), e.g., if the black hole has an accretion disk. The magnetic field near a stellar mass black hole may contain a contribution from the original magnetic field of the collapsed progenitor star. The dynamo mechanism in the plasma of the *accretion disk* might generate a regular magnetic field inside the disk. Such a field cannot 'cross' the conducting plasma region and is trapped in the vicinity of the black hole (see, e.g., the discussion in (Punsly 2008)).

Stellar mass and supermassive black holes often have jets, that is the collimated fluxes of relativistic plasma. It is generally believed that the *magnetohydrodynamics* (MHD) of the plasma in strong magnetic and gravitational fields of the black holes would allow one to understand the formation and energetics of the black hole jets (Punsly 2008). A magnetic field in the vicinity of the black holes might play an important role in the energy transfer from the accretion disk to jets. The existence of a regular magnetic field near the black holes is also required for the proper collimation of the plasma in the jets. It is believed that rotation of a black hole plays an important role in these processes. We discuss this problem in the next chapter. In this section we make a few remarks on properties of magnetized non-rotating black holes.

Suppose there exists an *axisymmetric magnetic field B* near the black hole. The spacetime local curvature created by it is of the order of B^2. It is comparable to the spacetime curvature near a black hole of mass M only if

$$B^2 \sim r_S^{-2} \sim M^{-2}. \tag{7.9.1}$$

This condition holds if

$$B \sim B_M \sim 10^{19}(M_\odot/M)\text{G}. \tag{7.9.2}$$

The magnetic fields that are expected near black holes are much smaller than B_M. This means that such a field can be considered as a test field in the given gravitational background. It practically does not affect the motion of neutral particles.

However, for charged particles, the acceleration induced by the *Lorentz force*, can be large. The acceleration is of the order of qB/m. Thus, the 'weakness' of the magnetic field B is compensated by the large value of the charge-to-mass ratio q/m, which, for example for electrons, is $e/m_e \approx 5.3 \times 10^{17}\,\text{g}^{-1/2}\text{cm}^{3/2-1}$.

Because the Schwarzschild metric is Ricci flat, the Killing vectors obey the equation (see Eq. (3.7.16))

$$\xi_{\alpha;\beta}{}^{;\beta} = 0. \tag{7.9.3}$$

This equation coincides with the Maxwell equation for a 4-potential A_α in the Lorentz gauge $A_\alpha{}^{;\alpha} = 0$.[3] The special choice

$$A^\alpha = \frac{B}{2}\xi_{(\phi)}^\alpha, \tag{7.9.4}$$

---

[3] In a generic spacetime the equation also contains the Ricci tensor $\xi_{\alpha;\beta}{}^{;\beta} + R_\alpha^\beta \xi_\beta = 0$ (see Eq. (3.7.16)). The vector potential A_α for the Maxwell field in the Lorentz gauge $A_\alpha{}^{;\alpha} = 0$ satisfies a similar equation but with the opposite sign before the Ricci tensor $A_{\alpha;\beta}{}^{;\beta} - R_\alpha^\beta A_\beta = 0$.

corresponds to a test magnetic field, which is homogeneous at the spatial infinity where it has the strength B (Wald 1984; Aliev and Galtsov 1989). In what follows, we assume that $B \geq 0$. The electric 4-potential Eq. (7.9.4) is static, that is it is invariant with respect to the isometry generated by the Killing vector $\xi_{(t)}^{\nu}$

$$(\mathcal{L}_{\xi_{(t)}} A)_{\mu} = A_{\mu,\nu}\xi_{(t)}^{\nu} + A_{\nu}\xi_{(t),\mu}^{\nu} = 0. \tag{7.9.5}$$

The magnetic field measured in the rest frame is

$$B_{\mu} = -\frac{1}{2}\varepsilon_{\mu\nu\lambda\rho}F^{\lambda\rho}\frac{\xi_{(t)}^{\nu}}{|\xi_{(t)}^2|^{1/2}}, \qquad \varepsilon_{0123} = \sqrt{-g}. \tag{7.9.6}$$

Here, $F_{\mu\nu} = A_{\nu,\mu} - A_{\mu,\nu}$ is the electromagnetic field tensor. The magnetic field corresponding to the 4-potential Eq. (7.9.4) is

$$B^{\mu}\partial_{\mu} = B\left(1 - \frac{r_S}{r}\right)^{1/2}\left[\cos\theta\frac{\partial}{\partial r} - \frac{\sin\theta}{r}\frac{\partial}{\partial\theta}\right]. \tag{7.9.7}$$

In the cylindrical coordinates (ρ, z, ϕ) at the spatial infinity it is directed along the z-axis.

If the size of the accretion disk is much larger than the gravitational radius, circular electric currents in it generate a magnetic field, which in the black hole vicinity can be approximated by Eq. (7.9.7). The magnetic field Eq. (7.9.7) coincides with the dipole approximation of the magnetic field of a current loop around a Schwarzschild black hole (Petterson 1974; Linet 1976). One can show that if the radius of the loop is much larger than the gravitational radius this approximation is quite good in the region between the horizon and the radii much smaller than the loop radius.

7.9.2 Charged-particle motion

The equation of motion for a charged particle is

$$m\frac{du^{\mu}}{d\tau} = qF^{\mu}{}_{\nu}u^{\nu}. \tag{7.9.8}$$

Here, $F_{\mu\nu}$ is the strength of the *electromagnetic field*, u^{μ} is the particle 4-velocity, $u^{\mu}u_{\mu} = -1$, q and m are its charge and mass, respectively. For the motion in the *magnetized black hole* there exist two conserved quantities associated with the Killing vectors: the energy $E > 0$ and the generalized azimuthal angular momentum $L \in (-\infty, +\infty)$,

$$E = -\xi_{(t)}^{\mu}P_{\mu} = m\left(1 - \frac{r_S}{r}\right)\frac{dt}{d\tau},$$

$$L = \xi_{(\phi)}^{\mu}P_{\mu} = mr^2\sin^2\theta\frac{d\phi}{d\tau} + \frac{qB}{2}r^2\sin^2\theta. \tag{7.9.9}$$

Here, $P_{\mu} = mu_{\mu} + qA_{\mu}$ is the generalized 4-momentum of the particle.

It is easy to check that the θ-component of Eq. (7.9.8) allows for a solution $\theta = \pi/2$. This is a motion in the equatorial plane of the black hole, which is orthogonal to the magnetic field. We restrict ourselves to this type of motion, for which the conserved quantities Eq. (7.9.9) are sufficient for the complete integrability of the dynamical equations. Let us denote

$$\mathcal{E} = \frac{E}{m}, \quad \ell = \frac{L}{mr_S}, \quad b = \frac{qBr_S}{2m}, \quad \rho = \frac{r}{r_S}, \quad \tilde{t} = \frac{t}{r_S}, \quad \sigma = \frac{\tau}{r_S}. \tag{7.9.10}$$

The first integrals of motion allow one to obtain the equation of motion in the first-order form

$$\left(\frac{d\rho}{d\sigma}\right)^2 = \mathcal{E}^2 - U, \quad U = \left(1 - \frac{1}{\rho}\right)\left[1 + \frac{(\ell - b\rho^2)^2}{\rho^2}\right], \tag{7.9.11}$$

$$\frac{d\phi}{d\sigma} = \frac{\ell}{\rho^2} - b, \quad \frac{d\tilde{t}}{d\sigma} = \frac{\mathcal{E}\rho}{\rho - 1}.$$

Here, U is an effective potential. According to the adopted convention, we have $b \geq 0$.

It should be emphasized that the magnetic field breaks the reflection symmetry $\phi \to -\phi$. The parameter ℓ can be positive or negative, $\ell = \pm|\ell|$. Trajectories with positive and negative ℓ are different. For $\ell > 0$ (sign $+$) the Lorentz force, acting on a charged particle, is repulsive, i.e. it is directed outward from the black hole, and for $\ell < 0$ (sign $-$) the Lorentz force is attractive, i.e. it is directed toward the black hole

Equations (7.9.11) are invariant under the following combined transformations:

$$b \to -b, \qquad \ell \to -\ell, \qquad \phi \to -\phi. \tag{7.9.12}$$

Thus, without loss of generality, one can assume that the charge q (and hence b) is positive. For a particle with a negative charge it is sufficient to make the transformation Eq. (7.9.12).

7.9.3 Trajectories

The effective potential U is positive in the black hole exterior, where $\rho > 1$. It vanishes at the black hole horizon, $\rho = 1$, and grows as $b^2\rho^2$ for $\rho \to +\infty$. The latter property implies that a particle never reaches the spatial infinity, i.e. its motion is always finite. One can easily understand why it happens. In the absence of the gravity, trajectories of charged particles in a magnetic field are circles of radius

$$r_c = \sqrt{2|L|/qB}. \tag{7.9.13}$$

Hence, they never reach infinity.

For small values of ℓ the effective potential is a monotonically increasing function of ρ (curve 1 in Figure 7.24). In this case, a particle starts its motion at the horizon, and, after reaching its radial turning point, it returns to the horizon. For ℓ larger than some critical value, the potential has one maximum and one minimum. For $\mathcal{E} > \mathcal{E}_{max}$ there is only one radial turning point, so that the motion is similar to the motion with small ℓ. One turning point exists also if $\mathcal{E} < \mathcal{E}_{max}$ and $\rho < \rho_{max}$. Besides the cases described above when the particle is finally trapped by the black hole, there exists only one other type of motion. This is a *bounded motion* with $\mathcal{E} < \mathcal{E}_{max}$ and $\rho \in (\rho_{max}, \rho_{min})$.

7.9.4 Circular orbits

Bounded orbits are of the most interest for astrophysical applications. Let us discuss their properties. A motion with $\mathcal{E} = \mathcal{E}_{min}$ is a *stable circular orbit*. As in the Schwarzschild geometry such orbits exist only in some domain of ρ. The radius of the *innermost stable*

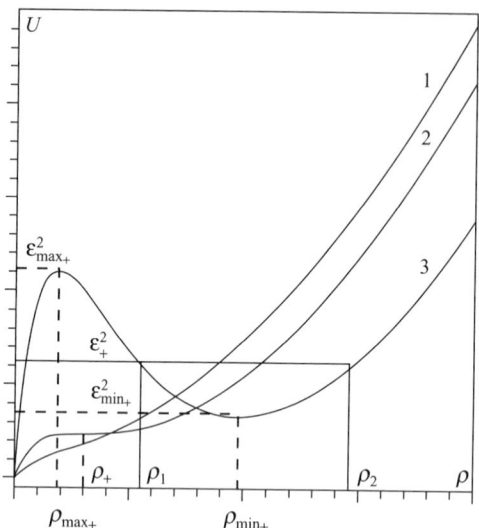

Fig. 7.24 Effective potential U as a function of ρ for fixed values of the magnetic field b and different values of the angular momentum parameter ℓ. For small values of ℓ the potential is a monotonically increasing function of ρ (curve 1). For sufficiently large value of ℓ it has a maximum (at $\rho_{\max}$) and a minimum (at $\rho_{\min}$) (curve 3). Curve 2 shows the potential for a critical value of ℓ, separating these two cases.

circular orbit is defined by equations

$$U_{,\rho} = U_{,\rho\rho} = 0. \tag{7.9.14}$$

Problem 7.3:   *Show that Eq. (7.9.14) implies the following parametric expressions for ℓ and b for the radius ρ of the innermost stable circular orbit*

$$\ell_{\pm} = \pm \frac{\rho_{\pm}(3\rho_{\pm} - 1)^{1/2}}{\sqrt{2}\left(4\rho_{\pm}^2 - 9\rho_{\pm} + 3 \pm \sqrt{(3\rho_{\pm} - 1)(3 - \rho_{\pm})}\right)^{1/2}},$$

$$b = \frac{\sqrt{2}(3 - \rho_{\pm})^{1/2}}{2\rho_{\pm}\left(4\rho_{\pm}^2 - 9\rho_{\pm} + 3 \pm \sqrt{(3\rho_{\pm} - 1)(3 - \rho_{\pm})}\right)^{1/2}}. \tag{7.9.15}$$

The signs $+$ and $-$ correspond to the motion with $\ell > 0$ and $\ell < 0$, respectively.

The second relation in Eq. (7.9.15) shows that in the absence of the magnetic field $\rho_{\pm} = 3$. This is the radius of the innermost stable circular orbit in the Schwarzschild geometry. When the value of the magnetic field b increases, both values $\rho_{\pm}$ become smaller. In the limit $b \to \infty$ they reach the limiting values $\rho_{+} = 1$ and $\rho_{-} = (5 + \sqrt{13})/4$, respectively. This means that in the presence of a strong magnetic field charged particles with a proper sign of the angular momentum can move around a circular orbit with the radius only slightly larger than a gravitational radius.

7.9.5   Bounded orbits

If the specific energy is higher than $\mathcal{E}_{\min}$ (but still smaller that $\mathcal{E}_{\max}$), the radial motion is an oscillation between the minimal ρ_1 and maximal ρ_2 radii. The right-hand side of the equation for the angle variable

$$\frac{d\phi}{d\sigma} = \frac{\ell}{\rho^2} - b \qquad (7.9.16)$$

is no langer constant, but it is modulated by the oscillation of the radial variable. Note that for $\ell < 0$ (attractive Lorentz force) the right-hand side of Eq. (7.9.16) is always negative, so that $\phi(\tau)$ is monotonically decreasing function. The particle is moving in the clockwise direction. For positive ℓ (repulsive Lorentz force) there exist different types of trajectories.

Let us mention another property of the effective potential, that we shall use later. Let us denote $\rho_* \equiv \sqrt{\ell/b}$, then simple calculations give $U_{,\rho}(\rho_*) = b/\ell$. This means that for bounded orbits and for positive ℓ one has $\rho_* > \rho_{min} \geq \rho_{max}$.

For sufficiently small amplitude of the radial oscillations, when $\rho_2 < \rho_*$, the trajectory remains 'smooth', but if the amplitude is greater than a critical one ($\rho_2 > \rho_*$), the particle moves along a 'curly' trajectory. For the critical case $\rho_2 = \rho_*$, separating these two different types of motion, the trajectory has cusps. All three types of motion ('smooth', 'critical', and 'curly') are illustrated in Figure 7.25 (plots a, b, and c, respectively).

To summarize: The main new features for the charged particle motion near a magnetized Schwarzschild black hole are:

1. No trajectory can reach infinity.

2. A particle is either finally trapped by the black hole, or is bounded.

3. The radius of the innermost stable circular orbit becomes smaller and for $b \to \infty$ and $\ell > 0$ it reaches r_S.

4. Bounded non-circular orbits can be either 'smooth', or 'curly'. The type of motion depends on the amplitude of radial oscillations.

5. A critical trajectory, separating these two different cases, has cusps.

6. The action of the gravitational field generates drift current.

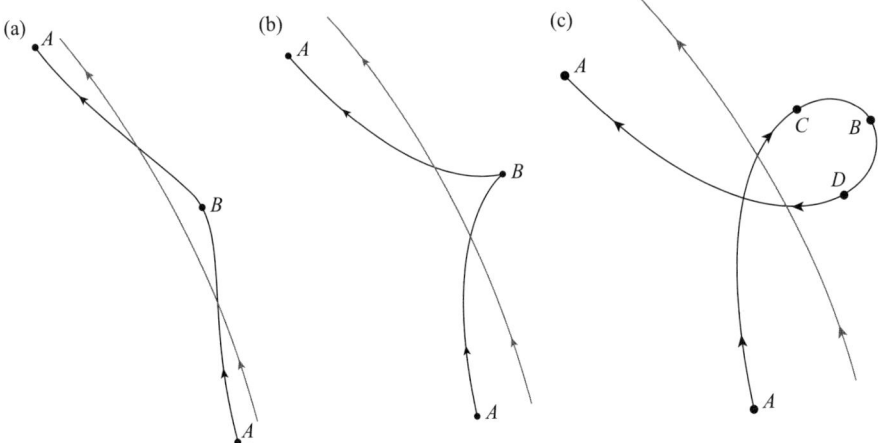

Fig. 7.25 Types of bounded trajectory corresponding to $\ell > 0$. The figures show segments of trajectories for 3 different types of motion, discussed in the text. Arrows illustrate the direction of motion of a charged particle. Circular arcs represent the stable circular orbit defined by $\rho = \rho_{min_+}$.

More details and references can be found in (Aliev and Galtsov 1989; Frolov and Shoom 2010).

7.10   Particle and Light Motion Near Higher-Dimensional Black Holes

7.10.1   Equation of motion

Let us discuss briefly some properties of particle and light motion in the gravitational field of a D-dimensional non-rotating black holes. The corresponding metric is (see Eq. (6.8.2))

$$ds^2 = -g\,dt^2 + \frac{dr^2}{g} + r^2\,d\omega_{D-2}^2\,, \qquad\qquad g = 1 - (r_S/r)^{D-3}. \tag{7.10.1}$$

Here, $d\omega_{D-2}^2$ is a line element of a round sphere S^{D-2} of a unit radius. Similarly to the four-dimensional case, it is easy to show that trajectories are plane. For simplicity, we chose a trajectory in the 'equatorial plane', so that the problem reduces to the 3-dimensional one. The corresponding 3D metric is (see Eq. (7.1.2))

$$ds^2 = -g\,dt^2 + \frac{dr^2}{g} + r^2 d\phi^2. \tag{7.10.2}$$

Two non-trivial integrals of motion Eq. (7.1.5), the energy E and the angular momentum L, are sufficient for the complete integrability of the dynamical equations. They can be written in the first-order form Eq. (7.1.6). By using the dimensional variables

$$ \mathcal{E} = E/m\,, \quad \zeta = r_S/r\,, \quad \ell = L/(mr_S)\,, \quad \sigma = \tau/r_S\,, \quad \tilde{t} = t/r_S, \tag{7.10.3}$$

one can write the equations of motion of a particle as follows (see Eq. (7.1.12))

$$ \zeta^{-4}\zeta'^2 = \mathcal{E}^2 - U\,, \qquad \phi' = \ell\zeta^2\,, \qquad \tilde{t}' = \mathcal{E}/g. \tag{7.10.4}$$

Here, $(\ldots)' = d(\ldots)/d\sigma$ and U is the effective potential

$$ U = (1 - \zeta^{D-3})(1 + \ell^2\,\zeta^2). \tag{7.10.5}$$

The effective potential is positive in the black hole exterior. It vanishes at the horizon (at $\zeta = 1$). For $D > 5$ its asymptotic at the spatial infinity (at $\zeta = 0$) is

$$ U \sim 1 + \ell^2\,\zeta^2. \tag{7.10.6}$$

In a special case $D = 5$ this asymptotic is

$$ U \sim 1 + (\ell^2 - 1)\,\zeta^2. \tag{7.10.7}$$

As stated earlier, it is sufficient to consider a case when $\ell \geq 0$.

7.10.2   Particle trajectories

Properties of gravity in higher dimensions are quite different from the four-dimensional case. The main important difference is that in a spacetime of a higher-dimensional ($D \geq 5$) black hole there are no stable bounded orbits.

To prove this let us assume the opposite, that is that the effective potential has a minimum at some finite value $0 < \zeta_{min} < 1$. For $\ell = 0$ the function U monotonically decreases from 1 at the infinity (at $\zeta = 1$) to 0 at the horizon (at $\zeta = 1$). Hence, there should exist some critical value $\ell = \ell_c$, so that for $\ell > \ell_c$ the minimum exists. Since $U(\zeta = 1) = 0$, the function U must also have a maximum at some point ζ_{max} between ζ_{min} and $\zeta = 1$. At the points of the maximum and minimum $U_{,\zeta} = 0$. At some point between ζ_{min} and ζ_{max} the function $U_{,\zeta\zeta}$ vanishes. When the parameter ℓ decreases, the points of maximum and minimum come closer to one another, so that at ℓ_c they coincide and one has

$$U_{,\zeta} = U_{,\zeta\zeta} = 0. \tag{7.10.8}$$

These conditions would determine a position of the *innermost stable circular orbit* and the critical parameter ℓ_*. These equations are of the form $(D = n + 2)$

$$U_{,\zeta} \equiv \zeta \left\{ \ell^2 \left[2 - (n+1)\zeta^{n-1} \right] - (n-1)\zeta^{n-3} \right\} = 0,$$

$$U_{,\zeta\zeta} \equiv \ell^2 \left[2 - n(n+1)\zeta^{n-1} \right] - (n-1)(n-2)\zeta^{n-3} = 0. \tag{7.10.9}$$

The first equation gives

$$\ell_c^2 = \frac{(n-1)\zeta^{n-3}}{2 - (n+1)\zeta^{n-1}}. \tag{7.10.10}$$

Substituting this expression into the second equaion in Eq. (7.10.9) and solving it one obtains the corresponding value ζ_c. For this value Eq. (7.10.9) determines ℓ_c. The obtained results are

$$\zeta_c^{n-1} = -\frac{(n-3)}{n+1}, \qquad \ell_c = \left(\frac{3-n}{n+1} \right)^{-\frac{(3-n)}{2(n+1)}}. \tag{7.10.11}$$

The solution ζ_c belongs to the interval $(0, 1)$ only when $n < 3$ $(D < 5)$, i.e. for four-dimensional spacetime. For $D \geq 5$ an innermost stable circular orbit does not exist. This contradiction means that there are no bounded orbits in the Tangherlini spacetime with the number of dimensions greater than 4. A similar result is valid in the Newtonian higher-dimensional gravity.

The obtained results also imply that the function U in the black hole exterior either has one maximum, or does not have it at all. As a result, only three types of orbits exist:

1. A particle propagates between the horizon and the infinity without radial turning points.

2. A particle propagates from the horizon to some radius where it has a turning point, after which the particle falls into the black hole.

3. A particle propagates from infinity, has a turning point at some radius, and than goes to infinity again.

In the last case, one has a *particle scattering by a black hole.*

Problem 7.4: *A D-dimensional physicist decided to study the interior of a D-dimensional Schwarzschild–Tangherlini black hole. How does he need to choose the parameters of his motion before crossing the event horizon $r = r_S$, in order to make maximal the proper time $\Delta \tau$ of his free fall from the horizon to the singularity? Calculate the value of $\Delta \tau$ in an arbitrary number of dimensions.*

7.10.3 Light propagation. Gravitational capture

By using variables

$$\zeta = r_S/r, \qquad \ell = L/(Er_S), \qquad \sigma = E\lambda/r_S, \qquad (7.10.12)$$

one can write the equations for *light propagation* in the field of a D-dimensional black hole in the form (see Eq. (7.1.14))

$$\zeta' = \pm\zeta^2\sqrt{1 - \ell^2\zeta^2 g}, \qquad g(\zeta) = 1 - \zeta^{D-3},$$

$$\phi' = \ell\zeta^2, \qquad t' = \frac{1}{g}. \qquad (7.10.13)$$

Here, ℓr_S is an *impact parameter*. It is sufficient to consider $\ell \geq 0$.
Solving the equation for radial turning points $1 - \ell^2\zeta^2(1 - \zeta^{D-3}) = 0$ one obtains

$$\ell = \frac{1}{\zeta\sqrt{1 - \zeta^{D-3}}}. \qquad (7.10.14)$$

Figure 7.26 shows plots of ℓ^{-1} as a function of ζ for $D = 4, 5, 6$ and 7. The behavior of the function $\ell(\zeta)$ is qualitatively the same in any number of dimensions. It grows infinitely at $\zeta = 0$ and $\zeta = 1$, and has a minimum inside this interval. Calculating the derivative of Eq. (7.10.14) one obtains

$$\ell' = \frac{(D-1)\zeta^{D-3} - 2}{2\zeta^2(1 - \zeta^{D-3})^{3/2}}. \qquad (7.10.15)$$

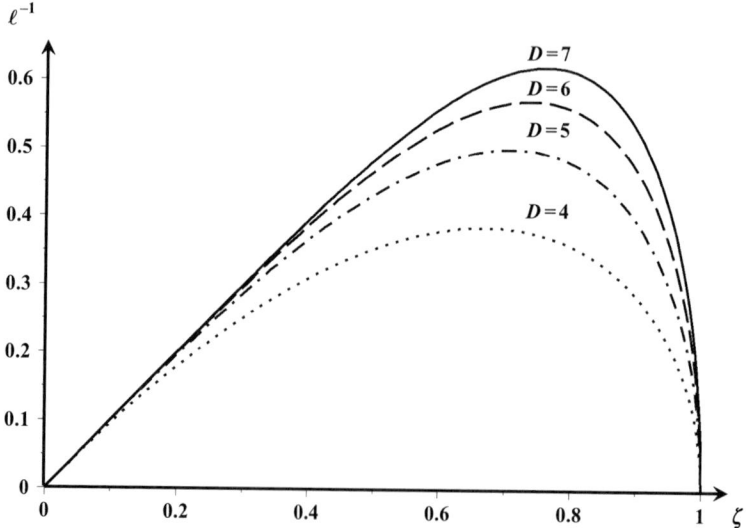

Fig. 7.26 The inverse impact parameter ℓ^{-1} as a function of the $\zeta = r_S/r$ for the dimensions $D = 4, 5, 6,$ and 7.

The position of the minimum is

$$\zeta_{\min} = \left(\frac{2}{D-1}\right)^{1/(D-3)}. \tag{7.10.16}$$

The corresponding critical value of the impact parameter is

$$\ell_{\min} = \left(\frac{D-1}{2}\right)^{1/(D-3)} \sqrt{(D-1)/(D-3)}. \tag{7.10.17}$$

For $D = 4$ these quantities are equal to (see page 200)

$$\zeta_{\min} = 2/3, \qquad \ell_{\min} = 3\sqrt{3}/2. \tag{7.10.18}$$

Null rays sent from infinity with the impact parameter $\ell < \ell_{\min}$ are captured by the black hole. The *capture cross-section* is

$$\sigma_{\text{capt}} = \mathcal{V}_{D-2}\, \ell_{\min}^{D-2}, \tag{7.10.19}$$

where $\mathcal{V}_n$ is the volume of an n-dimensional ball of radius r_S.

$$\mathcal{V}_n = \frac{\pi^{n/2}}{\Gamma\left(\frac{n}{2}+1\right)} r_S^n. \tag{7.10.20}$$

Finally, one has

$$\sigma_{\text{capt}} = \frac{r_S^{D-2}}{\Gamma(D/2)} \left[\frac{\pi(D-1)}{D-3}\right]^{\frac{D}{2}-1} \left(\frac{D-1}{2}\right)^{\frac{D-2}{D-3}}. \tag{7.10.21}$$

7.10.4   Light deflection

A null ray sent from infinity with the impact parameter $\ell > \ell_{\min}$ after reaching the radial turning point returns to infinity. The general solution $\phi(\zeta)$ for such scattering of a null ray is

$$\phi = \int \frac{d\zeta}{\sqrt{q^2 - \zeta^2 g}}, \qquad g(\zeta) = 1 - \zeta^{D-3}. \tag{7.10.22}$$

Here, $q = 1/\ell = E r_S/L$ is the inverse impact parameter. Similarly to the 4D case the deflection angle Θ can be written in the form

$$\Theta = -\pi + 2 \int_0^{\zeta_2} \frac{d\zeta}{\sqrt{q^2 - \zeta^2 g}}. \tag{7.10.23}$$

Here, ζ_2 corresponds to the turning point of the trajectory and is the solution of the algebraic equation

$$\zeta_2^2\, g(\zeta_2) = q^2. \tag{7.10.24}$$

After rescaling the integration variable $\zeta = \zeta_2 y$ we obtain

$$\Theta = -\pi + 2 \int_0^1 dy \frac{1}{\sqrt{1 - y^2 - \zeta_2^{D-3}(1 - y^{D-1})}}. \tag{7.10.25}$$

Since for large impact parameter $\ell \gg 1$ one has $\zeta_2 \simeq q \ll 1$ and therefore one can use an expansion of the integral in powers of a small parameter $\alpha = \zeta_2^{D-3}$

$$\Theta = -\pi + 2 \int_0^1 dy \frac{1}{\sqrt{1 - y^2}} + \zeta_2^{D-3} \int_0^1 dy \frac{1 - y^{D-1}}{(1 - y^2)^{3/2}} + \ldots. \tag{7.10.26}$$

The first two terms in this expression cancel and the third integral gives the leading contribution to the deflection angle. The result of the integration reads

$$\Theta = q^{D-3} \sqrt{\pi} \frac{\Gamma\left(\frac{D}{2}\right)}{\Gamma\left(\frac{D-1}{2}\right)} + O(q^{2D-6}), \tag{7.10.27}$$

where we have used the perturbative solution for $\zeta_2 = q + \frac{1}{2}q^{D-2} + O\left(q^{2D-4}\right)$. The gravitational radius r_S and the mass of the D-dimensional black hole are related according to Eq. (6.8.5)

$$r_S^{D-3} = \frac{4(D-3)\,\Gamma\left(\frac{D-3}{2}\right)}{(D-2)\,\pi^{\frac{D-3}{2}}} G^{(D)} M. \tag{7.10.28}$$

Thus, we can express the deflection angle explicitly in terms of the mass M of the black hole, the energy E, and the angular momentum L of the photon

$$\Theta \approx \frac{4\Gamma\left(\frac{D}{2} - 1\right)}{\pi^{\frac{D}{2}-2}} \left(\frac{E}{L}\right)^{D-3} G^{(D)} M. \tag{7.10.29}$$

This is a higher-dimensional generalization of the *Einstein formula for light deflection* (see Eq. (7.4.24)).

▌ **Problem 7.5:** *Obtain a generalization of Darwin's formula for a 5D non-rotating black hole.*

Solution: *The deflection angle Θ is*

$$\Theta = 2 \int_0^{\zeta_2} \frac{d\zeta}{\sqrt{q^2 - \zeta^2(1 - \zeta^2)}} - \pi. \tag{7.10.30}$$

Here, q is the inverse impact parameter and ζ_2 is the smaller of two roots of the expression inside the square-root. Let us change variables $\zeta^2 = y$, then we have

$$\Theta = \int_0^{y_2} \frac{dy}{\sqrt{y(y_2 - y)(y_3 - y)}} - \pi. \tag{7.10.31}$$

Here,

$$y_2 = \frac{1}{2}\left[1 - \sqrt{1 - 4q^2}\right], \qquad y_3 = \frac{1}{2}\left[1 + \sqrt{1 - 4q^2}\right]. \tag{7.10.32}$$

The integration over y leads to the complete elliptic integral **K**

$$\Theta = \frac{2}{\sqrt{y_3}} \mathbf{K}(k) - \pi , \qquad k = \sqrt{y_2/y_3}. \qquad (7.10.33)$$

For the critical trajectory one has $q = 1/2$, $\ell_{min} = 2$ *and* $k = 1$. *We have for the trajectories close to the critical one* $q = 1/2 - z$ *and*

$$k \simeq 1 - \sqrt{2z} , \qquad \mathbf{K}(k) \simeq \ln 2 - \frac{1}{4} \ln z , \qquad \Delta\ell = \ell - 2 \simeq 4z. \qquad (7.10.34)$$

Using these asymptotics we get

$$\Theta = 2\sqrt{2} \ln 2 - \frac{1}{\sqrt{2}} \ln z - \pi , \qquad \Theta = \theta + 2m\pi . \qquad (7.10.35)$$

And finally,

$$\Delta\ell = 64 \, e^{-\sqrt{2}[\theta + (2m+1)\pi]}, \qquad (7.10.36)$$

for any integer winding number m.

8

Rotating Black Holes

8.1 Kerr Spacetime

8.1.1 Astrophysical black holes are rotating

One can expect that the astrophysical black holes are rotating. Usually, a progenitor massive star has a non-vanishing angular momentum. Even if a considerable part of it is lost during the process of the black hole formation, the newly born black hole would be rotating. But even if a black hole is originally formed without angular momentum or with a small value of it, it will obtain angular momentum as a result of the interaction with the matter surrounding it. For example, accreting matter in a binary system that consists of a black hole and a normal star, forms an *accretion disk*. During the evolution of the disk matter from outer regions slowly moves to the inner region until it reaches the *innermost stable circular orbit*. Then, the matter falls into the black hole practically free, bringing to the black hole its mass and angular momentum. Thus, the accretion of the matter onto a black hole in a binary system speeds up its rotation. The calculation shows that this effect is strong and as a result of it the velocity of rotation of the black hole can reach a value close to the speed of light. (We discuss this *spinning up effect* in Section 8.4.) A similar mechanism may work for supermassive black holes as well.

Let us discuss the expected form of the metric for a *rotating black hole*. Consider the Schwarzschild metric for a non-rotating black hole and write it in the form

$$ds^2 = \xi_{(t)}^2 \, dt^2 + \xi_{(\phi)}^2 \, d\phi^2 + g^{-1}(r) \, dr^2 + r^2 \, d\theta^2. \tag{8.1.1}$$

We recall that this metric allows discrete symmetries

$$t \to -t, \qquad \phi \to -\phi, \qquad \theta \to \pi - \theta. \tag{8.1.2}$$

$\xi_{(t)}$ and $\xi_{(\phi)}$ are two mutually commuting and orthogonal Killing vectors. In fact, because of the special form of the function $\xi_{(\phi)}^2$ the metric allows two more Killing vectors and the symmetry group is $\mathbb{R} \times SO(3)$. When a massive object is rotating, its gravitational field has fewer symmetries. Namely:

- The presence of the axis of rotation breaks the spherical symmetry, so that only axial symmetry remains. As a result, the group of symmetry $SO(3)$ reduces to $SO(2) = U(1)$. In other words, the metric is axisymmetric.

- The metric is stationary but not static. Reflection of time $t \to -t$ changes the direction of rotation. However the metric is invariant under a joint reflection $(t, \phi) \to (-t, -\phi)$.

In the general case, a stationary axisymmetric metric has two commuting Killing vectors $\boldsymbol{\xi}_{(t)} = \partial_t$ and $\boldsymbol{\xi}_{(\phi)} = \partial_\phi$, but they are no longer orthogonal. One may expect that the symmetry of the reflection with respect to the 'equatorial plane' is still valid. Under these conditions the most general form of such a metric is

$$ds^2 = \xi_{(t)}^2 \, dt^2 + 2(\boldsymbol{\xi}_{(t)} \cdot \boldsymbol{\xi}_{(\phi)}) \, dt d\phi + \xi_{(\phi)}^2 \, d\phi^2 + B \, dr^2 + C \, d\theta^2. \tag{8.1.3}$$

The metric coefficients are functions of r and θ. (t, ϕ) are parameters along the Killing trajectories. We call (r, θ) *essential coordinates*, and (t, ϕ) *Killing coordinates*.

8.1.2 Kerr metric

A vacuum solution of the Einstein equations for a rotating black hole is the *Kerr metric*. This solution was found by Roy Kerr in 1963. In the *Boyer–Lindquist coordinates* it has the form

$$ds^2 = -\left(1 - \frac{2Mr}{\Sigma}\right) dt^2 - \frac{4Mra \sin^2 \theta}{\Sigma} \, dt \, d\phi + \frac{A \sin^2 \theta}{\Sigma} \, d\phi^2$$

$$+ \frac{\Sigma}{\Delta} \, dr^2 + \Sigma \, d\theta^2, \tag{8.1.4}$$

$$\Sigma = r^2 + a^2 \cos^2 \theta, \qquad \Delta = r^2 - 2Mr + a^2,$$

$$A = (r^2 + a^2)^2 - \Delta \, a^2 \sin^2 \theta. \tag{8.1.5}$$

As expected, this metric does not depend on t and ϕ, and hence $\boldsymbol{\xi}_{(t)} = \partial_t$ and $\boldsymbol{\xi}_{(\phi)} = \partial_\phi$ are two (commuting) Killing vectors. The metric also has the expected discrete symmetries

$$(t, \phi) \to (-t, -\phi), \qquad \theta \to \pi - \theta. \tag{8.1.6}$$

The metric contains two parameters M and a. At far distance $r \to \infty$ the metric has the form

$$ds^2 \approx -\left(1 - \frac{2M}{r}\right) dt^2 - \frac{4Ma \sin^2 \theta}{r} \, dt \, d\phi + dr^2 + r^2(d\theta^2 + \sin^2 \theta \, d\phi^2). \tag{8.1.7}$$

From this asymptotic form one can conclude that M is the *mass*, and $J = aM$ is the *angular momentum* of the black hole. The parameter a in the metric Eq. (8.1.4) is called the rotation parameter. Like the mass M it has the dimensionality of length. Their ratio is a dimensionless parameter of $\alpha = a/M$, which is called the *rotation rapidity*. Similarly to the case of the Schwarzschild black hole, one can use the mass M as a scale parameter and write the Kerr metric Eq. (8.1.7) in the form

$$ds^2 = M^2 d\bar{s}^2. \tag{8.1.8}$$

The dimensionless metric $d\bar{s}^2$ contains only one non-trivial dimensionless parameter, the rotation rapidity α.

Problem 8.1: *Check that in the absence of rotation, $a = 0$, the metric Eq. (8.1.4) coincides with the Schwarzschild metric.*

We present here the expressions for the metric determinant $\sqrt{-g}$ and for the non-vanishing components of the inverse metric that are often useful for calculation

$$g^{rr} = \Delta\Sigma^{-1}, \quad g^{\theta\theta} = \Sigma^{-1}, \quad g^{tt} = -A(\Delta\Sigma)^{-1}, \quad g^{t\phi} = -2Mra(\Delta\Sigma)^{-1},$$

$$g^{\phi\phi} = (\Delta\Sigma \sin^2\theta)^{-1}(\Delta - a^2 \sin^2\theta), \qquad \sqrt{-g} = \Sigma \sin\theta. \tag{8.1.9}$$

In order to check that Eq. (8.1.4) obeys the vacuum Einstein equations one needs to make standard calculations. Namely, to find the Christoffel symbols and the Ricci tensor. These calculations are straighforward, but quite long and time consuming. Fortunately, now there is another much faster way to do this. There exist computer programs adapted for the analytical calculations required in the tensor algebra and in the differential and Riemannian geometry. An example is a GRTensor package that can be found at the site: http://grtensor.phy.queensu.ca/. In particular, the inverse metric and the determinant g given in Eq. (8.1.9) can be obtained in fractions of a second. The GRTensor package can be used for the calculations in higher dimensions as well.

8.1.3   Zero-mass limit of the Kerr metric

In the absence of mass and angular momentum, that is when $M = J = 0$, the curvature vanishes and the spacetime is flat. If we take this limit keeping the ratio $J/M = a = $const for $M = 0$ the Kerr metric takes the form

$$ds^2 = -dt^2 + \left(r^2 + a^2 \cos^2\theta\right)\left[\frac{dr^2}{r^2 + a^2} + d\theta^2\right] + \left(r^2 + a^2\right)\sin^2\theta \, d\phi^2. \tag{8.1.10}$$

By the change of the coordinates

$$T = t, \qquad Z = r\cos\theta,$$

$$X = \sqrt{r^2 + a^2} \, \sin\theta \, \cos\phi, \tag{8.1.11}$$

$$Y = \sqrt{r^2 + a^2} \, \sin\theta \, \sin\phi,$$

the metric is transformed into the Minkowski metric

$$ds^2 = -dT^2 + dX^2 + dY^2 + dZ^2. \tag{8.1.12}$$

A surface $r = $ const is an *oblate ellipsoid of rotation*

$$\frac{X^2 + Y^2}{r^2 + a^2} + \frac{Z^2}{r^2} = 1. \tag{8.1.13}$$

8.1.4   Kerr–Schild form of the Kerr solution

Let $\eta_{\mu\nu}$ be a flat metric of the Minkowski spacetime, and (T, X, Y, Z) be Cartesian coordinates in it. Denote by r a function of X, Y, Z defined by Eq. (8.1.13), where a is a constant. Denote by $\boldsymbol{k}$ a vector

$$\boldsymbol{k} = (1, \frac{rX + aY}{r^2 + a^2}, \frac{rY - aX}{r^2 + a^2}, \frac{Z}{r}). \tag{8.1.14}$$

Let $g_{\mu\nu}$ be the following metric

$$g_{\mu\nu} = \eta_{\mu\nu} + 2Hk_\mu k_\nu, \qquad H = \frac{Mr^3}{r^4 + a^2 Z^2}. \qquad (8.1.15)$$

It is easy to check that k is a null vector in both metrics, η and g. For the chosen function H and the null vector Eq. (8.1.14), the metric Eq. (8.1.15) is a solution of the vacuum Einstein equations. It coinsides with the Kerr metric Eq. (8.1.4) after a proper change of the coordinates. When H and a null vector k are not fixed, expression Eq. (8.1.15) is called the *Kerr–Schild metric*.

8.1.5 Charged rotating black hole

There exists a solution, generalizing the Kerr metric to the case when a black hole is electrically charged. This solution is known as a *Kerr–Newman metric*. The parameter of the charge can be determined by measuring the electrostatic field of such a black hole at spatial infinity, similarly to how the mass and the angular momentum are defined in terms of the Komar integrals.

The Kerr–Newman metric is a solution of the Einstein–Maxwell equations. The Maxwell equations are

$$F^{\mu\nu}{}_{;\nu} = 4\pi J^\mu, \qquad F_{[\mu\nu;\alpha]} = 0. \qquad (8.1.16)$$

Let Σ be a global space-like surface. Then,

$$Q = \int_\Sigma J^\mu \, d\sigma_\mu \qquad (8.1.17)$$

is the electric charge located at this surface. Since $J^\mu{}_{;\mu} = 0$, Q does not depend on Σ provided during the deformation of Σ in the time-like direction there are no currents through the time-like boundary at infinity. Using Stokes' theorem one can rewrite the expression for the charge Q within a 3-dimensional volume Σ as the flux of the electric field $F_{\mu\nu}$ through its 2-dimensional boundary

$$4\pi Q = \int_\Sigma F^{\mu\nu}{}_{;\nu} \, d\sigma_\mu = \int_{\partial\Sigma} F^{\mu\nu} \, d\sigma_{\mu\nu}. \qquad (8.1.18)$$

This relation is the *Gauss law*.

The *Kerr–Newman metric* in the *Boyer–Lindquist coordinates* has the form

$$ds^2 = -\left(1 - \frac{2Mr - Q^2}{\Sigma}\right) dt^2 - \frac{(2Mr - Q^2)\, 2a \sin^2\theta}{\Sigma} \, dt\, d\phi$$

$$+ \frac{A \sin^2\theta}{\Sigma} \, d\phi^2 + \frac{\Sigma}{\Delta} \, dr^2 + \Sigma \, d\theta^2, \qquad (8.1.19)$$

$$\Sigma = r^2 + a^2 \cos^2\theta, \qquad \Delta = r^2 - 2Mr + a^2 + Q^2,$$

$$A = (r^2 + a^2)^2 - \Delta a^2 \sin^2\theta. \qquad (8.1.20)$$

The vector potential of the electromagnetic field reads

$$A_\alpha dx^\alpha = -\frac{Qr}{\Sigma}(dt - a \sin^2 \theta \, d\phi). \tag{8.1.21}$$

The horizon of the charged rotating black hole is located at

$$r_+ = M + \sqrt{M^2 - a^2 - Q^2}, \tag{8.1.22}$$

where M, $J = aM$, and Q are the mass, angular momentum and electric charge of the black hole, respectively. The surface area is

$$\mathcal{A} = 4\pi(r_+^2 + a^2). \tag{8.1.23}$$

For $a = 0$ the Kerr–Newman solution coincides with the Reissner-Nordström solution, describing a charged spherically symmetric black hole. The Kerr-Newman solution has many properties similar to the Kerr metric. The detailed discussion of this metric, including the global structure of the spacetime can be found e.g in the paper (Carter 1968) and the book (Heusler 1996).

8.2 Ergosphere. Horizon

8.2.1 Infinite redshift surface

From the form of the Kerr metric it is easy to conclude that

$$-\xi_t^2 = 1 - \frac{2Mr}{\Sigma}. \tag{8.2.1}$$

The *infinite redshift surface* Γ_+ is defined by the equation

$$\Sigma - 2Mr = r^2 - 2Mr + a^2 \cos^2 \theta = 0, \tag{8.2.2}$$

or

$$r = r_0 \equiv M + \sqrt{M^2 - a^2 \cos^2 \theta}. \tag{8.2.3}$$

This surface is also called an *ergosurface*.

The Killing vector ξ_t is time-like outside Γ_+, is null at Γ_+, and it is space-like inside Γ_+. In the absence of the rotation, that is in the Schwarzschild geometry, the surface Γ_+ is null and it coincides with the *event horizon*. In the presence of the rotation it is no longer true. In fact, Γ_+ is time-like. To prove this, consider a normal vector

$$n_\mu = (0, 1, -dr_0/d\theta, 0) \tag{8.2.4}$$

to the surface $r - r_0(\theta) = 0$. Its norm is

$$n^2 = g^{rr} + g^{\theta\theta}(dr_0/d\theta)^2 = \Sigma^{-1}[\Delta + (dr_0/d\theta)^2] \geq 0. \tag{8.2.5}$$

Thus, outside the axis of rotation the vector n^μ is space-like, and hence the surface Γ_+, to which it is normal, is time-like. A time-like surface cannot be a 'surface of no return'. It can be penetrated by particles and light rays in the both directions. Thus, in the Kerr spacetime the ergosurface is not the event horizon.

8.2.2 Ergosphere

If Γ_+ is not a horizon, what are the special properties of this surface? In order to answer this question, let us consider a particle that is moving at $r, \theta = \text{const}$ circles. In the general case, this motion is not geodesic. The four-velocity is

$$u^\mu = \eta^\mu \Big/ \left| \eta^2 \right|^{1/2}, \qquad \eta^\mu \equiv \xi_t^\mu + \omega \xi_\phi^\mu, \tag{8.2.6}$$

where ω is the angular velocity of a stationary observer at a circular orbit. To have an agreement with the Kerr metric we choose the direction of ϕ, which specifies the meaning of the rotation, so that $g_{t\phi} < 0$.

There exist special values of ω for which $\boldsymbol{u}$ becomes null. The condition $\boldsymbol{u}^2 = 0$ gives

$$\xi_\phi^2 \, \omega^2 + 2\omega \left(\xi_t, \xi_\phi \right) + \xi_t^2 = 0. \tag{8.2.7}$$

This equation has solutions

$$\omega_\pm = \frac{-g_{t\phi} \pm \sqrt{g_{t\phi}^2 - g_{tt}\, g_{\phi\phi}}}{g_{\phi\phi}}. \tag{8.2.8}$$

One can check that

$$g_{t\phi}^2 - g_{tt}\, g_{\phi\phi} = \Delta \sin^2 \theta. \tag{8.2.9}$$

For $\Delta > 0$ Eq. (8.2.7) has two roots $\omega_\pm$, while for $\Delta < 0$ there are no real roots.

The vector u^μ is time-like if

$$\omega_- < \omega < \omega_+, \tag{8.2.10}$$

Outside the infinite redshift surface, where $g_{tt} < 0$, ω_- is negative and ω_+ is positive. Thus, the observer can both corotate and counter-rotate with the black hole. Inside the infinite redshift surface all the observers are corotating with the black hole. The region between the infinite redshift surface and the surface where $\Delta = 0$ is called the *ergosphere*.

What happens if the particle inside the ergosphere moves in r and θ directions as well? Such a particle has the velocity

$$u^\mu \sim \eta^\mu + \alpha\, e_{\hat{r}}^\mu + \beta\, e_{\hat{\theta}}^\mu, \tag{8.2.11}$$

where $e_{\hat{r}}^\mu$ and $e_{\hat{\theta}}^\mu$ are unit vectors orthogonal to (t, ϕ) surface. The norm of this vector is

$$u^2 \sim \eta^2 + \alpha^2 + \beta^2. \tag{8.2.12}$$

The condition $\boldsymbol{u}^2 = 0$ gives an expression for $\omega_\pm$ similar to Eq. (8.2.8) with the change $g_{tt} \to \tilde{g}_{tt} = g_{tt} + \alpha^2 + \beta^2$. Thus, $\omega_- > 0$ inside the ergosphere even for motion with non-vanishing α and β.

Inside the ergosphere particles can move both with decrease and increase of r, but they must be necessarilay corotating with the black hole, that is to have a non-vanishing angular velocity $\omega \in (\omega_-, \omega_+) \, (\omega_- > 0)$.

8.2.3 Event horizon

The surface $\Delta = 0$ is the *event horizon*. At this surface

$$\omega_+ = \omega_- = \Omega. \tag{8.2.13}$$

Here,

$$\Omega = - \left. \frac{g_{t\phi}}{g_{\phi\phi}} \right|_{\mathrm{H}} = \frac{a}{2Mr_+} = \frac{a}{r_+^2 + a^2} \tag{8.2.14}$$

is the *angular velocity of the black hole*. The Killing vector η

$$\eta^\mu \equiv \xi_t^\mu + \Omega\, \xi_\phi^\mu, \tag{8.2.15}$$

is null at the horizon. The surface of the event horizon is also null and η is its null generator. A null surface where the Killing vector is also null is called a *Killing horizon*. In a stationary asymptotically flat spacetime the Killing horizon coincides with the *event horizon*.

The event horizon of a rotating black hole is located at $r = r_+$, where $\Delta(r_+) = 0$,

$$r = r_+, \qquad r_\pm = M \pm \sqrt{M^2 - a^2}. \tag{8.2.16}$$

The event horizon of the Kerr spacetime is a null 3-dimensional surface. Its spatial slices have the geometry of a 2-dimensional distorted sphere. The rotating black holes exist for $a \leq M$. For $a > M$ the Kerr solution does not have a horizon and it describes a naked singularity. It is generally believed that such a singularity does not arise in real physical processes, like gravitational collapse. The reason is that for rapidly rotating objects centrifugal forces can prevent the collapse. The collapse with a formation of a black hole is possible when the system loses enough of its angular momentum so that the condition $a/M \leq 1$ is satisfied. The event horizon is the inner boundary of the ergosphere. The infinite redshift surface is located outside the horizon and touches it only at two points, the 'north' and 'south' poles.

8.2.4 Surface gravity

Since the Killing vector η Eq. (8.2.15) is null and orthogonal to the hypersurface of the horizon, one has

$$\nabla^\mu (\eta^\alpha \eta_\alpha) = -2\kappa \eta^\mu. \tag{8.2.17}$$

The coefficient of proportionality κ here is a scalar function. If we take the Lie derivative of the previous equation then we obtain the property that κ is constant along the orbits of the Killing vector η

$$\mathcal{L}_\eta \kappa = 0. \tag{8.2.18}$$

In fact, the Einstein equations together with the *dominant energy condition* imply also that $\eta_{[\mu} \nabla_{\nu]} \kappa = 0$, i.e. κ is constant over the horizon. This constant has a meaning of a *surface gravity* of the horizon. In order to see this one can derive a few equivalent formulas for κ. First, from Eq. (8.2.17) and the fact that η is the Killing vector it trivially follows that

$$\eta^\alpha \nabla_\mu \eta_\alpha = -\eta^\alpha \nabla_\alpha \eta_\mu = -\kappa \eta_\mu. \tag{8.2.19}$$

Because η is orthogonal to the horizon, by the *Frobenius theorem* (see Wald 1984) we have

$$\eta_{[\alpha} \nabla_\beta \eta_{\gamma]} \Big|_{r=r_+} = 0 \tag{8.2.20}$$

and, hence,

$$\eta_\gamma \nabla_\alpha \eta_\beta \Big|_{r=r_+} = -2\eta_{[\alpha} \nabla_{\beta]}\eta_\gamma \Big|_{r=r_+}. \tag{8.2.21}$$

After multiplying this expression by $\nabla^\alpha \eta^\beta$, and twice using Eq. (8.2.19) we obtain

$$\kappa^2 = -\frac{1}{2}\eta^{\alpha;\beta}\eta_{\alpha;\beta}\Big|_{r=r_+}. \tag{8.2.22}$$

This formula for the *surface gravity* is valid for an arbitrary stationary black hole, provided its right-hand side is calculated on its horizon. For stationary black holes the surface gravity is constant everywhere on the horizon (Bardeen and Press 1973). In the particular case of the Kerr black hole it is

$$\kappa = \frac{\sqrt{M^2 - a^2}}{2M \left(M + \sqrt{M^2 - a^2}\right)} = \frac{r_+ - r_-}{2r_+(r_+ + r_-)}. \tag{8.2.23}$$

Problem 8.2: *Show that*

$$\kappa = \lim_{r \to r_+} \sqrt{w^\alpha w_\alpha} \sqrt{-\eta^\mu \eta_\mu}, \tag{8.2.24}$$

where

$$w_\alpha = \frac{1}{2}\nabla_\alpha \ln |\eta^\mu \eta_\mu| \tag{8.2.25}$$

is the 4-acceleration vector on the orbit η^μ and $\sqrt{-\eta^\mu \eta_\mu}$ is the redshift factor.

8.2.5   Geometry of a surface of revolution

Axisymmetric deformation of S^2

The surface of the Kerr black hole has the geometry of a 2-dimensional axisymmetric distorted sphere. In the general case, a metric of such a sphere can be written in the form

$$dl^2 = H(x)\,dx^2 + F(x)\,d\phi^2. \tag{8.2.26}$$

Here, $\boldsymbol{\xi} = \partial_\phi$ is a Killing vector field with closed trajectories and $F = \boldsymbol{\xi}^2$. The points where $F(x)$ vanishes are *fixed points* of the Killing vector $\boldsymbol{\xi}$. We assume that $F(x)$ is positive in the interval (x_1, x_2) and vanishes at its end points. Using the coordinate freedom we put $x_1 = -1$ and $x_2 = 1$, so that two fixed points of $\boldsymbol{\xi}$ are located at $x = \pm 1$. The surface area of the 2D surface with this metric is

$$\mathcal{A} = 4\pi B^2, \qquad B^2 = \frac{1}{2}\int_{-1}^{1} dx\sqrt{HF}. \tag{8.2.27}$$

Let us write the metric Eq. (8.2.26) in the form

$$dl^2 = B^2 dS^2. \tag{8.2.28}$$

The surface area for dS^2 is 4π. The scale factor B describes the size of a 2D deformed sphere, while the metric dS^2 contains information concerning its shape.

Introducing a new coordinate

$$z = B^{-2} \int_{-1}^{x} dx\sqrt{HF} - 1 \qquad (8.2.29)$$

one can write

$$dS^2 = f(z)^{-1}dz^2 + f(z)\,d\phi^2, \qquad f(z) = B^{-2}F(x). \qquad (8.2.30)$$

The new coordinate z also belongs to the interval $[-1, 1]$. At the end points of this interval the function $f(z)$ vanishes. In such a parametrization the geometry of a non-distorted round sphere is

$$dS^2 = \frac{dz^2}{1 - z^2} + (1 - z^2)\,d\phi^2. \qquad (8.2.31)$$

The metric Eq. (8.2.30) is regular (no conical singularities) at the points $x = \pm 1$ if $(df/dz)(\pm 1) = \mp 2$. The Ricci scalar for the metric Eq. (8.2.30) is (see Problem 3.8 on page 88)

$$R = -d^2f/dz^2. \qquad (8.2.32)$$

Instead of the scalar curvature R in two dimensions another quantity proportional to R is often used. It is known as the *Gaussian curvature*

$$\mathcal{K} = \frac{1}{2}R. \qquad (8.2.33)$$

These quantities give information about the local shape of the 2D surface. For a round unit sphere one has $\mathcal{K} = 1$.

Embedding

When we are working with a curved spacetime it is difficult to use our intuition. For this reason quite often one uses different tricks that help to understand better some of the features of the problem. The method of *embedding* is one such tool. For example, one can consider embedding of special slices of the spacetime into a flat space with a larger number of dimensions.

> It is well known that intrinsically defined Riemannian manifolds can be isometrically embedded in a flat space. The problem of isometric embedding of 2D manifolds in $\mathbb{E}^3$ is well studied. It is known that any compact surface embedded isometrically in $\mathbb{E}^3$ has at least one point of positive Gauss curvature. Any 2D compact surface with positive Gauss curvature is always isometrically embeddable in $\mathbb{E}^3$, and this embedding is unique up to rigid rotations. (For a general discussion of these results and for further references, see, e.g., (Berger 2003).)
>
> For the embedding of a 2D surface into the Euclidean space $\mathbb{E}^3$ there exists a simple relation between the Gaussian curvature $\mathcal{K}$ and the extrinsic curvature K_{ab} $(a, b = 1, 2)$,
>
> $$\mathcal{K} = \frac{1}{2}(K^2 - K^{ab}K_{ab}). \qquad (8.2.34)$$

This relation follows from the Gauss–Codazzi equation Eq. (B.1.5), when applied to a surface in a flat Euclidean space $\mathbb{E}^3$. Note that for the 2D surface the extrinsic curvature tensor can always be diagonalized simultaneously with the 2D metric. Its components in this diagonalized form are called principal curvatures k_1 and k_2,

$$K^a{}_b = diag(k_1, k_2). \tag{8.2.35}$$

Then,

$$K^2 = (k_1 + k_2)^2, \qquad K^{ab}K_{ab} = k_1^2 + k_2^2, \tag{8.2.36}$$

and, hence,

$$\mathcal{K} = k_1 k_2. \tag{8.2.37}$$

For axisymmetric distorted 2D spheres the problem of isometric embedding is quite simple. Consider a 3D Euclidean space in the cylindrical coordinates

$$dL^2 = dZ^2 + d\rho^2 + \rho^2 \, d\phi^2. \tag{8.2.38}$$

Choose a line $\rho = \rho(Z)$ at fixed ϕ and rotate it in the ϕ-direction around the Z-axis. The obtained surface is called a *revolution surface* (see Figure 8.1). The induced 2D geometry on it is

$$dS^2 = [1 + (d\rho/dZ)^2] \, dZ^2 + \rho^2 \, d\phi^2. \tag{8.2.39}$$

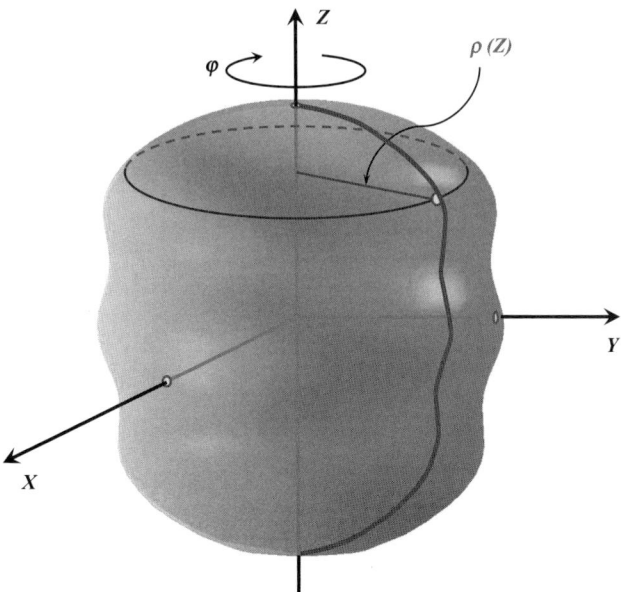

Fig. 8.1 A surface of revolution.

If one has a 2D axisymmetric metric, one says that it is embedded in $\mathbb{E}^3$ if its geometry is isometric to Eq. (8.2.39). For the metric Eq. (8.2.30) the embedding equations are

$$\rho(Z) = \sqrt{f(z)}, \qquad Z(z) = \pm \int_{-1}^{z} dz \frac{\sqrt{1 - 1/4(df/dz)^2}}{\sqrt{f(z)}}. \qquad (8.2.40)$$

Consider a vicinity of fixed points. For simplicity, we choose one of them, namely the 'northern' pole $z = 1$. (For the other point the result is the same.) Near this fixed point one has

$$f \sim 1 + 2(1 - z) - \mathcal{K}(1 - z)^2. \qquad (8.2.41)$$

Thus,

$$1 - 1/4(df/dz)^2 \sim 2\mathcal{K}(1 - z). \qquad (8.2.42)$$

The expression inside the square root in the nominator of the integrand in Eq. (8.2.40) becomes negative near $z = 1$ if at this point the Gaussian curvature is negative. This means that the embedding is possible only if at the fixed point $\mathcal{K} \geq 0$. (For further discussion see (Frolov 2006).)

8.2.6   Surface geometry of horizon and ergo surface

Geometry of Kerr horizon

The geometry of the 2D surface of the horizon in the Kerr spacetime is

$$dl^2 = \Sigma_+ \, d\theta^2 + \frac{\left(r_+^2 + a^2\right)^2 \sin^2 \theta}{\Sigma_+} \, d\phi^2, \qquad \Sigma_+ = r_+^2 + a^2 \cos^2 \theta. \qquad (8.2.43)$$

Let us introduce new parameters

$$B = \sqrt{r_+^2 + a^2}, \qquad \beta = \frac{a}{\sqrt{r_+^2 + a^2}}, \qquad (8.2.44)$$

then one has $B^{-2}\Sigma_+ = 1 - \beta^2 \sin^2 \theta$, and the metric on the surface of the horizon takes the form

$$dl^2 = B^2 \left[\left(1 - \beta^2 \sin^2 \theta\right) d\theta^2 + \frac{\sin^2 \theta}{1 - \beta^2 \sin^2 \theta} \, d\phi^2 \right]. \qquad (8.2.45)$$

The surface area of the horizon is

$$A = 4\pi B^2 = 4\pi \left(r_+^2 + a^2\right). \qquad (8.2.46)$$

Let us denote $z = \cos \theta$, then one has

$$dl^2 = B^2 dS^2, \qquad dS^2 = f^{-1} dz^2 + f \, d\phi^2, \qquad f = \frac{1 - z^2}{1 - \beta^2(1 - z^2)}. \qquad (8.2.47)$$

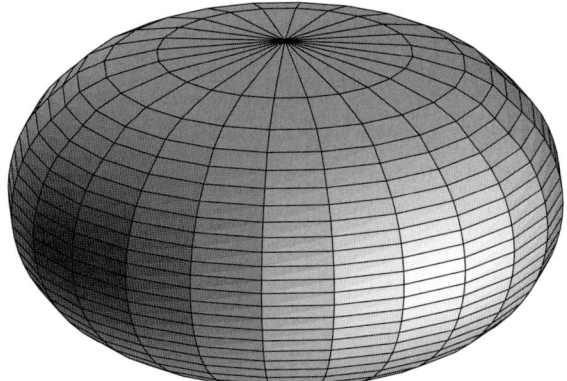

Fig. 8.2 The embedding diagram for a two-dimensional section of the event horizon of the Kerr black hole. The diagram is constructed for the critical value $a/M = \sqrt{3}/2$ of the rotation parameter so that the Gaussian curvature vanishes at the poles.

The length of the equatorial circle $\theta = \frac{\pi}{2}$ for the metric dS^2 is

$$L_1 = \frac{2\pi}{\sqrt{1 - \beta^2}}. \tag{8.2.48}$$

The length of the pole–pole closed geodesic (that is twice the length of the geodesic distance between the 'north' and 'south' poles) is

$$L_p = 2 \int_0^\pi d\theta \sqrt{1 - \beta^2 \sin^2 \theta} \le 2\pi \le \frac{2\pi}{\sqrt{1 - \beta^2}} = L_1. \tag{8.2.49}$$

Thus, the surface with the geometry Eq. (8.2.47) is an oblate surface of rotation.
Using Eq. (8.2.32) one obtains the Gaussian curvature of the horizon

$$\mathcal{K} = -\frac{d^2 f/dz^2}{2B^2} = \frac{1 - \beta^2 \left(1 + 3\cos^2 \theta\right)}{B^2 \left(1 - \beta^2 \sin^2 \theta\right)^3} = (r_+^2 + a^2) \frac{r_+^2 - 3a^2 \cos^2 \theta}{\left(r_+^2 + a^2 \cos^2 \theta\right)^3}. \tag{8.2.50}$$

Note that for a rapidly rotating black holes with $a > \sqrt{3}M/2$ (that is when $r_+^2 < 3a^2$) the Gaussian curvature becomes negative in the near-pole regions (Smarr 1973). The embedding diagram of the 2D surface of the Kerr horizon for the critical value $a/M = \sqrt{3}/2$ is shown in Figure 8.2. For larger values of the rotation rapidity parameter the regions near the poles cannot be embedded in $\mathbb{E}^3$ as a revolution surface.

Geometry of ergosurface

The surface geometry of the *ergosurface* can be obtained from the Kerr metric by putting there $t = \text{const}$ and $r = M + \sqrt{M^2 - a^2 \cos^2 \theta}$.

Problem 8.3: *Show that the 2D metric on the ergosurface can be written in the form*

$$dl^2 = 2M^2 \left[\frac{x(\theta)}{(x(\theta) - 1)^2} d\theta^2 + \sin^2 \theta [x(\theta) + \alpha^2 \sin^2 \theta] d\phi^2 \right], \tag{8.2.51}$$

where $\alpha = a/M$ is the rotation rapidity and

$$x(\theta) = 1 + \sqrt{1 - \alpha^2 \cos^2 \theta}. \tag{8.2.52}$$

The embedding diagrams for this 2D surface geometry can be found in the paper (Pelavas *et al.* 2001). It is interesting that this surface is smooth everywhere, except at the poles, where it has a *conical singularity*. Because of the symmetry $\theta \to \pi - \theta$ both of the singularities have the same *angle deficit*.

Problem 8.4: *Show that in the vicinity of the 'north' pole (for small θ) the metric Eq. (8.2.51) has the following form*

$$dl^2 \approx C^2 \left[d\theta^2 + (1 - \alpha^2) \theta^2 \, d\phi^2 \right], \qquad C^2 = 2M^2 \frac{1 + \sqrt{1 - \alpha^2}}{1 - \alpha^2}. \tag{8.2.53}$$

The circumference of a circle of given small θ is $l_\phi = 2\pi C \sqrt{1 - \alpha^2} \, \theta$, while the proper length to it from the pole is $l_\theta = C\theta$. For the rotating black hole the ratio $l_\phi/l_\theta = 2\pi \sqrt{1 - \alpha^2}$ differs from 2π. Near the pole the geometry is similar to the geometry of the cone with the angle deficit $\delta = 2\pi(1 - \sqrt{1 - \alpha^2})$ (Pelavas *et al.* 2001).

8.2.7 Einstein–Rosen bridge in the Kerr spacetime

For the Schwarzschild spacetime we also discussed the geometry of the *Einstein–Rosen bridge*. Due to the spherical symmetry it was sufficient to study the 2D section $\theta = \pi/2$, that is an intersection of the equatorial plane and the bridge surface. It is interesting to see how the rotation of the black hole deforms the Einstein–Rosen bridge. The metric on a $t =$const slice of the Kerr spacetime depends on r and θ, and it does not have spherical symmetry. Instead of a study of this quite complicated geometry, we focus on its equatorial slice. This 2D section $t = $ const of $\theta = \pi/2$ of the Kerr black hole can be embedded as a surface of revolution in $\mathbb{E}^3$. The 2D line element on this slice is

$$dl^2 = \left[\frac{\Sigma}{\Delta} dr^2 + \frac{A}{\Sigma} d\phi^2 \right]_{\theta=\pi/2} = \frac{r^2}{r^2 - 2Mr + a^2} dr^2 + \frac{r^3 + a^2 r + 2a^2 M}{r} d\phi^2. \tag{8.2.54}$$

The metric Eq. (8.1.4) is valid in the black hole exterior, where the radial coordinate r changes from r_+ (at the horizon) to $r = \infty$ (at the spatial infinity). The corresponding part of the Einstein–Rosen bridge is Σ_+. If one considers an analytical continuation of the Kerr metric, it also has an inner part that contains another branch Σ_- of the Einstein–Rosen bridge. The geometries of Σ_+ and Σ_- are isometric. For this reason it is sufficient to consider the external part Σ_+ only.

The surface with the geometry Eq. (8.2.54) can be embedded in a three-dimensional flat space

$$dL^2 = dZ^2 + d\rho^2 + \rho^2 d\phi^2. \tag{8.2.55}$$

It is convenient to parametrize the embedding surface as $Z(r)$, $\rho(r)$, where r is the coordinate that enters Eq. (8.2.54). In this parametrization the induced geometry is

$$dl^2 = \left[(Z')^2 + (\rho')^2 \right] dr^2 + \rho^2 d\phi^2, \tag{8.2.56}$$

where $(\ldots)'$ is the derivative with respect to r. By comparing the metrics Eq. (8.2.54) and Eq. (8.2.56) one obtains the following equations of the embedding

$$\rho = \sqrt{\frac{r^3 + a^2 r + 2a^2 M}{r}},$$

(8.2.57)

$$(Z')^2 + (\rho')^2 = \frac{r^2}{r^2 - 2Mr + a^2}.$$

We can always choose the coordinate Z to be zero on the horizon $r = r_+$. Then,

$$Z(r) = \pm \int_{r_+}^{r} dr \sqrt{\frac{r^2}{r^2 - 2Mr + a^2} - (\rho')^2}.$$

(8.2.58)

Substituting here the expression Eq. (8.2.57) one gets an explicit integral formula for the shape of the surface. Though this formula looks quite messy, numerically it is easy to plot the revolution surface for the embedding of the equatorial slice of the Einstein–Rosen bridge. For a non-rotating black hole $a = 0$ we reproduce the Einstein–Rosen bridge.

In Figure 8.3 we have depicted an upper half $Z \geq 0$ of the embedding surfaces (corresponding to Σ_+) for different values of the rotation parameter a. The closer the rotation parameter to the critical one the longer the 'throat' of the bridge. The lower halves of the Einstein–Rosen bridges are just symmetrical reflections of the upper ones.

The case of the extremally rotating Kerr black hole, when $a/M = 1$ is a special one. In this case, the throat of the bridge becomes infinitely long. It does not mean though that one can not reach the horizon in a finite proper time. Even in an extremal case all time-like geodesics reach the horizon in a finite proper time. For null geodesics its takes a finite value of the affine parameter to reach the event horizon (see discussion in the next section).

Problem 8.5: *Show that for a given mass M all the embedding surfaces have the same value $\rho = 2M$ at the horizon. Show that the ergosurface is located at $r = 2M$ or, equivalently, at $\rho = \sqrt{4M^2 + 2a^2}$.*

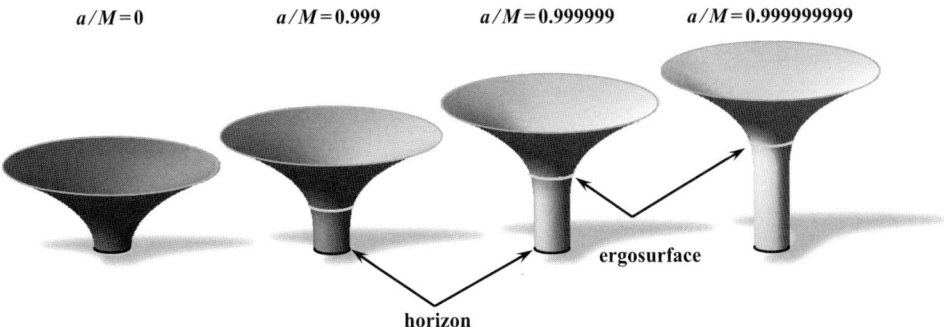

$a/M = 0$ $a/M = 0.999$ $a/M = 0.999999$ $a/M = 0.999999999$

ergosurface

horizon

Fig. 8.3 The embedding diagram for a two-dimensional section of equatorial plane ($t = 0$, $\theta = \pi/2$) of the Kerr black hole. The diagram depicts embedding surfaces for a non-rotating $a/M = 0$ black hole and three near-extremal black holes, with a close to M.

8.2.8 Near-extremal Kerr spacetime

Properties of the extemal Kerr black hole

The Kerr solution describes a black hole for $M > a$, while for $M < a$ there is no horizon and this solution describes a naked singularity. The critical solution with $M = a$ that separates these two different phases is called an *extremal Kerr metric*. Let us discuss some properties of the near-extremal and the extremal Kerr spacetimes.

For the exremal black hole M is the only parameter that completely specifies the metric. One has

$$r_+ = M, \qquad \Omega = \frac{1}{2M}, \qquad \mathcal{A} = 8\pi M^2, \qquad \kappa = 0. \qquad (8.2.59)$$

In this limit $\Delta = (r - M)^2$. As a result, a proper distance to the horizon along space-like geodesics (for $t = $ const) is infinite. On the other hand, it is possible to show that dispite this divergence the proper time that a particle takes to reach the horizon remains finite. For a null geodesic the affine parameter is also finite at the horizon.

> **Problem 8.6:** *Show that in the $a = M$ limit the Kerr metric Eq. (8.1.4) at the axis of symmetry $\theta = 0, \pi$ reduces to the following 2D metric*
>
> $$dS^2 = -\left(1 - \frac{2Mr}{r^2 + M^2}\right)dt^2 + \frac{r^2 + M^2}{(r - M)^2}dr^2. \qquad (8.2.60)$$
>
> *Show that the proper distance (for $t = $ const) from a point r_0 to the horizon is logarithmically divergent. Using a geodesic equation for a particle motion in this 2D metric, show that the proper time required to reach the horizon along the symmetry axis is finite. Show that the affine parameter for the null geodesic for the metric Eq. (8.2.60) also remains finite at the horizon.*

Near-horizon geometry

The space near a horizon of the extremal Kerr metric has a 'tube-like' geometry. It is clear that r is not a good choice of the coordinate in this region. To study the near-horizon geometry of the extremal Kerr spacetime we make the following coordinate transformation

$$r = M + \lambda\rho, \qquad t = T/\lambda, \qquad \phi = \varphi + t/(2M\lambda). \qquad (8.2.61)$$

This transformation for small λ 'zooms in' the near-horizon region. Writing the extremal Kerr metric in the new coordinates $(T, \rho, \theta, \varphi)$ and taking the limit $\lambda \to 0$ one obtains the following metric (Bardeen and Horowitz 1999)

$$ds^2 = \frac{1}{2}(1 + \cos^2\theta)\left[-(\rho/\rho_0)^2 dT^2 + (\rho_0/\rho)^2 d\rho^2 + \rho_0^2 d\theta^2\right]$$

$$+ \frac{2\rho_0^2 \sin^2\theta}{1 + \cos^2\theta}[d\varphi + (\rho/\rho_0^2)dT]^2. \qquad (8.2.62)$$

Here, $\rho_0 = \sqrt{2}M$. Since the mass M enters as a global scale parameter it is sufficient to consider the metric Eq. (8.2.62) for a special value. In particular, it is convenient to put $M = 1/\sqrt{2}$, so that $\rho_0 = 1$ The form of this metric can be slightly simplified by means of the following transformation

$$\rho = \sqrt{1 + y^2}\,\cos\tau + y, \qquad T = \rho^{-1}\sqrt{1 + y^2}\,\sin\tau,$$

$$\varphi = \Phi + \ln\left|\frac{\cos\tau + y\sin\tau}{1 + \sqrt{1 + y^2}\,\sin\tau}\right|. \tag{8.2.63}$$

The corresponding metric is (Bardeen and Horowitz 1999)

$$ds^2 = \frac{1}{2}(1 + \cos^2\theta)\left[-(1 + y^2)d\tau^2 + \frac{dy^2}{1 + y^2} + d\theta^2\right]$$

$$+ \frac{2\sin^2\theta}{1 + \cos^2\theta}(d\Phi + y\,d\tau)^2. \tag{8.2.64}$$

The symmetry of the near-horizon metric Eq. (8.2.63) is enhanced. Namely, it has four Killing vectors, instead of two Killing vectors of the Kerr spacetime. These vectors are

$$\xi_1 = \partial_\Phi,$$

$$\xi_2 = \frac{y\cos\tau + \sqrt{1 + y^2}}{\sqrt{1 + y^2}}\partial_\tau + \sin\tau\sqrt{1 + y^2}\,\partial_y + \frac{\cos\tau}{\sqrt{1 + y^2}}\partial_\Phi,$$

$$\xi_3 = \frac{y\sin\tau}{\sqrt{1 + y^2}}\partial_\tau - \cos\tau\sqrt{1 + y^2}\,\partial_y + \frac{\sin\tau}{\sqrt{1 + y^2}}\partial_\Phi, \tag{8.2.65}$$

$$\xi_4 = \frac{\cos\tau + y\sin^2\tau\sqrt{1 + y^2}}{2\sqrt{1 + y^2}(y + \cos\tau\sqrt{1 + y^2})}\partial_\tau + \frac{1}{2}\sin\tau\sqrt{1 + y^2}\,\partial_y - \frac{\cos\tau}{2\sqrt{1 + y^2}}\partial_\Phi.$$

8.3 Particle and Light Motion in Equatorial Plane

In this and subsequent sections we discuss particle motion and light propagation in the Kerr geometry. These problems are of special interest in connection with possible astrophysical applications. Let us indicate some of these problems:

1. A distant observer obtains information concerning a rotating black hole by studying light emitted from matter in its vicinity.

2. Matter in the *accretion disk* surrounding a black hole is moving on the almost circular Keprlerian orbits.

3. Fluctuations of the radiation emitted by inhomogenaties of the disk are modulated by the Keplerian frequency.

4. To explain the broading of the sharp Fe α lines emitted by Fe ions in the accretion disk one needs to know both the characteristics of the Keplerian orbits and of the light propagation in the Kerr spacetime.

5. Information concerning the motion of extended objects in the gravitational field of a black hole is required for study of the tidal disruption of stars encountering a supermassive black hole.

These and other problems require study of time-like and null geodesics in the Kerr geometry. Additional information on this subject can be found in the books (Misner *et al.* 1973; Chandrasekhar 1983) and review articles (Bardeen 1973; Carter 1973; Wilkins 1972; Teo 2003). The next chapter contains a discussion of the test-field propagation.

8.3.1 Integrals of motion

When a black hole is rotating, there exists a preferable direction in the space, i.e. the axis of the rotation. In such a gravitational field the generic orbits of particle and light are no longer planar. Motion in the equatorial plane is a special case. Because of the reflection symmetry $\theta \to \pi - \theta$, this plane is a geodesic submanifold. This means that the motion of a particle or light with the initial data $\theta = \pi/2$, $\dot{\theta} = 0$ is restricted to the equatorial plane. Let us consider this type of motion first (Bardeen 1973). The reduced metric in the equatorial plane is

$$ds^2 = g_{tt}\, dt^2 + 2g_{t\phi}\, dt d\phi + g_{\phi\phi}\, d\phi^2 + g_{rr}\, dr^2, \tag{8.3.1}$$

$$g_{tt} = -\left(1 - \frac{2M}{r}\right), \qquad g_{t\phi} = -\frac{2aM}{r},$$

$$g_{\phi\phi} = r^2 + a^2 + \frac{2a^2 M}{r}, \qquad g_{rr} = \left(1 - \frac{2M}{r} + \frac{a^2}{r^2}\right)^{-1}. \tag{8.3.2}$$

We choose the affine parameter λ along the trajectory so that $p^\mu = \dot{x}^\mu = dx^\mu/d\lambda$. The conserved quantities, the energy E and angular momentum L, are

$$E = -\xi_{(t)\mu}\, p^\mu = -g_{tt}\, \dot{t} - g_{t\phi}\, \dot{\phi},$$
$$L = \xi_{(\phi)\mu}\, p^\mu = g_{t\phi}\, \dot{t} + g_{\phi\phi}\, \dot{\phi}. \tag{8.3.3}$$

The normalization condition $p^2 = -m^2$ takes the form

$$g_{tt}\, \dot{t}^2 + 2g_{t\phi}\, \dot{t}\dot{\phi} + g_{\phi\phi}\, \dot{\phi}^2 + g_{rr}\, \dot{r}^2 = -m^2. \tag{8.3.4}$$

Using Eq. (8.3.3) to find $\dot{t}$ and $\dot{\phi}$ and substituting these expressions in Eq. (8.3.4) one obtains

$$\dot{r} = \pm r^{-3/2}\sqrt{P},$$

$$\dot{\phi} = \frac{(r - 2M)L + 2aME}{r\Delta}, \tag{8.3.5}$$

$$\dot{t} = \frac{Er(r^2 + a^2) - 2aM(L - aE)}{r\Delta}.$$

Here,

$$P = E^2\left(r^3 + a^2\, r + 2a^2 M\right) - 4aMEL - (r - 2M)\, L^2 - m^2\, r\Delta,$$

$$\Delta = r^2 - 2Mr + a^2. \tag{8.3.6}$$

In the absence of rotation (for $a = 0$) these equations coincide with Eq. (7.1.10).

8.3.2   Particle motion. Radial equation

Equation 8.3.5 are written in the form that is valid for both cases, particle and light motion. Let us discuss now the motion of particles. In this case it is convenient to use the proper time parameter $\tau = m\lambda$ and instead of the energy E and the angular momentum L to use *specific energy* $\mathcal{E} = E/m$ and *specific angular momentum* $\mathcal{L} = L/m$. Equation 8.3.5 take the form

$$\frac{dr}{d\tau} = \pm r^{-3/2}\sqrt{\mathcal{P}},$$

$$\frac{d\phi}{d\tau} = \frac{(r-2M)\mathcal{L} + 2aM\mathcal{E}}{r\Delta},$$

$$\frac{dt}{d\tau} = \frac{\mathcal{E}r(r^2+a^2) - 2aM(\mathcal{L}-a\mathcal{E})}{r\Delta},$$

$$\mathcal{P} = \mathcal{E}^2\left(r^3 + a^2\,r + 2a^2M\right) - 4aM\mathcal{E}\mathcal{L} - (r-2M)\,\mathcal{L}^2 - r\Delta.$$

$\mathcal{P}$ is a cubic polynomial in r. The radial motion has turning points at $\mathcal{P} = 0$. At these points

$$\mathcal{E} = V_\pm, \tag{8.3.8}$$

where $V_\pm$ are solutions of the equations $\mathcal{P} = 0$ for $\mathcal{E}$

$$V_\pm = \frac{2aM\mathcal{L} \pm X}{r^3 + a^2 r + 2a^2 M},$$

$$X^2 = r\left[r^3 + (\mathcal{L}^2 + a^2)r + 2a^2 M\right]\Delta. \tag{8.3.9}$$

For the Schwarzschild spacetime $V_\pm = \pm\sqrt{U}$, where U is the effective potential Eq. (7.1.10). We call $V_\pm$ a *potential function*.

> Quite often in the literature $V_\pm$ is also called an effective potential, both for the Schwarzschild and Kerr metrics. We prefer to keep the name of the effective potential for U. To distiguish $V_\pm$ from U, we use another name for it, namely a potential function. Another remark might also be useful. Similarly to the Schwarzschild case, in the Kerr metric it is also convenient to use dimensionless quantities. But instead of normalizing the variables by using the size of the event horizon, r_+, it is much more convenient to use for the normalization the parameter of mass M. For this, mainly historical, reason the dimensionless quantities in the Kerr geometry contain an additional factor 2 as compared with the Schwarzschild case.

For given parameters of the black hole, M and a, the potential functions $V_\pm$ depend on r and the integral of motion $\mathcal{L}$. The motion of particles with the specific energy $\mathcal{E}$ is possible only in the regions where either $\mathcal{E} \geq V_+$ or $\mathcal{E} \leq V_-$. One can show that the function $\mathcal{P}$ is invariant under the discrete transformation $\mathcal{E} \to -\mathcal{E}$, $\mathcal{L} \to -\mathcal{L}$, and the regions, where the motion with these parameters is possible, are interchanged. In the Schwarzschild black hole the particle energy outside the horizon is always non-negative and, hence, the region with $\mathcal{E} \leq V_-$ is to be excluded because in this case $V_- \leq 0$. Potential functions $V_\pm$ for different values of the rotation parameter are shown in Figures 8.4–8.7.

The asymptotic values of the potential functions at infinity and at the event horizon are

$$V_\pm(r=\infty) = \pm 1, \qquad V_\pm(r=r_+) = a\mathcal{L}/(2Mr_+) = \Omega\mathcal{L}. \tag{8.3.10}$$

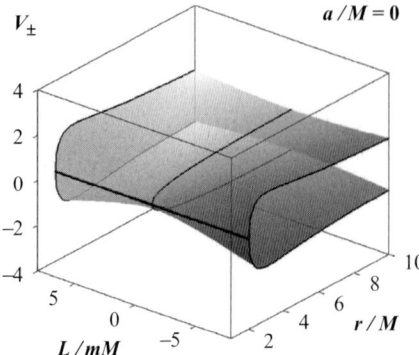

Fig. 8.4 The potential functions $V_\pm$ for $a/M = 0$.

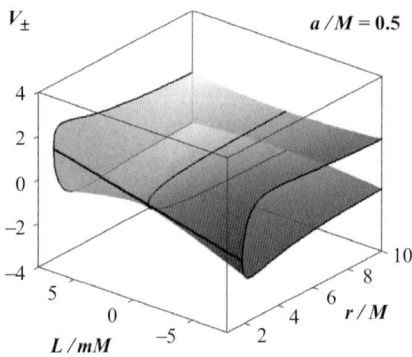

Fig. 8.5 The potential functions $V_\pm$ for $a/M = 0.5$.

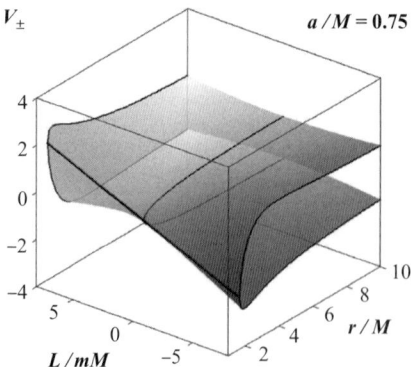

Fig. 8.6 The potential functions $V_\pm$ for $a/M = 0.75$.

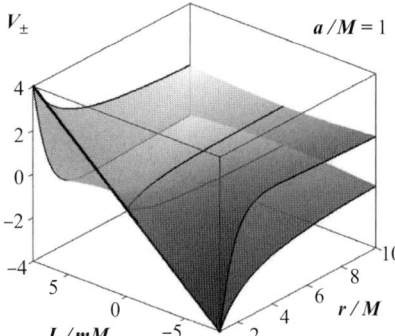

Fig. 8.7 The potential functions $V_\pm$ for $a/M = 1$.

Here, $\Omega = a/(2Mr_+)$ is the angular velocity of rotation of the Kerr black hole. The first of these relations shows that a particle can reach infinity only if its specific energy $\mathcal{E} \geq 1$. For $\mathcal{E} > 1$ (*hyperbolic motion*) the particle has a non-vanishing velocity at infinity. For $\mathcal{E} = 1$ (*parabolic motion*) its asymptotic velocity vanishes. For $\mathcal{E} < 1$ the outgoing particle has a radial turning point, after which it starts its incoming motion.

8.3.3 Circular orbits

In the vicinity of any point r_0 one can write a Taylor expansion of a function $\mathcal{P}$, Eq. (8.3.7) as follows

$$\mathcal{P} = P_0 + P_1(r - r_0) + P_2(r - r_0)^2 + \dots . \tag{8.3.11}$$

If r_0 is a turning point, then $P_0 = 0$. A *circular orbit* is a special type of the *bounded motion* when two of its turning points coincide with one another and one has

$$P_0 = P_1 = 0. \tag{8.3.12}$$

In the vicinity of the circular orbit the radial equation of motion takes the form

$$\frac{dr}{d\tau} = \pm r_0^{-3/2} \sqrt{P_2} \, |r - r_0|. \tag{8.3.13}$$

For negative P_2 the only possible solution is $r = r_0$ and the circular orbit is stable, while $P_2 > 0$ corresponds to the unstable orbits.

Two equations Eq. (8.3.12) establish relations between three quantities, r_{circ}, $\mathcal{E}_{\mathrm{circ}}$ and $\mathcal{L}_{\mathrm{circ}}$, which characterize the circular motion. It is convenient to choose r_{circ} as an 'independent' parameter. For this parametrization the other two parameters can be written in the following explicit form

$$\mathcal{E}_{\text{circ}} = \frac{r^2 - 2Mr \pm a\sqrt{Mr}}{r\left(r^2 - 3Mr \pm 2a\sqrt{Mr}\right)^{1/2}},$$

$$\mathcal{L}_{\text{circ}} = \frac{\pm\sqrt{Mr}\left(r^2 \mp 2a\sqrt{Mr} + a^2\right)}{r\left(r^2 - 3Mr \pm 2a\sqrt{Mr}\right)^{1/2}}. \qquad (8.3.14)$$

The signs $+$ and $-$ in these relations correspond to the prograde and retrograde motion, respectively. Circular orbits exist only if

$$\left(r^2 - 3Mr \pm 2a\sqrt{Mr}\right) \geq 0. \qquad (8.3.15)$$

The angular velocity of the Keplerian circular motion is

$$\omega_{\text{circ}} = \frac{d\phi}{dt} = \pm\frac{\sqrt{Mr}}{r^2 \pm a\sqrt{Mr}}. \qquad (8.3.16)$$

Stable circular orbits exist in the domain $r > r_{\text{ISCO}}$. The radius r_{ISCO} of the *innermost stable* or *marginally stable* circular orbit can be found as a solution of the equations

$$P_0 = P_1 = P_2 = 0. \qquad (8.3.17)$$

Substituting Eq. (8.3.14), which give a solution of the first two equations, into the third equation $P_2 = 0$ one obtains the following equation for r_{ISCO}

$$r^2 - 3a^2 - 6Mr \pm 8a\sqrt{Mr} = 0. \qquad (8.3.18)$$

A solution of this equation for $r > r_+$ can be written in the following parametric form

$$r_{\text{ISCO}} = M\{3 + Z_2 \mp [(3 - Z_1)(3 + Z_1 + 2Z_2)]^{1/2}\}, \qquad (8.3.19)$$

where

$$Z_1 = 1 + (1 - a^2/M^2)^{1/3}\left[(1 + a/M)^{1/3} + (1 - a/M)^{1/3}\right],$$

$$Z_2 = (3a^2/M^2 + Z_1^2)^{1/2}. \qquad (8.3.20)$$

As we have already mentioned, one can use M as a scale parameter. Equation (8.3.19) shows that r_{ISCO}/M is a function of the rotation rapidity parameter $\alpha = a/M$. The quantities $\mathcal{E}_{\text{circ}}$ and $\mathcal{L}_{\text{circ}}/M$ for the innermost stable circular orbit are functions of α as well.

The equality in Eq. (8.3.15) is reached in the case of the circular null rays, one solution for the direct motion and one for the retrograde motion. The radius of the (unstable) circular orbit closest to the black hole (the motion along it being at the speed of light) is

$$r_{\text{photon}} = 2M\left\{1 + \cos\left[\frac{2}{3}\arccos\left(\mp\frac{a}{M}\right)\right]\right\}. \qquad (8.3.21)$$

For $a = 0$, $r_{\text{photon}} = 3M$, while for $a = M$, $r_{\text{photon}} = M$ (direct motion) or $r_{\text{photon}} = 4M$ (retrograde motion). Because null rays in the equatorial plane cannot have more than one

Table 8.1 Specific energy $\mathcal{E}$, specific binding energy $1 - \mathcal{E}$, and specific angular momentum $|\mathcal{L}|/M$ of a test particle at the last stable circular orbit.

	$a = 0$	$a = M$	
		$\mathcal{L} > 0$	$\mathcal{L} < 0$
$\mathcal{E}$	$\sqrt{8/9}$	$\sqrt{1/3}$	$\sqrt{25/27}$
$1 - \mathcal{E}$	0.0572	0.4236	0.0377
$|\mathcal{L}|/M$	$2\sqrt{3}$	$2/\sqrt{3}$	$22/3\sqrt{3}$

turning point any small perturbation will kick off the massless particle from the circular orbit to infinity or to the black hole. So, these circular null trajectories are unstable and mark the closest possible circular orbits for any ultrarelativistic particle. At larger radii there are circular orbits for massive relativistic particles that are also unstable because they correspond to the maximum rather than minimum of the effective potential. At even larger radii the circular orbits become stable.

The specific energy $\mathcal{E}$ at the *innermost stable circular orbit* for different values of the angular momentum is given in Table 8.1. This table demonstrates, in particular, that the efficiency of the extremely rotating black hole can reach 42.4%. This means, in the gravitational field of such a rotating black hole a particle moving at slightly changing circular Keplerian orbits can lose up to $0.42mc^2$ of its proper energy before it reaches the innermost stable circular orbit. After this, the orbit becomes unstable and the particle falls into the black hole. This efficiency is much higher than the efficiency of the non-rotating black holes (about 5.7%).

For massive particles the circular orbits with $r > r_{\text{photon}}$ and $E/m \geq 1$ are unstable. A small perturbation directed outward forces this particle to leave its orbit and escape to infinity on an asymptotically hyperbolic trajectory. The unstable circular orbit on which $E_{\text{circ}} = m$ is given by the expression

$$r_{\text{bind}} = 2M \mp a + 2M^{1/2}(M \mp a)^{1/2}. \tag{8.3.22}$$

These values of the radius are the minima of periastra of all parabolic orbits. If a particle, moving in the equatorial plane, comes in from infinity (where its velocity $v_\infty \ll c$) and passes within a radius r_{bind}, it will be captured.

The quantities r_{ISCO}, r_{photon}, and r_{bind} as functions of the rotation parameter a/M are shown in Figure 8.8. Note that as $a \to M$, the invariant distance from a point r to the horizon r_+, equal to

$$\int_{r_+}^{r} \frac{r' \, dr'}{\Delta^{1/2}(r')},$$

diverges. As a result, it is not true that all three orbits coincide in this limit and lie at the horizon even though for $L > 0$ the radii r of all three orbits tend to the same limit r_+ (Bardeen *et al.* 1972).

Problem 8.7: *What is the minimum value for the parameter a at which the innermost stable circular orbit r_{ISCO} is inside the ergosphere?*

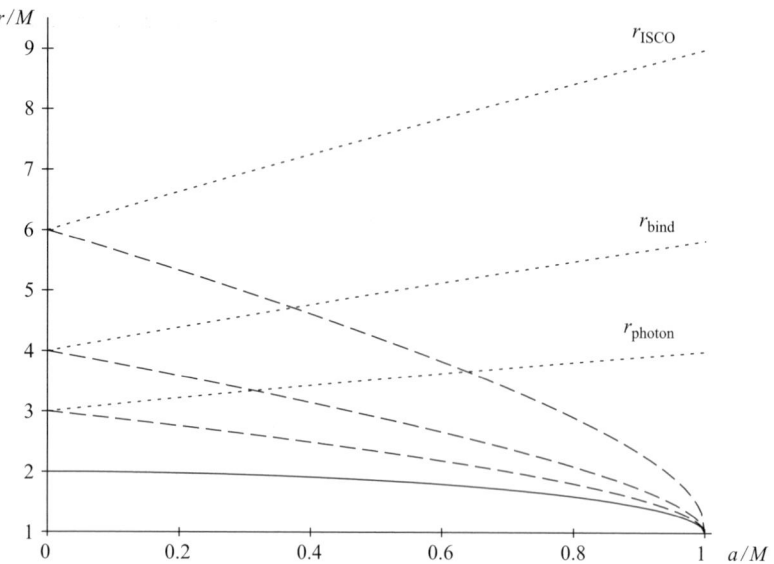

Fig. 8.8 r_{ISCO}, r_{photon}, and r_{bind} as functions of the rotation parameter a/M. The quantities corresponding to the direct and retrograde motions are shown by dashed and dotted lines, respectively. A solid line shows the horizon.

▌**Solution:** $a/M = 2\sqrt{2}/3 \simeq 0.9428$.

▌**Problem 8.8:** *What is the minimum value for the parameter a at which the closest (unstable) circular orbit r_{photon} is inside the ergosphere?*

▌**Solution:** $a/M = 1/\sqrt{2} \simeq 0.7071$.

▌**Problem 8.9:** *What is the minimum value for the parameter a at which r_{bind} is inside the ergosphere?*

▌**Solution:** $a/M = 2(\sqrt{2} - 1) \simeq 0.8284$.

Negative-energy orbits

If in Eq. (8.3.9) $L < 0$ and $|2aML| > X$, then the energy of a particle or a photon may be negative $E < 0$. This happens only when they move inside the ergosphere. This property is related to the fact that $\xi^{\mu}_{(t)}$ is not time-like inside the ergosphere. Since the corresponding momentum p_{μ} is a future-directed time-like or null vector, the energy of a particle inside the ergosphere,

$$E = -p_{\mu} \xi^{\mu}_{(t)}, \tag{8.3.23}$$

may be both positive and negative. The energy may become negative not only on the equatorial plane, but outside it as well, that is everywhere inside the ergosphere.

This property of the ergosphere makes possible what is known as the *Penrose process*. Let a particle 0 with energy E_0 enter the ergosphere and decay there into two particles, 1 and 2,

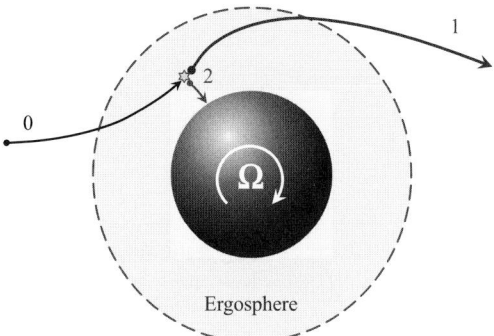

Fig. 8.9 Penrose process. A particle 0 enters the ergosphere and decays there into two particles, 1 and 2. One of them with a negative energy (2) falls into the black hole. The other one (1) escapes the ergosphere with an energy exceeding an energy of the original particle.

so that the energy of one of them, say 2, is negative, $E_2 < 0$. Then, the other particle can leave the ergosphere and propagate to infinity carrying the energy $E_1 > E_0$. In the process the energy $\Delta E = E_1 - E_0 > 0$ is extracted from the black hole.

8.3.4 Twin paradox in the Kerr spacetime

A free motion of objects along a circular trajectory gives an interesting example of the twin paradox in general relativity. Consider two observers, Alice and Bob, moving in opposite directions along a circular orbit of the same radius r in the equatorial plane of the Kerr black hole. Let us assume that Bob is moving in the direction of rotation of the black hole ($\epsilon = +1$) and Alice is counter-rotating ($\epsilon = -1$). Because the black hole drags into rotation the space around it, Alice should move faster than Bob to stay at the same circular orbit. Therefore, Alice covers more distance than Bob after their first encounter till the next one. Her higher speed in combination with other relativistic effects leads to the slower proper time pace for Alice. So Bob grows old faster than Alice, in spite of the fact that both move geodesically along the same orbit. The rotation of the black hole makes a big difference.

Let us calculate how big this effect is. We assume that Alice and Bob meet one another at the initial moment $t = 0$ at the point with $\phi = 0$. Using Eq. (8.3.16) we can write

$$\phi_\epsilon = \frac{\epsilon}{\rho^{3/2} + \epsilon\alpha}(t/M). \tag{8.3.24}$$

Here, $\rho = r/M$ and $\alpha = a/M$. They meet again at the moment Δt when (see Figure 8.10)

$$\phi_+ - \phi_- = 2\pi. \tag{8.3.25}$$

This condition implies

$$\Delta t/M = \pi(\rho^{3/2} - \alpha^2\rho^{-3/2}), \tag{8.3.26}$$

and, therefore,

$$\phi_+ = \pi - \pi\alpha\rho^{-\frac{3}{2}}, \qquad \phi_- = -\pi - \pi\alpha\rho^{-\frac{3}{2}}. \tag{8.3.27}$$

One can see that $|\phi_-| \geq |\phi_+|$, since $\alpha \geq 0$.

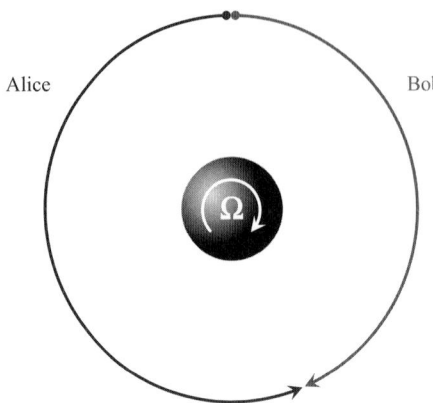

Fig. 8.10 Alice and Bob are orbiting around the Kerr black hole along the same orbit but in opposite directions. Bob is corotating with the black hole while Alice is counter-rotating. For Bob it takes more proper time of waiting between every meeting on the orbit than for Alice.

To calculate the proper times τ_ϵ between two sequential events of the meetings as measured individually by Alice and by Bob, we write the interval Eq. (8.3.1) along the trajectory $r = \text{const}$

$$ds^2 = M^2 \left[-\left(1 - \frac{2}{\rho}\right) d(t/M)^2 - \frac{4\alpha}{\rho} d(t/M)d\phi + \frac{\rho^3 + \alpha^2(\rho + 2)}{\rho} d\phi^2 \right]. \qquad (8.3.28)$$

The interval s calculated along the particle trajectory is connected with the proper time τ as follows $\tau^2 = -s^2$. Calculations give

$$\Delta_\epsilon \tau = \frac{\pi M \sqrt{(\rho^{3/2} - 3\rho^{1/2} + 2\epsilon\alpha)}(\rho^{3/2} - \epsilon\alpha)}{\rho^{3/4}}, \qquad (8.3.29)$$

where $\epsilon = +1$ corresponds to Bob measurements while $\epsilon = -1$ to Alice's ones. By construction, $\Delta_\epsilon \tau \geq 0$ for all circular orbits. On the other hand, one can easily check that

$$(\Delta_+ \tau)^2 - (\Delta_- \tau)^2 = 4\pi^2 M^2 \alpha (3\rho^2 + \alpha^2)\rho^{-3/2} \geq 0. \qquad (8.3.30)$$

This means that

$$\Delta\tau = \Delta_+ \tau - \Delta_- \tau \geq 0, \qquad (8.3.31)$$

and each time when Alice meets Bob, Alice becomes younger than Bob by the time interval Eq. (8.3.31). For orbits with large radii

$$\Delta\tau = \frac{6\alpha\pi M}{\rho} \left(1 + \frac{3}{2\rho}\right) + O(\rho^{-3}). \qquad (8.3.32)$$

The closer the orbit is to the black hole, the bigger is the relative time delay. Of course, the effect is more pronounced for faster-rotating black holes. In the extremal limit $a = M$ and for the closest possible circular orbit (though unstable), when Alice moves with the ultrarelativistic speed at $r \approx 4M$ we have the strongest effect. After their first encounter at $\phi = 0$ they meet again at $\phi \approx 7\pi/8$. For Bob it takes $\Delta_+ \tau \approx 7\pi M/\sqrt{2}$, while for Alice it happens practically in no time $\Delta_- \tau \approx 0$ because she moves almost at the speed of light.

Let us emphasize that both of them are moving freely, that is their worldlines are geodesic. The curvature of the spacetime created by the rotating black hole makes possible the existence of two different geodesics that start at the same point and meet one another again. This is more evidence that the global causal structure of a spacetime can differ from the flat one.

8.3.5 Null geodesics

We finish this section with brief remarks concerning *null-ray propagation* in the equatorial plane of the *Kerr black hole*. As a starting point we again use Eq. (8.3.5). We put there $m = 0$ and introduce dimensionless parameters

$$\rho = \frac{r}{M}, \quad \tau = \frac{t}{M}, \quad \sigma = \frac{E\lambda}{M}, \quad \alpha = \frac{a}{M}, \quad \ell = \frac{L}{EM}. \tag{8.3.33}$$

The parameter ℓ is the impact parameter measured in M units,[1] ρ is the radius in the same units, and σ is a new affine parameter. In these notations the equations for a light ray take the form

$$\frac{d\rho}{d\sigma} = \pm\rho^{-3/2}\sqrt{\bar{P}},$$

$$\frac{d\phi}{d\sigma} = \frac{(\rho - 2)\ell + 2\alpha}{\rho\,(\rho^2 - 2\rho + \alpha^2)}, \tag{8.3.34}$$

$$\frac{d\tau}{d\sigma} = \frac{\rho\,(\rho^2 + \alpha^2) - 2\alpha(\ell - \alpha)}{\rho\,(\rho^2 - 2\rho + \alpha^2)},$$

where

$$\bar{P} = \rho^3 + (\alpha^2 - \ell^2)\rho + 2(\alpha - \ell)^2. \tag{8.3.35}$$

Every null geodesic has no more than one turning point. If a turning point exists, $\bar{P}$ vanishes at this point. This condition determines the position of the turning point as a function of the momentum parameter ℓ. This is the solution of the cubic equation $\bar{P} = 0$ for ρ. On the other hand, one can consider this condition as the equation for ℓ, which is quadratic and has two solutions (see Figure 8.11)

$$\ell = \frac{\pm\rho\sqrt{\rho^2 - 2\rho + \alpha^2} - 2\alpha}{\rho - 2}. \tag{8.3.36}$$

For every rotation rapidity parameter α one can find two critical values of ℓ corresponding to prograde and retrograde circular orbits of photons.

The null-ray trajectories can be classified as follows:

1. the geodesics that come from infinity and then are bounced back to infinity;
2. the geodesics that come from the horizon and are bounced back to the black hole;

---

[1]Note, that, as was already mentioned, in the Kerr metric we use M as a scale parameter. This normalization differs from the normalization adopted in the Schwarzschild metric by a factor of 2.

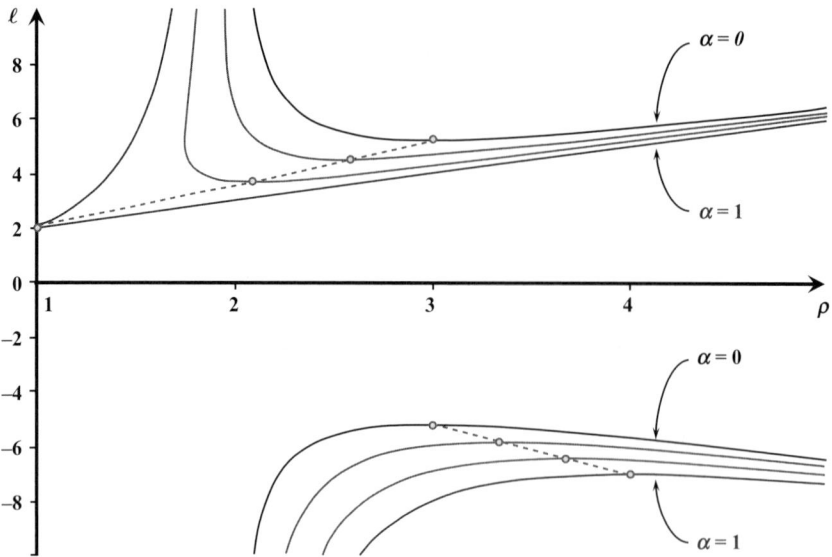

Fig. 8.11 The functions $\ell(\rho)$ corresponding to the turning points are depicted for $\alpha = 0$, $\alpha = 1/3$, $\alpha = 2/3$, and $\alpha = 1$. For all rotation parameters the ergosurface is located at $\rho = 2$. Two different branches correspond to positive and negative impact parameters ℓ. For both of these families $|\ell| \to \infty$ at the ergosurface. One can see from these plots that for every fixed ℓ there exist no more than two intersections of the fixed ℓ with functions $\ell_{\pm}(\rho)$. Each such intersection corresponds to the radial turning point. Any given null ray has no more than one radial turning point. In the case when there is no radial turning point light propagates between the horizon and infinity. When there are 2 turning points, $\rho_1 < \rho_2$, a null ray either is scattered by the black hole ($\rho \geq \rho_2$), or it moves from the black hole and after it reaches ρ_1 it moves back. When there is only one solution, this corresponds to the unstable circular orbit. All the retrograde circular orbits ($\ell < 0$) are located outside the ergosphere in the range of $3 \leq \rho \leq 4$.

3. unstable circular orbits corresponding to the critical values of ℓ separating the above two families.

The radii of unstable circular orbits of photons can be found from the condition $d\ell/d\rho = 0$, which results in the cubic equation with only two solutions that lie outside the black hole

$$\rho_{\text{photon}} = 2 + \cos \psi \mp \sqrt{3} \sin \psi, \qquad \psi = \frac{2}{3} \arcsin \alpha. \qquad (8.3.37)$$

Note that this formula is just another form of Eq. (8.3.21). These stationary points for the whole range of rotation rapidity parameter $0 \leq \alpha \leq 1$ are shown in Figure 8.11.

8.4   Spinning up the Black Hole

8.4.1   Matter accretion and black hole parameters change

Black holes in many astrophysical objects like quasars or binary systems are accompanied by an accretion disk. The matter that is gradually falling into the black hole from the accretion disk carries angular momentum and, therefore, speeds up the rotation of the black hole

(Lynden-Bell 1969). In reality, the process is quite complicated and its study requires the knowledge of the disk structure, the properties of the matter interaction in the disk, the flux of the matter supplied by a companion of the black hole, the magnetic field, etc. But it happens that some properties of the process of spinning up the black hole are quite robust and do not depend much on the details of the particular system.

For example, let us try to answer the question: What is the dependence of the rotation rapidity parameter of the black hole $\alpha = J/M^2$ on the mass of the black hole, during the process of accretion of dust-like matter from a thin disk? We assume that the accretion is quasistationary, and therefore the gravitational radiation is negligible. We also neglect its electromagnetic radiation. The influence of the gravitational field of the disk itself on the trajectories of matter in the disk is also assumed to be negligible as compared to the role of the gravity of Kerr black hole. Note that these assumptions do not restrict the total amount of matter that can be accreted to the black hole. Let us consider the case when the accretion disk lies in the equatorial plane of the Kerr black hole. The matter in the *accretion disk* is moving mostly along the circular orbits. Interaction of particles in the disk leads to the effective friction between its layers and results in the transfer of the energy and the angular momentum between the circular layers. This process depends on the details of the matter interaction in the disk. At the same time after the particles reach the *innermost stable circular orbit* they continue their motion by freely falling into the black hole.

The radius r_{ISCO} of the innermost stable circular orbit is given by the relation Eq. (8.3.19). By substituting the expression in Eq. (8.3.14), one obtains the specific energy $\mathcal{E}_{\text{circ}}$ and specific angular momentun $\mathcal{L}_{\text{circ}}$ for a particle at this orbit. It is worth noting that these quantities are uniquely related to each other. As soon as the particle leaves this orbit it does not interact with the rest of the disk matter any longer and falls into the black hole, preserving its uniquely defined energy and angular momentum. Because at the innermost stable circular orbit $dE_{\text{circ}}/dr = 0$ and $dL_{\text{circ}}/dr = 0$ the energy and angular momentum of particles on this and at the nearby orbits are the same to a high precision.

Denote by δM and δJ the change of the mass and angular momentum of the black hole as a result of this process. Their ratio can be written in the form

$$\frac{\delta J}{\delta M} = Mf(\alpha). \tag{8.4.1}$$

(See discussion on page 262.) Thus, one has the following universal equation for the change of the parameters of the black hole as a result of the accretion of the matter onto it

$$2\frac{\delta J}{\delta (M^2)} = f\left(\frac{J}{M^2}\right). \tag{8.4.2}$$

8.4.2   Evolution in the black hole parameter space

In order to calculate the function $f(\alpha)$ it is enough to substitute $r = r_{\text{ISCO}}$ (see Eq. (8.3.19)) to $\delta M = E_{\text{circ}}$ and $\delta J = L_{\text{circ}}$, where E_{circ} and L_{circ} are given by Eq. (8.3.14). The result looks quite messy. However, one can simplify it by a proper choice of variables. Let us introduce new variables

$$x = \sqrt{r_{\text{ISCO}}/M}\,, \qquad \alpha = \frac{J}{M^2} = \frac{a}{M}. \tag{8.4.3}$$

The rotation rapidity factor $\alpha = 0$ for a non-rotating black hole, $\alpha = \pm 1$ for maximally rotating black holes. The condition $dE_{\mathrm{circ}}/dr = 0$ defining the innermost stable orbits Eq. (8.3.18) then can be rewritten as

$$x^4 - 6x^2 + 8\epsilon\,\alpha x - 3\alpha^2 = 0. \tag{8.4.4}$$

$\epsilon = +1$ corresponds to a prograde motion, while $\epsilon = -1$ corresponds to orbits with a retrograde motion. Technically, it is convenient to use x instead of α as the parameter describing the black hole. So we have to express α in terms of x

$$\alpha = \frac{x}{3}\left(4\epsilon \pm \sqrt{3x^2 - 2}\right). \tag{8.4.5}$$

This relation besides the sign parameter ϵ contains $\pm$ signs for the two roots of the quadratic equation. The conditions $0 \leq \alpha \leq 1$ and $x \geq 1$ exclude the half of the solutions. As a result, one has that for the prograde orbits ($\epsilon = +1$) $1 \leq x \leq \sqrt{6}$, while for the retrograde motion ($\epsilon = -1$) $\sqrt{6} \leq x \leq 3$. The corresponding value of α is

$$\alpha = \epsilon\,\frac{x}{3}\left(4 - \sqrt{3x^2 - 2}\right). \tag{8.4.6}$$

The variation of the angular momentum due to the accretion of a particle from the innermost stable orbit is

$$\frac{1}{M}\frac{\delta J}{\delta M} = \frac{1}{M}\frac{L_{\mathrm{circ}}}{E_{\mathrm{circ}}}\bigg|_{r=r_{\mathrm{ISCO}}} = \epsilon\,\frac{2x\left(6x^2 - 5 - \sqrt{3x^2 - 2}\right)}{3\left(3x^2 - 2 - \sqrt{3x^2 - 2}\right)}. \tag{8.4.7}$$

So the function $f(\alpha)$ in Eq. (8.4.2) is determined by the formula

$$f(\alpha) = \epsilon\,\frac{2x\left(6x^2 - 5 - \sqrt{3x^2 - 2}\right)}{3\left(3x^2 - 2 - \sqrt{3x^2 - 2}\right)}, \tag{8.4.8}$$

where $x = x(\alpha)$ is implicitly defined by Eq. (8.4.6). Taking into account the relation $J = \alpha M^2$ we can rewrite Eq. (8.4.7) in the form

$$\frac{\delta\alpha}{\delta(\ln M)} = f(\alpha) - 2\alpha. \tag{8.4.9}$$

Substitution of Eq. (8.4.6) in this relation leads to a much simpler equation for $x(M)$

$$\frac{\delta x}{\delta(\ln M)} = -x, \tag{8.4.10}$$

which has an evident solution $x = CM^{-1}$. The constant C here is defined by the initial conditions. For example, if initially the black hole with mass M_0 is not rotating and the accretion disk has been formed later, then $x_0 = \sqrt{r_{\mathrm{ISCO}\,0}/M} = \sqrt{6}$. Therefore, $C = \sqrt{6}\,M_0$ and

$$x = \sqrt{6}\,\frac{M_0}{M}, \tag{8.4.11}$$

or equivalently

$$r_{\text{ISCO}} = 6 \frac{M_0^2}{M}. \tag{8.4.12}$$

Substitution of this solution into Eq. (8.4.6) gives the answer for the dependence of the rotation rapidity factor of the Kerr black hole during the accretion from the disk (Bardeen 1970)

$$\alpha = \frac{\sqrt{2}}{\sqrt{3}} \frac{M_0}{M} \left(4 - \sqrt{18 \frac{M_0^2}{M^2} - 2} \right), \qquad 1 \le \frac{M}{M_0} \le \sqrt{6},$$

$$\alpha = \frac{\sqrt{2}}{\sqrt{3}} \frac{M_0}{M} \left(\sqrt{18 \frac{M_0^2}{M^2} - 2} - 4 \right), \qquad \frac{\sqrt{2}}{\sqrt{3}} \le \frac{M}{M_0} \le 1, \tag{8.4.13}$$

$$\alpha = 1, \qquad \frac{M}{M_0} \ge \sqrt{6}.$$

Looking at Eq. (8.4.12) we arrive at the conclusion that during the process of spinning up the black hole by the matter falling from the accretion disk the product of the mass of the Kerr black hole to the radius of the innermost stable circular orbit (the inner edge of the disk) remains constant

$$M r_{\text{ISCO}} = 6 M_0^2. \tag{8.4.14}$$

This evolution law remains valid until the black hole reaches the critical value of rotation $\alpha = 1$. After this moment its mass continues to grow while the black hole remains maximally rotating and the inner edge of the disk is at $r_{\text{ISCO}} = M$.

Problem 8.10: *Show that the Schwarzshild black hole with mass M_0, after the spin-up by a thin dust-like accretion disk, reaches maximum rotation rapidity factor $\alpha = 1$ when its mass becomes equal to $\sqrt{6} M_0$.*

Problem 8.11: *What is the rest mass Δm of the dust from the disk accreted by a black hole during this spin-up process? Assume that initially the black hole is not rotating and has the mass M_0 and the final mass is M.*

Solution:

$$\Delta m = 3 M_0 \left[\arcsin \left(\frac{M}{3 M_0} \right) - \arcsin \left(\frac{1}{3} \right) \right], \qquad 1 \le \frac{M}{M_0} \le \sqrt{6},$$

$$\Delta m = 3 M_0 \left[\arcsin \left(\sqrt{\frac{2}{3}} \right) - \arcsin \left(\frac{1}{3} \right) \right] \tag{8.4.15}$$

$$+ \sqrt{3} \left[M - \sqrt{6} M_0 \right], \qquad \frac{M}{M_0} \ge \sqrt{6}.$$

Problem 8.12: *Show that the maximally rotating black hole $\alpha = 1$ with mass M_1, after the spin-down by a thin dust-like accretion disk rotating in the opposite direction, stops rotating $\alpha = 0$ when its mass is equal $M_0 = \sqrt{1.5} M_1$.*

The described model of accretion neglects the effects of radiation emitted by the disk. Kip Thorne (1974) demonstrated that these effects prevent spin-up of the black hole beyond the limit $\alpha_{\lim} \approx 0.998$.

8.5   Geodesics in Kerr Spacetime: General Case

8.5.1   Separation of variables in the Hamilton–Jacobi equation

The equations of motion of particles and light out of the equatorial plane are much more complicated. However, the good news is that these equations are completely integrable. This means that the motion is regular (not chaotic). The evident symmetries of the Kerr spacetime are not sufficient to guarantee the complete integrability. In fact, one has 2 conserved quantities, energy E and azimuthal angular momentum L_z. They, together with $m^2 = -g_{\mu\nu}p^\mu p^\nu$, give 3 integrals of motion. One more integral of motion is required for the complete integrability. This integral of motion, discovered by Carter (1968), is quadratic in momenta. It is generated by the second-rank *Killing tensor* responsible for a *hidden symmetry*. The expression for this Killing tensor can be obtained by separating variables in the *Hamilton–Jacobi equations* for particle and light motion.[2]

The *Hamilton–Jacobi equation* for the *Jacobi action* $\tilde{S}$ has the form (see Eqs. (4.1.22) and (4.1.21)

$$\tilde{S} = S + \frac{1}{2}m^2\lambda, \tag{8.5.1}$$

$$g^{\mu\nu}\partial_\mu S\partial_\nu S + m^2 = 0. \tag{8.5.2}$$

A solution of the Hamilton–Jacobi equation that contains as many arbitrary constants as the number of independent coordinates is called a *complete integral*

$$\tilde{S} = \tilde{S}(\lambda, x^\mu, \alpha_\mu). \tag{8.5.3}$$

Solving the equation

$$\frac{\partial \tilde{S}}{\partial \alpha_\mu} = \beta^\mu, \tag{8.5.4}$$

one obtains x^μ as functions of λ, α_μ and β^μ. Relations $p_\mu = \partial \tilde{S}/\partial x^\mu$ determine the dependence of the momentum on λ.

For a particle in the Kerr spacetime we write the Jacobi function S in the form

$$S = -Et + L_z\phi + S_r(r) + S_\theta(\theta). \tag{8.5.5}$$

This form includes the information that the coordinates t and ϕ are cyclic variables. As earlier, E is the energy. A conserved quantity L_z is a component of the angular momentum along the axis of the rotation. For the motion in the equatorial plane we denoted this parameter as L.

---

[2]The Kerr–Newman geometry, like the Kerr black hole, obeys the hidden symmetries that lead to separability of the Hamilton–Jacobi equations and make completely integrable the equations of motion of charged particles (Carter 1968).

Let us demonstrate (following Carter (1968)), that this action allows the complete separation of variables. Using the expressions Eq. (8.1.9) for the upper components of the Kerr metric we present the equation

$$g^{\mu\nu} S_{,\mu} S_{,\nu} = -m^2 \tag{8.5.6}$$

in the form

$$- \left[\frac{(r^2 + a^2)^2}{\Delta} - a^2 \sin^2\theta \right] E^2 + \frac{4Mar}{\Delta} EL_z + \left[\frac{1}{\sin^2\theta} - \frac{a^2}{\Delta} \right] L_z^2$$

$$+ \Delta S_{,r}^2 + S_{,\theta}^2 = -m^2 \left[r^2 + a^2 \cos^2\theta \right]. \tag{8.5.7}$$

We can rewrite this equation as

$$- \frac{1}{\Delta} \left[\left(r^2 + a^2 \right)^2 E^2 - 4MarEL_z + a^2 L_z^2 \right]$$

$$+ \Delta S_{,r}^2 + m^2 r^2 + 2aL_z E + a^2 \sin^2\theta E^2 + \frac{L_z^2}{\sin^2\theta} \tag{8.5.8}$$

$$+ S_{,\theta}^2 + m^2 a^2 \cos^2\theta - 2aL_z E = 0.$$

From this equation it is evident that the ansatz Eq. (8.5.5) does work and

$$K \equiv \left(Ea\sin\theta - \frac{L_z}{\sin\theta} \right)^2 + S_{,\theta}^2 + m^2 a^2 \cos^2\theta \tag{8.5.9}$$

is a separation constant.

8.5.2 Equations of motion in the first-order form

The obtained separated Hamilton–Jacobi equation are

$$\begin{cases} \left(\dfrac{dS_\theta}{d\theta} \right)^2 + \left(aE\sin\theta - \dfrac{L_z}{\sin\theta} \right)^2 + m^2 a^2 \cos^2\theta = K, \\[3mm] \Delta \left(\dfrac{dS_r}{dr} \right)^2 - \dfrac{\left[(r^2 + a^2) E - aL_z \right]^2}{\Delta} + m^2 r^2 = -K. \end{cases}$$

Let us introduce quantities

$$\mathcal{R} = \left[E(r^2 + a^2) - aL_z \right]^2 - \left(m^2 r^2 + K \right) \Delta, \tag{8.5.10}$$

$$\Theta = K - m^2 a^2 \cos^2\theta - \left(Ea\sin\theta - \frac{L_z}{\sin\theta} \right)^2. \tag{8.5.11}$$

The *Jacobi action* for our problem is

$$\tilde{S} = \frac{1}{2}m^2\lambda - Et + L_z\phi + S_r + S_\theta,$$

$$S_r = \int^r \frac{\sqrt{\mathcal{R}}}{\Delta}\, dr, \qquad S_\theta = \int^\theta \sqrt{\Theta}\, d\theta.$$

(8.5.12)

It is a function of four integrals of motion E, L_z, K, and m^2, and thus this is a complete integral of the *Hamilton–Jacobi equations*. Note that the signs of $\sqrt{\mathcal{R}}$ and $\sqrt{\Theta}$ can be chosen independently, but once the choice has been made, it must be used further consistently in all equations. The change of the signs occurs at the turning points of the corresponding equations, that is when either $\mathcal{R} = 0$ or $\Theta = 0$. The lower limit of integration in Eq. (8.5.12) can be chosen independently in each integral since only variations of the action are physically meaningful.

The integrated form of the geodesic equations can be obtained by using the fact that partial derivatives of the Jacobi action with respect to the integrals of motion are constants. These relations give (Misner *al* 1974)

$$\int^r \frac{dr}{\sqrt{\mathcal{R}}} = \int^\theta \frac{d\theta}{\sqrt{\Theta}},$$

$$t = \int^r \frac{(r^2 + a^2)\left[E\left(r^2 + a^2\right) - L_z a\right]}{\Delta \sqrt{\mathcal{R}}}\, dr - \int^\theta \frac{a\left(aE\sin^2\theta - L_z\right)}{\sqrt{\Theta}}\, d\theta,$$

$$\phi = \int^r \frac{a\left[E\left(r^2 + a^2\right) - L_z a\right]}{\Delta \sqrt{\mathcal{R}}}\, dr - \int^\theta \frac{aE\sin^2\theta - L_z}{\sin^2\theta \sqrt{\Theta}}\, d\theta,$$

(8.5.13)

$$\lambda = \int^r \frac{r^2\, dr}{\sqrt{\mathcal{R}}} + \int^\theta \frac{a^2\cos^2\theta}{\sqrt{\Theta}}\, d\theta.$$

These equations allow one to obtain the following set of first-order equations for particles and light motion in the Kerr spacetime (Carter 1968; Bardeen 1973, Misner *et al.* 1974)

$$\Sigma\frac{dr}{d\lambda} = \sqrt{\mathcal{R}},$$

(8.5.14)

$$\Sigma\frac{d\theta}{d\lambda} = \sqrt{\Theta},$$

(8.5.15)

$$\Sigma\frac{d\phi}{d\lambda} = -\left(aE - \frac{L_z}{\sin^2\theta}\right) + \frac{a}{\Delta}\left(E(r^2 + a^2) - L_z a\right),$$

(8.5.16)

$$\Sigma\frac{dt}{d\lambda} = -a\left(aE\sin^2\theta - L_z\right) + \frac{r^2 + a^2}{\Delta}\left[E(r^2 + a^2) - L_z a\right].$$

(8.5.17)

For the light, one has to substitute $m = 0$ into $\mathcal{R}$ and Θ. Then, λ plays the role of an affine parameter.

8.5.3 Particle motion

At large distance r the radial function is of the form $\mathcal{R} \sim (E^2 - m^2)r^4$. A particle can 'travel' to infinity (hyperbolic motion) only when $E > m$. In the opposite case, the particle always has a radial turning point r_{turn}, and the motion is restricted to the domain $r \leq r_{\text{turn}}$ (elliptic type of motion). The case $E = m$ is marginal: the particle that reaches infinity has zero velocity there (parabolic motion). The detailed analysis and the classification of different types of the radial motion is quite an involved problem, since $\mathcal{R}$ is a quartic polynomial of the general type.

> **Problem 8.13:** *Show that in the absence of rotation, $a = 0$, Eqs. (8.5.14)–(8.5.17) reproduce Eq. (7.8.4) in the tilted spherical coordinates for the particle motion in the Schwarzschild spacetime, with the following identification*
>
> $$\mathcal{L}_z = L_z/m, \qquad \mathcal{L}^2 = K/m^2. \tag{8.5.18}$$

Equation (8.5.18) implies that in the limiting case $a = 0$ the Carter's integral of motion K is proportional to the square of the total angular momentum of the particle. In the Kerr metric, when the spherical symmetry is broken, the total angular momentum is no longer a conserved quantity. The symmetry, which is responsible for this fourth integral of motion K, is the *hidden symmetry* of the Kerr spacetime (see discussion in Section 8.7).

8.5.4 Solutions of the θ-equation

Equation (8.5.15) shows that the angle θ can change in the region where $\Theta(\theta) \geq 0$. The function $\Theta(\theta)$ is symmetric with respect to the reflection with respect to the equatorial plane, $\theta \to \pi - \theta$. It is interesting that this function does not depend on the mass M of the Kerr metric. This means that it has the same form even when $M = 0$, that is in the flat-space limit. The study of motion in the θ-direction is relatively simple. If one denotes $z = \sin^2 \theta$, then the function Θ is a quadratic in z polynomial divided by z. Values of θ where $\Theta(\theta) = 0$ (the roots of the quadratic in z polynomial) correspond to the turning points of the angular motion. For the classification of types of angular motion it is convenient to denote

$$Q = K - (Ea - L_z)^2, \tag{8.5.19}$$

and to rewrite Eq. (8.5.11) in the form

$$\Theta = Q - \cos^2 \theta [a^2(m^2 - E^2) + L_z^2/\sin^2 \theta]. \tag{8.5.20}$$

Possible types of θ-motion can be classified as follows (Carter 1968)

1. $Q > 0$: The function $\Theta(\theta)$ is positive at the equatorial plane $\theta = \pi/2$. For $L_z \neq 0$ it tends to $-\infty$ at the symmetry axes, $\theta = 0, \pi$. This means that it passes through zero at the points θ_0 and $\pi - \theta_0$, and is positive within this interval. In its motion between θ_0 and $\pi - \theta_0$ the particle crosses the equatorial plane, where $\Theta = Q$. The interval of the angle extends to poles if and only if $L_z = 0$ and $Q + a^2(E^2 - m^2) \geq 0$.

2. $Q = 0$: In this case $\theta = \pi/2$ is a solution. This is a motion in the equatorial plane. A non-trivial solution when θ changes with time exists only when $a^2(E^2 - m^2) > L_z^2$. For such motion a particle trajectory touches the equatorial plane on one side or the other. The range of θ extends to the axis of symmetry if and only if $L_z = 0$.

3. $Q < 0$: A real solution exists only when $Q \geq -(a\sqrt{E^2 - m^2} - |L_z|)$. If the equality is valid, a solution is $\theta = \text{const}$. For $L_z = 0$ it lies on the symmetry axis. For $L_z \neq 0$, the fixed value of θ is between the symmetry axis and the equatorial plane. Finally, when the inequality is strict, θ varies between two values, and a particle does not cross the equatorial plane.

8.6   Light Propagation

8.6.1   Null geodesics

In the previous section we studied particle motion in the gravitational field of a black hole. This problem is important for understanding the behavior of matter in its vicinity and an accreation disk structure. Most of the information about black holes and surrounding matter is obtained by studying the electromagnetic radiation. When the wavelength of the electromagnetic waves is much smaller than the size of the black hole, one can use the *geometrical-optics approximation* to describe it. In this approximation photons are propagating along null geodesics. In this section we consider null rays in the Kerr geometry.

Denote

$$\tilde{\lambda} = E\lambda, \qquad \ell = L_z/E, \qquad \mathbb{Q} = Q/E^2. \qquad (8.6.1)$$

Then, the equations of motion Eqs. (8.5.14)–(8.5.17) for the light are simplified and take the form

$$\Sigma \frac{dr}{d\tilde{\lambda}} = \sqrt{\mathcal{R}}, \qquad (8.6.2)$$

$$\Sigma \frac{d\theta}{d\tilde{\lambda}} = \sqrt{\Theta}, \qquad (8.6.3)$$

$$\Sigma \frac{d\phi}{d\tilde{\lambda}} = -a + \frac{\ell}{\sin^2 \theta} + \frac{a}{\Delta}\left(r^2 + a^2 - \ell a\right), \qquad (8.6.4)$$

$$\Sigma \frac{dt}{d\tilde{\lambda}} = -a\left(a \sin^2 \theta - \ell\right) + \frac{r^2 + a^2}{\Delta}\left[r^2 + a^2 - \ell a\right], \qquad (8.6.5)$$

where

$$\mathcal{R} = \left(r^2 + a^2 - a\ell\right)^2 - [\mathbb{Q} + (a - \ell)^2]\Delta, \qquad (8.6.6)$$

$$\Theta = \mathbb{Q} + \cos^2 \theta(a^2 - \ell^2/\sin^2 \theta) = \mathbb{Q} + (a - \ell)^2 - (a \sin \theta - \ell/\sin \theta)^2.$$

Let us discuss some properties of $\mathcal{R}$. This is a polynomial of r of the fourth order. Its value at the horizon

$$\mathcal{R}|_{r=r_+} = (r_+^2 + a^2 - a\ell)^2 \qquad (8.6.7)$$

is positive. At the infinity $r \to \infty$ it grows as r^4. The equation

$$\mathcal{R} = 0 \qquad (8.6.8)$$

determines the radial turning points of the photon trajectory.

Problem 8.14:   *Show that a photon trajectory in the Kerr spacetime has no more than one radial turning point in the black hole exterior.*

Solution: *Since $\mathcal{R}$ is the polynomial of the fourth order in r, it has no more than 4 real roots. Suppose the number of solutions of Eq. (8.6.8) is 4. Denote these solutions as $r_1 \geq r_2 \geq r_3 \geq r_4$. Since the polynomial $\mathcal{R}$ does not contain the third power of r, the roots must obey the condition*

$$r_1 + r_2 + r_3 + r_4 = 0. \tag{8.6.9}$$

This means that at least one of them, r_4, is negative, while the other 3 may be positive. Since the function $\mathcal{R}$ is positive both on the horizon and at the infinity, this implies that outside the horizon there exists not more than 2 real roots $r_+ \leq r_2 < r_1$ of Eq. (8.6.8). If real roots are absent, a trajectory extends from the horizon to infinity and a solution describes a capture of the photon by the black hole. For 2 different roots the function $\mathcal{R}$ is negative in the interval (r_2, r_1). Such a photon propagates either from the horizon to a turning point, and then back to the horizon, or it comes from infinity and after the turning point goes back to infinity. The latter case decribes the photon scattering. In any case the photon has no more than one radial turning point.

In a critical case, when 2 roots of $\mathcal{R}$ coincide, one has

$$\mathcal{R} = 0, \qquad \frac{d\mathcal{R}}{dr} = 0. \tag{8.6.10}$$

Denote a solution of these two equations by r_{photon}. The radius $r = r_{\text{photon}}$ corresponds to unstable 'spherical' orbits. The orbits that lie in the equatorial plane are the circles with r_{photon} given by Eq. (8.3.21). The generic orbits satisfying Eq. (8.6.10) outside the equatorial plane are quite complicated three-dimensional curves but these curves are located at fixed radius. In the vicinity of this radius one has $r = r_{\text{photon}} + x$, $\mathcal{R} \sim x^2$ and

$$\frac{d\phi}{dx} \sim x^{-1}. \tag{8.6.11}$$

The photon trajectory that is on the verge of having a turning point at r_{photon} circles many times around the black hole and then either falls into the black hole or escapes to infinity.

8.6.2   Impact parameters

For light rays the parameters ℓ and $\mathbb{Q}$ are independent. For rays that reach the spatial infinity, these parameters have a simple physical meaning. Suppose an observer is located at a point $(r_0, \theta_0, \phi = 0)$ with large r_0. We call θ_0 an *inclination angle*. Assume that the observer measures directions of photons. At large distance r_0 Eqs. (8.6.3)–(8.6.5) give

$$\frac{d\phi}{dt} \sim \frac{\ell}{r_0^2 \sin^2 \theta_0}, \qquad \frac{d\theta}{dt} \sim \pm \frac{1}{r_0^2} \sqrt{\mathbb{Q} + a^2 \cos^2 \theta_0 - \ell^2 \cot^2 \theta_0}. \tag{8.6.12}$$

The angles of displacement of the photon in the ϕ and θ directions are $r_0 \sin \theta_0 \, d\phi/dt$ and $r_0 \, d\theta/dt$, respectively. These angles decrease as r_0^{-1}. By multiplying these quantities by r_0 one obtains impact parameters

$$x = -\frac{\ell}{\sin \theta_0},$$
$$\tag{8.6.13}$$
$$y = \pm \sqrt{\mathbb{Q} + a^2 \cos^2 \theta_0 - \ell^2 \cot^2 \theta_0}.$$

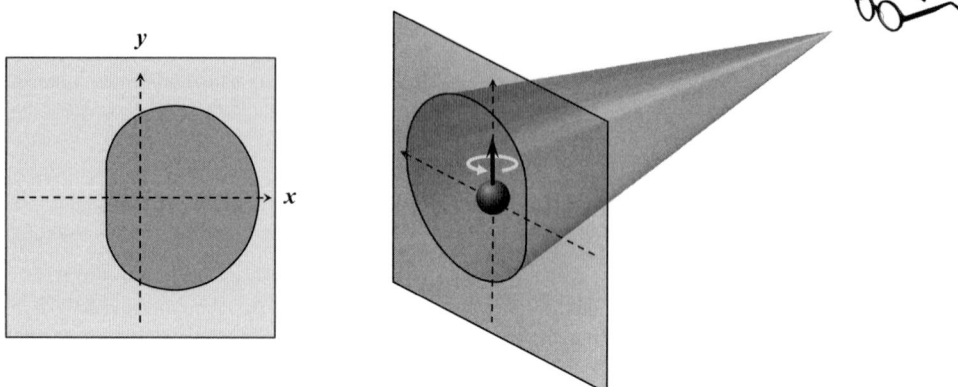

Fig. 8.12 The observer is looking in the direction of the black hole. On the background of the sky the black hole will be visible as a black spot ('shadow') with the shape of a deformed disk that is slightly shifted in the direction of rotation of the black hole (to the right). For an extremal black hole $a/M = 1$, when observed from the equatorial plane, the left edge of the shadow is a vertical straight line between the points $(-2M, -\sqrt{3}M)$ and $(-2M, \sqrt{3}M)$. The right edge is at $x = 7M$.

Using these equations one obtains

$$x^2 + y^2 = \mathbb{Q} + \ell^2 + a^2 \cos^2 \theta_0. \qquad (8.6.14)$$

Each photon that reaches a distant observer determines a point on the (x, y)-plane of the image of the sky visible through the telescope. In Figure 8.12 we depicted the orientation of the image and the axes x and y relative to the rotating black hole.

8.6.3   Black hole shadow

Now imagine that behind the rotating black hole there is a source of light. Let its angular size be much larger than the angular size of the black hole. Then, a distant observer will see a dark spot on the (x, y)-plane, which is an apparent image of the black hole. Certainly, our assumption about the properties of the light source is an idealization. In reality, one can expect a similar shadow effect when the photons are emitted from the parts of the accreting disc, located behind the black hole, and that on their way to a distant observer propagate in the black hole's vicinity. Suppose a photon from the source, which reaches the distant observer, has the impact parameters (x, y). Let us trace back in time the trajectory of this photon. In its backward motion such a photon must pass near the black hole and go to infinity. Hence, it has a radial turning point. A trajectory without a radial turning point cannot be emitted by the source and reach the distant observer. As a result, the black hole will be seen as a dark region on the background of the bright source. This is what is called the *black hole shadow*. One might expect that such a shadow can be found, for example, in direct observations by future interferometers of the black hole *Sgr A** in the center of the Milky Way. Let us study the black hole shadow effect in more detail.

The rim of this shadow corresponds to photons that are marginally trapped by the black hole. They revolve around the black hole many times before finally reaching the distant observer. The size and shape of the rim depends on the black hole parameters (M and a) as well as on the inclination angle θ_0 between the direction to the distant observer and the axis of symmetry. The size and the shape of the rim of the black hole shadow can be obtained from Eq. (8.6.10). It is convenient to rewrite these equations in the equivalent form

$$r^{-1}\mathcal{R} = d\left(r^{-1}\mathcal{R}\right)/dr = 0. \tag{8.6.15}$$

Written in an explicit form Eq. (8.6.15) gives

$$[r(r^2 + a^2) + 2a^2 M] - 4aM\ell - (r - 2M)\ell^2 - r^{-1}(r^2 - 2Mr + a^2)Q = 0,$$
$$3r^2 + a^2 - \ell^2 - r^{-2}(r^2 - a^2)Q = 0. \tag{8.6.16}$$

One can use one of these equations to find the parameter r and to use the other one to obtain the relation between ℓ and Q, which determines the position of the shadow rim on the screen. Instead of this, it is more convenient to use these equations to determine ℓ and Q as functions of the parameter r. These expressions give the relation between ℓ and Q in the parameteric form. We shall use this approach.

Equations (8.6.16) are linear in Q and quadratic in ℓ. Excluding Q one obtains a single quadratic equation for ℓ. This equation has two solutions. The first one is

$$\ell = (r^2 + a^2)/a, \qquad Q = -r^4/a^2. \tag{8.6.17}$$

Here, r is just a parameter. The explicit relation between ℓ and Q we are looking for comes from these formulas and reads

$$Q + (a - \ell)^2 = 0. \tag{8.6.18}$$

Then, the condition $\Theta \geq 0$ and Eq. (8.6.6) lead to a solution $\theta = \theta_0 = \text{const}$, where $\sin^2 \theta_0 = \ell/a$. These null rays are characterized by two conserved quantities

$$\ell = a \sin^2 \theta_0, \qquad Q = -a^2 \cos^4 \theta_0, \tag{8.6.19}$$

and propagate at $\theta = \theta_0$ and $r = a \cos \theta_0$. Evidently these orbits are inside the horizon except for the one in the equatorial plane $\theta_0 = \pi/2$ of an extremal black hole. No photons with these values of parameters can reach the observer. In the extremal case $a = M$ the observer in the equatorial plane could see the dark point (Hioki and Maeda 2009) corresponding to this degenerate trajectory. But this dark point happens to be inside the dark shadow anyway (see the dark point $x/M = -1, y = 0$ in the right picture of Figure 8.13).

The other (physically meaningful) solution of Eq. (8.6.16) gives

$$\ell = -\frac{r^3 - 3Mr^2 + a^2(r + M)}{a(r - M)},$$
$$Q = \frac{r^3}{a^2(r - M)^2}[4a^2 M - r(r - 3M)^2]. \tag{8.6.20}$$

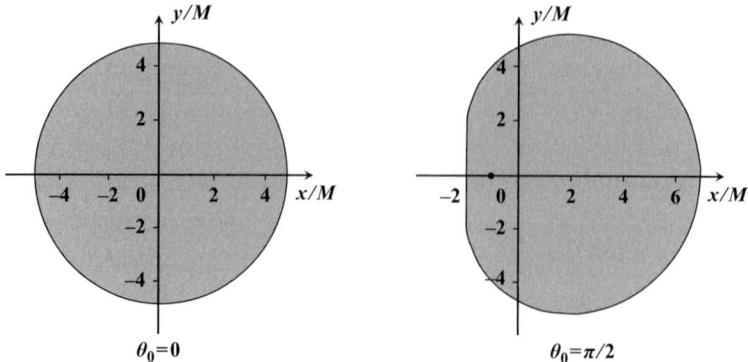

Fig. 8.13 The shadows of the extremely rotating black hole $a = M$ for the view from the pole $\theta_0 = 0$ (picture on the left) and from the equator $\theta_0 = \pi/2$ (picture on the right).

The function $\ell(r)$, determined by Eq. (8.6.20), has the derivative

$$\frac{d\ell}{dr} = -2\frac{(r-M)^3 + M(M^2 - a^2)}{a(r-M)^2}. \tag{8.6.21}$$

It vanishes at

$$r = M - [M(M^2 - a^2)]^{1/3}. \tag{8.6.22}$$

Problem 8.15: *Prove that any constant-r orbit either lies in the equatorial plane or crosses it. In the former case $\mathbb{Q} = 0$, while in the latter one $\mathbb{Q} > 0$.*

Proof: If the orbit is in the equatorial plane, then $\theta = \pi/2$ must be the root of the function Θ. This is so only if $\mathbb{Q} = 0$. Suppose now that the light ray moving at $r =$const never crosses the equatorial plane. In this case, the function Θ must have roots $\theta_1 \geq \theta_2$ located either between 0 and $\pi/2$ or between $\pi/2$ and π. Since the function Θ is symmetric with respect to the reflection $\theta \to \pi - \theta$ it is sufficient to consider only the former case.

Let us introduce new variables

$$\rho = r/M, \quad p = (a/M)^2, \quad s = \cos^2\theta. \tag{8.6.23}$$

We substitute Eq. (8.6.20) into the expression Eq. (8.6.6) for Θ. This gives

$$\frac{\Theta}{M^2} = -\frac{p^2(\rho-1)^2 s^2 + 2p\rho(\rho^3 - 3\rho + 2p)s + \rho^3(\rho^3 - 6\rho^2 + 9\rho - 4p)}{p(\rho-1)^2(1-s)}. \tag{8.6.24}$$

Under our assumption there exist two roots of Θ, $0 < s_1 \leq s_2 \leq 1$. According to Eq. (8.6.24) the sum of the roots is

$$Z = \frac{1}{2}(s_1 + s_2) = -\frac{\rho(\rho^3 - 3\rho + 2p)}{p(\rho-1)^2}. \tag{8.6.25}$$

We show now that in the black hole exterior, where $\rho > \rho_+ = 1 + \sqrt{1-p}$, Z is negative for any value $p \in [0,1]$. It is sufficient to show that in this domain

$$z = \rho^3 - 3\rho + 2p \qquad (8.6.26)$$

is positive. The function z has a maximum at $\rho = -1$ and minimum at $\rho = 1$. For $\rho > 1$ it is a monotonically growing function of ρ At the horizon

$$z = (1-p)(1 + \sqrt{1-p}) > 0. \qquad (8.6.27)$$

Hence, it is positive everywhere outside the horizon.

Thus, we proved that Z is negative in the black hole exterior, and so that the lowest root s_1 must also be negative. This contradicts our assumption that the motion is totally localized in the upper semi-sphere. This means that if the constant-r orbit is not in the equatorial plane it always crosses it. At the equatorial plane $\Theta = \mathbb{Q}$ (see Eq. (8.6.6)). On the other hand, for $r =$ const light orbits $\rho_{photon_-} < \rho < \rho_{photon_+}$ (see Eq. (8.3.37)) and $\Theta > 0$, hence, for these trajectories we have $\mathbb{Q} > 0$.

The rim of the black hole shadow can be found by substituting the parametric equations Eq. (8.6.20), relating $\mathbb{Q}$ and ℓ, into Eq. (8.6.13). It is easy to check that

$$(x - a\sin\theta_0)^2 + y^2 = \frac{4M(2r^3 - 3Mr^2 + Ma^2)}{(r-M)^2}. \qquad (8.6.28)$$

It is interesting that the right-hand side of this equation does not depend on the inclination angle θ_0.

8.6.4 Special cases

Let us calculate now the size and the shape of the *black hole shadow* in the two limiting cases: (i) $a \ll M$ and (ii) $a = M$.

Slowly rotating black hole

For simplicity we put $\theta_0 = \pi/2$. In this case

$$x = -\ell, \qquad y = \pm\sqrt{\mathbb{Q}}. \qquad (8.6.29)$$

Keeping the terms up to the first order in $\alpha = a/M$ in Eq. (8.6.16) one gets two equations for the constant-r trajectories

$$r^3 - (\mathbb{Q} + \ell^2)r + 2M(\mathbb{Q} + \ell^2 - 2a\ell) = 0,$$
$$3r^2 - \mathbb{Q} - \ell^2 = 0. \qquad (8.6.30)$$

We use the second relation to obtain $\mathbb{Q}$, substitute this solution into the first equation, and solve this equation perturbatively. As a result we get

$$r = 3M - \frac{2a\ell}{9M}. \qquad (8.6.31)$$

This allows us to find $\mathbb{Q}$

$$\mathbb{Q} = 27M^2 - \ell^2 - 4a\ell. \qquad (8.6.32)$$

The condition that y in Eq. (8.6.29) is real gives

$$\ell \in [\ell_-, \ell_+], \qquad \ell_\pm = \pm 3\sqrt{3}M - 2a. \qquad (8.6.33)$$

The shadow rim is described by the equation

$$(x - 2a)^2 + y^2 = 27M^2. \qquad (8.6.34)$$

This relation for $a = 0$ correctly reproduces the result for the Schwarzschild black hole, where the critical impact parameter is $3\sqrt{3}M$. Equation (8.6.34) also shows that for a slowly rotating black hole the rim of the shadow remains approximately a circle of radius $3\sqrt{3}M$ shifted in the direction of the rotation by a constant $2a$. This shift effect can be easily understood. For the Kerr black hole corotating circular photon orbits are closer to the horizon than the counter-rotating ones (see Figure 8.8). Therefore, the prograde photons that have reached the observer, are visible from lesser angular distance from the black hole than the retrograde ones. Altogether, this results in the shift of the image in the direction of rotation of the black hole.

For $\theta_0 \neq 0$ the result is similar: the shadow is again a circle of radius $3\sqrt{3}M$ shifted along the x-axis by the distance $2a \sin \theta_0$. No shift is present when the observer is located at the symmetry axis.

Extremely rotating black hole

For the extremal $a = M$ Kerr metric the relations Eq. (8.6.20) simplify and give

$$\ell = -M^{-1}[(r - M)^2 - 2M^2], \qquad Q = M^{-2}r^3(4M - r). \qquad (8.6.35)$$

The first of these equations can be used to obtain r as a function of ℓ

$$r = M + \sqrt{M(2M - \ell)}. \qquad (8.6.36)$$

The second solution, for which $r < M$, is unphysical. Substituting Eq. (8.6.36) into the expression for Q one gets

$$Q = M^{-2}\left(M + \sqrt{M(2M - \ell)}\right)^3 \left(3M - \sqrt{M(2M - \ell)}\right). \qquad (8.6.37)$$

This equation determines the position of the rim of the black hole shadow in the (x, y)-plane.

In the simplest case, when the observer is in the equatorial plane, $\theta_0 = \pi/2$, one has $Q = y^2$ and $\ell = -x$, so that the equation for the shape of the shadow is

$$y^2 = M^{-2}\left(M + \sqrt{M(2M + x)}\right)^3 \left(3M - \sqrt{M(2M + x)}\right). \qquad (8.6.38)$$

Since y^2 must be real and non-negative, the impact parameter x changes in the interval $(-2M, 7M)$.

Note that $\ell = 2M$ is a degenerate solution. It corresponds to $r = M$. At this value of r the coefficients of the term Q in Eq. (8.6.16) vanish. Hence, Q can have an arbitrary value, as soon as $\ell = 2M$. For this value the rim of the shadow on the (x, y)-plane is a vertical straight line connecting the points $(-2M, -\sqrt{3}M)$ and $(-2M, \sqrt{3}M)$. As we shall see in the next section, this 'flattening' of the shadow rim is a generic property of the rotating black holes.

For an observer looking at the black hole along the symmetry axis, $\theta_0 = 0$, a shadow is a circle. For the rays propagating along this direction the impact parameter $\ell = 0$. Using Eq. (8.6.37) one finds $\mathbb{Q} = (11 + 8\sqrt{2})M^2$. The radius of the circle can be found by using the relation Eq. (8.6.14)

$$R = \sqrt{x^2 + y^2} = 2(1 + \sqrt{2})M \approx 4.83M. \tag{8.6.39}$$

This radius is slightly less that the shadow radius for the Schwarzschild black hole of the same mass, $R_S = 3\sqrt{3}M \approx 5.20M$.

8.6.5   Generic properties of the rotating black hole shadows

Asymmetry

Figures 8.15 show the curves for the rim of the black hole shadow for different values of the rotation parameter a and different values of the *inclination angle* θ_0. For small values of the rotation parameter the deformation of the *apparent shape of the black hole* is small. The main effect is the shift of the shadow region in the (x, y)-plane to the right. This shift is more profound for large values of the rotation parameter. This effect can be easily explained as follows. For rotating black holes the radius of corotating circular orbits descreases when a grows. For counter-rotating orbits the effect is opposite. Similarly, the photons with positive impact parameters (corotating with the black hole), reach the (x, y)-plane at the left end point of the shadow, which is moved to the right. For the counter-rotating photons (with $\ell < 0$) the left end point of the shadow moves to the right as well. As a result, the shadow is moved as a whole to the right, that is in the direction of the rotation of the black hole (see Figure 8.14). This effect of the *asymmetry of the black hole shadow* is a common feature of the rotating black holes.

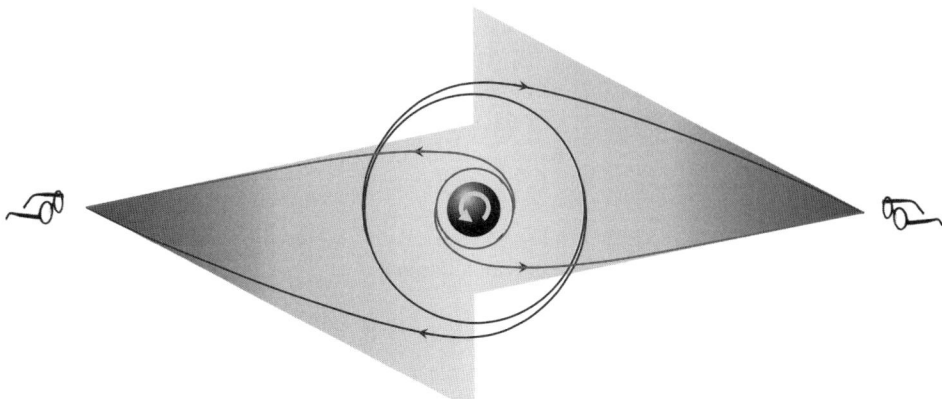

Fig. 8.14 The rim of the shadow corresponds to photons that come from infinity, rotate around the black hole many times near the unstable spherical orbit (direct or retrograde) and then reach the observer. Different observers see the black hole shadow shifted in different directions.

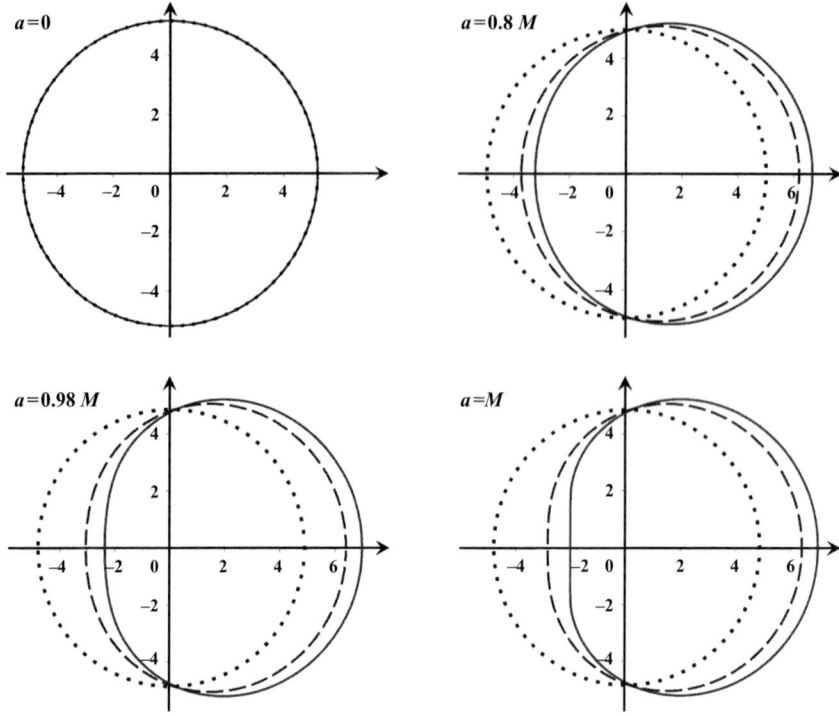

Fig. 8.15 The shadows for different inclination angles: $\theta_0 = \pi/2$ – solid line, $\theta_0 = \pi/4$ – dashed line, $\theta_0 = 0$ – dotted line. Evidently, for the non-rotating black hole the shadow is independent of the inclination angle and all three projections coincide. All shadows visible from the pole $\theta_0 = 0$ are circles.

Flattening

Another generic feature is the flattening of the left side of the shadow of a rotating black hole. The black hole is assumed to rotate counter-clockwise as seen from the north pole. We have already mentioned this effect for the extremally rotating black hole. Let us discuss it in more detail. For the inclination angle $\theta_0 = \pi/2$ Eq. (8.6.13) takes the form

$$x = -\ell, \qquad y = \pm\sqrt{\mathbb{Q}}, \qquad (8.6.40)$$

where ℓ and $\mathbb{Q}$ are related according to Eq. (8.6.20). Points where the shadow rim crosses the x-axis are defined by the equation $\mathbb{Q} = 0$, which implies

$$r^3 - 6Mr^2 + 9M^2r - 4a^2M = 0. \qquad (8.6.41)$$

We use this equation to find a^2 as a function of r

$$a^2 = \frac{1}{4M}r(3M - r)^2. \qquad (8.6.42)$$

The rotation parameter vanishes for $r = 3M$. For other values of a, there exist 2 roots of Eq. (8.6.41). When a changes from 0 to M the lower root $r_{(l)}$ monotonically decreases from

$3M$ to M, while the bigger one $r_{(r)}$ changes from $3M$ to $4M$. Let $\ell_{(l,r)}$ be corresponding values of the impact parameter

$$\ell_{(l,r)} = \mp \frac{1}{2}\sqrt{r/M}(r+3M).$$ (8.6.43)

The impact parameter $\ell_{(l)}$ corresponds to a left point of where the shadow rim crosses the x-axis, and $\ell_{(r)}$ is for a right point.

> **Problem 8.16:** *Show that at the left and right points of the intersection of the rim curve Eq. (8.6.40) with the x-axis one has*
>
> $$\left.\frac{d^2x}{dy^2}\right|_{(l,r)} = \frac{1}{2Mr_{(l,r)}^2}(r_{(l,r)} - M)\sqrt{Mr_{(l,r)}}.$$ (8.6.44)

The plot of the quantity d^2x/dy^2 a function of the rotation parameter is shown in Figure 8.16. The upper line on this plot corresponds to the right point $r_{(r)}$, and the lower line is for $r_{(l)}$. One can see that the curvature of the right part of the shadow rim curve practically remains the same for the values of the rotation parameter $a \in [0, M]$. The curvature of the left part decreases rapidly when $a \to M$. This means that for a rapid rotation of the black hole the left side of the shadow is always flattened.

To summarize, observations of the black hole shadow and study of its size and shape would allow one to obtain direct information about the mass and the rotation parameter of the black hole.

8.6.6 Principal null geodesics

In the Kerr spacetime there exist special types of photon trajectories, called *principal null rays*, which play an important role. For these rays the conserved quantities are

$$\ell = a\sin^2\theta_0, \qquad Q = -a^2\cos^4\theta_0,$$ (8.6.45)

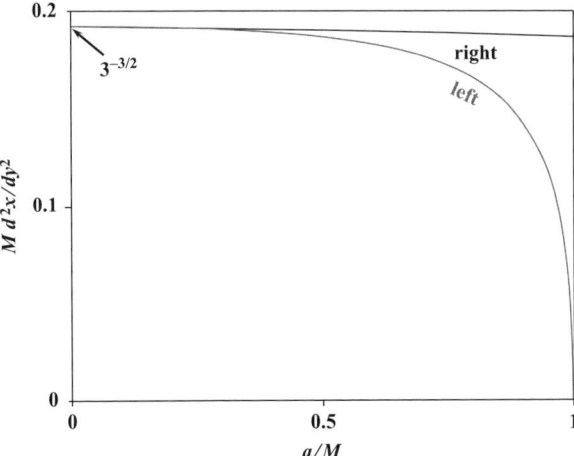

Fig. 8.16 The curvature of the shadow at its left and right sides as a function of the rotation parameter a.

so that the function $\Theta(\theta_0)$ vanishes. The trajectory of principal null rays is on the 'cone' of the constant value of $\theta = \theta_0$. The expression for $\mathcal{R}$ also simplifies, and one gets $dr/d\tilde{\lambda} = \pm 1$. Using Eqs. (8.6.4) and (8.6.5), one finds the other components of the null vector $l_{\pm}^{\mu} = dx^{\mu}/d\tilde{\lambda}$

$$l_{\pm}^{\mu}\partial_{\mu} = \frac{r^2 + a^2}{\Delta}\partial_t \pm \partial_r + \frac{a}{\Delta}\partial_{\phi}. \tag{8.6.46}$$

The integral lines of these vectors are *principal null geodesics*. These vectors obey the relation

$$C_{\alpha\beta\gamma[\delta}l_{\pm\epsilon]}l_{\pm}^{\beta}l_{\pm}^{\gamma} = 0, \tag{8.6.47}$$

where $C_{\alpha\beta\gamma\delta}$ is the *Weyl tensor*. When Eq. (8.6.47) has two independent solutions the spacetime belongs to the *Petrov type D*. One can also show that the principal null vectors are eigenvectors of the antisymmetric tensor $\xi_{(t)\mu;\nu}$

$$\xi_{(t)\mu;\nu}l_{\pm}^{\nu} = \pm\frac{1}{2}F_{,r}l_{\pm\mu}. \tag{8.6.48}$$

Here, $\xi_{(t)}$ is the Killing tensor, and

$$F = -\xi_{(t)}^2 = 1 - \frac{2Mr}{\Sigma}. \tag{8.6.49}$$

8.7   Hidden Symmetries of Kerr Spacetime

8.7.1   Killing tensor

We already mentioned that the 'miraculous' properties of the Kerr spacetime are connected with the *hidden symmetry*, which, in particular, implies the *complete integrability of geodesic equations*. In the general case, according to the *Liouville theorem*, the complete integrability is a property of the Hamiltonian system with $2m$-dimensional phase space, when there exist m independent integrals of motion in the involution (see Section 4.3). In the relativistic theory, if the integrals of motion are monomials in the momentum p_{μ}, the existence of the hidden symmetry can be formulated as a property of the spacetime manifold. Each such monomial of the power q is connected to the Killing tensor of the same rank q (see Section 4.3). For the complete integrability of geodesic equations it is sufficient to have 4 independent Killing vectors and/or Killing tensors. One of the Killing tensors, which always exists, is the metric. The corresponding conserved quantity is

$$g^{\mu\nu}p_{\mu}p_{\nu} = -m^2. \tag{8.7.1}$$

In the Kerr metric there exist two more integrals of motion, energy E and azimuthal angular momentum L_z, connected with the Killing vectors $\xi_{(t)}$ and $\xi_{(\phi)}$, generating the spacetime symmetries

$$E = -\xi_{(t)}^{\mu}p_{\mu}, \qquad L_z = \xi_{(\phi)}^{\mu}p_{\mu}. \tag{8.7.2}$$

An additional integral of motion (*Carter's constant*) is quadratic in the momentum. It can be written as

$$K = K^{\mu\nu} p_\mu p_\nu = \left[p_0\, a \sin\theta + \frac{p_3}{\sin\theta} \right]^2 + p_2^2 - a^2 \cos^2\theta\, g^{\mu\nu} p_\mu p_\nu. \qquad (8.7.3)$$

The corresponding *Killing tensor* is

$$K^{\mu\nu} = a^2 \sin^2\theta\, \xi^\mu_{(t)} \xi^\nu_{(t)} + 2a\, \xi^{(\mu}_{(t)} \xi^{\nu)}_{(\phi)} + \frac{1}{\sin^2\theta} \xi^\mu_{(\phi)} \xi^\nu_{(\phi)} + \delta^\mu_\theta \delta^\nu_\theta - a^2 \cos^2\theta\, g^{\mu\nu}. \qquad (8.7.4)$$

It obeys the equation

$$K_{(\mu\nu;\varepsilon)} = K_{\mu\nu;\varepsilon} + K_{\varepsilon\mu:\nu} + K_{\nu\varepsilon;\mu} = 0. \qquad (8.7.5)$$

Problem 8.17: *Let $\xi^{(1)}_\mu$ and $\xi^{(2)}_\nu$ be two Killing vectors. Show that a tensor product $\xi^{(1)}_{(\mu} \xi^{(2)}_{\nu)}$ is a rank-2 Killing tensor.*

A Killing tensor of rank q, which can be presented as a linear combination of the other Killing tensors of the same rank and/or of a properly symmetrized tensor product of the Killing tensors of the lower rank, is called *reducible* . The corresponding conserved quantity for such a tensor is a function of the other conserved quantities. Hence, it does not give a new independent integral of motion. To construct a complete set of conserved quantities it is sufficient to consider only *irreducible Killing tensors* .

It is easy to see that the above four integrals of motion are generically functionally independent. Moreover, they are in involution (see Eq. (4.3.32)). The latter property follows from the following relations

$$[\boldsymbol{\xi}_{(t)}, \boldsymbol{\xi}_{(\phi)}] = 0,$$
$$\mathcal{L}_{\boldsymbol{\xi}_{(t)}} g_{\mu\nu} = \mathcal{L}_{\boldsymbol{\xi}_{(t)}} K_{\mu\nu} = \mathcal{L}_{\boldsymbol{\xi}_{(\phi)}} g_{\mu\nu} = \mathcal{L}_{\boldsymbol{\xi}_{(\phi)}} K_{\mu\nu} = 0, \qquad (8.7.6)$$
$$[g, K] = 0.$$

In the last relation we used the *Schouten–Nijenhuis bracket* Eq. (4.3.29). This relation is simply the Killing equation Eq. (8.7.5). The conditions Eq. (8.7.6) provide the complete integrability of the geodesic equations in the Kerr spacetime.

Problem 8.18: *In the limit $M = 0$ (for finite a) the Kerr spacetime becomes flat and its metric is given by Eq. (8.1.10). The Killing tensor, written in the form Eq. (8.7.4), depends on the mass M only through the term, proportional $g^{\mu\nu}$, and hence it has a well-defined limit at $M \to 0$. Using Eq. (8.1.11), write the expressions for K and Q (for $M = 0$) (see Eq. (8.5.19)) in the Cartesian coordinates and show that the corresponding conserved quantities can be written in the form*

$$K = L^2 - 2aEL_Z + a^2(E^2 - p_Z^2),$$
$$Q = L^2 - L_Z^2 - a^2 p_Z^2, \qquad L^2 = L_X^2 + L_Y^2 + L_Z^2. \qquad (8.7.7)$$

Here, L_i and p_i are Cartesian components of the angular momentum and momentum of the particle.

For the scattering problem in the Kerr spacetime, one can use Eq. (8.7.7) to determine the value of K in the asymptotic region, where the spacetime is practically flat. In the process of scattering of a particle by the rotating black hole, the value of K remains constant. This means that if a particle reaches the spatial infinity again, this conserved quantity can be written again in the form Eq. (8.7.7). For incoming and outgoing motion the form of K is the same, but the values that enter this expression are different asymptotic (in- and out-) data. The corresponding (global) conservation law is a non-trivial property of the Kerr spacetime, and, in the general case, it is not valid for the processes in the gravitational field of rotating bodies, different from the black holes. For more detailed discussion of the hidden symmetries see Appendix D.

It is interesting that in the near-extremal limit when $a/M \to 1$ the hidden symmetry connected with the Killing tensor K becomes explicit. Namely, for the metric Eq. (8.2.64) it takes the form

$$K^{\mu\nu} = 2\xi_1^{(\mu}\xi_2^{\nu)} + \xi_2^{\mu}\xi_2^{\nu} - \xi_3^{\mu}\xi_3^{\nu} - \frac{1}{2}g^{\mu\nu}, \tag{8.7.8}$$

where ξ_i^{μ} are the Killing vectors defined in Eq. (8.2.65). This means that the Killing tensor becomes redundant in this limit, and it can be expressed in terms of the Killing vectors ξ_i responsible for the explicit symmetry of the spacetime.

8.7.2 Kerr metric in the canonical form

The Kerr metric given by Eq. (8.1.4) is written in special, *Boyer–Lindquist coordinates*. These coordinates are a natural generalization on the Schwarzschild coordinates to the case of a rotating black hole. We present now another form of the Kerr metric, in which its hidden symmetry is more evident. Let us make the following coordinate transformations

$$y = a\cos\theta, \qquad \psi = \phi/a, \qquad \tau = t - a\phi. \tag{8.7.9}$$

We call (τ, r, y, ψ) Darboux (or canonical) coordinates. The Kerr metric Eq. (8.1.4) in the *Darboux coordinates* takes the form

$$ds^2 = \frac{1}{r^2 + y^2}\left[-\Delta_r(d\tau + y^2 d\psi)^2 + \Delta_y(d\tau - r^2 d\psi)^2\right]$$

$$+(r^2 + y^2)\left[\frac{dr^2}{\Delta_r} + \frac{dy^2}{\Delta_y}\right], \tag{8.7.10}$$

where

$$\Delta_r = r^2 - 2Mr + a^2, \qquad \Delta_y = a^2 - y^2. \tag{8.7.11}$$

■ **Problem 8.19:** *Prove this.*

Equation (8.7.10) is a *canonical form* of the Kerr metric. In the canonical (Darboux) coordinates the Killing tensor Eq. (8.7.4) reads[3]

$$K^{\mu}{}_{\nu} = \begin{pmatrix} 0 & 0 & 0 & -r^2 y^2 \\ 0 & -y^2 & 0 & 0 \\ 0 & 0 & r^2 & 0 \\ -1 & 0 & 0 & r^2 - y^2 \end{pmatrix}. \tag{8.7.12}$$

[3] In Appendix D hidden symmetries are described in a more systematic way. The Killing tensor K_{Carter} used for separation of variables and given by Eqs. (8.7.4) and (8.7.12) differs only by sign from the Killing tensor $K_{canonical}$ given by Eqs. (D.5.1) and (D.7.6) in Appendix D.

Let us emphasize that the Killing tensor Eq. (8.7.12) does not depend on the parameters of the Kerr metric. Moreover, it is a Killing tensor for any metric of the form Eq. (8.7.10), which contains 2 arbitrary functions of one variable, $\Delta_r(r)$ and $\Delta_y(t)$ (see Appendix D).

The *principal null vectors* $l_\pm$ in the canonical coordinates are

$$l_\pm^\mu = \frac{r^2}{\Delta_r}\partial_\tau \pm \partial_r + \frac{1}{\Delta_r}\partial_\psi. \tag{8.7.13}$$

It is easy to see that $l_\pm$ are eigenvectors of the Killing tensor Eq. (8.7.12)

$$K^\mu{}_\nu l_\pm^\nu = -y^2 l_\pm^\mu. \tag{8.7.14}$$

This means that the principal null vectors $l_\pm$ are eigenvectors of the operator K, and $-y^2$ are its eigenvalues.

Note that the principal null vectors are related to the *normalized Darboux basis* vectors Eq. (D.7.9) according to

$$l_\pm^\mu \partial_\mu = -\sqrt{\frac{r^2 + y^2}{\Delta_r}}\left(n_1^\mu \pm \bar{n}_1^\mu\right). \tag{8.7.15}$$

It is possible to show that the orthonormal vectors, orthogonal to the (l_+, l_-) plane are also eigenvectors of K with the eigenvalues proportional to r^2. This means that the essential canonical coordinates (r, y) are related to the Killing tensor K responsible for the hidden symmetry. This property has a wide-ranging generalization in the case of the higher-dimensional rotating black hole metrics. This subject, which goes far beyond the scope of this book, is very briefly discussed at the end of Appendix D.

8.8 Energy Extraction from a Rotating Black Hole

8.8.1 Reversible process

Consider the Penrose process (see Figure 8.9) in more detail. Let us recall that in this process the incoming particle 0 decays inside the ergosphere into two particles, 1 and 2. Local conservation of energy-momentum requires $p_0^\mu = p_1^\mu + p_2^\mu$. For every particle conserved quantities are

$$\varepsilon_i = -p_i^\mu \xi_{(t)\mu}, \qquad j_i = p_i^\mu \xi_{(\phi)\mu}. \tag{8.8.1}$$

Denote by l^μ the null generator of the horizon

$$l^\mu = \xi_{(t)}^\mu + \Omega \xi_{(\phi)}^\mu. \tag{8.8.2}$$

Here, Ω is the angular velocity of the black hole

$$\Omega = \frac{a}{r_+^2 + a^2} = \frac{a}{2Mr_+}. \tag{8.8.3}$$

We assume that particle 2 has the negative energy and it crosses the horizon. Since a scalar product of the null and time-like vectors is non-positive, provided both vectors are future-directed. Thus, we have

$$0 \geq l^{\mu} p_{2\mu} = -\varepsilon_2 + \Omega j_2 \rightarrow j_2 \leq \varepsilon_2/\Omega. \tag{8.8.4}$$

The last relation implies

$$j_1 - j_0 = -j_2 \geq -\frac{\varepsilon_2}{\Omega} \geq 0. \tag{8.8.5}$$

This means that the extraction of the energy is always accompanied by the extraction of the angular momentum.

As a result of this process the black hole absorbs some amount of (negative) energy and angular momentum. Because of the backreaction its parameters change

$$\delta M \geq \Omega \, \delta J, \qquad \delta M = \varepsilon_2, \qquad \delta J = j_2. \tag{8.8.6}$$

The 'most economic' process in which $\delta M = \Omega \, \delta J$ is called *reversible*. Let us show that for the reversible process the area of the black hole remains the same.

The surface area of a rotating black hole is

$$A = 4\pi (r_+^2 + a^2) = 8\pi M r_+$$

$$= 8\pi M \left(M + \sqrt{M^2 - a^2} \right) = 8\pi \left(M^2 + \sqrt{M^4 - J^2} \right). \tag{8.8.7}$$

For the variation of the mass, δM, and the angular momentum, δJ, the area of the black hole changes by

$$\delta A = 16\pi \left(M + \frac{M^3}{\sqrt{M^4 - J^2}} \right) [\delta M - \Omega \, \delta J]. \tag{8.8.8}$$

For the reversible process $\delta M - \Omega \, \delta J = 0$ and hence, the black hole area remains constant.

This is a special case of the general *Hawking's area theorem*, which reveals that *in the absence of naked singularities (i.e. singularities visible from infinity) the surface area of a black hole is a non-decreasing function of time, provided the weak-energy condition is satisfied.* (See Section 10.2.3.)

8.8.2 Irreducible mass

The quantity $M_{\mathrm{ir}} = (A/16\pi)^{1/2}$ is called the *irreducible mass*. In the absence of rotation $M_{\mathrm{ir}} = M$. For the extremely rotating black hole $(M = a)$ $M_{\mathrm{ir}} = M/\sqrt{2}$. Using the definition of the irreducible mass one obtains

$$M_{\mathrm{ir}}^2 = \frac{1}{2} \left(M^2 + \sqrt{M^4 - J^2} \right). \tag{8.8.9}$$

This relation allows one to find the mass of the black hole as a function of its irreducible mass and angular momentum

$$M^2 = M_{\mathrm{ir}}^2 + \frac{J^2}{4M_{\mathrm{ir}}^2} \geq M_{\mathrm{ir}}^2. \tag{8.8.10}$$

It is easy to see that the mass of the black hole is never less than its irreducible mass. It is equal to the latter only if the black hole is non-rotating.

Consider a rotating black hole with mass M_0 and angular momentum J_0, and let $M_{ir}(M_0, J_0)$ be its irreducible mass. In the energy-extraction process the final mass M_1 cannot be smaller than the irreducible mass. The most efficient process occurs when, during this process, M_{ir} remains the same. The maximal energy extracted for this reducible process is

$$\Delta M = M_0 - M_{ir}(M_0, J_0). \tag{8.8.11}$$

For the extremal black hole

$$\Delta M = \left(1 - \frac{1}{\sqrt{2}}\right) M_0 = 0.29 M_0. \tag{8.8.12}$$

This is a part of the black hole energy, which is connected with its rotation. For a stellar mass black hole of $10 M_\odot \sim 2 \times 10^{34}$ g the maximal rotating energy

$$\Delta E \approx 0.6 \times 10^{55} \text{ erg.} \tag{8.8.13}$$

8.8.3 Generalization

The above results allow a generalization to the case of a charged rotating black hole. In this case, the Hawking area theorem implies

$$\delta M \geq \Omega \, \delta J + \Phi \, \delta Q, \tag{8.8.14}$$

where $r_+ = M + \sqrt{M^2 - a^2 - Q^2}$, and

$$\Phi = -A_\mu l^\mu \big|_{r=r_+} = -(A_t + \Omega A_\phi)\big|_{r=r_+} = \frac{Q r_+}{r_+^2 + a^2}, \qquad \Omega = \frac{a}{r_+^2 + a^2} \tag{8.8.15}$$

are the *electrostatic potential* at the horizon of the charged black hole and its angular velocity of rotation. l^μ is the null generator of the horizon Eq. (8.8.2).

Let us apply Eq. (8.8.14) to the case when a black hole absorbs a particle of charge e and spin s and with the initial energy ε, falling along the axis of symmetry. In this case

$$\delta Q = e, \qquad \delta J = \sigma s \hbar, \qquad \delta M \leq \varepsilon. \tag{8.8.16}$$

The last inequality reflects the fact that part of the initial energy of the particle can be emitted to infinity in the form of the gravitational radiation. Combining Eq. (8.8.14) and Eq. (8.8.16) one obtains

$$\varepsilon > e\Phi + \sigma s \Omega. \tag{8.8.17}$$

The sign function $\sigma = 1$ when the direction of the spin coincides with the direction of the angular momentum, and $\sigma = -1$ in the opposite case. Equation (8.8.17) shows that if a charged particle falls onto the black hole it must have more energy ε in the case when $eQ > 0$ than for the opposite sign of the charge. The explanation is simple: There is an electric repulsion between the charges of the same sign and the attraction between the charges of the opposite sign. A similar conclusion can be made concerning the term $\sigma s \Omega$ describing the *spin–spin interaction*. Namely, there is a repulsive force acting on a particle that has the same direction of the spin and the angular momentum of the black hole. For the opposite orientation this force is an attraction.

8.8.4 Superradiance

There is an analog of the Penrose process for the field scattering by a rotating black hole. This effect is known as the *superradiance*. We shall discuss in detail the field propagation and interaction with the spacetime of a black hole in the next chapter. Here, we just make simple remarks connected with energy aspects of this problem. Consider a bosonic field $\varphi_\bullet$ in the Kerr spacetime. The index $\bullet$ means that the field may have several components, for example to be a vector or tensor one. The nature of the field is not important at the moment. Since the spacetime is stationary and axisymmetric we can decompose the field into the modes

$$\varphi_\bullet \sim f_\bullet(r,\theta)\, e^{-i\omega t + im\phi}. \tag{8.8.18}$$

One can consider such a classical field as a special correlated coherent state in which there is a condensate of quanta with fixed values of the energy ε and the projection of the angular momentum on the axis of the rotation j

$$\varepsilon = \hbar\omega, \qquad j = \hbar m, \qquad q = e. \tag{8.8.19}$$

We include here the parameter of the charge q in the case when the field is charged. As a result of the absorption of the field the parameters of the black hole change and one has

$$\frac{\delta J}{\delta M} = \frac{m}{\omega}, \qquad \frac{\delta Q}{\delta M} = \frac{e}{\hbar\omega}. \tag{8.8.20}$$

The Hawking area theorem implies

$$\delta M - \Omega\, \delta J - \Phi\, \delta Q \geq 0,$$

so that we arrive at the following inequality

$$\frac{\delta M}{\hbar\omega}\, [\,\hbar\omega - \hbar m\Omega - e\Phi\,] \geq 0. \tag{8.8.21}$$

This shows that for

$$\hbar\omega \leq \hbar m\Omega + e\Phi \tag{8.8.22}$$

one has $\delta M \leq 0$ and hence the energy is extracted from the black hole. In this case the scattered wave has more energy. Since the frequency remains the same, its amplitude is greater than the amplitude of the incident wave. This effect of wave amplification in the scattering by the black hole is called *superradiance*.

> *This effect is of a general nature. Suppose we have a rotating dielectric or conducting (absorbing) cylinder of the radius R. For a wave $\varphi_\bullet \sim e^{-i\omega t + im\phi}$ the condition of the fixed phase is $\omega t = m\phi$ and ω/m determines the phase velocity. If ΩR (velocity of the surface motion) is greater than the linear phase velocity $R\omega/m$, absorbtion is replaced by amplification. This effect is similar to the anomalous Doppler effect.*

Anomalous Doppler effect. *Let us consider a transparent medium with the refraction coefficient* $n(\omega)$ *and an oscillator or atom moving in it with a constant velocity v. We assume that this system (oscillator or atom) can radiate at frequency ω_0 (as measured in its own reference frame). Then, the frequency ω of the radiation from the moving system, measured by an observer that is at rest with respect to the medium, is*

$$\omega(\theta) = \frac{\omega_0 \sqrt{1 - v^2}}{1 - v\,n(\omega)\,\cos\theta}. \tag{8.8.23}$$

For the emission outside the Cherenkov *cone, when* $[v\,n(\omega)\,\cos\theta < 1$, *this formula describes an ordinary (normal) Doppler effect. For a superluminal motion,* $v > n^{-1}$, *and for the direction inside the Cherenkov cone, when* $v\,n(\omega)\,\cos\theta > 1$, *the radiation with positive ω is possible only when $\omega_0 < 0$; that is, when the system emitting photons becomes existed as a result of emission. This effect is known as the* anomalous Doppler effect *(see, e.g., (Ginzburg 1985)).*

8.9 Black Holes in External Magnetic Field

8.9.1 Black holes as membranes

Black holes are very different from other astrophysical compact objects. The latter have a boundary that separates its 'inner' matter from the empty 'outer' space. A black hole does not have such a boundary. In the absence of matter, the Kerr black hole is a vacuum solution of the Einstein equations. One may say that the black hole is plenty of nothing. At the same time, for an external observer it has a boundary, the *event horizon*. A free-falling observer will see nothing special, when he/she crosses it. As we know, the main property of the horizon is that it works as a one-way membrane. It 'allows' everything to fall inside, but it does not allow anything to go out. This happens because the regular horizon is a null surface.

Nevertheless, for outside observers the black hole in many aspects looks like an extended compact body. It is possible to describe black holes as some kind of 'bodies' with special physical properties. Such a formalism was developed by Kip Thorne and collaborators and is called the *membrane formalism* (see the book (Thorne *et al.* 1986). The main idea of this approach is to 'substitute' a real black hole with a fictitious 'body', with well-defined physical properties. These properties can be chosen so that they guarantee that the physics in the black hole exterior and outside the 'body' are the same. In section 2.4.3 in our discussion of electrostatics in the Rindler space we have already seen that the horizon is similar to a conducting surface. It is easy to show that for the consistency of the Maxwell equations in the Rindler frame it is sufficient to assume that the horizon is a conducting surface with a universal surface resistance

$$R_H = 4\pi \approx 377 \text{ Ohm.} \tag{8.9.1}$$

This conclusion remains valid for any null surface in a curved spacetime, and, in particular, for the event horizon of a stationary black hole.

Similarly, one can introduce other characteristics of the 'horizon membrane', such as a surface temperature, an entropy, surface densities of charge and electric current, a surface viscosity, and so on. In the framework of the membrane paradigm, black holes are described as extended bodies that not only have a mass, a charge, and an angular momentum, but possess a surface with very special properties. Using this paradigm one arrives at the same answeres to the problems, which involve black holes, as the standard 4D curved spacetime approach. The appealing feature of the membrane approach is that it allows one to use intuition, based on the

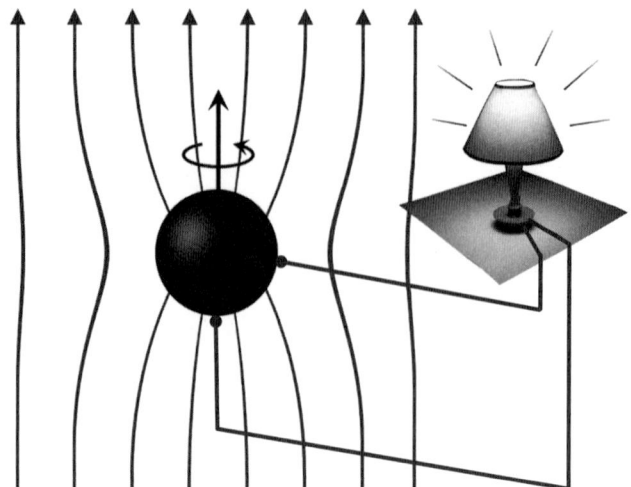

Fig. 8.17 Black hole as a unipolar inductor. Rotation of a black hole in an external magnetic field generates the electric potential difference between the pole and the equator.

knowledge of the 'usual' physical laws, to 'predict' different properties of systems, containing black holes.

As an important example consider the the following problem: *What happens when a rotating black hole is inserted in an originally homogeneous magnetic field parallel to the axis of its rotation?*

The membrane paradigm teaches us that the black hole is similar to a conducting body. But when a rotating conducting body is immersed in the magnetic field, it works like an electric generator. Thus, one can expect that a rotating black hole surrounded by the conducting plasma in the magnetic field generates currents. In this process the rotation energy of the black hole is transformed into the energy of currents in the external space. The idea to use this mechanism to explain the energetics of black holes and jet formations was proposed by Blandford and Znajek (1977) (for the review see also (Thorne *et al.* 1986; Lee *et al.* 2000)). Figure 8.17 illustrates this mechanism.

*Assume a conducting body moves with the velocity **v** with respect to the magnetic field **B**. As a result of the Lorentz transformation, in a reference frame comoving with the body there exists an electric field $E = [v \times B]$. This field generates an electric currect, which results in the separation of charges. This process continues until the electric field induced by these charges compensates the field **E**. As a result, in the original frame there will be a non-trivial electric field inside the conducting body, and, as a consequence, the electric-potential difference between different parts of the rotating body. If outside the body there exists an electric circuit connecting the points with the different potentials, the induced voltage generates an electic current.*

This effects is quite general. Magnetism can make, or induce, electric currents whenever the conductor that carries the current and the lines of magnetic force move so that they cut across each other. The device that contains a rotating body in the magnetic field is well known in physics. It has different names, such as a unipolar generator, a homopolar generator, a disk dynamo. The unipolar generator was developed first by Michael Faraday during his experiments in 1831 in which he used a rotating conducting disc. It is sometimes called the Faraday disc in his honor. It was the beginning of modern dynamos that is, electrical generators that operate using a magnetic field.

8.9.2 Black hole in a homogeneous magnetic field

In this section we demonstrate how the rotating black hole becomes a *unipolar inductor*. We shall use a 'standard' 4D form of general relativity. However, let us first make a few general remarks concerning electrodymanics of black holes. An isolated stationary black hole in the asymptotically flat spacetime is uniquely characterized by the parameters[4] M, J, and Q. In this sense, the electromagnetic properties of the charged black holes are quite trivial. In particular, dipole and higher electric and magnetic moments are uniquely determined by its parameters $\{M, J, Q\}$. Nevertheless, the *black hole electrodynamics* is quite non-trivial if we deal not with an isolated black hole, but with a black hole surrounded by matter. Namely, these situations are of the most astrophysical interest. In a generic case a black hole is surrounded by accreting matter, which forms an *accretion disk*. This matter is usually hot enough and is in the form of plasma. The plasma is an ideal conductor and the magnetic field (if it is present) is frozen into it. In the presence of the rotating plasma the well-known *dynamo mechanism* can amplify the magnetic field. Its inhomogeneities fall down to the black hole, but the magnetic field cannot escape to infinity, being captured by the plasma of the acreting disk. As a result of this mechanism a magnetic field in the black hole vicinity is generated, which can produce observable effects.

In Section 7.9 we have already discussed non-rotating black holes inserted in a test magnetic field. As before, we assume that the energy density of the magnetic field is small enough, so that its backreaction effects can be neglected. We also exclude the plasma from our consideration by 'moving' the region it occupies far enough from the black hole. We use the approximation in which test the magnetic field is stationary, axisymmetric, and homogeneous at infinity. We first obtain a solution of the Maxwell equations in the Kerr spacetime background obeying these boundary conditions and that is regular at the event horizon. Then we study the properties of the solution. We focus mainly on the effects related to the black hole rotation.

Instead of solving the Maxwell equations directly we use the same trick as in Section 7.9. In the Kerr spacetime there exist two Killing vectors $\boldsymbol{\xi}_{(t)}$ and $\boldsymbol{\xi}_{(\phi)}$. As mentioned in Section 7.9, in the vacuum spacetime the Killing vector obeys the equation

$$\xi^{\alpha;\beta}{}_{;\beta} = 0, \tag{8.9.2}$$

which is identical to the Maxwell equation in the Lorentz gauge. Let us define two antisymmetric tensors

$$F_{(a)\mu\nu} = -2\xi_{(a)\mu;\nu} = \xi_{(a)\nu;\mu} - \xi_{(a)\mu;\nu}, \tag{8.9.3}$$

where $a = t$ or ϕ. According to Eq. (8.9.2), $F_{(a)\mu\nu}$ can be interpreted as solutions of the source-free Maxwell equation, $\xi_{(a)\mu}$ being identified with the electromagnetic vector potential in the Lorentz gauge $\xi^{\mu}_{(a);\mu} = 0$. At the asymptotic infinity in the cylindrical coordinates one has

---

[4]If there exist magnetic monopoles, a black hole can also have a magnetic charge P. We shall not discuss this case here.

$$ds^2 \approx -dt^2 + dz^2 + d\rho^2 + \rho^2\, d\phi^2, \quad \xi_{(t)\mu} \approx -\delta_\mu^t, \quad \xi_{(\phi)\mu} \simeq \rho^2\, \delta_\mu^\phi. \tag{8.9.4}$$

One can check that in this region

$$F_{(t)\mu\nu} = \xi_{(t)\mu;\nu} \approx 0,$$
$$F_{(\phi)\mu\nu} \approx \rho(\delta_\mu^\rho\, \delta_\nu^\phi - \delta_\mu^\phi\, \delta_\nu^\rho). \tag{8.9.5}$$

Hence,

$$B^i = -\frac{1}{2} e^{ijk} F_{(\phi)jk} = 2\delta_z^i. \tag{8.9.6}$$

Thus, asymptotically **B** is a homogeneous magnetic field of value 2 directed along the *z*-axis.

Both fields $F_{(\phi)\mu\nu}$ and $F_{(t)\mu\nu}$ have a monopole (Coulomb) component of an electric field. The corresponding electric charges are related to the mass *M* and the angular momentum of the black hole as follows

$$4\pi Q_{(t)} = \int F_{(t)}{}^{\mu\nu}\, d\sigma_{\mu\nu} = -2 \int \xi_{(t)}{}^{\mu;\nu}\, d\sigma_{\mu\nu} = -8\pi M,$$
$$4\pi Q_{(\phi)} = \int F_{(\phi)}{}^{\mu\nu}\, d\sigma_{\mu\nu} = -2 \int \xi_{(\phi)}{}^{\mu;\nu}\, d\sigma_{\mu\nu} = 16\pi aM. \tag{8.9.7}$$

Thus, one has $Q_{(t)} = -2M$ and $Q_{(\phi)} = 4aM$. Thus, if the rotating black hole with charge *Q* is in the homogeneous at infinity magnetic field *B*, the corresponding vector potential is

$$A^\mu = \frac{B}{2}\xi_{(\phi)}{}^\mu + (Ba - \frac{Q}{2M})\xi_{(t)}{}^\mu. \tag{8.9.8}$$

For this field there exists the electrostatic potential difference

$$\Delta\Phi = \frac{Q - 2aMB}{2M}, \tag{8.9.9}$$

between the event horizon and an infinitely distant point. In the presence of plasma, the particles with the proper sign of their charge fall into the black hole until this 'chemical potential' vanishes. As a result, the black hole will get charge $Q = 2aMB$, and the electric potential takes the form

$$A^\mu = \frac{1}{2} B\xi_{(\phi)}{}^\mu. \tag{8.9.10}$$

8.9.3 Black hole as a unipolar inductor

Consider a stationary 'observer' with angular velocity Ω^F

$$u^\mu \sim \xi_{(t)}{}^\mu + \Omega^F \xi_{(\phi)}{}^\mu. \tag{8.9.11}$$

The electric potential U^F (as measured by this observer) is

$$U^F = A_\mu \left(\xi_{(t)}{}^\mu + \Omega^F \xi_{(\phi)}{}^\mu\right) = \frac{1}{2} B\left(g_{t\phi} + \Omega^F g_{\phi\phi}\right). \tag{8.9.12}$$

Let us calculate the difference of electric potential ΔU between the pole ($\theta = 0$) and equator ($\theta = \pi/2$) of the horizon $r = r_+$ for the field Eq. (8.9.10)

$$\Delta U = U^F(r_+, \theta = 0) - U^F(r_+, \theta = \pi/2). \tag{8.9.13}$$

In the Kerr metric

$$g_{t\phi} = -\frac{2Mra\sin^2\theta}{\Sigma}, \qquad g_{\phi\phi} = \left[\left(r^2 + a^2\right)^2 - \Delta a^2 \sin^2\theta\right]\frac{\sin^2\theta}{\Sigma}. \tag{8.9.14}$$

Since $\Delta = 0$ at the horizon one has

$$g_{t\phi}(r_+, 0) = 0, \qquad\qquad g_{\phi\phi}(r_+, 0) = 0,$$

$$g_{t\phi}(r_+, \pi/2) = -a, \qquad g_{\phi\phi}(r_+, \pi/2) = r_+^2 + a^2. \tag{8.9.15}$$

Hence,

$$\Delta U^F = \frac{1}{2} B \left(r_+^2 + a^2\right)\left(\Omega^H - \Omega^F\right). \tag{8.9.16}$$

Thus, if $\Omega^F \neq \Omega^H$, there is a non-vanishing difference of the electric potentials between the pole and the horizon. This result is in agreement with our expectations, based on the analogy with the *unipolar inductor*.

For estimations, let us assume that

$$\Omega^F \sim \frac{1}{2}\Omega^H = \frac{1}{2}\frac{a}{r_+^2 + a^2}. \tag{8.9.17}$$

Then,

$$\Delta U^F \simeq \frac{1}{4} Ba, \qquad P = \dot{E} \sim B^2 a^2,$$

$$P \approx 10^{39} \frac{\text{erg}}{\text{s}} \left(\frac{M}{10^6 \, M_\odot}\right)^2 \left(\frac{a}{a_{\text{max}}}\right)^2 \left(\frac{B}{10^4 \, \text{G}}\right)^2. \tag{8.9.18}$$

This mechanism is used to explain the energetics of the powerful jets emission by black holes.

The detailed study of processes in the black hole vicinity in the presence of the magnetic field and plasma requires the complete theory of the *magnetohydrodynamics* in the strong gravitational field, which is quite complicated. There exist plenty of publications on this subject, in which different models are considered. Quite a lot of work was done in the numerical simulations. However, at the moment we are still far from the complete understanding of these processes. (For further discussion and references see, e.g., the book (Punsly 2008).)

9

Classical and Quantum Fields near Black Holes

9.1 Introduction

In the previous chapters we discussed null rays or 'photon trajectories' in the vicinity of black holes. In this description we used the *geometrical-optics approximation*, which is valid when the wavelength of the 'photon' λ is much smaller than the scale at which the characteristics of the gravitational field change. For a black hole of mass M the curvature of the spacetime is $\mathcal{R} \sim M/r^3$. Near the black hole $\mathcal{R} \sim M^{-2}$. Thus, one can expect that the geometrical optics works well everywhere outside the black hole when the wavelength is much smaller than its gravitational radius, $\lambda/M \ll 1$. When this condition is broken, wave aspects of the light propagation, e.g. its diffraction, become important. In the next chapter we shall discuss gravitational-wave radiation by black holes. This is an important example, where it naturally happens. For example, a newly formed black hole is usually non-stationary, and it relaxes to the stationary state by emitting the gravitational waves. Since the only scale for such a problem is the gravitational radius, one can expect that the wavelength of the emitted radiation is also of the order of M. A similar situation occurs in a completely different case of the quantum radiation of a black hole (see Section 9.6). These and other problems require study of the solutions of the wave equations in the black hole spacetime. This is the main subject of the present chapter.

Two types of fields are of special interest, electromagnetic and gravitational. The electromagnetic field is described by the *Maxwell equations*. The gravitational waves are described as perturbations over a chosen gravitational background. Both of these fields are massless and have spin, which is equal to 1 and 2 for the electromagnetic and gravitational waves, respectively. In most cases the backreaction of the fields on the background geometry can be neglected. Thus, one can consider the fields as test ones. In this approximation the fields obey linear partial differential equations with coefficients depending on the background geometry.

The analysis of the propagation of massless fields with a spin in the rotating black hole background is greatly simplified due to the wonderful properties of the Kerr metric, connected with its hidden symmetries. Namely, it was shown that the massless field equations in the Kerr spacetime can be decoupled, that is reduced to one *master equation*. This second-order equation can be solved by the separation of variables. We briefly discuss this property in Section 9.4. (For detailed discussion and references see e.g. Chandrasekhar 1983; Frolov and Novikov 1998).

In Section 8.8 we mentioned that there exists an interaction of the spin of a particle with a rotating black hole. This effect is of a general nature. Gravitational spin–spin and spin–orbit interactions are common features for the field propagation in the black hole background. For some problems this interaction may be important. For example, the Hawking radiation of the black hole (Section 9.7) depends essentially on the spin of the field. For other problems one can neglect spin effects. In this case, the problem of the field propagation in the black hole geometry is greatly simplified. In this approximation one can use a model of a *scalar massless field*. It will allow one to clarify many interesting features of the problem. We shall use this approximation. We also focus mainly on the case of a spherically symmetric black hole.

9.2 Static Field in the Schwarzschild Spacetime

9.2.1 Scalar field of a static source

The most important for astrophysical applications is a case when a source of the field in moving in the background of the black hole spacetime. For example, when a black hole of relatively small mass is moving in the field of a large-mass black hole. In this case, one can approximate a small black hole by a point-like particle and study the gravitational waves generated by the motion of this particle in the gravitational field of the large-mass black hole. We consider similar problems later. For the moment we focus on a simpler problem. Namely, we study a field created by a static source. In our model we assume that a source is kept at rest by means of some external force.

The action of a scalar massless field φ in a curved spacetime is

$$S[\varphi] = -\frac{1}{2} \int d^4x \sqrt{|g|} \, g^{\mu\nu} \, \partial_\mu \varphi \partial_\nu \varphi + S_j \,, \qquad S_j = \int d^4x \sqrt{|g|} \, j\varphi. \qquad (9.2.1)$$

The second term in the action describes an interaction of a source $j(x)$ with the scalar field. The field equation is

$$\Box\varphi \equiv \frac{1}{\sqrt{|g|}} \partial_\mu \left(\sqrt{|g|} \, g^{\mu\nu} \partial_\nu \varphi \right) = -j. \qquad (9.2.2)$$

In the Schwarzschild spacetime it is possible to obtain exact static asymmetric solutions of this equation. We assume that both the field φ and the source j do not depend on time t and the angle ϕ. Let us make the coordinate transformation from the standard coordinates (r, θ) to new coordinates

$$\rho = \sqrt{r(r - r_S)} \sin\theta \,, \qquad z = (r - r_S/2) \cos\theta. \qquad (9.2.3)$$

The Schwarzschild metric in these coordinates takes the form

$$ds^2 = -e^{2U} dt^2 + e^{-2U} \left[e^{2V} (d\rho^2 + dz^2) + \rho^2 d\phi^2 \right]. \qquad (9.2.4)$$

The concrete form of the functions U and V is not important. In the new coordinates $\sqrt{|g|} = e^{-2(U-V)}\rho$ and the field equation Eq. (9.2.2) for a static source has the form

$$\left[\frac{1}{\rho} \partial_\rho(\rho \, \partial_\rho) + \partial_z^2 \right] \varphi = -J \,, \qquad J = e^{-2(U-V)} j. \qquad (9.2.5)$$

In order to solve this equation, let us consider a 3D flat spacetime with the metric

$$dl^2 = \sum_{i=1}^{3} dX_i^2 = dz^2 + d\rho^2 + \rho^2 d\psi^2. \tag{9.2.6}$$

The 3D flat Laplacian is

$$^{(3)}\Delta = \frac{1}{\rho} \partial_\rho (\rho\, \partial_\rho) + \partial_z^2 + \frac{1}{\rho^2} \partial_\psi^2. \tag{9.2.7}$$

The 3D flat Green function $^{(3)}G$, that is a decreasing-at-infinity solution of the equation

$$^{(3)}\Delta \,^{(3)}G = -\frac{\delta(\rho - \rho')}{\rho} \delta(z - z')\delta(\psi - \psi'), \tag{9.2.8}$$

is

$$^{(3)}G = \frac{1}{4\pi\,|\mathbf{X} - \mathbf{X}'|} = \frac{1}{4\pi\sqrt{(z - z')^2 + \rho^2 + \rho'^2 - 2\rho\rho'\cos(\psi - \psi')}}. \tag{9.2.9}$$

Consider now the following equation in the flat spacetime

$$^{(3)}\Delta\, \Phi = -\hat{J}. \tag{9.2.10}$$

Its solution is

$$\Phi(\rho, z, \psi) = \int \rho' d\rho' dz' d\psi' \,^{(3)}G(\rho, z, \psi; \rho', z', \psi')\, \hat{J}(\rho', z', \psi'). \tag{9.2.11}$$

If $\hat{J}$ does not depend on ψ the equation takes the form

$$\left[\partial_\rho(\rho\, \partial_\rho) + \partial_z^2 \right] \Phi = -\hat{J}. \tag{9.2.12}$$

Let us note that this equation is identical to Eq. (9.2.5). Denote

$$^{(3)}\tilde{G}(\rho, z; \rho', z') = \int_0^{2\pi} d\psi' \,^{(3)}G(\rho, z, \psi; \rho', z', \psi'). \tag{9.2.13}$$

Taking this integral one has

$$^{(3)}\tilde{G}(\rho, z; \rho', z') = \frac{1}{\pi R} \mathbf{K}(k),$$

$$R^2 = (\rho - \rho')^2 + (z - z')^2, \qquad k = 2\sqrt{\frac{\rho\rho'}{R}}. \tag{9.2.14}$$

Here, $\mathbf{K}(k)$ is the complete elliptic integral of the first kind. Thus, a solution of the initial problem Eq. (9.2.5) is

$$\varphi(\rho, z) = \int_{-\infty}^{\infty} dz' \int_0^{\infty} \rho' d\rho' \,^{(3)}\tilde{G}(\rho, z; \rho', z')\, J(\rho', z'). \tag{9.2.15}$$

If the source J is located along the z-axis ($\theta' = 0$) the solution is simplified. For $\rho' = 0$, where $k = 0$, one has $\mathbf{K}(0) = \pi/2$. The Green function Eq. (9.2.14) takes the form

$$^{(3)}\tilde{G}(\rho, z; 0, z') = \frac{1}{2R}, \tag{9.2.16}$$

where

$$\begin{aligned}
R^2 &= \rho^2 + (z - z')^2 \\
&= (r - M)^2 + (r' - M)^2 - 2(r - M)(r' - M)\cos\theta - M^2 \sin^2\theta.
\end{aligned} \tag{9.2.17}$$

Here, $M = r_S/2$. The field of a point-like scalar charge located at the arbitrary point (r', θ', ϕ') in the Schwarzschild spacetime is given by Eqs. (9.2.16) and (9.2.17), where θ must be replaced by λ

$$\cos\lambda = \cos\theta\cos\theta' + \sin\theta\sin\theta'\cos(\phi - \phi'). \tag{9.2.18}$$

9.2.2   Electric field of a point-like charge

It is interesting that a similar trick can be used to find an explicit solution for the electric field of a static point-like charge. We shall not repeat the steps similar to these presented in the previous subsection, and just give the final expression for the vector-potential A_μ

$$A_\mu(r, \theta, \phi) = -q\delta_\mu^0 \frac{1}{rr'}\left(M + \frac{\Pi}{R}\right). \tag{9.2.19}$$

As earlier, we assume that the charge q is located at a point (r', θ', ϕ'). The function R is given by Eqs. (9.2.17) and (9.2.18), and

$$\Pi = (r - M)(r' - M) - M^2 \cos\lambda. \tag{9.2.20}$$

Problem 9.1:   *Check by a direct substitution that Eq. (9.2.19) is a solution of the Maxwell equations in the Schwarzschild spacetime with a point-like electric charge q at rest.*

9.2.3   Flat spacetime limit

Consider the Schwarzschild metric

$$ds^2 = -g\,dt^2 + g^{-1}\,dr^2 + r^2\left(d\theta^2 + \sin^2\theta\,d\phi^2\right), \quad g = 1 - r_S/r. \tag{9.2.21}$$

In order to study the properties of the field near the horizon let us make the following coordinate transformation

$$r = r_S + \frac{\rho^2}{4r_S}, \quad t = 2r_S a\tau,$$

$$\tan\left(\frac{\theta}{2}\right) = \frac{1}{2r_S}\sqrt{x^2 + y^2}, \quad \tan\phi = \frac{y}{x}. \tag{9.2.22}$$

Here, a is an arbitrary parameter with the dimensionality of $[\text{Length}]^{-1}$. In the above relations the dimensionality of t and τ is the same. In these coordinates the metric takes the form

$$ds^2 = -\frac{\rho^2 a^2}{F} d\tau^2 + F d\rho^2 + F^2 \frac{dx^2 + dy^2}{1 + (x^2 + y^2)/(2r_S)^2},$$ (9.2.23)

$$F = 1 + (\rho/2r_S)^2.$$

If we fix the value of ρ, x, and y and take the limit $r_S \to \infty$ we obtain

$$ds^2 \sim -\rho^2 a^2 d\tau^2 + d\rho^2 + dx^2 + dy^2.$$ (9.2.24)

This is the *Rindler metric* (see Section 2.4). This means that when one considers a finite-size domain in the vicinity of the black hole horizon and takes the limit $r_S \to \infty$ one can use the flat-spacetime approximation. The coordinate ρ has the meaning of the distance to the horizon. An observer located at the point with fixed value of (ρ, x, y) is moving with constant acceleration ρ^{-1}. At the normalization point, where $g_{\tau\tau} = -1$, this acceleration is a. The fact that the corresponding limiting metric is flat is not surprising. In fact, the curvature of the spacetime calculated at the horizon is of the order of $1/r_S^2$. In the limit $r_S \to \infty$ it vanishes.

Let us now find out how the solution Eq. (9.2.19) for the electromagnetic field looks in this limit. Note that

$$\mathbf{A} = A_\mu dx^\mu = A_t dt = A_\tau d\tau.$$ (9.2.25)

Thus, $A_\tau = 2r_S a A_t$. The contribution of the first term in the brackets in Eq. (9.2.19) to A_τ is of the form $r_S^2/(rr')$. Near the horizon and in the limit $r_S \to \infty$ it gives a constant that can be neglected. One also has

$$\Pi \sim \frac{S}{8}, \qquad R^2 \sim \frac{1}{16 r_S^2} \left(S^2 - 4\rho^2 \rho'^2 \right),$$ (9.2.26)

$$S = \rho^2 + \rho'^2 + x^2 + y^2.$$

Thus, if the source is at $\rho' = a^{-1}$ then one has

$$A_\tau \sim -qa \frac{S}{\sqrt{S^2 - 4\rho^2 a^{-2}}}, \qquad S = \rho^2 + x^2 + y^2 + a^{-2}.$$ (9.2.27)

This expression coincides with the expression Eq. (2.4.15) for the electric potential in the Rindler spacetime.

9.3 Dimensional Reduction

9.3.1 Spherical reduction

When a source of the field is time dependent the above-described trick does not work. A natural way to study such problems is to exploit the symmetry of the gravitational field. Let us demonstrate that in the spherically symmetric spacetime the field equations can be reduced to a set of 2D equations. Let us write a spherically symmetric metric in the form

$$ds^2 = d\gamma^2 + r^2 \, d\Omega^2,$$

$$d\gamma^2 = \gamma_{AB} \, dx^A \, dx^B, \quad d\Omega^2 = d\theta^2 + \sin^2\theta \, d\phi^2,$$

(9.3.1)

where $A, B = 0, 1$ and $d\Omega^2$ is the metric on a unit 2D sphere. Let us write the field φ as

$$\varphi = \frac{\psi}{r}.$$

(9.3.2)

Then, the field equation

$$\Box\varphi = -j$$

(9.3.3)

gives

$$^{(2)}\Box\psi - \frac{^{(2)}\Box r}{r}\psi + \frac{1}{r^2}\triangle\psi = -rj.$$

(9.3.4)

Here, $^{(2)}\Box$ is a 2D box in the 2D metric $d\gamma^2$ and $\triangle$ is a 2D Laplace operator on a unit sphere $d\Omega^2$

$$^{(2)}\Box = \frac{1}{\sqrt{|\gamma|}}\partial_A\left(\sqrt{|\gamma|}\,\gamma^{AB}\partial_B\right), \quad \triangle = \frac{1}{\sin\theta}\partial_\theta(\sin\theta\,\partial_\theta) + \frac{1}{\sin^2\theta}\partial_\phi^2.$$

(9.3.5)

We express a complete set of eigenfunctions $\hat{Y}_{lm}(\theta,\phi)$ of $\triangle$ in terms of the associated Legendre polynomials $P_l^m(x)$

$$\hat{Y}_{lm}(\theta,\phi) = \frac{1}{\sqrt{\pi}}P_l^m(\cos\theta)\begin{cases} \frac{1}{\sqrt{2}}, & m = 0, \\ \cos(m\phi), & 0 < m \le l, \\ \sin(m\phi), & -l < m < 0. \end{cases}$$

(9.3.6)

They obey an equation

$$\triangle\hat{Y}_{lm} = -l(l+1)\,\hat{Y}_{lm}.$$

(9.3.7)

These eigenfunctions are just the real and imaginary parts of the standard spherical harmonics $Y_{lm}(\theta,\phi)$. They obey the following normalization conditions

$$\int_0^{2\pi} d\phi \int_0^\pi d\theta \sin\theta \, \hat{Y}_{l'm'}(\theta,\phi)\,\hat{Y}_{lm}(\theta,\phi) = \delta_{ll'}\,\delta_{mm'}.$$

(9.3.8)

In a spherically symmetric spacetime one can decompose both the field φ and the source j into modes

$$\varphi = \sum_{l=0}^\infty \sum_{m=-l}^{l} \frac{\varphi_{lm}(x^A)}{r}\hat{Y}_{lm}(\theta,\phi),$$

$$j = \sum_{l=0}^\infty \sum_{m=-l}^{l} \frac{j_{lm}(x^A)}{r}\hat{Y}_{lm}(\theta,\phi).$$

(9.3.9)

The field equation Eq. (9.3.4) results in the following mode equations

$$^{(2)}\Box \varphi_{lm} - W_l \varphi_{lm} = -j_{lm}, \qquad W_l = \frac{l(l+1)}{r^2} + \frac{^{(2)}\Box r}{r}. \tag{9.3.10}$$

Let us substitute the decomposition Eq. (9.3.9) into the action Eq. (9.2.1). Note that

$$\sqrt{|g|} = r^2 \sqrt{|\gamma|} \sqrt{\Omega}, \qquad g^{\mu\nu} \varphi_{,\mu} \varphi_{,\nu} = \gamma^{AB} \varphi_{,A} \varphi_{,B} + \frac{1}{r^2} \Omega^{ij} \varphi_{,i} \varphi_{,j},$$

$$\int d\Omega \, \Omega^{ij} \, \hat{Y}_{lm,i} \, \hat{Y}_{l'm',j} = l(l+1) \, \delta_{l,l'} \delta_{m,m'}, \tag{9.3.11}$$

where $d\Omega = \sin\theta \, d\theta \, d\phi$. The latter relation is obtained by integrating by parts. Using these relations after some simplifications one obtains

$$S = \sum_{l=0}^{\infty} \sum_{m=-l}^{l} S_{lm}, \tag{9.3.12}$$

$$S_{lm} = -\frac{1}{2} \int d^2x \sqrt{|\gamma|} \left[\gamma^{AB} \partial_A \varphi_{lm} \partial_B \varphi_{lm} + W_l \varphi_{lm}^2 - 2 j_{lm} \varphi_{lm} \right].$$

Thus, in a spherically symmetric case the original 4D theory is effectively reduced to an infinite set of 2D theories enumerated by the indices (l, m). The expression Eq. (9.3.12) gives the action of the corresponding reduced 2D theory with a field variable $\varphi_{lm}(x)$. The adopted procedure is known as *dimensional reduction*. This is a general procedure that is useful when a theory is considered in a spacetime that has some symmetries. Let us note that there exists a degeneracy. Namely, the function W_l does not depend on m. Hence, in the absence of the source the 2D actions for a given l with different m are identical.

In the case when the background geometry is described by the Schwarzschild metric,

$$d\gamma^2 = -g dt^2 + dr^2/g, \qquad g = 1 - r_S/r, \tag{9.3.13}$$

the 2D action Eq. (9.3.12) allows further simplification. Denote as earlier by

$$r_* = \int \frac{dr}{g} = r + r_S \ln\left(\frac{r - r_S}{r_S}\right) \tag{9.3.14}$$

a *tortoise coordinate* in the Schwarzschild geometry. The expression of the coordinate r as a function of r_* is $r = r_S \left\{ 1 + W\left(\exp\left[\frac{r_*}{r_S} - 1 \right] \right) \right\}$. It is given in terms of a so-called *Lambert W-function* that is defined as the solution of the equation $W(z) \exp[W(z)] = z$. Using this notation one can rewrite Eq. (9.3.12) in the form

$$S_{lm} = \int dt \, L_{lm},$$

$$L_{lm} = \frac{1}{2} \int dr_* \left[(\partial_t \varphi_l)^2 - (\partial_{r_*} \varphi_l)^2 - U_l \varphi_l^2 - 2j_{lm}\varphi_{lm} \right], \tag{9.3.15}$$

$$U_l(r_*) = g(r)W_l = \left(1 - \frac{r_S}{r}\right)\left(\frac{l(l+1)}{r^2} + \frac{r_S}{r^3}\right), \qquad r = r(r_*).$$

We used here the relation $^{(2)}\Box r = r_S/r^2$. It is easy to see that in the absence of a source the action Eq. (9.3.15) coincides with the action Eq. (F.4.1) for the $(1+1)$-dimensional theory considered in Appendix F, provided the *tortoise coordinate* r_* is identified with the flat space coordinate x. In the absence of a source the field equation for φ_l is

$$-\partial_t^2 \varphi_l + \partial_{r_*}^2 \varphi_l - U_l \varphi_l = 0. \tag{9.3.16}$$

The form of the effective potential U_l in this equation (see Eq. (9.3.15)) looks similar to the form of the effective potential U_ℓ for null-ray motion in the Schwarzschild geometry (see Section 7.1.4). This is not a coincidence. The light-ray equation can be obtained from the field equation Eq. (9.3.16) in the geometrical-optics approximation. The condition of the applicability of this approximation is $l \gg 1$ and the wave frequency ω must be high. Let us write $\varphi_l \sim e^{-i\omega(t-S)}$. By substituting this expression in Eq. (9.3.16) and keeping the highest order in ω terms one obtains

$$\left(\frac{dS}{dr_*}\right)^2 = 1 - \frac{U_l}{\omega^2}. \tag{9.3.17}$$

This is an analog of the Hamilton–Jacobi equation for the radial motion. Substituting $S_{,r_*} = p_{r_*} = dr_*/dt$ one obtains the radial equation in the form

$$g^{-2}(dr/dt)^2 = 1 - \frac{U_l}{\omega^2}. \tag{9.3.18}$$

The radial equation for null rays follows from Eq. (7.1.7) with mass $m = 0$ and $E = \omega$

$$g^{-2}(dr/dt)^2 = 1 - g\frac{L^2}{\omega^2 r^2}. \tag{9.3.19}$$

By comparing these relations we see that they are identical if $L^2 = l(l+1)$.
This is an expected result. For the motion in the spherically symmetric potential the geometrical-optics approximation for the angular motion requires that $L = l + 1/2$, and the momentum is large. To summarize, one correctly reproduces the null-ray equations from the field equation by means of the following transformation

$$l + \frac{1}{2} \to \omega r_S \, \ell. \tag{9.3.20}$$

If one substitutes this relation into the expression Eq. (9.3.15) for U_l and omits terms that do not depend on ℓ in the second brackets, one obtains the effective potential for the null rays.

9.3.2 Green function

In order to solve the inhomogeneous equation Eq. (9.2.2) it is sufficient to know its Green function G^{ret}

$$\Box G^{\text{ret}}(x, x') = -\delta^4(x, x').$$
(9.3.21)

Here, $\delta^4(x, x')$ is a 4-dimensional invariant delta function determined by a relation

$$\int d^4x \sqrt{|g|}\, \delta^4(x, x') f(x) = f(x'),$$
(9.3.22)

for any scalar function f. The *invariant delta function* is related to a *coordinate delta function* $\delta^4(x - x')$ as follows

$$\delta^4(x, x') = \frac{1}{\sqrt{|g|}}\, \delta^4(x - x').$$
(9.3.23)

Since we are looking only for the solutions that are generated by j and propagated to the future we impose the following boundary condition

$$G^{\text{ret}}(x, x') = 0, \qquad \text{for } t < t'.$$
(9.3.24)

It is a so-called *retarded Green function*.

In a spherically symmetric spacetime one can use the decompositions Eq. (9.3.9). Substituting these decompositions in Eq. (9.3.3) one obtains

$$\left[{}^{(2)}\Box - W_l \right] \varphi_{lm} = -j_{lm}.$$
(9.3.25)

One can solve this equations using 2D retarded Green functions D obeying the equation

$$\left[{}^{(2)}\Box - W_l \right] D_l(x^A, x^{A'}) = -\frac{\delta^2(x^A - x^{A'})}{\sqrt{|\gamma|}},$$
(9.3.26)

and a boundary condition

$$D(x^A, x^{A'}) = 0, \qquad \text{for } t < t'.$$
(9.3.27)

In the Schwarzschild spacetime Eq. (9.3.26) takes the form

$$[-\partial_t^2 + \partial_{r_*}^2 - U_l]\, D_l(t, r_*; t', r'_*) = -\delta(t - t')\delta(r_* - r'_*).$$
(9.3.28)

Since the background metric is static D_l depends only on the time difference

$$D_l(t, r_*; t', r'_*) = D_l(t - t', r_*, r'_*).$$
(9.3.29)

The boundary condition for this retarded Green function is

$$D_l(t, r_*, r'_*) = 0, \qquad \text{for } t < 0.$$
(9.3.30)

Using this Green function one can write a solution of Eq. (9.3.25) in the form

$$\varphi_{lm}(t, r_*) = \int_{-\infty}^{t} dt' \int_{-\infty}^{\infty} dr'_*\, D_l(t - t', r_*, r'_*) j_{lm}(t', r'_*).$$
(9.3.31)

The 4D retarded Green function $G^{\text{ret}}(x, x')$ is

$$G^{\text{ret}}(x, x') = \sum_{l=0}^{\infty} \sum_{m=-l}^{l} \frac{D_l(t, r_*, r'_*)}{rr'} \hat{Y}_{lm}(\theta, \phi) \hat{Y}_{lm}(\theta', \phi'). \tag{9.3.32}$$

Because G^{ret} is real, the 2D Green function D_l is also real. Since D_l does not depend on m, one can make a summation in Eq. (9.3.32) over m and get

$$G^{\text{ret}}(x, x') = \sum_{l=0}^{\infty} (2l + 1) \frac{D_l(t, r_*, r'_*)}{rr'} P_l(\cos \lambda). \tag{9.3.33}$$

Here, P_l is the Legendre polynomial and

$$\cos \lambda = \cos \theta \cos \theta' + \sin \theta \sin \theta' \cos(\phi - \phi'). \tag{9.3.34}$$

The parameter λ is a 2D interval between the points (θ, ϕ) and (θ', ϕ') of the unit sphere.

9.4 Quasinormal Modes

9.4.1 Modes in the frequency domain

Let us make the Fourier transform

$$\varphi_{lm}(t, r_*) = \frac{1}{2\pi} \int_{-\infty}^{\infty} d\omega \, e^{-i\omega t} \tilde{\varphi}_{lm}(\omega, r_*),$$

$$j_{lm}(t, r_*) = \frac{1}{2\pi} \int_{-\infty}^{\infty} d\omega \, e^{-i\omega t} \tilde{j}_{lm}(\omega, r_*). \tag{9.4.1}$$

Substituting these representations into Eq. (9.3.31) we obtain

$$\tilde{\varphi}_{lm}(\omega, r_*) = \int_{-\infty}^{\infty} dr'_* \tilde{D}_l(\omega, r_*, r'_*) \tilde{j}_{lm}(\omega, r'_*). \tag{9.4.2}$$

Here,

$$\tilde{D}_l(\omega, r_*, r'_*) = \int_{0^-}^{+\infty} dt \, e^{i\omega t} D_l(t, r_*, r'_*). \tag{9.4.3}$$

This transform is well defined for $\Im(\omega) \geq 0$, and the inverse transform is given by

$$D_l(t, r_*, r'_*) = \frac{1}{2\pi} \int_{-\infty+ic}^{+\infty+ic} d\omega e^{-i\omega t} \tilde{D}_l(\omega, r_*, r'_*), \tag{9.4.4}$$

where c is a positive number. The change of the integration variable $\omega = is$ in Eq. (9.4.3) and Eq. (9.4.4) shows that these relations are simply the direct and inverse *Laplace transforms*. Using ω instead of s appears to be more convenient for our purposes.

Since $D_l(t, r_*, r'_*)$ is a real function, its transform has the following property

$$\tilde{D}_l(-\omega, r_*, r'_*) = \tilde{D}_l^*(\omega, r_*, r'_*). \tag{9.4.5}$$

Substituting Eq. (9.4.4) into Eq. (9.3.28) gives

$$\left[\frac{d^2}{dr_*^2} + \omega^2 - U\right]\tilde{D}(\omega, r_*, r_*') = -\delta(r_* - r_*').$$

(9.4.6)

In order to make expressions more compact we omit the subscript l in U_l and $\tilde{D}_l$.

9.4.2 Green function in the frequency domain

The ordinary differential operator on the left-hand side of Eq. (9.4.6) is well known from quantum mechanics. In fact, $\tilde{D}$ coincides with the Green function for a standard 1-dimensional Schrödinger equation with the energy $E = \omega^2$. This quantum-mechanical problem on the infinite line with a potential U is well known and is discussed in any textbook on quantum mechanics. In our case, a potential U_l is positive and its shape depends on l (see Figure 9.1).

Some of the features of this problem are discussed in Appendix F. Let us note that now we are using frequency ω instead of the wavelength k. It does not create any problems since for the massless field $\omega = k$. It is convenient to use the solutions $\tilde{u}_{in}$ and $\tilde{u}_{up}$ that differs from basic solutions u_{in} and u_{up} (see Appendix F) by a normalization factor

$$\tilde{u}_{in,\omega} = \frac{\sqrt{2\pi}}{T_\omega} u_{in,\omega}, \qquad \tilde{u}_{up,\omega} = \frac{\sqrt{2\pi}}{t_\omega} u_{up,\omega}.$$

(9.4.7)

These solutions of the equation

$$\left[\frac{d^2}{dr_*^2} + \omega^2 - U\right]\tilde{u} = 0,$$

(9.4.8)

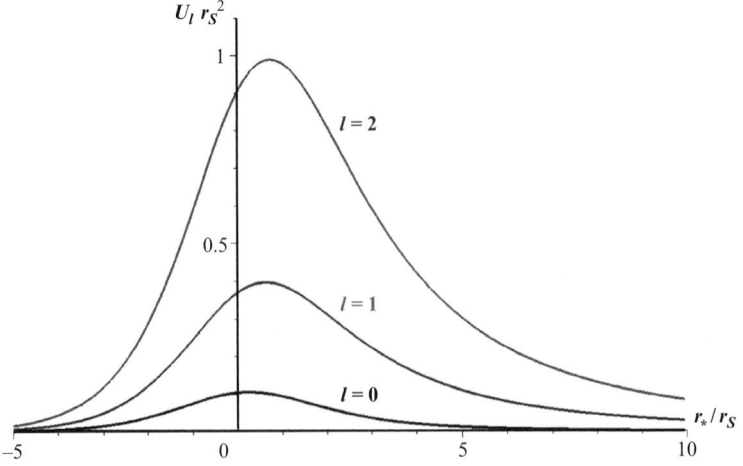

Fig. 9.1 Potential $U_l(r_*)$.

are singled out by the following asymptotics

$$
\tilde{u}_{\text{in},\omega}(r_*) = \begin{cases} e^{-i\omega r_*} \,, & r_* \to -\infty \,, \\[2mm] A_{\text{in},\omega}e^{-i\omega r_*} + A_{\text{out},\omega}e^{i\omega r_*} \,, & r_* \to +\infty \,; \end{cases} \tag{9.4.9}
$$

$$
\tilde{u}_{\text{up},\omega}(r_*) = \begin{cases} B_{\text{out},\omega}e^{i\omega r_*} + B_{\text{in},\omega}e^{-i\omega r_*} \,, & r_* \to -\infty \,, \\[2mm] e^{i\omega r_*} \,, & r_* \to +\infty \,. \end{cases} \tag{9.4.10}
$$

The scattering coefficients $A_{\text{in,out}}$ and $B_{\text{in,out}}$ are known in quantum mechanics as the *Jost functions*. There exist simple relations between the Jost functions and the transmission and reflection amplitudes (see Appendix F)

$$
T_\omega = \frac{1}{A_{\text{in},\omega}} \,, \qquad R_\omega = \frac{A_{\text{out},\omega}}{A_{\text{in},\omega}} \,, \qquad t_\omega = \frac{1}{B_{\text{out},\omega}} \,, \qquad r_\omega = \frac{B_{\text{in},\omega}}{B_{\text{out},\omega}} \,. \tag{9.4.11}
$$

The *Wronskian*

$$
W[v_1, v_2] = v_1 (dv_2/dr_*) - v_2 (dv_1/dr_*), \tag{9.4.12}
$$

calculated for any two solutions v_1 and v_2 of Eq. (9.4.8), is constant (independent of r_*). Calculating the values of the Wronskian for $\tilde{u}_{\text{in},\omega}$, $\tilde{u}_{\text{out},\omega}$ and their complex conjugate at $r_* = \pm\infty$ allows one to obtain the following relations

$$
|A_{\text{in},\omega}|^2 = 1 + |A_{\text{out},\omega}|^2 \,, \qquad |B_{\text{out},\omega}|^2 = 1 + |B_{\text{in},\omega}|^2 \,,
$$

$$
B_{\text{out},\omega} = A_{\text{in},\omega} \,, \qquad B_{\text{in},\omega}^* = -A_{\text{out},\omega} \,. \tag{9.4.13}
$$

These relations are equivalent to the relations Eq. (F.4.12) and Eq. (F.4.15) for the reflection and transmission amplitudes. One also has

$$
W[\tilde{u}_{\text{in},\omega}, \tilde{u}_{\text{up},\omega}] = 2i\omega A_{\text{in},\omega}. \tag{9.4.14}
$$

The Green function $\tilde{D}(\omega, r_*, r_*')$ can be written as follows

$$
\tilde{D}(\omega, r_*, r_*') = \frac{1}{2i\omega A_{\text{in},\omega}} \tilde{u}_{\text{in},\omega}(r_*^<) \tilde{u}_{\text{up},\omega}(r_*^>). \tag{9.4.15}
$$

The notation $r_*^<$ means r_* if $r_* < r_*'$ and r_*' in the opposite case. The quantity $r_*^>$ has a similar meaning. This Green function is symmetric with respect to its variables r_* and r_*'. It describes the propagation of the modes $\sim e^{-i\omega v}$ to the horizon and modes $\sim e^{-i\omega u}$ to infinity. Here, $u = t - r_*$ and $v = t + r_*$ are the *retarded* and *advanced time*, respectively.

In order to have a real Green function $D(t, r_*, r_*')$ one needs to consider both positive and negative frequency ω and impose the condition Eq. (9.4.5). Using the expressions Eq. (9.4.9) and Eq. (9.4.10) for both positive and negative frequencies it is easy to check that if one changes $\omega \to -\omega$ and applies the complex conjugation to the function, the functions $\tilde{u}_{\text{in},\omega}$ and $\tilde{u}_{\text{out},\omega}$ transform into themselves

$$
\tilde{u}_{\text{in},-\omega}^* = \tilde{u}_{\text{in},\omega} \,, \qquad \tilde{u}_{\text{up},-\omega}^* = \tilde{u}_{\text{up},\omega} \,. \tag{9.4.16}
$$

This means that

$$A^*_{\text{in},-\omega} = A_{\text{in},\omega}, \quad A^*_{\text{out},-\omega} = A_{\text{out},\omega}, \quad B^*_{\text{in},-\omega} = B_{\text{in},\omega}, \quad B^*_{\text{out},-\omega} = B_{\text{out},\omega}. \quad (9.4.17)$$

These conditions guarantee that $\tilde{D}$, defined by Eq. (9.4.15), obeys the relation Eq. (9.4.5).

9.4.3   Field at infinity

We assume that the source of the field is localized in some region in space so that $\tilde{j}(\omega, r'_*) = 0$ outside that region. Outside the source one has

$$\tilde{\varphi}(\omega, r_*) = \frac{J(\omega)}{2i\omega A_{\text{in},\omega}} \tilde{u}_{\text{up},\omega}(r_*), \quad J(\omega) = \int_{-\infty}^{\infty} dr'_* \tilde{u}_{\text{up},\omega}(r'_*) \tilde{j}(\omega, r'_*). \quad (9.4.18)$$

At $r_* \to \infty$ we get

$$\tilde{\varphi}(\omega, r_*) \sim \frac{J(\omega)}{2i\omega A_{\text{in},\omega}} e^{i\omega r_*}, \quad (9.4.19)$$

and

$$\varphi(t, r_*) \sim \frac{1}{4i\pi} \int_{-\infty+ic}^{\infty+ic} d\omega e^{-i\omega u} \frac{J(\omega)}{\omega A_{\text{in},\omega}}. \quad (9.4.20)$$

We have omitted (l, m)-indices in the above relations. Equation (9.4.20) allows one to calculate the scalar field created at large distance by a source moving near the Schwarzschild black hole. A similar expression can be easily obtained for the scalar field crossing the horizon.

9.4.4   Ringing radiation

Equation (9.4.20) formally solves the problem of radiation. One needs only:

- to calculate accurately $J(\omega)$ for different l and m;
- to take the integrals over ω in Eq. (9.4.20);
- to substitute the obtained value φ_{lm} in Eq. (9.3.9);
- to perform summation over l and m.

This procedure is very similar to finding the scattering data and cross-sections in standard quantum mechanics. But we know from quantum mechanics that knowledge of the *resonances* is very important for understanding of the qualitative features of the scattering problem. Resonances are almost bounded states. If a falling particle has energy E close to the resonance energy E_R, a particle can be trapped for some period of time in a *metastable state*. This results in the sharp changes of the cross-section near the resonance energy. The resonance states are connected with the poles of the *Jost function* in the lower part of the complex plane of ω. To study this effect and to understand better other properties of the scattering it is very useful to consider the analytical properties of the scattering data in the complex ω-plane.

Such an analysis can be applied to the problem of our interest. The resonance modes of the black hole are known as *quasinormal modes*. In the WKB approximation the field propagation is related to null geodesics ('photons'). For a black hole of mass M there exists a critical value of the impact parameter $l_{\min}r_s \approx 3\sqrt{3}M$, for which photons are almost captured and spend

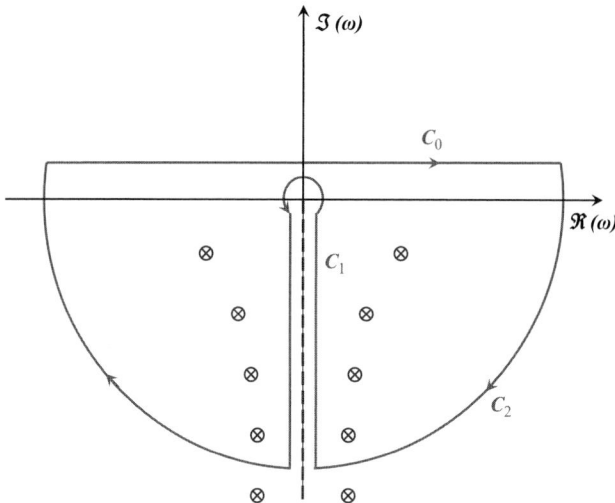

Fig. 9.2 Analytic properties of the Jost function $A_{in,\omega}$. Crosses show the position of the poles. The branch cut is located along the negative part of the $\Im(\omega)$ axis. C is the contour of integration in Eq. (9.4.20).

some time near a circular unstable orbit. This is an analog of the resonance effect in the wave description of the photon scattering.

For a study of the quasinormal modes let us consider the properties of the Jost function $A_{in,\omega}$ in the complex ω-plane. Its analytic properties depend on the form of the potential U_l. The Jost function is analytic in the upper half of the ω-plane and has an infinite set of discrete poles in the lower half-plane. They are shown by crosses in Figure 9.2. Because of the property Eq. (9.4.17) these poles are located symmetrically with respect to the $\Im(\omega)$ axis. These poles correspond to the *quasinormal modes*. Besides poles, the Jost function has a branch cut along the negative part of the $\Im(\omega)$ axis. This cut makes the Jost function single-valued.

To single out the contribution of the quasinormal modes to the asymptotic field at infinity let us consider a new contour of integration in Eq. (9.4.20), which consists of three parts C_0, C_1, and C_2 shown in Figure 9.2. One has

$$\int_{C_0} + \int_{C_1} + \int_{C_2} = \frac{1}{2\pi i} \sum_n \text{Res}_n. \tag{9.4.21}$$

This schematic formula means that the original integral over C_0 in Eq. (9.4.20) can be reduced to the integrals over C_1 and C_2 plus the contribution of the complex poles of the Jost function enumerated by a number n. The latter contribution can be calculated using the *Cauchy formula*. Namely, for any holomorphic function $f(z)$ one has

$$f(a) = \frac{1}{2\pi i} \oint \frac{f(z)}{z - a} dz. \tag{9.4.22}$$

The contour integral is taken over the counter-clockwise contour γ surrounding a point a (see Figure 9.3)

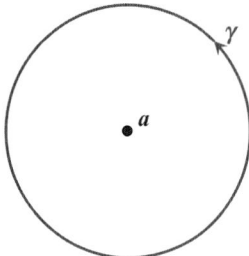

Fig. 9.3 Illustration of the Cauchy formula.

Let us study the contribution of a quasinormal mode n to the radiated wave. In the vicinity of the pole ω_n one has

$$A_{\text{in},\omega} \approx (\omega - \omega_n) \left. \frac{dA_{\text{in},\omega}}{d\omega} \right|_{\omega=\omega_n} \equiv (\omega - \omega_n)\alpha_n. \tag{9.4.23}$$

Thus, one obtains

$$\varphi_n(t, r_*) \sim \frac{1}{2} \frac{e^{-i\omega_n u}}{\omega_n \alpha_n} J(\omega_n). \tag{9.4.24}$$

Since all the poles are located in the lower half-plane of ω, these modes do not grow with the retarded time u. This property reflects *black hole stability*. The contribution of the quasinormal mode n has the form of oscillations in the retarded time u of the wave φ_n with a decreasing (damping) amplitude. Because of these features such radiation is called a *ringing radiation*.

In quantum mechanics the scattering resonances usually arise for energies close to the top of the potential barrier

$$\Re(\omega_0) \approx \sqrt{U_l^{\text{max}}} \approx \frac{1}{3\sqrt{3M}} (l + 1/2). \tag{9.4.25}$$

The imaginary part of the frequency, which determines the lifetime of the resonance, is connected with the curvature of the potential at its top

$$\Im(\omega_0) \approx -\frac{1}{2} \left| \frac{1}{2U_l} \frac{d^2 U_l}{dr_*^2} \right|^{1/2} \approx -\frac{\sqrt{3}}{18M}. \tag{9.4.26}$$

The *quality factor*

$$Q = \frac{1}{2} \left| \frac{\Re(\omega)}{\Im(\omega)} \right| \tag{9.4.27}$$

is of order of l. This factor shows how many oscillations a system makes before its significant damping.[1] Quasinormal modes with the smallest value of n have the smallest value of the imaginary part of the frequency. Thus, they have the lowest rate of decay and dominate.

---

[1] For atomic systems the quality factor $Q \sim 10^6$ is achievable.

It is interesting that the quasinormal modes with large n for the Schwarzschild black hole have a universal asymptotic behavior

$$8\pi M\omega_n \sim \ln(3) - i\pi(2n+1), \qquad n \to \infty. \qquad (9.4.28)$$

The discussion of this subject can be found in the review (Berti et al. 2009).

Equation (9.4.21) shows that besides the contribution of the quasinormal modes there are two additional terms in the asymptotic expression for $\varphi_n(t, r_*)$ given by integrals over contours C_1 a C_2. The first one (the integral along the branch cut) describes the power-law tail at late time. The second one (the integral along the high-frequency arc) gives the high-frequency contribution that might be related with the 'early-time radiation'.

9.4.5 Further remarks

The quasinormal modes are related to long-living resonances and reflect the presence of nearly circular critical orbits of zero-mass particles in the Schwarzschild geometry. One can expect that this phenomenon is of a general nature and should exist not only for the scalar massless field. This expectation is correct. The quasinormal modes were found for electromagnetic field (spin $s = 1$) and for gravitational perturbations (spin $s = 2$) as well. It was demonstrated by Price (1972a,b) that for any massless field of spin s and each radiative multipole l in the Schwarzschild geometry there exists a scalar function $\Phi_l^{(s)}$ that can be used to reconstruct all the components of the original field. This mode function obeys the same equation Eq. (9.4.8)

$$\left[\frac{d^2}{dr_*^2} + \omega^2 - U_l^{(s)} \right] \Phi_l^{(s)} = 0 \qquad (9.4.29)$$

as the scalar field with the only difference that the potential in this equation is

$$U_l^{(s)} = \left(1 - \frac{r_S}{r}\right) \left[\frac{l(l+1)}{r^2} + \frac{r_S(1 - s^2)}{r^3} \right]. \qquad (9.4.30)$$

The quasinormal modes exist in the rotating black holes as well. As we have already mentioned, the massless field equations allow a complete separation of variables in the Kerr spacetime. The quasinormal modes in the Kerr geometry are also connected with the poles of the Jost functions for the separated radial equation for the corresponding field. One can find more about the quasinormal modes and their properties in the reviews (Kokkotas and Schmidt 1999; Berti *et al.* 2009; Konoplya and Zhidenko 2011).

The quasinormal modes are deeply connected with the black hole properties. Gravitational radiation from black holes contains very special features determined by the quasinormal modes. In fact, these modes are 'dynamical vibrations' of the spacetime geometry and they contain direct information about spacetime regions close to the event horizon. In principle, the study of these 'vibrations' not only allows one to prove that these waves are generated by a black hole, but also would allow one to confirm the validity of the Einstein equations in the strong-field regime.

9.5 Massless Fields in the Kerr Spacetime

The analysis of the previous section can be generalized to the case of the fields with the spin propagating in the Kerr black hole background. Namely, the massless field equations in the Kerr metric can be decoupled, that is to be reduced to a single master equation. The latter equation allows the complete separation of variables. This result was obtained by Press and Teukolsky (1973). (See also the books (Chandrasekhar 1983; Frolov and Novikov 1998) and references therein.) For illustration, we discuss in this section the case of the *Klein–Gordon equation* in the Kerr spacetime that obeys the equation[2]

$$[\Box - \mu^2]\Phi = 0. \tag{9.5.1}$$

The solution of this equation can be written as follows

$$\Phi(t, r, \theta, \phi) = \int d\omega \sum_{l,m} R_{lm}(r, \omega)\, Z_{lm}^\omega(\theta, \phi)\, e^{-i\omega t}. \tag{9.5.2}$$

Here, (t, r, θ, ϕ) are the *Boyer–Lindquist coordinates*. Radial mode functions $R_{lm}(r, \omega)$ are solutions of the radial solution

$$\frac{d}{dr}\left(\Delta \frac{dR}{dr}\right) + \left[\frac{Q^2}{\Delta} - \mu^2 r^2 - \lambda\right] R = 0, \tag{9.5.3}$$

where

$$Q \equiv (r^2 + a^2)\,\omega - a\,m, \qquad \lambda \equiv E - 2a\,m\,\omega + a^2\omega^2. \tag{9.5.4}$$

The functions

$$Z_{lm}^\omega(\theta, \phi) = (2\pi)^{-1/2} S_{lm}^\omega(\theta)\, e^{im\phi} \tag{9.5.5}$$

are the *spheroidal harmonics*. The functions $S_{lm}^\omega(\theta)$ are solutions of the angular equation

$$\frac{1}{\sin\theta}\frac{d}{d\theta}\left(\sin\theta\frac{dS}{d\theta}\right) + \left[a^2\omega^2\cos^2\theta - \frac{m^2}{\sin^2\theta} + E - \mu^2 a^2\cos^2\theta\right] S = 0, \tag{9.5.6}$$

obeying the regularity condition at the singular point of this equation, $\theta = 0$ and $\theta = \pi$.

> *The functions $S_{lm}(\theta)$ are solutions of a Sturm–Liouville eigenvalue problem for the separation constant E. The imposed regularity conditions lead to a discrete set of possible eigenvalues of E. According to Sturm–Liouville theory, the eigenfunctions form a complete, orthogonal set on the interval $0 \le \theta \le \pi$ for each combination of a ω, and m. This infinite set of eigenfunctions is labelled by integer number l.*

The spheroidal harmonics were studied in detail by Flammer (1957). For the scalar field they are expressed in terms of spheroidal wave functions (Abramowitz and Stegun 1972). The spheroidal harmonics are normalized by the condition

---

[2]We include the mass term in this equation since it does not break the separability.

$$\int_0^{2\pi} d\phi \int_0^{\pi} d\theta \sin\theta \, \bar{Z}_{l'm'}^{\omega}(\theta,\phi) \, Z_{lm}^{\omega}(\theta,\phi) = \delta_{ll'} \, \delta_{mm'}, \qquad (9.5.7)$$

which means that

$$\int_0^{\pi} |S_{lm}(\theta)|^2 \sin\theta \, d\theta = 1. \qquad (9.5.8)$$

Moreover, because of the symmetries of Eq. (9.5.6), one has

$$Z_{lm}^{\omega}(\pi - \theta, \phi) \cong Z_{lm}^{\omega}(\theta, \phi), \qquad \bar{Z}_{lm}^{\omega}(\theta, \phi) \cong Z_{l-m}^{-\omega}(\theta, \phi). \qquad (9.5.9)$$

We use the symbol $\cong$ to indicate that a relation is valid up to some arbitrary multiplicative constant. For the chosen fixed normalization (9.5.7) this constant is just a phase factor.

A decomposition similar to Eq. (9.5.2) can be written for massless fields with the spin in the Kerr spacetime. This was demonstrated by Press and Teukolsky (1973). See also the books (Chandrasekhar 1983; Frolov and Novikov 1998). In the presence of spin the angular harmonics are described by *spin-weighted spheroidal harmonics*. For the fields with spin at the Schwarzschild spacetime (where $a = 0$), the required solutions are the spin-weighted spherical harmonics Y_{lm}. They can be expressed in terms of standard spherical harmonics (Goldberg *et al.* 1967). (For the properties of the spin-weighted spheroidal harmonics, see, e.g., review by Berti *et al.* 2006).)

9.6   Black Hole in a Thermal Bath

9.6.1   Thermal radiation in a static spacetime

Let us discuss now interaction of a black hole with a thermal radiation. We start with general remarks concerning properties of the thermal radiation in a static spacetime. Let ξ^{μ} be a time-like *Killing vector*. Denote the *redshift factor* α,

$$\alpha = \sqrt{-\xi^2}, \qquad \xi^2 = \xi^{\epsilon}\xi_{\epsilon}. \qquad (9.6.1)$$

We also denote

$$w_{\mu} = \nabla_{\mu} \ln\alpha. \qquad (9.6.2)$$

The Killing vector in a static spacetime obeys the following relation (see Eq. (3.7.18))

$$\xi_{\mu;\nu} = \xi_{\mu}w_{\nu} - \xi_{\nu}w_{\mu}. \qquad (9.6.3)$$

Denote by $u^{\mu} = \xi^{\mu}/\alpha$ a unit time-like vector. This vector is a four-velocity of a *Killing observer*, that is the observer that is at rest in the static spacetime. Using Eq. (9.6.3) we get

$$w^{\mu} = u^{\nu}u^{\mu}{}_{;\nu}, \qquad u^{\nu}{}_{;\nu} = 0. \qquad (9.6.4)$$

The first relation means that w^{μ} coincides with the 4-acceleration of the static observer.

A fluid in thermodynamic equilibrium is described by the following stress-energy tensor

$$T_{\mu\nu} = (\rho + p)\, u_{\mu}u_{\nu} + p\, g_{\mu\nu}. \qquad (9.6.5)$$

Here, ρ is the energy density and p is the pressure. They are related by an equation of state $p = p(\rho)$, which depends on the type of fluid. For example, for the thermal radiation the equation of state is

$$p = \frac{1}{3}\rho. \tag{9.6.6}$$

For a gas of non-relativistic particles (dust) it takes the form $p = 0$.

The covariant conservation law

$$T_{\mu\nu}{}^{;\nu} = 0 \tag{9.6.7}$$

gives a relation

$$0 = T_{\mu\nu}{}^{;\nu} = (\rho + p)_{;\nu}\, u_\mu u^\nu + (\rho + p)\, u^\nu u_{\mu;\nu} + (\rho + p)\, u_\mu u^\nu{}_{;\nu} + p_{;\mu}. \tag{9.6.8}$$

In the static spacetime the first and third terms in the right-hand side vanish. Using Eq. (9.6.4) we find

$$p_{;\mu} = -(\rho + p)\, w_\mu. \tag{9.6.9}$$

This equation has a simple meaning. A static equilibrium is possible if and only if the gravitational force (the right-hand side) is compensated by the gradient of the pressure. This relation is similar to the condition of the equilibrium in the gravitational field in the Newtonian theory. Note that in the Newtonian limit, when the motion of constituents of the media is non-relativistic, $p \ll \rho$, and in the right-hand side of Eq. (9.6.9) instead of $\rho + p$ one has ρ.

9.6.2 Tolman's temperature

For the thermal radiation the pressure and density are connected by Eq. (9.6.6). Substituting this relation into Eq. (9.6.9) we get $\rho_{;\mu} = -4\rho w_\mu$. Integration of this equation gives $\rho = \rho_0/\alpha^4$. Since the energy density of the thermal radiation is proportional to the fourth power of the temperature $\rho \sim \Theta^4$ one obtains the following condition of the equilibrium of the thermal radiation in a static spacetime

$$\alpha\Theta = \text{const.} \tag{9.6.10}$$

For the standard normalization of the Killing vector, when $\alpha = 1$ at infinity,

$$\Theta = \frac{\Theta_0}{\alpha}, \qquad \alpha = \sqrt{-\xi^2} = \sqrt{-g_{00}}. \tag{9.6.11}$$

Here, Θ_0 is the temperature measured at infinity (the normalization point of the Killing vector). Thus, in thermal equilibrium the local temperature depends on a point. This local temperature given by Eq. (9.6.11) is called *Tolman's temperature*.

The condition of thermal equilibrium Eq. (9.6.11) is quite natural. In fact, when a thermal photon propagates in a static gravitational field its frequency ω obeys the relation $\omega\alpha = \text{const.}$ When such a photon arrives at a region with a smaller value of the redshift factor α it is blueshifted. But it must remain in thermal equilibrium with the surrounding photons at this point. This is possible only if the local temperature is also blueshifted in a similar way. Tolman's law Eq. (9.6.11) reflects this fact.

Equation (9.6.10) also follows from the general conditions of thermal equilibrium. Let us recall the thermodynamical definition of the temperature

$$\Theta^{-1} = \frac{dS}{dE}. \tag{9.6.12}$$

Here, S is the entropy of the system and E is its energy. The entropy depends only on the internal states of the system and remains unchanged if one adiabatically puts the system in the static gravitational field. In this process the energy is redshifted and the derivative of the entropy with respect to the redshifted energy remains constant. Hence, according to Eq. (9.6.12) one reproduces the relation Eq. (9.6.10).

9.6.3 Accretion of thermal radiation on the black hole

If we try to apply the Tolman formula Eq. (9.6.11) to the case of the thermal radiation in the presence of a black hole we immediately have a problem. The redshift factor α at the surface of the Schwarzschild black hole vanishes. As a result, the Tolman local temperature diverges at the horizon. The local energy density of the thermal radiation, which is proportional to Θ^4, grows infinitely as well, so that the backreaction of the thermal radiation would destroy the black hole itself. This means that such a state is pathological. In fact, in order to keep the thermal radiation in equilibrium, one must provide conditions that prevent the free fall of the radiation into the black hole. This happens, for example, when the black hole is surrounded by a screen. Suppose the screen is spherically symmetric. We assume that the screen surface is heated and emits photons exactly with the same thermal properties as photons of the thermal bath. This occurs when the screen temperature is equal to the local Tolman temperature at its surface. This guarantees the thermal equilibrium of the system. When the radius of the screen tends to the gravitational radius, an infinitely growing force is required to keep it static. In other words, the limit of the infinite Tolman temperature is not physical.

What happens when there is no screen? Evidently, some of the photons from the thermal bath will fall into the black hole, increasing its mass. For a short interval of time, when the mass change is small, one may consider this effect as the *accretion* of the thermal radiation on a static black hole with a given mass M.

Let us calculate the rate of this accretion. Consider a black hole of mass M in the thermal bath of radiation, which has the temperature Θ_0 at infinity. Far away from the black hole the number of thermal photons with the momentum $(\mathbf{k}, \mathbf{k} + \Delta\mathbf{k})$ in a small space volume Δv is

$$\Delta N = g\frac{\Delta v d^3 k}{(2\pi)^3} n_k. \tag{9.6.13}$$

Here, $g = 2$ is the number of polarization states of a photon, and

$$n_k = \frac{1}{e^{k/\Theta_0} - 1} \tag{9.6.14}$$

is a number density of photons with the energy $\omega = k$.

Consider a sphere of large radius R surrounding the black hole. We assume that the photons do not interact, so that those photons that are moving in the direction of the black hole and have impact parameter less than the critical one will be absorbed by the black hole. Since the redshifted energy of a photon remains constant along its trajectory, such a photon after falling into a black hole brings energy k with it. Consider the spatial volume between spheres of radius R and $R + \Delta R$. The condition that a photon from this region will be absorbed by the black hole is $0 \le \theta \le \theta_*$, where

$$R \sin \theta_* = b_* \equiv l_{\min} \, r_S. \tag{9.6.15}$$

Here, $r_S = 2M$ is the gravitational radius, θ is the angle between the direction of the photon motion and the direction of the black hole, and b_* is the critical impact parameter (see Figure 9.4). Using the results of Section 7.4.1 one has

$$\sin \theta_* = \frac{3\sqrt{3}}{2} \frac{r_S}{R}. \tag{9.6.16}$$

The energy ΔE that is brought to the black hole by the photons of the thermal bath from the layer between the spheres is

$$\Delta E = \frac{1}{\pi^2} R^2 \Delta R \mathcal{Q}, \qquad \mathcal{Q} = 2\pi \int_0^{\theta_*} d\theta \sin \theta \int_0^\infty dk k^3 n_k. \tag{9.6.17}$$

Since for large R the angle θ_* is very small one has

$$\int_0^{\theta_*} d\theta \sin \theta \approx (\theta_*)^2 / 2 = \frac{27}{8} \left(\frac{r_S}{R} \right)^2. \tag{9.6.18}$$

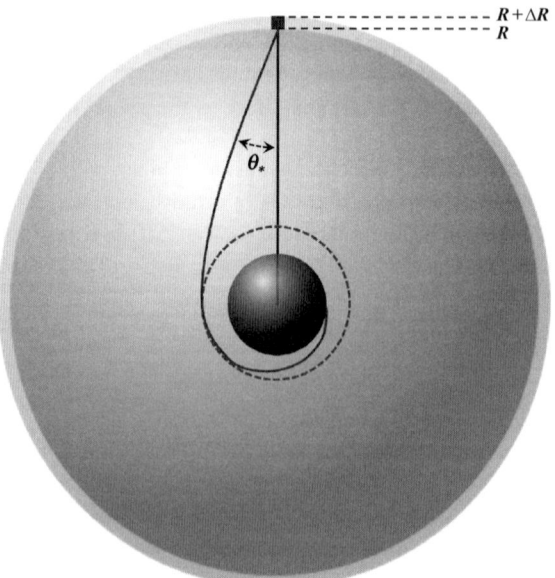

Fig. 9.4 Accretion of thermal radiation into a black hole.

One also has

$$\int_0^\infty dk k^3 n_k = A \, \Theta_0^4, \qquad A = \int_0^\infty \frac{dx x^3}{e^x - 1} = \frac{\pi^4}{15}. \qquad (9.6.19)$$

In order to reach the black hole the photon must move practically in the direction of the black hole. For such a photon $\Delta R = \Delta t$. At far distance from the black hole the temperature is Θ_0. Combining these results one gets

$$\dot{E} = \frac{\Delta E}{\Delta t} = \frac{9}{20} \pi^3 r_S^2 \, \Theta_0^4 = 27 \pi r_S^2 \, \sigma \, \Theta_0^4. \qquad (9.6.20)$$

Here, $\sigma = \pi^2/60$ is the *Stefan–Boltzmann constant*. One can rewrite this expression in the form

$$\dot{E} = \pi b_*^2 \, \rho, \qquad (9.6.21)$$

where $\rho = 4\sigma \, \Theta_0^4$ is the energy density of the thermal radiation at infinity, and $b_* = 3\sqrt{3} \, r_S/2$ is the critical impact parameter.

9.6.4 Thermofield calculations

To calculate the rate of accretion of the thermal radiation into the black hole we considered photons as zero-mass particles propagating along null geodesics. In other words we used the *geometrical-optics approximation*. This approximation is valid when the characteristic wavelength of the waves is much smaller than the scale at which the spacetime characteristics change essentially. If $\mathcal{R} \sim L^{-2}$ is the spacetime curvature the corresponding length scale is L. The characteristic wavelength λ of photons in the thermal bath with the temperature Θ is $\lambda \sim \Theta^{-1}$. Thus, the required condition means that $\Theta \gg 1/L$. If this condition is not satisfied, more accurate consideration is required.

It is instructive to repeat the calculations of the accretion rate of the thermal radiation into the black hole without using the geometrical-optics approximation. To simplify calculations we consider a scalar massless field φ. After dimensional reduction the scalar massless field theory in the Schwarzschild spacetime is reduced to an infinite set of 2D theories (see Eq. (9.3.12)). Each of the mode φ_{lm} propagates in a space with the effective potential U_l given by Eq. (9.3.15). In the tortoise coordinates this problem is equivalent to the scattering problem considered in Appendix F.4. In such a description the thermal equilibrium state corresponds to the case when there exist reservoirs of thermal radiation with the same temperature at both spatial infinities. If the thermal bath at $r_* = -\infty$ is absent, in the corresponding *non-equilibrium thermal state* there exists the energy flux directed to $r_* \to -\infty$, which describes the accretion of the thermal radiation into the black hole. For the modes with angular momentum l this flux is

$$\dot{E}_l = \frac{1}{2\pi} \int_0^\infty d\omega \, \omega \, \frac{|T_{\omega l}|^2}{e^{\beta \omega} - 1}, \qquad \beta = \Theta_0^{-1}. \qquad (9.6.22)$$

Here, $T_{\omega l}$ is the *transmission coefficient*. This relation is similar to Eq. (F.4.44)[3]

> *Equation (9.6.22) has a simple interpretation. $(e^{\beta\omega} - 1)^{-1}$ is the number density of quanta with frequency ω sent from infinity $r = \infty$. These quanta have energy ω. The graybody factor $|T_{\omega l}|^2$ is the probability of the penetration though the potential barrier. Thus, the integrand in Eq. (9.6.22) gives the energy of those quanta that reach the black hole horizon.*

Summation over the modes gives

$$\dot{E} = \sum_{l=0}^{\infty}(2l+1)\dot{E}_l = \frac{1}{2\pi}\int_0^{\infty}\frac{d\omega\,\omega\Gamma_\omega}{e^{\beta\omega}-1}.$$

$$\Gamma_\omega = \sum_{l=0}^{\infty}(2l+1)|T_{\omega l}|^2. \tag{9.6.23}$$

The quantity Γ_ω has the meaning of the cross-section of the black hole at the given frequency ω.

In the geometrical-optics approximation one identifies $l + 1/2$ with a dimensionless impact parameter $r_S\omega\ell$ (see Eq. (9.3.20)). For large l the term $1/2$ can be omitted. The condition that all photons, with the impact parameter less than the critical one $r_S\ell_{\min}$ (see Section 7.4.1), are captured by the black hole reads

$$|T_{\omega l}|^2 \approx \theta(3\sqrt{3}M\omega - l), \quad l_{\min} = 2M\omega\,\ell_{\min}, \quad \ell_{\min} = 3\sqrt{3}/2. \tag{9.6.24}$$

Here, $\theta(x)$ is the *Heaviside step function*. The angular momentum quantum number l and the parameter ℓ are related according to Eq. (9.3.20). In this approximation (proposed by DeWitt (1975) and called the *DeWitt approximation*) one has

$$\Gamma_\omega \approx \sum_{l=0}^{\infty}(2l+1)\theta(3\sqrt{3}M\omega - l) \approx 27M^2\omega^2, \tag{9.6.25}$$

and hence

$$\dot{E} \approx \frac{27M^2}{2\pi}\int_0^{\infty}\frac{d\omega\,\omega^3}{e^{\beta\omega}-1} = \frac{9}{40}\pi^3 r_S^2\Theta_0^4. \tag{9.6.26}$$

This expression is in agreement with the result Eq. (9.6.20) of the previous subsection. The difference in the factor 2 is because the number of polarization states of the photon is $g = 2$ (see Eq. (9.6.13)), that is twice the corresponding factor for the scalar field.

9.7 Hawking Effect

9.7.1 Quantum radiation of black holes

Till now in considering the interaction of matter and fields with a black hole we neglected their quantum properties. In this description a black hole is an absolutely 'cold' black object

---

[3] Since the field is massless we put $k = \omega$. Another difference is that the source of the thermal radiation with temperature Θ_0 is at $r_* = \infty$, while the flux is calculated at $r_* = -\infty$, that is the horizon of the black hole.

that absorbs everything and emits nothing. The quantum theory dramatically changes this paradigm. As it was demonstrated by Hawking (1974, 1975) in the presence of a black hole the vacuum is unstable. As a result of the vacuum decaying, the black hole of mass M emits radiation. This radiation is thermal and it has the *Hawking temperature*

$$\Theta_H = \frac{\hbar c^3}{8\pi G k_B M}. \tag{9.7.1}$$

We restored all the dimensional constants in this relation to demonstrate that this is a remarkable relation that besides the black hole parameter (mass M) contains the Planck's constant $\hbar$, the speed of light c and the gravitation coupling constant G. This happens because the effects connected with this temperature are determined by the *quantum* nature of the radiation, described by the *relativistic* quantum theory interacting with the *gravitational* background. As a result of the *Hawking process* the isolated black hole emits thermal radiation with the thermal spectrum and the temperature (as measured at infinity) equal to Θ_H.

> *The vacuum in the quantum theory is defined as a state without real propagating particles. A stable vacuum has the lowest energy. This makes it similar to the classical vacuum. However, due to the uncertainty relations a quantum field in the vacuum state has non-vanishing quantum fluctuations. Namely, these, so-called zero-point fluctuations make the properties of the quantum vacuum quite non-trivial and different from properties of the classical vacuum. Under a non-adiabatic change of parameters of the quantum system some field modes, which were initially in the ground state can be excited. In such a process 'real' particles are created. In the classical theory, if the field initially is in the classical ground state, the effect of the parametric excitation does not work.*
>
> *Energy conservation requires that the energy of a pair created from the vacuum in a stationary external field must be the same as the energy of the vacuum. Since a particle of the created pair that reaches infinity, has positive energy, its 'companion' should have negative energy. That is, such a particle must be created either inside the black hole, where the Killing vector is space-like, or, in the case of a rotating black hole inside the ergosphere.*
>
> *Only a fraction of the created particles reach a distant observer, while the other fraction is either created inside the black hole or is absorbed by it. As a result of such a process, the distant observer would have access only to a part of the total system. As is well known from quantum mechanics such a subsystem of observed particles, which form the Hawking radiation, is described by the* density matrix. *It happens even in the case when the initial state was the vacuum, that is a* pure quantum state. *The thermal properties of the Hawking radiation mean that the corresponding density martix is thermal.*

Hawking (1974, 1975) theoretically discovered the effect of the quantum radiation of black holes by studying the quantum field theory in the external gravitational field of the black hole (see also, e.g., (DeWitt 1975; Frolov and Novikov 1998) and references therein). Instead of reproducing these calculations we shall demonstrate that the main property of the quantum radiation, namely its thermality, in fact follows from the Euclidean formulation of the quantum field theory in the black hole background. As earlier, we consider a simple model of a scalar massless field propagating near the Schwarzschild black hole.

9.7.2 Euclidean black hole and Hawking temperature

Let us recall some properties of the *Euclidean black hole*, which we have discussed in Section 6.2.4. This metric is of the form

$$ds_E^2 = d\gamma_E^2 + r^2\, d\Omega^2, \qquad d\gamma_E^2 = g\, dt_E^2 + \frac{dr^2}{g}, \qquad g = 1 - \frac{r_S}{r}. \tag{9.7.2}$$

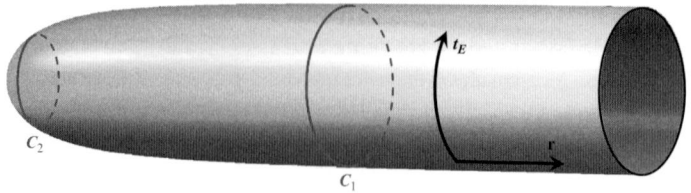

Fig. 9.5 Embedding diagram for 2D (t, r)-section of the Gibbons–Hawking instanton.

It is regular at the Euclidean horizon $r = r_S$ only if the time coordinate t_E is periodic and has the period $2\pi/\kappa$, where $\kappa = 1/(2r_S)$ is the *surface gravity*. The *embedding diagram* for the metric $d\gamma_E^2$ is shown in Figure 9.5. The Euclidean metric Eq. (9.7.2) is known as the *Gibbons–Hawking instanton*.

The requirement of the periodicity with respect to t_E determines the global structure and the topology of the Euclidean black hole. The topology of the 2D sector is $\mathbb{R}^2$. At large r the 2D metric $d\gamma_E^2$ has the geometry of a flat 2D cylinder with the Euclidean time direction being compact. Asymptotically, the geometry Eq. (9.7.2) is the same as for 4D flat metric

$$ds_E^2\Big|_{r\to\infty} \sim dt_E^2 + dr^2 + r^2 d\Omega^2, \tag{9.7.3}$$

with the fixed periodicity property of t_E. The topology of 4D space of the Euclidean black hole is $\mathbb{R}^2 \times S^2$.

In the Euclidean formulation of quantum field theory one uses Wick's rotation $t \to it_E$. The period $2\pi/\kappa$ of t_E is fixed by the condition of the regularity of the Gibbons–Hawking instanton at the horizon. To make the Euclidean field theory well defined we have to impose the regularity condition of the field at the Euclidean horizon. This condition implies that the corresponding field automatically has the same period $2\pi/\kappa$[4] with respect to the Euclidean time t_E. In particular, the Euclidean Green functions also have this property.

The quantum field theory in the physical spacetime is restored by performing the inverse Wick's rotation $t_E \to -it$. Then, the Euclidean Green functions are transformed into the *Feynman propagators* (see Appendices F.2 and F.3 for details). A so-defined quantum field theory has 'memory' about its Euclidean origin. In particular, its propagators are periodic with respect to the imaginary time, so that they correspond to a special choice of the quantum state. The state that is singled out by this condition is a thermal one, described by the thermal density matrix with the temperature Θ_H. This is the equilibrium state of the quantum field in the black hole background, where the thermal field has the temperature measured at infinity equal to the Hawking temperature Θ_H.

Consider, for example, a quantum scalar massless field φ in the Schwarzschild spacetime. After the dimensional reduction one obtains an infinite set of 2D quantum theories enumerated by the index (l, m). After the inverse Wick's rotation of the Euclidean theory, each of these 2D fields is also in the state of thermal equilibrium without energy fluxes.[5] As we know in each (l, m)-sector the corresponding theory is equivalent to the 2D theory in the flat

---

[4]For fermions this is a condition of anti-periodicity.
[5]In the general case this equilibrium is unstable.

spacetime in the presence of the potential U_l. But we also know that the state of the thermal equilibrium in such a 2D theory is possible only if besides a thermal radiation from $r_* = \infty$ there exists a thermal radiation from $r_* = -\infty$ (see Appendix F.4). This means that a regular thermal equilibrium state in a spacetime of the black hole is possible when there exists the thermal radiation emitted by the black hole, provided its redshifted temperature is equal to the Hawking temperature. This thermal radiation 'compensates' the thermal flux of the radiation from infinity.

9.7.3 Hawking radiation

In the case when there is no thermal bath at infinity, one has a *non-equilibrium thermal state*. It can be shown that in each of the (l, m)-sectors there exists a net flux of the thermal radiation from the black hole to infinity (see Appendix F.4). This is the *Hawking radiation*. The temperature of the radiation is universal and is given by Eq. (9.7.1).

Expression (F.4.44) gives a contribution of the (l, m)- modes in the flux. Summing over the modes we obtain

$$\dot{E} = \frac{1}{2\pi} \int_0^\infty d\omega \frac{\omega}{e^{\omega/\Theta_H} - 1} \sum_{l=0}^\infty (2l + 1)|T_{\omega,l}|^2. \tag{9.7.4}$$

Note that the expression Eq. (9.7.4) is similar to the relation Eq. (9.6.22) for the rate of energy absorption of the thermal radiation by the classical black hole for the special choice of the temperature $\Theta_0 = \Theta_H$. In particular, the graybody factor $|T_{\omega,l}|^2$ is the same in both equations. This happens because for the scattering problem the probability of the penetration through the potential barrier for the right- and left-moving waves is the same.

In order to estimate the rate of energy emission by the black hole one can use the *DeWitt approximation* Eq. (9.6.25) for the transmission coefficients. In this approximation the rate of emission Eq. (9.7.4) coincides with the rate of absorption Eq. (9.6.26), provided the temperature $\Theta_0 = \Theta_H$

$$\dot{E} \approx \frac{9}{10 \times 8^4 \times \pi} \frac{1}{M^2}. \tag{9.7.5}$$

Note that in the case of a scalar field the main contribution (91%) comes from the s-mode ($l = 0$–mode). To calculate $\dot{E}$ more accurately one must use the numerical calculations of the *graybody factor* $|T_{\omega,l}|^2$. These calculations show that the estimation Eq. (9.7.5) differs from the exact result at the level of 6% accuracy. This difference is explained by the violation of the WKB approximation for low-frequency modes with $\omega \sim \Theta_H$. It is surprising that the difference is so small despite the fact that these frequencies give the main contribution to the Hawking radiation.

Problem 9.2: *An approximation for the graybody factor $\Gamma_\omega = \sum_{l=0}^\infty (2l + 1)|T_{\omega,l}|^2$, which is more accurate at low frequency ω, was obtained by Sánchez (1978) and is of the form*

$$\Gamma_\omega \approx 27M^2\omega^2 \left(1 - \frac{2\sqrt{2}}{27M\pi\omega} \sin(2\sqrt{27}\pi\omega M)\right). \tag{9.7.6}$$

Calculate the rate of emission $\dot{E}$ in this approximation.

9.7.4 General case

The expression for the energy rate of the Hawking emission Eq. (9.7.4) has a simple interpretation. The thermal factor

$$n_\omega = \frac{1}{e^{\omega/\Theta_H} - 1},$$ (9.7.7)

is the number density of thermal photons emitted by the black hole at the redshifted frequency ω. The factor $|T_{\omega,l}|^2$ gives the probability of the photon in the mode (ω, l) emitted by the black hole to reach infinity. The factor ω in the nominator is the energy (frequency) of the photon. This expression also contains the summation over the discrete quantum numbers (l) and the integration over the continuous spectrum of ω. Direct calculations show that this formula can be easily generalized to the case of the quantum emission of the massless field by an arbitrary black hole that besides mass M has an angular velocity J and electric charge Q. The massless field is characterized by its spin s, which is equal to $0, 1/2, 1, 2$ for the scalar field, neutrino, photons, and gravitons, respectively. All these fields allow the separation of variables in the Kerr–Newman geometry. We denote by J a collective index 'enumerating' the modes. This index includes the frequency ω, the analog of the angular momentum l, the azimuthal quantum number m, and the polarization $P = \pm 1$

$$J = \{\omega, l, m, P\}.$$ (9.7.8)

We also denote

$$\varpi_J = \omega - m_J \Omega_H - q_J |e| \Phi_H, \qquad \sigma_J = \text{sign}(\varpi_J).$$ (9.7.9)

Here, $|e|$ is the absolute value of the charge of the emitted particle, q_J is its sign, Ω_H and Φ_H are the angular velocity and the electric potential of the black hole

$$\Omega_H = \frac{a}{r_+^2 + a^2}, \qquad \Phi_H = \frac{Qr_+}{r_+^2 + a^2},$$

$$r_+ = M + \sqrt{M^2 - a^2 - Q^2}.$$ (9.7.10)

Here, $a = J/M$ is the *rotation parameter*. The Hawking temperature is

$$\Theta_H = \frac{\kappa}{2\pi} = \frac{1}{2\pi} \frac{r_+ - M}{r_+^2 + a^2}.$$ (9.7.11)

Consider the rate of emission of particles and let f_J be one of its quantum numbers: energy ω, azimuthal quantum number m, or charge e. We denote by $\mathcal{F}$ the flux of the corresponding parameter, the mass M, the angular momentum J, or the electric charge Q. Then, one has

$$\dot{\mathcal{F}} = \frac{1}{2\pi} \int_0^\infty d\omega \sum_J \frac{f_J \sigma_J |T_J|^2}{\exp(\varpi_J/\Theta_H) - \epsilon}.$$ (9.7.12)

Here, $\epsilon = 1$ for bosons and $\epsilon = -1$ for fermions. A similar formula is also valid for the massive particle with the mass μ. In the latter case, the integration over the frequency must start from μ.

The classical effect of the superradiance has an evident quantum analog. For bosons there exist states with $\sigma_J < 0$ that obey the superradiance condition $\varpi_J < 0$. For these modes the denominator in Eq. (9.7.12) becomes negative as well. However, this change of sign of the denominator is compensated by the change of the sign of σ_J in the numerator, so that the total sign of the integrand remains the same. More accurate study of the transmission amplitude T_J shows that at the point where the change of the signs occurs this coefficient vanishes, so that the integrand remains finite at this point.

9.8 Quantum Fields in the Rindler Spacetime

9.8.1 Thermal radiation near the horizon

At infinity the Euclidean proper time τ_E coincides with coordinate time t_E. Hence, the proper length of the circle C_1 shown in Figure 9.5 coincides with the inverse Hawking temperature Θ_H^{-1}. It is evident that the circles $r =$const similar to C_2, which are located closer to the black hole, have smaller proper length. Indeed, their length, or the period in the proper time τ_E, is

$$\Theta^{-1} = \sqrt{g}\,\Theta_H^{-1}. \tag{9.8.1}$$

This relation, rewritten for the local temperature $\Theta = \Theta_H/\sqrt{g}$ is simply *Tolman's law*. The proper length of the circle vanishes at the horizon. As we have already learned, this means that the local temperature grows infinitely near the horizon.

At first glance we have a paradox. The energy density ε of the thermal radiation at the temperature Θ is

$$\varepsilon \sim \Theta^4. \tag{9.8.2}$$

For any given temperature at infinity, including the Hawking temperature Θ_H, the local temperature diverges at the horizon. On the other hand, the thermal state at the Hawking temperature is regular since the horizon is a regular surface of the Gibbons–Hawking instanton and the fields are regular there. The resolution of this paradox is the following. Equation (9.8.2) is valid only if the temperature is almost constant and/or the wavelength of thermal photons is much less than the characteristic scale where the external fields change significantly. For the black hole one has

$$\left|\frac{\Delta\Theta}{\Theta}\right| = \left|\frac{\Delta g}{2g}\right| = \frac{r_S}{2r^2}\frac{\Delta r}{g}. \tag{9.8.3}$$

Since near the horizon $r \approx r_S$ and $g \approx \Delta r/r_S$, Eq. (9.8.3) shows that $|\Delta\Theta/\Theta| \sim 1$. This means that the approximation used to derive the law Eq. (9.8.2) is not valid. This relation should be modified. The corresponding additional terms are connected to the *vacuum polarization effect*. In fact, for the thermal state with the Hawking temperature these additional terms exactly compensate the apparent divergencies of the thermal energy density at the horizon.

To illustrate the origin of the vacuum polarization effect let us consider the massless scalar field theory in the flat spacetime (see Appendix F.3). For the plane-wave decomposition (F.3.26) the quantity $|\Phi_k(X)|^2$ is proportional to the probability to 'meet' a quanta at the point X. Summing over the modes with the frequency in the interval $(\omega, \omega + d\omega)$ one finds the distribution of the density of states $n(\omega)d\omega \sim \omega^2 d\omega$. This result does not depend on the choice of the modes used in the field decomposition. The density distribution does not depend on the point X. This is not surprising

since the flat spacetime is homogeneous and anisotropic. The energy density of the zero-point
fluctuations is

$$\varepsilon \sim \int_0^{\omega_{\text{cut-off}}} \omega^3 d\omega \sim \omega_{\text{cut-off}}^4. \tag{9.8.4}$$

It also does not depend on the choice of the point X and is divergent and proportional to the fourth
power of the cut-off parameter $\omega_{\text{cut-off}}$. This divergence is considered as non-physical and one simply
subtracts it. In the presence of an external field the distribution of the density of states becomes
dependent on the point, $n(\omega, X)d\omega$, so that after renormalization (subtraction of the zero-point
fluctuations in the absence of the external field), there remains a finite contribution to the vacuum
energy density. This contribution, which is a functional of the external field, is known as the vacuum
polarization.

9.8.2 Thermal radiation in the Rindler space

Note that the leading divergence of the energy density of the thermal field Eq. (9.8.2) does
not depend on the curvature. For this reason, to study the role of the vacuum polarization
effect we neglect for a moment the curvature effects. In other words, we use the *Rindler space*
approximation.

Consider a scalar massless field φ in the *Rindler metric* (see Eq. (2.4.3))

$$ds^2 = -a^2\rho^2 d\tau^2 + d\rho^2 + dy^2 + dz^2. \tag{9.8.5}$$

As earlier, we use a normalization point $\rho = a^{-1}$. At this point τ is a proper time, and a is the
acceleration of the Rindler observer. The field equation

$$\Box\varphi = 0 \tag{9.8.6}$$

in Rindler coordinates takes the form

$$\left(-a^{-2}\rho^{-2}\partial_\tau^2 + \rho^{-1}\partial_\rho(\rho\,\partial_\rho) + \partial_y^2 + \partial_z^2\right)\varphi = 0. \tag{9.8.7}$$

The field φ can be decomposed into modes

$$\varphi_k = e^{i(k_y y + k_z z)}\Phi_k(\tau, \rho), \tag{9.8.8}$$

where

$$\left[-a^{-2}\rho^{-2}\partial_\tau^2 + \rho^{-1}\partial_\rho(\rho\partial_\rho) - k^2\right]\Phi_k = 0, \qquad k = \sqrt{k_y^2 + k_z^2}. \tag{9.8.9}$$

Using the decomposition Eq. (9.8.8) one effectively reduces the 4D theory to the infinite set of 2D
theories Eq. (9.8.9), 'enumerated' by the continuous index (k_y, k_x). One can obtain these equations by
taking a limit $r_S \to \infty$ in the dimensionally reduced equations in the Schwarzschild geometry. In this
limit, when the gravitational radius becomes very large, the black hole surface can be approximated
by a 2D plane. Thus, a 2D Laplace operator on such a surface can be approximated by a 2D flat
Laplacian with a continuous spectrum.

Denote by r_* a *tortoise coordinate* in the Rindler space

$$r_* = a^{-1}\ln(a\rho). \tag{9.8.10}$$

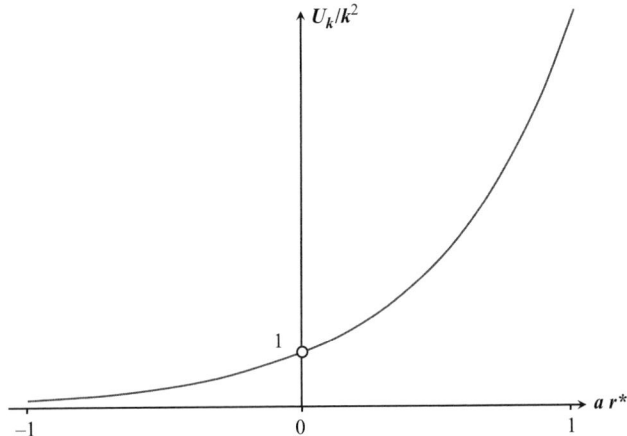

Fig. 9.6 Effective potential $U_k(r_*)$ in the Rindler space.

This tortoise coordinate changes from $-\infty$ at the horizon $\rho = 0$ to $+\infty$ (at $\rho = +\infty$). Equation (9.8.9) takes the form

$$\left(-\partial_\tau^2 + \partial_{r_*}^2 - U_k\right)\Phi_k = 0, \qquad U_k = k^2\, e^{2ar_*}. \tag{9.8.11}$$

The effective potential U_k is shown in Figure 9.6. The reduced theory is equivalent to the theory in the 2D flat spacetime with the effective potential U_k considered in Appendix F.4. However, there is a difference: the effective potential U_k grows infinitely at $r_* \to \infty$.

Let us write

$$\Phi \sim e^{-i\omega\tau}\, \varphi_{\omega k}. \tag{9.8.12}$$

One has

$$\left(\frac{d^2}{dr_*^2} + \omega^2 - U_k\right)\varphi_{\omega k} = 0. \tag{9.8.13}$$

A solution $\varphi_{\omega k}$ tends to zero at large r_* for any ω and $k > 0$. In other words, the transmission coefficient $t_{\omega k}$ vanishes, while the absolute value of the reflection coefficient is $|r_{\omega k}| = 1$. One can always choose the phase of the upgoing mode in such a way that the corresponding solution is real. The conformal diagram for these modes is shown in Figure 9.7.

9.8.3 Thermal Green functions

Consider the thermal radiation in the Rindler space. The outgoing thermal flux from the Rindler horizon, being reflected back by the potential, generates the incoming flux. The combination of these two fluxes results in a thermal equilibrium state. Since in an external gravitational field the local temperature depends on the position, one needs to choose a special normalization point for its identification. We use $\rho = a^{-1}$ as such a point and denote by Θ_0 the temperature there.

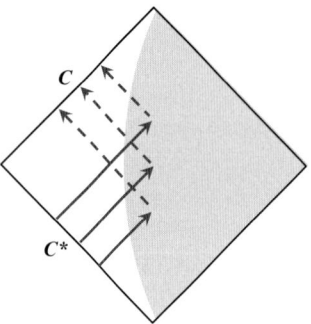

Fig. 9.7 Conformal diagram for Rindler modes.

The detailed information about the local observables can be obtained if one knows the thermal Green function. In the case of the thermal radiation in the Rindler spacetime this Green function can be found in an explicit form. Consider the Euclidean Rindler space

$$ds^2 = a^2 \rho^2 d\tau_E^2 + d\rho^2 + dy^2 + dz^2, \tag{9.8.14}$$

and impose the periodicity condition on the Euclidean time τ_E. Namely, we assume that the length of a circle $\rho = a^{-1}$ is β. In the general case, such a space has a cone-like singularity at $\rho = 0$. The corresponding *embedding diagram* for the 2D (τ, ρ)-section is shown in Figure 9.8. The singularity is absent only when

$$\beta = 2\pi/a. \tag{9.8.15}$$

For the quantum field theory the periodicity in the imaginary time corresponds to the thermal state with the temperature

$$\Theta_0 = \beta^{-1}. \tag{9.8.16}$$

In the absence of the conical singularity the temperature Θ_0 is

$$\Theta_a = \frac{a}{2\pi}. \tag{9.8.17}$$

This is called the *Unruh temperature*.

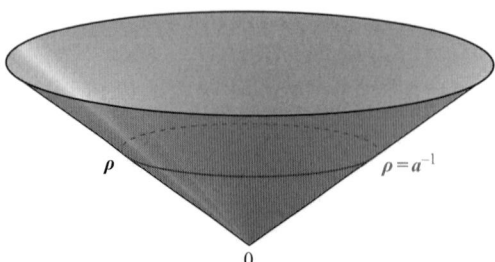

Fig. 9.8 The embedding diagram for 2D (τ, ρ)-section of the Euclidean Rindler space.

The Euclidean Green function is a decreasing-at-infinity solution of the equation

$$\Box_E G_E(x, x') = -\delta^4(x, x'), \tag{9.8.18}$$

where $\Box_E$ is the 4D Laplace operator. Denote by G_E^0 the usual Euclidean Green function in a flat 4D Euclidean space. In the Cartesian coordinates X^μ it has the form

$$G_E^0(X, X') = \frac{1}{4\pi^2} \frac{1}{|X - X'|^2}. \tag{9.8.19}$$

Making the transformation to the Rindler coordinates one obtains

$$G_E^0(x, x') = \frac{1}{8\pi^2 \rho \, \rho'} \frac{1}{\cosh \lambda - \cos[(\tau_E - \tau_E')a]},$$

$$\cosh \lambda = \frac{\rho^2 + \rho'^2 + (y - y')^2 + (z - z')^2}{2\rho \, \rho'}. \tag{9.8.20}$$

This Green function depends on $\Delta \tau_E = \tau_E - \tau_E'$ and has the period $\Delta \tau_E = 2\pi/a$. From the point of view of the Rindler observer it describes the thermal state of 'Rindler quanta', with the *Unruh temperature*.[6]

The Euclidean Green function on the cone space, where the period of time τ_E differs from $a/(2\pi)$, can be obtained by means of the *Sommerfeld transformation*. This is an integral transformation that changes the periodicity of Green functions. This procedure is described in detail in (Dowker 1977). Here, we just present the final result

$$G_E(x, x') = \frac{\Theta_0}{4\pi \, a \rho \, \rho' \sinh \lambda} \frac{\sinh(2\pi \, \Theta_0 \, \lambda/a)}{\cosh(2\pi \, \Theta_0 \, \lambda/a) - \cos[2\pi \, \Theta_0 \, (\tau_E - \tau_E')]}. \tag{9.8.21}$$

One can check directly that this is the required Green function. It is evident that it has the period Θ_0^{-1} in the Euclidean time τ_E. For $\Theta_0 = a/(2\pi)$ it coincides with Eq. (9.8.20). The difference, the *renormalized Green function*,

$$G_E^{\text{ren}}(x, x') = G_E(x, x') - G_E^0(x, x') \tag{9.8.22}$$

is a regular function in the limit of coincident points $x \to x'$, so that $G_E(x, x')$ has the same singular behavior in this limit as $G_E^0(x, x')$. Outside this point it is a regular decreasing-at-infinity solution of Eq. (9.8.18).

Analytical continuation of the Euclidean propagator G_E, which is periodic in the Euclidean time with the period β, to the physical spacetime with the Minkowskian signature gives the Feynman propagator at finite temperature $\Theta = 1/\beta$ (see Appendix F)

$$G_{F,\beta}(t, \mathbf{x}; t', \mathbf{x}') = iG_E(\tau, \mathbf{x}; \tau', \mathbf{x}')\big|_{\tau \to -it}. \tag{9.8.23}$$

For the calculation of the stress-energy tensor one needs to know the finite-temperature Hadamard function $G_\beta^{(1)}$ rather than the Feynman propagator. However, outside the light cone they are proportional to each other, namely, if we take $\mathbf{x}$ and $\mathbf{x}'$ separated by a space-like interval, that is when $(t - t')^2 < (\mathbf{x} - \mathbf{x}')^2$, then

---

[6]We discuss the meaning of the 'Rindler quanta' in Section 9.8.5.

$$G_\beta^{(1)}(t, \mathbf{x}; t', \mathbf{x}') = -2iG_{F,\beta}(t, \mathbf{x}; t', \mathbf{x}') = 2G_E(-it, \mathbf{x}; -it', \mathbf{x}'). \tag{9.8.24}$$

This relation gives us the recipe for the calculation of the thermal averages using the Euclidean Green function. The renormalized Green functions are related in exactly the same way.

9.8.4   Local observables

In order to calculate local observables in the thermal state we can use the renormalized thermal Hadamard Green function $G_{\text{ren},\beta}^{(1)}(x, x') = 2G_E^{\text{ren}}(x, x')$ for space-like separated[7] points x and x'. For example, the stress-energy tensor is then defined as follows

$$\langle T_{\mu\nu}(x) \rangle_\beta = \lim_{x' \to x} \left[\partial_{(\mu} \partial_{\nu')} - \frac{1}{2} g_{\mu\nu} g^{\lambda\epsilon'} \partial_\lambda \partial_{\epsilon'} \right] G_E^{\text{ren}}(x, x'). \tag{9.8.25}$$

Here, ∂_μ and $\partial_{\nu'}$ are the partial derivatives with respect to x and x', respectively. After performing differentiations and taking the coincidence limit $x \to x'$ one obtains

$$\langle T_\mu^\nu(x) \rangle_\beta = \frac{\pi^2}{90(\rho a)^4} \left[\Theta_0^4 - \left(\frac{a}{2\pi} \right)^4 \right] \text{diag}(-3, 1, 1, 1). \tag{9.8.26}$$

Let us note that $\langle T_\mu^\nu(x) \rangle_\beta$ is invariant under Wick's rotation, so that the quantity Eq. (9.8.26), calculated in the Euclidean space can be used without any changes in the 'physical' spacetime.

By definition, Θ_0 and $\Theta_a = a/(2\pi)$ are the temperature of the radiation and the Unruh temperature measured at the same reference point $\rho = a^{-1}$ of the Rindler frame. Since the redshift factor in the Rindler space is

$$\alpha = a\rho, \tag{9.8.27}$$

one can define the corresponding Tolman local temperatures at an arbitrary point ρ as

$$\Theta(\rho) = \frac{\Theta_0}{a\rho}, \qquad \Theta_U(\rho) = \frac{\Theta_a}{a\rho} = \frac{1}{2\pi\rho}. \tag{9.8.28}$$

One can rewrite Eq. (9.8.26) in the form

$$\langle T_\mu^\nu(x) \rangle_\beta = \frac{\pi^2}{90} \left[\Theta^4(\rho) - \Theta_U^4(\rho) \right] \text{diag}(-3, 1, 1, 1). \tag{9.8.29}$$

Note that $\varepsilon = \pi^2 \Theta^4/30$ coincides with the energy density of the thermal radiation with the temperature Θ in the Minkowski spacetime.

For $\Theta_0 = \Theta_a$ the cancellation of the two terms in the square brackets takes place for any ρ. The corresponding state is the *Minkowski vacuum*. In other words, in the Rindler frame the Minkowski vacuum corresponds to a thermal state with the *Unruh temperature*. From the point of view of the Rindler observer the relation Eq. (9.8.29) allows the following interpretation. Besides the thermal contribution into $\langle T_\mu^\nu(x) \rangle_\beta$, proportional to $\Theta^4(\rho)$, it also contains another

---

[7] The easiest way is to take $t = t'$ and $\mathbf{x} \to \mathbf{x}'$. When the required quantity is expressed in terms of the Euclidean Green function then the same result can be obtained also by taking the limit $x \to x'$ along any curve in the Euclidean coordinate space.

contribution, proportional to Θ_U^4. The latter does not depend on the temperature. It describes the energy density and pressure of the *vacuum polarization*.

> *The quantity $|\varphi_{\omega k}(x)|^2$ has the meaning of the probability for a Rindler particle in the mode (ω, k_y, k_z) to be near a point x. For the Minkowski particles in a flat spacetime for a plane-wave mode **k** this probability does not depend on the point. For Rindler particles this is not so. As a result, the contribution to the stress-energy tensor of a single mode also depends on its position. After summing over the modes one obtains the contribution $\sim a^4$. In the general case, the effect of an external field on the mode propagation results in the dependence of the vacuum observables on the point.*

9.8.5 Uniformly accelerated detectors

The *Unruh temperature* is an effective temperature measured by a uniformly accelerated detector. If a is the value of the proper acceleration of the particle detector, the corresponding Unruh temperature is

$$\Theta_U = \frac{\hbar}{k_B c} \frac{a}{2\pi}. \tag{9.8.30}$$

The *Boltzman factor* k_B in the dimensionful expression is required if the temperature is measured in kelvins rather than in the energy units. For a detailed discussion of the Unruh effect and quantum effects in the Rindler spacetime see, e.g., the reviews by (Takagi 1986; Ginzburg and Frolov 1987)

> *In order to measure the temperature one needs to use a special physical device such as a thermometer. In the general case, such a device has internal degrees of freedom that interact with the thermal media. As a result of this interaction, the internal degrees of freedom will be thermally exited with the same temperature as the media. Registering the change of the internal parameters of the thermometer one measures the temperature of the media.*

To demonstrate how the temperature of the thermal bath can be measured, let us consider the following simple model of the 'thermometer'. Consider a quantum field $\hat{\varphi}$ in a flat space-time, and a point-like detector interacting with this field. Such a detector must be localized in a small spatial volume near a worldline representing its motion. It must have internal degrees of freedom such that an interaction of the detector with the quantum field results in the transitions between the energy levels of the internal degrees of freedom. We assume that interaction between the detector and the field φ is of the simplest form

$$S_{\text{int}}[\varphi, q] = \lambda \int_\gamma d\tau \, q(\tau) \, \varphi(x^\mu(\tau)). \tag{9.8.31}$$

Here, the integration is taken along the worldline γ of the detector, τ is the proper time on this line, λ is the coupling constant, $q(\tau)$ describes the internal degrees of freedom of the detector.

> *In Appendix F.3.6 we discuss the case when a quantum oscillator is used as a 'detector'. It is assumed that the oscillator is at rest in a flat spacetime, and it is interacting with a thermal radiation of the field φ. It is demonstrated that the equilibrium distribution of such a 'detector' is thermal with its temperature equal to the temperature of the bath. The main details of the calculations do not depend on the nature of the degrees of freedom of the detector q. It can be an oscillator, as we used for the calculations in Appendix F.3.6, or a two-level system, the model that is often used to simplify the calculations.*

For a static detector in the thermal bath the calculations show that the probability of the transition between the detector levels, in the lowest in λ approximation, is proportional to the proper time τ. The probability of transition between the levels 1 and 2 per unit time can be written in the form

$$\dot{w}_{1 \to 2} = \lambda^2 Q(1,2) F(|\omega_1 - \omega_2|) \tilde{G}_\beta(\omega_2 - \omega_1). \tag{9.8.32}$$

Here, $\lambda^2 Q(1, 2)$ is the 'cross-section' of the interaction of the detector with the field φ. Besides the coupling constant λ it depends on the structure of the detector, that is of the form of the detector action and the type of interaction described by S_{int}. The factor $F(|\omega|)$ is a phase-space factor, which is proportional to the density of the number of states of the quanta of the frequency $(\omega, \omega + d\omega)$ in the vicinity of the detector. For the detector in the thermal bath $\tilde{G}_\beta(\omega_2 - \omega_1)$ is the Fourier transform of the Matsubara function for an oscillator of frequency ω in the thermal bath with the inverse temperature β

$$\tilde{G}_\beta(\omega) = \frac{\omega}{2\pi} \frac{1}{e^{\beta\omega} - 1}, \qquad \tilde{G}_\beta(-\omega) = \frac{\omega}{2\pi} \frac{e^{\beta\omega}}{e^{\beta\omega} - 1}. \tag{9.8.33}$$

Note that only this factor depends on the sign of the frequency ω, while the other factors in Eq. (9.8.32) depend only on the absolute value $|\omega|$. This is why the ratio $\dot{w}_{1\to2}/\dot{w}_{2\to1}$ is universal. As a result, one has (c.s. Eq. F.3.78)

$$\frac{\text{Probability of } 1 \to 2 \text{ transition per a unit time}}{\text{Probability of } 2 \to 1 \text{ transition per a unit time}} = e^{-\beta\omega}. \tag{9.8.34}$$

Here, $\omega = \omega_2 - \omega_1$ is the difference of the energies of the levels 2 and 1, and β is the inverse temperature. This relation is known as the detailed balance relation. It should be emphasized that the probability of the transition itself depends on the coupling constant λ, the nature of the internal degrees of freedom $q(\tau)$, and the phase volume of thermal quanta of the field φ near the position of the detector. These factors account for the difference of a 'relaxation time', that is the characteristic time that is required for the detector to reach its thermal equilibrium state. This time is not universal. For example, for small λ it can be large. For the case of a thermal quantum field in the presence of the potential U, considered in Appendix F.4, the relaxation time depends on the spatial position of the detector.

Similar calculations can be repeated for a *uniformly accelerated detector* in the flat spacetime interacting with the quantum (massless) field φ in the Minkowski vacuum state (see (Takagi 1986) for the review). The vacuum state $|M\rangle$ is a state without any quanta as measured by an inertial observer. The worldline γ of a uniformly accelerated observer is an integral line of the boost Killing vector field ξ_τ. One obtains for the probability of transition of the detector between the levels 1 and 2 an expression similar to Eq. (9.8.32). The factor $\lambda^2 Q(1, 2)$ is the cross-section of the detector. It depends on the detector structure. If the acceleration does not deform the detector essentially, this factor remains the same as for the inertial detector. The factor $F(|\omega|)$ describes the density of states of the Rindler quanta with the Rindler frequency ω measured at the position of the detector. Both of these factors depend on $|\omega|$, and cancel in the ratio Eq. (9.8.34).

The factor that determines this ratio contains the Fourier transform of the Wightman Green function (see Eq. (F.3.33)) of the quantum field φ

$$G^+(x, x') = \langle M|\varphi(x)\varphi(x')|M\rangle. \tag{9.8.35}$$

Let ω_1 and ω_2 be the Rindler energies of the detector levels 1 and 2, respectively. Then the rate of transitions of the accelerated detector per unit of Rindler proper time τ for these levels is

$$\dot{w}_{1 \to 2} \propto \int_{-\infty}^{\infty} d\tau \, e^{-i\omega\tau} G^+ \left(x(t - \frac{\tau}{2}), x(t + \frac{\tau}{2}) \right). \tag{9.8.36}$$

The Wightman function in the Minkowski vacuum state $|M\rangle$ is

$$G^+(X, X') = -\frac{1}{4\pi^2} \frac{1}{(T - T' - i\epsilon)^2 - (\mathbf{X} - \mathbf{X}')^2}. \tag{9.8.37}$$

This function calculated on the trajectory of the detector

$$T = \frac{1}{a} \sinh(at), \qquad X = \frac{1}{a} \cosh(at), \qquad Y = Z = 0, \tag{9.8.38}$$

and written in the Rindler coordinates takes the form

$$G^+ \left(x(t - \frac{\tau}{2}), x(t + \frac{\tau}{2}) \right) = -\frac{1}{8\pi^2} \frac{a^2}{\cosh(a(\tau - i\epsilon)) - 1}. \tag{9.8.39}$$

Remarkably, the form of this expression is the same as the form of the similar thermal Wightman function of the quantum field in the Minkowski spacetime for the inverse temperature $\beta = 2\pi/a$. This can be checked by comparison of Eq. (9.8.39) with Eq. (F.3.50). Substituting this Green function into Eq. (9.8.36) we get

$$\frac{\dot{w}_{1 \to 2}}{\dot{w}_{2 \to 1}} = e^{-\beta(\omega_2 - \omega_1)}, \qquad \beta = \frac{2\pi}{a}. \tag{9.8.40}$$

This is simply the detailed balance relation that guarantees the final thermal distribution over the (Rindler) energy levels of the uniformly accelerated detector interacting with the quantum field φ in the Minkowski vacuum state $|M\rangle$. This result is universal, that is it does not depend on the nature of the degrees of freedom of the detector and its structure. Moreover, this relation obtained for the scalar massless field, remains valid for other quantum fields.

This property can be formulated as follows: *from the point of view of a uniformly accelerated observer the Minkowski vacuum 'looks like' a thermal bath with the Unruh temperature*

$$\Theta_U = \frac{\hbar}{k_{\mathrm{B}} c} \frac{a}{2\pi}. \tag{9.8.41}$$

It should be emphasized that in a general case the phase factor $F(|\omega_{1 \to 2}|)$ in the Rindler frame differs from a similar factor for the similar inertial detector at rest in the thermal bath with the same temperature Θ_U. Thus, while the thermal nature of the detector excitation Eq. (9.8.40) is universal, the rate of excitation in these two cases (the accelerated and inertial detectors) can be different. To perform the calculations of the transition rates of the detector per unit time detailed information is required concerning both the type of the field, and the detector properties.

Similar arguments can be applied to a static detector in the vicinity of the static black hole. Let us assume that the quantum field is in the *Hartle–Hawking state*, that is the thermal Hawking radiation of the black hole is in equilibrium with the thermal radiation from the thermal reservoir at infinity. For the interaction Eq. (9.8.31) such a detector will measure a local *Tolman temperature* at the point of the observation. The locally measured inverse temperature is equal to the length of a circle at the position of the detector r, i.e. $\beta(r) = 2\pi\sqrt{|g_{00}|}/\kappa$ (see line C_2 in Figure 9.5), where κ is the horizon surface gravity. In the vicinity of the tip of the *Gibbons–Hawking instanton*, where the distance from the horizon is much smaller than the gravitational radius, the Rindler approximation is valid. In this limit, the local Hawking temperature coincides with the Unruh temperature measured by the Rindler observer with the same proper acceleration.

Let us make a few general remarks concerning the *Unruh effect*. First, the very concept of *Rindler particles* and thermal properties of the Minkowski vacuum arise only when one describes the corresponding processes (such as the accelerated detector behavior) in the *uniformly accelerated frame*. It is always possible to 'come back' to the 'inertial observer point of view' and to perform the required calculations in the inertial frame. If one addresses the effects that have a covariant description, the results of the calculations do not depend on the choice of the frame used for the calculations. For example, the distribution of the uniformly accelerated detector over its internal states can be easily calculated in the Minkowski frame. However, the concept of a thermal bath of the Rindler particles allows one to make quite general 'predictions' of the properties of accelerated systems with internal degrees of freedom. This is a consequence of the universality of the thermal properties of the Minkowski vacuum in an accelerated frame. Another aspect of the Unruh effect is a better understanding of quantum effects in the vicinity of the black hole horizon, in the regime when the curvature effects are not important and the Rindler approximation can be used.

9.8.6 Unruh effect and observations

The *Hawking temperature*

$$\Theta_H = \frac{\hbar c^3}{k_B G} \frac{1}{8\pi M} \approx 1.55 \times \left(\frac{M_\odot}{M}\right) \times 10^{-6}\text{K} \tag{9.8.42}$$

is extremely small for the stellar mass and supermassive black holes. The effect of quantum evaporation is important for mini-black holes, such as small-mass primordial black holes (Section 10.2.6) and hypothetical micro-black holes in the models with large extra dimensions (Section 10.5). At the moment we do not have an indication of their existence. An interesting option is to test the Hawking effect in special condensed-matter systems, imitating properties of black holes. We discuss this option in Section 10.4.

The Unruh effect is closely related to the Hawking effect. Let us discuss whether one can use accelerated particles to test the Unruh effect. Let us estimate what acceleration a is required to provide the Unruh temperature, say 1 K. Simple calculations give

$$a = 2.466 \times 10^{22} \left(\frac{\Theta_U}{1\,\text{K}}\right) \frac{\text{cm}}{\text{s}^2}. \tag{9.8.43}$$

The Unruh temperature is quite small for accelerations that are achievable in the laboratory experiments. One can say also vice versa that in order to observe the effects related to the Unruh temperature one needs such extreme accelerations that would destroy most physical objects. Here are some examples that provide an intuition on the characteristic scales of proper acceleration in different experiments.

For the acceleration on the surface of the Earth, which is $982\,\text{cm}\,\text{s}^{-2}$, the Unruh temperature would be $3.98 \times 10^{-20}\,\text{K}$. This temperature is too small to be detected in any feasible setup of the experiment. The frequency of thermal photons in this case is about one year^{-1}. Electrons in linear accelerators provide an example of a system where much higher accelerations can be achieved. For example, at the Stanford linear accelerator the electric field is about $10^5\,\text{V}\,\text{cm}^{-1}$. The acceleration of electrons in this field is about $2 \times 10^{20}\,\text{cm}\,\text{s}^{-2}$. The corresponding Unruh temperature is $\sim 7 \times 10^{-3}\,\text{K}$. Unfortunately, in the experiments with linear accelerators there is not enough time for the system of electrons to reach a thermal equilibrium. The typical time for the polarizations to relax to the equilibrium distribution is many orders of magnitude larger than the time of the experiment itself.

For circular accelerators and storage rings the case looks more promising. An achievable acceleration experienced by electrons in existing storage rings is about $3 \times 10^{25}\,\text{cm}\,\text{s}^{-2}$. If the direction of the acceleration were constant in time, the corresponding Unruh temperature would be around 1200 K. Electrons with the spins directed along to the external magnetic field and in the opposite direction have different energies. So, at zero temperature all spins, would be relaxed to one direction. Thermal effects will partially depolarize the electron spins, which is a measurable effect. The polarization equilibrium can be reached in a reasonable time of a few hours or even minutes. Unfortunately, the vector of acceleration for circular trajectories rotates and the direct comparison with the Unruh effect is not possible. However, for this case one can make quantum calculations similar to the Unruh calculations and find out qualitatively similar terms in the probability of excitation of the detector. In the reference frame of the electrons the spacetime is stationary but the vacuum does not look like exactly thermal any longer, contrary to the linear acceleration case. Nevertheless, the effect of spontaneous excitation of the stationary detector is still present. The probability of excitation should also be of the same order of magnitude as that of the detector in the thermal bath with the Unruh temperature. For example, consider electrons moving in a constant magnetic field. Because of the interaction of the spin of the electrons with the magnetic field the classical equilibrium state of the electrons is when all electrons are 100% polarized along the magnetic field. Quantum calculations similar to the calculations of the Unruh effect show that there should be spontaneous spin flips of the electrons accompanied by the photon radiation. As a result, the equilibrium distribution is not 100% polarized but rather a mixture of up- and down-polarized electrons. In the ideal case, for typical characteristics of the electron storage rings the polarization reaches the value of $\sim 92\%$. This property of depolarization of electrons is called the *Sokolov–Ternov effect*. The Sokolov–Ternov effect has been observed in real accelerators.

The proton accelerators are not as good as electron storage rings because protons are more massive, and hence the accelerations are smaller and the characteristic time of relaxation to the equilibrium is much longer. The Large Hadron Collider (LHC) is the world's largest and highest-energy particle accelerator. Its radius is $R = 4.3$ km. The projected top energy of protons is 7 TeV, which corresponds to the gamma factor $\gamma \equiv (1 - v^2/c^2)^{-1/2} \simeq 7500$. The acceleration experienced by protons then can be estimated from the formula $a = (\gamma^2 - 1)c^2/R$. The temperature corresponding to this acceleration is only 0.76 K. The smallness of this temperature though is not the main obstacle to measuring the depolarization effect. The problem is that the equilibrium can not be achieved in a lifetime of the experiment.

In the electric fields in the atomic nuclei one could expect much stronger accelerations. Hypothetically, if the electron is placed in the homogeneous electric field with a strength equal to that of the atom nuclei with the charge $Q = Ze$ at the closest orbits of electrons $r = \hbar^2/(Ze^2 m_e)$, then the Unruh temperature is $k_B \Theta_U/m_e c^2 = (Z\alpha)^3/2\pi$, where m_e is the electron mass and $\alpha = e^2/\hbar c \simeq 137.036^{-1}$ is the fine-structure constant. For small Z the energy of the thermal quanta is much less than the $m_e c^2$ and may become of the same order only for electric fields corresponding to $Z \sim 1/\alpha$.

Another possibility to reach large accelerations is to use ultra-intense lasers. In contemporary lasers by focusing an ultrashort laser pulse to a very small spot an irradiance of 10^{22} W cm^{-2} has been achieved. This irradiance is very high but still less than the critical one, when electron–positron pairs start to be produced from the vacuum by the electric field of the laser beam. This irradiance 10^{29} W cm^{-2} would correspond to a Schwinger limiting electric field of $E \sim 10^{16}$ V cm^{-1}. In real conditions by focusing an ultrashort pulse of the petawatt laser the irradiance 10^{23} W cm^{-2} or more could be achieved. Then, the characteristic accelerations are $\sim 10^{28}$ cm s^{-2} and the corresponding Unruh temperature $\sim 10^6$ K. For these intensities the electric field in the pulse is so strong that during half a period of the laser wave the electrons under its action become ultrarelativistic. The Larmor radiation of an accelerated electron in this laser pulse can be easily calculated. The electron backreacts to the Larmor radiation. This backreaction leads to an additional 'quivering radiation' that is quantum in nature and is closely related to the Unruh effect. In this setup of the problem, the degrees of freedom transverse to the acceleration of the electron effectively play the role of internal degrees of freedom of the Unruh detector. The classical Larmor radiation of the electron in the magnetic node of a linearly polarized standing electromagnetic wave vanishes in the direction of the electric field while the intensity of the quantum ('Unruh') part of the radiation is not zero in this direction. So, in a narrow range of angles where the Larmor radiation has a blind spot the quantum radiation may even dominate. Evidently, it is true only in an ideal setup of an experiment that is quite difficult to implement in real conditions.

The Unruh effect is a natural property of the quantum field theory in flat spacetime and its experimental testing is simply testing of the quantum field theory. What is appealing in the Unruh effect is the non-trivial interpretation of relativistic quantum effects from the point of view of non-inertial observers. More detailed discussion of experimental applications of the Unruh effect can be found in the review (Crispino 2008).

9.9 Black Hole Thermodynamics

9.9.1 Black holes and thermodynamics

Wheeler seems to have been the first to notice that the very existence of a black hole in the classical theory of gravitation contradicts the law of non-decreasing entropy. Indeed, imagine that a black hole swallows a hot body possessing a certain amount of entropy. Then, the observer outside of it finds that the total entropy in the part of the world accessible to his observation has decreased. This disappearance of entropy could be avoided in a purely formal way if we simply were to assign the entropy of the ingested body to the inner region of the black hole. In fact, this 'solution' is patently unsatisfactory because any attempt by an 'outside' observer to determine the amount of entropy 'absorbed' by the black hole is doomed to failure. Quite soon after the absorption, the black hole becomes stationary and completely 'forgets', as a result of 'balding', such 'details' as the structure of the ingested body and its entropy.

If we are not inclined to forgo the law of non-decreasing entropy because a black hole has formed somewhere in the universe, we have to conclude that any black hole *by itself* possesses a certain amount of entropy and that a hot body falling into it not only transfers its mass, angular momentum and electric charge to the black hole, but its entropy S as well. As a result, the entropy of the black hole increases by at least S. Bekenstein (1972, 1973) noticed that the properties of one of the black hole characteristics – its area $\mathcal{A}$ – resemble those of entropy. Indeed, Hawking's area theorem (see Section 10.2) implies that the area $\mathcal{A}$ does not decrease in any classical processes; that is, it behaves as entropy does. It was found, in fact, that the analogy of black hole physics to thermodynamics is quite far-reaching. It covers both gedanken experiments with specific thermodynamic devices (like the heat engine) and the general laws of thermodynamics, each of which has an analog in black hole physics.

An arbitrary black hole, like a thermodynamic system, reaches an equilibrium (stationary) state after the relaxation processes are completed. In this state, it is completely described by the small number of parameters: M, J, Q.

9.9.2 Four laws of black hole thermodynamics

The surface gravity κ, the angular velocity of the black hole Ω, and the electric potential Φ are constant on the event horizon of a stationary black hole (see, e.g., (Frolov and Novikov 1998) and references therein). Denote by $\mathcal{A}$ the surface area of the black hole of mass M. According to the thermodynamic analogy in black hole physics, the quantities

$$\Theta = \Theta_H = \frac{\hbar \kappa}{2\pi k_B c}, \quad S^H = \frac{\mathcal{A}}{4\,l_{Pl}^2}, \quad E = Mc^2 \tag{9.9.1}$$

play the role of *temperature*, *entropy*, and *internal energy* of the black hole, respectively. The angular velocity of the black hole Ω and the electric potential Φ of the horizon play the role of potentials conjugated to the total angular momentum and to the charge Q of the black hole.

Bardeen *et al.* (1973) formulated the four laws of black hole physics, which are similar to the four laws of thermodynamics.

Zeroth law. *The surface gravity κ of a stationary black hole is constant everywhere on the surface of the event horizon.*

Thermodynamics does not permit equilibrium when different parts of a system are at different temperatures. The existence of a state of thermodynamic equilibrium and temperature is postulated by the zeroth law of thermodynamics. The zeroth law of black hole physics plays a similar role and has been proved (Bardeen and Press 1973) under the assumption that the *dominant energy condition* (see Section 5.2.5) is satisfied.

First law. *When the system incorporating a black hole switches from one stationary state to another, its mass changes by*

$$dM = \Theta \, dS^H + \Omega \, dJ + \Phi \, dQ + \delta q, \tag{9.9.2}$$

where dJ and dQ are the changes in the total angular momentum J and electric charge Q of the black hole, respectively, and δq is the contribution to the change in the total mass due to the change in the stationary matter distribution outside the black hole.

Second law. *In any classical process, the area of a black hole, A, and hence its entropy S^H, do not decrease:*

$$\Delta S^H \geq 0. \tag{9.9.3}$$

This analog of the second law of thermodynamics is a consequence of Hawking's area theorem, which is valid provided the *weak-energy condition* is satisfied. In both cases (thermodynamics and black hole physics), the second law signals the irreversibility inherent in the system as a whole, and thus singles out the direction of the time arrow. In thermodynamics, the law of non-decreasing entropy signifies that the part of the internal energy that cannot be transformed into work grows with time. Quite similarly, the law of non-decreasing area of a black hole signifies that the fraction of a black hole's internal energy that cannot be extracted grows with time. As in thermodynamics, the quantity S^H stems from the impossibility of extracting any information about the structure of the system (in this case, the structure of the black hole).

Third law. In thermodynamics, the third law has been formulated in a variety of ways.[8]
Two (essentially equivalent) formulations, due to Nernst, state that:

- Isothermal reversible processes become isentropic in the limit of zero temperature.
- It is impossible to reduce the temperature of any system to the absolute zero in a finite number of operations.
- A stronger version, proposed by Planck, states that: The entropy of any system tends, as $\Theta \to 0$, to an absolute constant, which may be taken as zero.

Bardeen *et al.* (1973) formulated the analog of the **third law** for black holes in the following form: *It is impossible by any procedure, no matter how idealized, to reduce the black hole temperature to zero by a finite sequence of operations.*
Since the black hole temperature Θ vanishes simultaneously with κ, this is only possible if an isolated stationary black hole is extremal: $M^2 = a^2 + Q^2$. The impossibility of transforming a black hole into an extremal one is closely related to the impossibility of realizing a state with $M^2 < a^2 + Q^2$ in which a naked singularity would appear and the 'cosmic censor' principle would be violated. An analysis of specific examples (Wald 1974) shows that the closer the state of the black hole comes to the extremal one, the more restrictive are the conditions on the possibility of making the next step.
Israel (1986a) emphasized that it is difficult to define the meaning of an 'finite sequence of operations' considering only quasistatic processes that were analyzed by Bardeen *et al.* (1973). Israel proposed and proved the following version of the **third law**: *A non-extremal black hole cannot become extremal at a finite advanced time in any continuous process in which the stress-energy tensor of accreted matter stays bounded and satisfies the weak-energy condition in the neighborhood of the outer apparent horizon.*

---

[8]The third law of thermodynamics was formulated in 1906 by Nernst. In 1911 it was extended and reformulated by Planck. The precise conditions under which this law is valid are quite subtle, and there were many attempts to clarify them (see Wilson (1957), Callon (1960), Lewis and Randall (1961), Kastin (1968)). For a discussion of the third law in connection with black hole physics, see Israel (1986b).

9.9.3 Generalized second law

Quantum effects violate the condition for the applicability of Hawking's area theorem. Thus, quantum evaporation reduces the area of black holes, and inequality (9.9.3) is violated. On the other hand, black hole radiation is thermal in nature, and the black hole evaporation is accompanied by a rise in entropy in the surrounding space. One may expect that the so-called *generalized entropy* $\tilde{S}$, defined as the sum of the black hole entropy S^H and the entropy of the radiation and matter outside the black hole, S^m,

$$\tilde{S} = S^H + S^m, \tag{9.9.4}$$

does not decrease. In fact, we note that the rate of increase (by the clock of a distant observer) of mass and entropy of matter in the black hole exterior, because of Hawking radiation of a massless spin-s field, can be written in the form

$$\frac{dM^m}{dt} = -\frac{dM}{dt} = \frac{1}{4} \sigma_s \, h_s \, \Sigma_s \, \Theta_H^4, \qquad \frac{dS^m}{dt} = \frac{1}{3} \sigma_s \, B_s \, h_s \, \Sigma_s \, \Theta_H^3, \tag{9.9.5}$$

where h_s is the number of polarizations of the field; $\sigma_s = \pi^2/30$ for bosons and $7\pi^2/240$ for fermions; Σ_s is the effective cross-section of the black hole; Θ is its temperature, and B_s is a dimensionless coefficient of order of unity. On the other hand, the change in the entropy S^H of a non-rotating black hole is related to the change in its mass M by the formula

$$dS^H = \Theta_H^{-1} \, dM. \tag{9.9.6}$$

Comparing Eq. (9.9.5) and Eq. (9.9.6), we find (Zurek 1982)

$$R \equiv -\frac{dS^m}{dS^H} = \frac{4}{3} B_s. \tag{9.9.7}$$

Numerical results obtained (Zurek 1982; Page 1983) demonstrate that the coefficient B_s is always greater than 3/4; hence, the generalized entropy $\tilde{S}$ increases when an isolated black hole emits radiation. It can be shown (Zurek 1982) that if there is black-body radiation at a temperature $\tilde{\Theta}$ outside the black hole, the generalized entropy again increases, except in the case when $\tilde{\Theta} = \Theta$. In this special case, the increase in entropy in the space around the black hole due to its evaporation is exactly compensated for by a decrease in the entropy in this space, due to the accretion of thermal radiation onto the black hole. These arguments give sufficient grounds for assuming that the following law holds:[9]

Generalized second law. *In any physical process involving black holes, the generalized entropy $\tilde{S}$ does not decrease.*

$$\Delta \tilde{S} = \Delta S^H + \Delta S^m \geq 0. \tag{9.9.8}$$

The fact that the generalized second law includes, on an equal footing, the seemingly very different quantities, S^m (which characterizes the 'degree of chaos' in the structure of the physical matter) and S^H (which is a geometric characteristic of the black hole), is a new

---

[9]The generalized second law in this formulation was first suggested by Bekenstein (1972, 1973, 1974) before the quantum radiation of black holes was discovered.

indication of their profound similarity. In fact, the very possibility of this relation is rooted in Einstein's equations, which relate the physical characteristics of matter with the geometric characteristics of spacetime.

> *Thermal properties of ordinary thermodynamical systems are macroscopic reflections of their micro-scopic nature. From the microscopic point of view entropy is the logarithm of the number of micro-scopic states of the system with given macroscopic properties. Temperature describes the average energy of constituents of a system. The great success of statistical mechanics is the microscopic expla-nation of thermodynamical properties. In the presence of black holes thermodynamics is inconsistent without prescribing very specific thermodynamical properties such as entropy and temperature to the black holes themselves. Therefore, one can expect that there should exist a statistical mechanical derivation of the black hole entropy as well. Then we are in big trouble, since a stationary black hole is a vacuum solution of the Einstein equations. The gravitational field is in its ground vacuum state without excitations of the field. It seems that in gravitational theory there is nothing to play the role of microscopical degrees of freedom.*
>
> *A hint to a possible resolution of this paradox lies in an observation that the Bekenstein–Hawking entropy depends both on Planck's constant ħ and the gravitational constant G. So, the degrees of freedom that may describe the statistical mechanics of black holes, have to be related to quantum gravity. There is a 'natural' explanation of the black hole entropy if the gravity is an emergent phenomenon. Namely, there exists a more fundamental theory of some 'heavy' constituents, which are dynamical at the high energy comparable with the Planckian one. In the low-energy regime only some collective degrees of freedom of these constituents can be excited. These collective low-energy degrees of freedom are described by the gravitational field. To calculate the black hole entropy one needs to calculate the number of states of the heavy constituents corresponding to the fixed configuration of the gravitational field describing the black hole.*
>
> *Until now there is no consensus about what these degrees of freedom are. For example, in the string theory, the fundamental degrees of freedom are connected with string excitations. There are interesting examples where the entropy of the string states gives the correct value of the black hole entropy (see, e.g, the recent review by Sen (2008) and references therein).*
>
> *There are many very different microscopic models that provide a correct final formula for the black hole entropy. Here is an incomplete list of approaches: Weakly coupled strings and branes in string theory; ADS–CFT correspondence; various induced gravity models; entanglement entropy; fuzzballs, loop quantum gravity, etc. (see (Carlip 2009) for a review). All these approaches, which have very different candidates for the fundamental constituents of quantum gravity, reproduce the expression for Bekenstein–Hawking entropy of a black hole. In this sense, the law $S_{\mathrm{BH}} = \mathcal{A}/4$ is universal. An interesting question is: "Why is everyone getting the same answer?" (Carlip 2009)*

9.10   Higher-Dimensional Generalizations

9.10.1   Dimensional reduction

Dimensional reduction of field equations

Let us discuss briefly the problem of the field propagation in the spacetime of higher-dimensional black holes. Following the adopted strategy we indicate only a general approach to this problem and demonstrate its similarity and differences with respect to the 4D case. As earlier, we use a simplest possible case of a spherically symmetric black hole. We also use a model of a scalar massless field in order to simplify calculations. The action for the field φ is

$$S = -\int d^D x \sqrt{|g|} \left[\frac{1}{2} g^{\mu\nu} \partial_\mu \varphi \, \partial_\nu \varphi - j\varphi \right]. \tag{9.10.1}$$

The spherically symmetric vacuum solution of the D-dimensional vacuum Einstein equations is the *Tangherlini metric* , which is a straightforward generalization of the 4D Schwarzschild solution. It has the form (see Section 6.8)

$$ds^2 = d\gamma^2 + r^2\, d\Omega_n^2, \qquad d\gamma^2 = \gamma_{AB}\, dx^A\, dx^B, \tag{9.10.2}$$

where $n = D - 2$, $A, B = 0, 1$ and $d\Omega_n^2$ is a metric on a unit n-dimensional sphere S^n. For the Tangherlini metric in the 'standard' (t, r) coordinates

$$d\gamma^2 = -g\, dt^2 + \frac{1}{g}dr^2, \qquad g = 1 - \frac{r_S^{n-1}}{r^{n-1}}. \tag{9.10.3}$$

We shall use this expression later, but at the moment we discuss the massless field theory in a general spherically symmetric spacetime with the metric Eq. (9.10.2).
Let us write the field φ as

$$\varphi = \frac{\psi}{r^{n/2}}. \tag{9.10.4}$$

Then, the field equation

$$\Box\varphi = -j \tag{9.10.5}$$

in the metric Eq. (9.10.2) takes the form

$$^{(2)}\Box\psi - \frac{n}{2}\left[\frac{^{(2)}\Box r}{r} + \left(\frac{n}{2} - 1\right)\frac{(\nabla r)^2}{r^2}\right]\psi + \frac{1}{r^2}{}^{(n)}\triangle\psi = -r^{n/2}j. \tag{9.10.6}$$

Here, $(\nabla r)^2 = \gamma^{AB}\partial_A r\partial_B r$, $^{(2)}\Box$ is a box in the 2D metric $d\gamma^2$, and $^{(n)}\triangle$ is a Laplace operator on a unit n-dimensional sphere S^n

$$^{(2)}\Box = \frac{1}{\sqrt{|\gamma|}}\partial_A\left(\sqrt{|\gamma|}\gamma^{AB}\partial_B\right), \qquad {}^{(n)}\triangle = \frac{1}{\sqrt{\Omega}}\partial_i(\sqrt{\Omega}\,\Omega^{ij}\partial_j). \tag{9.10.7}$$

Eigenfunctions of $^{(n)}\triangle$

The complete set of eigenfunctions of $^{(n)}\triangle$ can be obtained as follows (Rubin and Ordóñez 1984). Let X^c, $c = 1, \ldots, n + 1$ be the Cartesian coordinates in $(n + 1)$-dimensional Euclidean space $\mathbb{R}^{n+1}$. Consider functions of the form

$$\mathcal{Y}_l = R^{-l}C_{c_1\ldots c_l}X^{c_1}\ldots X^{c_l}, \tag{9.10.8}$$

where $C_{c_1\ldots c_l}$ is a symmetric traceless rank-l tensor and

$$R^2 = \sum_{c=1}^{n+1}(X^c)^2. \tag{9.10.9}$$

The number of linearly independent components of the coefficients $C_{c_1\ldots c_l}$ is

$$d_0(n, l) = \frac{(l + n - 2)!\,(2l + n - 1)}{l!(n - 1)!}. \tag{9.10.10}$$

We shall use the index $q = 1, \ldots, d_0(n, l)$ to enumerate these independent components and write them in the form $C^q_{c_1 \ldots c_l}$. We also use the notation

$$\mathcal{Y}_{lq} = R^{-l} C^q_{c_1 \ldots c_l} X^{c_1} \ldots X^{c_l}. \tag{9.10.11}$$

The functions Eq. (9.10.11) are in fact functions on a unit sphere since they depend only on the unit vectors $n^i = X^i/R$, $\sum_{i=1}^{n+1} n^{i2} = 1$. These harmonics are eigenfunctions of the invariant Laplace operator $^{(n)}\triangle$ on a unit sphere S^n with eigenvalues $-l(n + l - 1)$

$$^{(n)}\triangle \mathcal{Y}_{lq} = -l(n + l - 1)\mathcal{Y}_{lq}. \tag{9.10.12}$$

They form a complete set. One can choose these harmonics so that they satisfy the following normalization conditions

$$\int_{S^n} d\Omega_n \sqrt{\Omega_n} \, \mathcal{Y}_{lq} \mathcal{Y}_{l'q'} = \delta_{ll'} \delta_{qq'}. \tag{9.10.13}$$

Reduction to 2D problem

As in the 4D case we decompose both the field ϕ and the source j into modes and write

$$\varphi = \sum_{l=0}^{\infty} \sum_{q} \frac{\varphi_{lq}(x^A)}{r^{n/2}} \mathcal{Y}_{lq}, \qquad j = \sum_{l=0}^{\infty} \sum_{q} \frac{j_{lq}(x^A)}{r^{n/2}} \mathcal{Y}_{lq}. \tag{9.10.14}$$

The field equation Eq. (9.10.6) results in the following mode equations

$$^{(2)}\Box \varphi_{lq} - W_l \varphi_{lq} = -j_{lq},$$

$$W_l = \frac{n}{2} \left[\frac{^{(2)}\Box r}{r} + \left(\frac{n}{2} - 1 \right) \frac{(\nabla r)^2}{r^2} \right] + \frac{l(l + n - 1)}{r^2}. \tag{9.10.15}$$

By substituting the decomposition Eq. (9.10.14) into the action Eq. (9.10.2) one obtains

$$S = \sum_{l=0}^{\infty} \sum_{q} S_{lq},$$

$$S_{lq} = -\frac{1}{2} \int d^2x \sqrt{|\gamma|} \left[\gamma^{AB} \partial_A \varphi_{lq} \partial_B \varphi_{lq} + W_l \varphi_{lq}^2 - 2 j_{lq} \varphi_{lq} \right]. \tag{9.10.16}$$

Thus, as earlier, the original D-dimensional theory in a spherically symmetric spacetime is effectively reduced to an infinite set of 2D theories enumerated by the index (l, q). The expression for S_{lq} in Eq. (9.10.16) gives the action of the corresponding reduced 2D theory with a field variable $\varphi_{lq}(x)$. Let us note that the function W_l does not depend on q. Hence, in the absence of the source the 2D actions for a given l with different q are identical.

Spherical reduction in the Tangherlini metric

In the case when the background geometry is described by the *Tangherlini metric*,

$$d\gamma^2 = -g \, dt^2 + dr^2/g, \qquad g = 1 - (r_S/r)^{n-1}, \tag{9.10.17}$$

the 2D action Eq. (9.10.16) allows a further simplification. Denote as earlier by r_* a *tortoise coordinate* that now is

$$r_* = \int \frac{dr}{g(r)} = rF\left(1, -\frac{1}{n-1}; \frac{n-2}{n-1}; (r_S/r)^{n-1}\right), \tag{9.10.18}$$

where $F(a, b; c; z)$ is a hypergeometric function. Using this notation, one can rewrite Eq. (9.10.16) in the form

$$S_{lq} = \int dt \, L_{lq},$$

$$L_{lq} = \frac{1}{2} \int dr_* \left[(\partial_t \varphi_{lq})^2 - (\partial_{r_*} \varphi_{lq})^2 - U_l \varphi_{lq}^2 + 2g(r) j_{lq} \varphi_{lq}\right],$$

$$U_l(r_*) = g(r) \, W_l \tag{9.10.19}$$

$$= \left(1 - \frac{r_S^{n-1}}{r^{n-1}}\right)\left[\left(\frac{n}{2}\right)^2 \frac{r_S^{n-1}}{r^{n+1}} + \frac{(n/2)(n/2 - 1) + l(l + n - 1)}{r^2}\right].$$

We used here the expressions

$$\frac{{}^{(2)}\Box r}{r} = \frac{(n-1)r_S^{n-1}}{r^{n+1}}, \qquad \frac{(\nabla r)^2}{r^2} = \frac{1}{r^2} - \frac{r_S^{n-1}}{r^{n+1}}. \tag{9.10.20}$$

For $n = 2$ the action Eq. (9.10.19) coincides with the reduced action Eq. (9.3.12) for the Schwarzschild metric. It is easy to see that in the absence of a source the action Eq. (9.10.19) has the form of the action Eq. (F.4.1) for the $(1 + 1)$-dimensional theory considered in Appendix F.4, provided the *tortoise coordinate* r_* is identified with the flat space coordinate x.

9.10.2   Quasinormal modes and Hawking radiation

After reduction of the D-dimensional field theory to a set of 2D problems one can apply most of the relevant results discussed in the present chapter to the D-dimensional case. In D dimensions the characteristic form of the effective potential $U_l(r_*)$ qualitatively has the same behavior as in the 4D case. Namely, it takes a zero value at the horizon, $r_* = -\infty$, reaches a maximal value near $r_* \sim r_S$ and falls down to zero value at infinity $r_* = \infty$. The characteristic form of the potential indicates that for the modes propagating in the vicinity of the top of the potential one may expect the existence of resonances (*quasinormal modes*). The numerical calculations confirm this expectation. These modes, when excited by time-dependent sources, are responsible for the *ringing radiation*.

In the application to the quantum effects in the higher-dimensional black holes, the obtained result allows one immediately to write the following expression for the energy rate of emission for quantum (Hawking) radiation by a non-rotating D-dimensional black hole

$$\dot{E} = \frac{1}{2\pi} \int_0^\infty d\omega \frac{\omega}{e^{\omega/\Theta_H} \mp 1} \sum_{l=0}^\infty d_0(n, l) |T_{\omega,l}|^2. \tag{9.10.21}$$

The minus sign in the denominator corresponds to bosons, while the sign plus stands for fermions. This formula is similar to Eq. (9.7.4). $|T_{\omega,l}|^2 = |t_{\omega,l}|^2$ is the graybody factor for the potential U_l given by Eq. (9.10.19). The exact value of the evaporation rate of the black hole requires numerical estimations of the graybody factors for every mode. But one can easily estimate the evaporation rate using the fact that the only parameter describing the system is the gravitational radius or, equivalently, the Hawking temperature that in $D = 2 + n$ dimensions is

$$\Theta_H = \frac{\kappa}{2\pi} = \frac{n-1}{4\pi r_S}. \tag{9.10.22}$$

To estimate the rate of emission of the higher-dimensional black hole we note that its surface area is $A \sim r_S^n$, while the thermal energy density $\varepsilon \sim \Theta_H^{n+2}$. Thus, the rate of the energy emission by such a black hole can be estimated as $\dot{E} \sim \varepsilon A$, which gives

$$\dot{E} = q_n \frac{1}{r_S^2}, \tag{9.10.23}$$

where the coefficient q_n depends on the number of dimensions and on the spin of the field.

Exact calculations show that this dependence is not very strong and every polarization of the quantum field contributes roughly the same value to the radiation rate of the higher-dimensional black hole. Scalar particles contribute a little more than other fields (per number of polarizations) but this property is common to all dimensions. Qualitatively, quantum evaporation of the higher-dimensional black hole looks very much like the evaporation of the four-dimensional one.

Problem 9.3: *Calculate the rate of accretion of a thermal radiation onto a static D-dimensional black hole.*

Solution: *This problem is a higher-dimensional generalization of the problem we discussed in Section 9.6.3. Let us give here first expressions for a volume V_n and the surface area S_{n-1} of an n-dimensional ball of radius R*

$$V_n = C_n R^n, \quad S_{n-1} = n C_n R^{n-1}, \quad C_n = \frac{\pi^{n/2}}{\Gamma(n/2 + 1)}. \tag{9.10.24}$$

Denote by g the number of polarizations of a 'photon', and $d\Gamma$ their phase volume

$$d\Gamma = \frac{d^{D-1} x \, d^{D-1} k}{(2\pi)^{D-1}}. \tag{9.10.25}$$

For the thermal distribution with temperature Θ_0 the number density is

$$n_k = \frac{1}{e^{k/\Theta_0} - 1}. \tag{9.10.26}$$

Thus, one has for the energy of photons in this phase volume

$$\Delta E = g \, d\Gamma \, k \, n_k. \tag{9.10.27}$$

As in Section 9.6.3 we consider those photons in the spherical shell of radius R and width ΔR, which are absorbed by the black hole during time $\Delta t = \Delta R$. As earlier, we take the radius R large enough so that we can neglect the gravitational field at this distance and use 'flat space' quantities. For the corresponding spatial volume $d^{D-1} x$ one has

$$d^{D-1} x = \Delta R \, S_{D-2} = \frac{2\pi^{(D-1)/2}}{\Gamma((D-1)/2)} R^{D-2} \Delta R. \tag{9.10.28}$$

To find the momentum volume $d^{D-1}k$ near the momentum k we choose the spherical coordinates in the momentum space so that the axis of symmetry $\theta = 0$ passes through k. One has

$$d^{D-1}k = dk\, k^{D-2} d\Omega, \quad d\Omega = S_{D-3} \sin^{D-3}\theta \, d\theta. \tag{9.10.29}$$

$d\Omega$ is a solid angle corresponding to the momentum volume $d^{D-1}k$. The condition that photons are captured by the black hole gives the following restriction on the angle of their momentum with respect to the symmetry axis

$$0 \le \theta \le \theta_* = \frac{\ell_{\min} r_S}{R}. \tag{9.10.30}$$

Here, $\ell_{\min}$ is the critical value of the impact parameter which is (see Eq. (7.10.17))

$$\ell_{\min} = \left(\frac{D-1}{2}\right)^{1/(D-3)} \sqrt{(D-1)/(D-3)}. \tag{9.10.31}$$

Since $r_S/R \ll 1$ one has

$$\int_0^{\theta_*} d\theta \, \sin^{D-3}\theta = \frac{1}{D-2} \frac{\ell_{\min}^{D-2} r_S^{D-2}}{R^{D-2}}. \tag{9.10.32}$$

Since the selected photons move practically along the radius, during the time interval Δt their radius changes by ΔR.
Combining the obtained relations one gets

$$\dot{E} = g \frac{1}{\pi \Gamma(D-1)} \ell_{\min}^{D-2} r_S^{D-2} J_D,$$

$$J_D = \int_0^\infty \frac{dk \, k^{D-1}}{e^{k/\Theta_0} - 1}. \tag{9.10.33}$$

To obtain this result we used the following property of Γ-functions

$$\Gamma((D-2)/2)\Gamma((D-1)/2) = \frac{8\sqrt{\pi}}{2^D} \Gamma(D-2). \tag{9.10.34}$$

The integral over the momentum k can be taken explicitly

$$J_D = \Theta_0^D \Gamma(D)\zeta(D), \tag{9.10.35}$$

where ζ is the Riemann zeta-function.
Finally, for the rate of energy accretion by the black hole we obtain

$$\dot{E} = g(D-1)\pi^{-1}\zeta(D)\, \Theta_0^D \, \ell_{\min}^{D-2} r_S^{D-2}, \tag{9.10.36}$$

where $\ell_{\min}$ is given by Eq. (9.10.31).

Problem 9.4: *Calculate the rate of energy accretion $\dot{E}$ of the thermal radiation with redshifted temperature Θ_0 for a D-dimensional static black hole using the DeWitt approximation.*

Solution: *Like in the case of a 4-dimensional non-rotating black hole, the rate of emission can be calculated as follows (cf. Eq. (9.6.23))*

$$\dot{E}_{\mp} = \frac{g}{2\pi} \int_0^\infty d\omega \frac{\omega \Gamma_\omega}{e^{\beta\omega} \mp 1},$$

$$\Gamma_\omega = \sum_{l=0}^\infty d_o(D,l)|T_{\omega,l}|^2, \quad d_o(D,l) = \frac{(l+D-4)!(2l+D-3)}{l!(D-3)!}.$$

(9.10.37)

Here, g is the number of polarization states. The sign $(-)$ in the integrand is for bosons, while the sign $(+)$ stands for fermions.

In the DeWitt approximation we have

$$|T_{\omega,l}|^2 \approx \theta(l_o - l),$$

(9.10.38)

where $l_o = \omega r_S \ell_{min}$, and ℓ_{min} is given by Eq. (7.11.17). This gives

$$\Gamma_\omega \approx \sum_{l=0}^\infty d_o(D,l)\theta(l_o - l) = \frac{(l_o + 1)(2l_o + D - 2)(l_o + D - 3)!}{(D-2)!(l_o + 1)!}.$$

(9.10.39)

For $l_o \gg D$ we have

$$\Gamma_\omega \approx \frac{2\omega^{D-2} r_S^{D-2} \ell_{min}^{D-2}}{(D-2)!}.$$

(9.10.40)

Substituting this into the expression for the energy rate we derive

$$\dot{E}_{\mp} \approx g \frac{r_S^{D-2} \ell_{min}^{D-2}}{\pi(D-2)!} \Theta_0^D \int_0^\infty \frac{x^{D-1} dx}{e^x \mp 1}.$$

(9.10.41)

The integral on the right-hand side gives

$$\int_0^\infty \frac{x^{D-1} dx}{e^x \mp 1} = B_{\mp} \Gamma(D)\zeta(D),$$

(9.10.42)

where $B_- = 1$ for bosons, $B_+ = 1 - 2^{-(D+1)}$ for fermions. The emission rate is

$$\dot{E}_{\mp} = g B_\pm (D-1)\pi^{-1} \zeta(D) \Theta_0^D \ell_{min}^{D-2} r_S^{D-2}.$$

(9.10.43)

For bosons, this result coincides with Eq. (9.10.36), as it must.

In a special case, when the temperature Θ_0 coincides with the Hawking temperature

$$\Theta_H = \frac{D-3}{4\pi r_S},$$

(9.10.44)

the same expression Eq. (9.10.43) can be used to find the energy rate for the Hawking radiation of the higher-dimensional black hole in the DeWitt approximation. It is easy to check that this result correctly reproduces the universal relation $\dot{E} = q_n r_S^{-2}$ (see Eq. (9.10.23)), is valid for any number of dimensions and determines the coefficient q_n in this expression.

10
Black Holes and All That Jazz

10.1 Asymptotically Flat Spacetimes

10.1.1 Introduction

A Kerr black hole is a stationary vacuum solution of the Einstein equations. This solution is uniquely specified by its mass and angular momentum. Natural questions are:

1. Do there exist non-stationary black holes?
2. Do there exist stationary black holes, different from the Kerr one?

The answer to the first question is 'Yes'. A collapse of an ideal spherical body creates a static Schwarzschild black hole. But if a body is not spherical or/and rotates, the black hole will be non-static. However, such a non-stationary black hole emits gravitational waves. These waves partly propagate to infinity and partly are scattered back and absorbed by the black hole. As a result of this process an isolated black hole relaxes to a final stationary state. In this sense, a time-dependent black hole can be considered as an 'excited state' that evolves after relaxation to a stationary 'ground state'. That is why study of stationary black holes is such an important part of black hole physics.

The answer to the second question is 'No'. Any regular stationary solution of the vacuum Einstein equations, which contains a black hole in an asymptotically flat spacetime, necessarily is the Kerr metric or, in the absence of rotation the Schwarzschild one. This is a result summarizing now-classical theorems on general properties of black holes. In the next section we discuss some of these results. The comprehensive presentation of this material can be found in the books (Misner *et al.* 1973; Hawking and Ellis 1973, Thorne *et al.* 1986, Wald 1984; Frolov and Novikov 1998).

It should be emphasized that it is not so simple to address these kind of problems without using developed mathematical tools. That is why special assumptions are often present in the formulation of the theorems on black holes. Usually, these assumptions are quite 'natural'. We focus mainly on vacuum black hole solutions, which are of most importance for applications.

Our first goal is to give a general definition of a black hole. It is quite natural to define a black hole as a *region of a spacetime from where no information-carrying signals can escape to infinity*. The picture we have in mind is an isolated compact object with a strong gravitational field. An observer is located at a far distance from this object, where the gravitational field is weak and the spacetime is practically flat. When we say that the information cannot escape from some region to infinity, we understand that all the far-distant observers agree with this conclusion. It looks that at least in the case of an isolated black hole the definition is clear. However, already at this first step there is a problem: What does it mean *asymptotically flat spacetime* and which set of observers do we include in our consideration? The questions are mainly related to the general covariance of the Einstein theory. The property that is valid in one coordinate system may be wrong in another, until one uses geometrically invariant and well-defined description. For this reason, let us first give a definition of an *asymptotically flat spacetime*.

> *To illustrate the 'problem of infinity' let us discuss a simple example. Consider a two-dimensional flat plane $\mathbb{R}^2$. Denote by (r, ϕ) polar coordinates on it, connected with the standard Cartesian coordinates as follows*
>
> $$x = r \cos \phi, \qquad y = r \sin \phi. \tag{10.1.1}$$
>
> *Let us make the following coordinate transformation $r = \tan(\theta/2)$. It is easy to see that it brings points of the 'coordinate infinity' $r = \infty$ to a finite coordinate distance. Namely, any point of $\mathbb{R}^2$ has finite coordinates (θ, ϕ) inside the coordinate disc $0 \le \theta \le \pi$, $0 \le \phi < 2\pi$. The points at infinity, where $r = \infty$, are mapped on the boundary $\theta = \pi$ of the disc. The flat plane metric $dl^2 = dx^2 + dy^2 = dr^2 + r^2 d\phi^2$ in the new coordinates is*
>
> $$dl^2 = \Omega^2 d\tilde{l}^2, \quad d\tilde{l}^2 = d\theta^2 + \sin^2 \theta d\phi^2, \quad \Omega^2 = \frac{1}{4\cos^4(\theta/2)}. \tag{10.1.2}$$
>
> *Here, $d\tilde{l}^2$ is a metric on a unit round sphere. The 'infinity' of $\mathbb{R}^2$ is mapped to the 'south pole' of the unit sphere. At this point the conformal factor Ω becomes infinite. This is a natural price one needs to pay for bringing 'infinity to a finite distance'. As a result of the coordinate transformations combined with the conformal one, one can reduce study of asymptotic behavior at $r \to \infty$ of functions, for example solutions of the Laplace equations, to a local problem of their properties at a single point (the south pole) of the regular 2D space (unit sphere).*

In a curved spacetime a similar approach was developed by Penrose (1968) who proposed a definition of an asymptotically flat spacetime. To illustrate this definition we first discuss *conformal infinity* of the Minkowski spacetime. The main idea is to make first a coordinate transformation 'bringing' infinity to finite values of the coordinate parameters, and after this to find a proper conformal factor that maps the points of infinity to some boundary points of a regular geometry. The conformal map preserves the null geodesics. Thus, the conformal image has the same causal structure as the original spacetime. The basic idea of the Penrose approach is that the conformal structure of the asymptotically flat spacetime at infinity is similar to the structure of the flat spacetime. We also introduce so-called *Carter–Penrose conformal diagrams*, which provide one with powerful tools to study global causal structure of a spacetime. Later, we show how this approach allows one to formulate the general definition of a black hole in purely geometrical terms.

10.1.2   Infinity structure of the Minkowski spacetime

Let us consider the Minkowski spacetime and write its metric

$$ds^2 = -dT^2 + dX^2 + dY^2 + dZ^2, \tag{10.1.3}$$

in the spherical coordinates $X = R \sin \theta \sin \phi$, $Y = R \sin \theta \cos \phi$, $Z = R \cos \theta$

$$ds^2 = -dT^2 + dR^2 + R^2 d\omega^2. \tag{10.1.4}$$

Here, $d\omega^2 = d\theta^2 + \sin^2\theta\, d\phi^2$ is the metric on a unit sphere S^2. The radial coordinate R changes in the interval $[0, \infty)$. Let us make the following change of coordinates

$$T \pm R = \tan\frac{1}{2}(\psi \pm \xi). \tag{10.1.5}$$

The coordinates (ψ, ξ) change in the domain

$$0 \le \xi \le \pi, \qquad -\pi + \xi \le \psi \le \pi - \xi. \tag{10.1.6}$$

In these coordinates the interval ds^2 takes the form

$$ds^2 = \Omega^{-2}\, d\tilde{s}^2, \qquad d\tilde{s}^2 = -d\psi^2 + d\xi^2 + \sin^2\xi\, d\omega^2, \tag{10.1.7}$$

where

$$\Omega = 2\cos\frac{1}{2}(\psi + \xi)\cos\frac{1}{2}(\psi - \xi). \tag{10.1.8}$$

The conformal factor Ω vanishes at the 'boundary' of the Minkowski spacetime. This boundary consists of the following pieces:

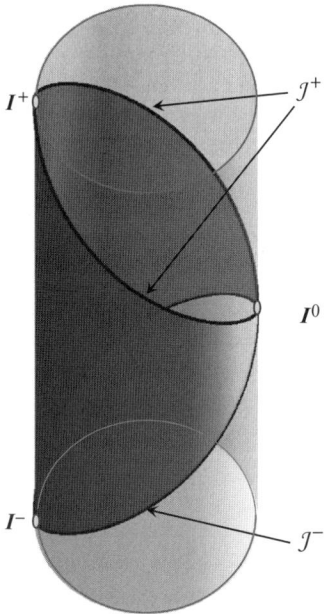

Fig. 10.1 Carter–Penrose diagram of the Minkowski spacetime.

1. The past null infinity $\mathcal{J}^-$, where $\psi = -\pi + \xi$ and $0 < \xi < \pi$.

2. The future null infinity $\mathcal{J}^+$, where $\psi = \pi - \xi$ and $0 < \xi < \pi$. These boundaries are null surfaces in the metric $d\tilde{s}^2$.

3. The spatial infinity I^0 where $\xi = \pi$.

4. The future, I^+, and the past I^- time-like infinities, where $\psi = \pi$ and $\psi = -\pi$, respectively.

The coordinate transformation Eq. (10.1.5) is chosen so that it brings the points of infinity to the finite coordinate 'distance'. Points at infinity in the Minkowski spacetime correspond to $\psi + \xi$ and $\psi - \xi$ with the values $\pm\pi$. At these values, the metric ds^2 becomes meaningless, but the metric $d\tilde{s}^2$, conformal to ds^2, is regular[1]. At a fixed value of ψ the metric Eq. (10.1.7) is

$$d\tilde{l}^2 = d\xi^2 + \sin^2\xi\, d\omega^2. \tag{10.1.9}$$

This is a metric on a unit 3D sphere S^3. Hence Eq. (10.1.7) is the metric of a static *Einstein universe*. In fact the Minkowski spacetime is mapped onto a part of this space. Figure 10.1 shows a 2D-sector $\theta = \pi/2, \phi = 0$, where the conformal metric is

[1] The metric $d\tilde{s}^2$ has removable coordinate singularities at $\xi = 0$ and $\xi = \pi$.

$$d\tilde{s}^2 = -d\psi^2 + d\xi^2. \tag{10.1.10}$$

Equations $\psi \pm \xi = $ const give null lines. A shadowed region is an image of the Minkowski spacetime.

To illustrate the properties of the Minkowski spacetime in the new coordinates we consider its 3D section $Z = 0$. On this section $\theta = \pi/2$ and the line element is

$$ds^2 = -dT^2 + dR^2 + R^2 d\phi^2 = \Omega^2 d\tilde{s}^2, \quad d\tilde{s}^2 = -d\psi^2 + d\xi^2 + \sin^2 \xi d\phi^2. \tag{10.1.11}$$

Figures 10.2–10.4 show the coordinate domain for (ψ, ξ, ϕ). We choose ξ as a radial coordinate, and ϕ is the corresponding polar angle. Let us emphasize that these figures are not embedding diagrams. We use them to illustrate the coordinate form of different surfaces and lines of the original Minkowski spacetime. Namely, Figure 10.2 shows surfaces of constant radius R of the Minkowski spacetime, Figure 10.3 shows surfaces of the fixes Minkowski time T, and Figure 10.4 shows a set of null cones passing through the center $R = 0$.

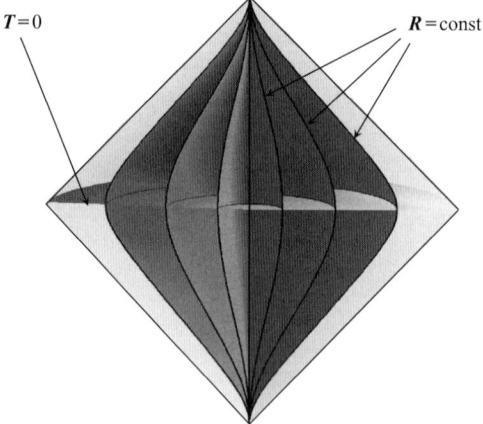

Fig. 10.2 Carter–Penrose diagram of the Minkowski spacetime. This figure shows time-like surfaces $R = $ const.

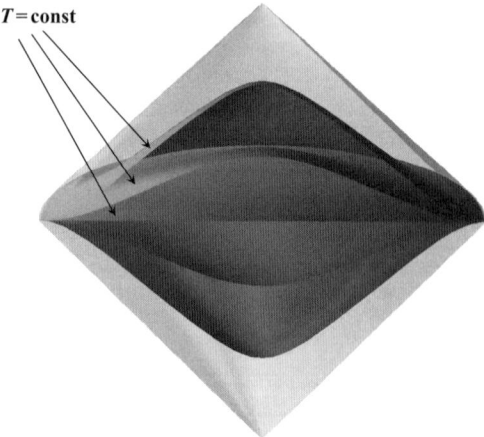

Fig. 10.3 Carter–Penrose diagram of the Minkowski spacetime. This figure shows space-like surfaces $T = $ const.

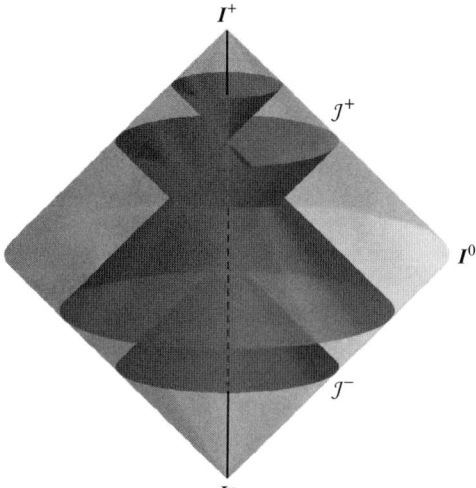

Fig. 10.4 Carter-Penrose diagram of the Minkowski spacetime. This figure shows null cones passing through the origin $R = 0$ at different moments of time T.

Let us discuss now behavior of null rays in the conformal space in more detail (we still keep $Z = 0$ for simplicity). Consider a null plane in the direction of the X-axis and passing through $X = 0$. Its equation is $T = X$. We call a null line $Y = b$ on this plane a generator and b its *impact parameter*. Let us find an equation of the null plane in the (ψ, ξ, ϕ) coordinates. For a generator with the impact parameter b one has

$$Y = R\cos\phi = b, \qquad\qquad R = \frac{b}{\cos\phi},$$

$$X = R\sin\phi = b\tan\phi, \qquad \phi \in (-\pi/2, \pi/2). \tag{10.1.12}$$

Denote $W_\pm \equiv T \pm R = b(\sin\phi \pm 1)/\cos\phi$. Then,

$$\psi = \arctan(W_+) + \arctan(W_-), \quad \xi = \arctan(W_+) - \arctan(W_-). \tag{10.1.13}$$

Equations (10.1.13) determine the null generators in a parametric form (as a function of the angular coordinate). The surface $T = X$ and its generators are shown in Figure 10.5. One can see that all the null generators start at a single point of the $\mathcal{J}^-$ and end at a single point of the $\mathcal{J}^+$. The basic generator $b = 0$ is a straight line going through $\mathcal{J}^-$ to $\mathcal{J}^+$ and passing through the origin $\xi = 0$.

Problem 10.1: *Let η be a null geodesic in the metric $g_{\mu\nu}$. Show that after a conformal transformation $g_{\mu\nu} = \Omega^2 \tilde{g}_{\mu\nu}$ it remains a null geodesic in the conformal metric $\tilde{g}_{\mu\nu}$.*

Solution: *Denote $\sigma = \ln\Omega$. The Christoffel symbols for the conformally related spaces are connected as follows*

$$\Gamma^\lambda_{\mu\nu} = \tilde{\Gamma}^\lambda_{\mu\nu} + \gamma^\lambda_{\mu\nu}, \qquad \gamma^\lambda_{\mu\nu} = \sigma_{,\mu}\delta^\lambda_\nu + \sigma_{,\nu}\delta^\lambda_\mu - \tilde{g}_{\mu\nu}\tilde{g}^{\lambda\alpha}\sigma_{,\alpha}. \tag{10.1.14}$$

Denote by l^μ a tangent vector to η. It is null in both conformally related metrics. Using Eq. (10.1.14) one obtains

$$l^\mu\nabla_\mu l^\lambda = l^\mu\tilde{\nabla}_\mu l^\lambda + 2l^\mu\sigma_{,\mu}l^\lambda. \tag{10.1.15}$$

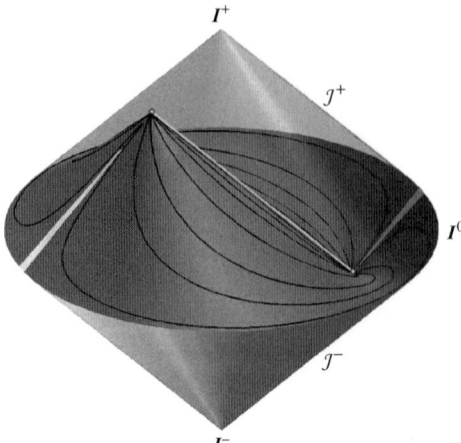

Fig. 10.5 Carter–Penrose diagram of the Minkowski spacetime. This figure shows a null plane described by the equation $T = X$.

Here, $\tilde{\nabla}_\mu$ is the covariant derivative in the metric $\tilde{g}_{\mu\nu}$. The geodesic equation

$$l^\mu \nabla_\mu l^\lambda = 0 \qquad (10.1.16)$$

implies that $l^\mu \tilde{\nabla}_\mu l^\lambda = -2 l^\mu \sigma_{,\mu} l^\lambda$. This is a geodesic equation. It can be put into the canonical form similar to Eq. (10.1.16) after proper change of the parametrization along the null curve η.

Problem 10.2: *The conformal metric $d\tilde{s}^2$ in Eq. (10.1.11) is* ultrastatic. *In the ultrastatic spacetime a projection of a geodesic on the constant-time surface is a geodesic of the 3D spacial metric (see Problem on page 77). Show that spatial lines $(\xi(\phi), \phi)$ determined by Eq. (10.1.13) are large circles on a unit sphere Eq. (10.1.9).*

Let us summarize. The conformal map is constructed in two steps. First, one makes the coordinate transformation that brings the coordinate infinity of the Minkowski spacetime to the finite coordinate 'volume'. Secondly, one accompanies these transformations by a conformal map. As a result, one obtains a new regular 'unphysical' space $\tilde{M}$, so that the original Minkowski spacetime M is its part, and the conformal factor Ω is chosen so that the conformal metric $d\tilde{s}^2$ is regular at the image of 'infinity'. Time-like geodesics begin at the infinite time-like past I^- and end at the infinite time-like future I^+ (see Figure 10.2). Null geodesics propagate from $\mathcal{J}^-$ (*past null infinity*) to $\mathcal{J}^+$ (*future null infinity*). All the null generators of the fixed-null plane start at a single point of $\mathcal{J}^-$ and end at a single point $\mathcal{J}^+$ (see Figure 10.3). Space-like geodesics 'propagate' between the points of the *spatial infinity I^0*. The null boundary of the conformal image of the Minkowski spacetime is formed by two null 'cones' $\mathcal{J}^-$ an $\mathcal{J}^+$, the past-directed null cone from the future infinity I^+ and the future-directed null cone from the past infinity I^- (see Figure 10.4). In addition, the boundary includes spatial infinity I^0, and 'points' $I^\pm$.

We have already mentioned that at the 'boundary' the conformal factor Ω vanishes. However, its gradient is finite at $\mathcal{J}^\pm$. In the coordinates $(\psi, \xi, \theta, \phi)$ one has

$$\Omega_{,\mu}\big|_{\mathcal{J}^+} = -\sin\xi \cdot (1, 1, 0, 0),$$
$$\Omega_{,\mu}\big|_{\mathcal{J}^-} = \sin\xi \cdot (1, -1, 0, 0). \qquad (10.1.17)$$

The vectors $\Omega_{,\mu}$ are regular null vectors on $\mathcal{J}^\pm$, tangent to these surfaces.

10.1.3   Conformal diagram of the Schwarzschild spacetime

Let us construct a conformal diagram of the spacetime of an eternal spherically symmetric black hole. As the first step, one makes the coordinate transformation that brings the infinity to a finite coordinate distance. We already discussed such coordinates in Section 6.3. For the Kruskal metric ($r_S = 2M$) (6.3.14)

$$ds^2 = -\frac{4r_0^3}{r} e^{-(r/r_S-1)}\, dVdU + r^2(d\theta^2 + \sin^2\theta\, d\phi^2),$$

$$-UV = \left(\frac{r}{r_S} - 1\right)\exp\left(\frac{r}{r_S} - 1\right),$$

$$(10.1.18)$$

the new coordinates (ς, η) are connected with (U, V) by the relations

$$U = \sinh[\tan(\varsigma)], \qquad V = \sinh[\tan(\eta)]. \qquad (10.1.19)$$

In these coordinates the infinities $U = \pm\infty$ correspond to the finite values $\varsigma = \pm\pi/2$, and the infinities $V = \pm\infty$ correspond to the finite values $\eta = \pm\pi/2$ (see Figure 10.6). The event horizon H^+ ($U = 0$) is given by the equation $\varsigma = 0$, while the past horizon H^- ($V = 0$) is given by $\eta = 0$. Near the horizons one has $U \approx \varsigma$ and $V \approx \eta$.

The Kruskal metric in these coordinates is given by Eq. (6.3.22), and it can be written in the form

$$ds^2 = r^2\, d\tilde{s}^2,$$

$$d\tilde{s}^2 = \frac{4(r/r_S - 1)}{(r/r_S)^3\, \tanh[\tan(\varsigma)]\, \tanh[\tan(\eta)]} \cdot \frac{d\varsigma\, d\eta}{\cos^2\varsigma\, \cos^2\eta} + d\omega^2.$$

$$(10.1.20)$$

The boundary points $(-\pi/2 < \varsigma < 0,\ \eta = \pi/2)$ and $(-\pi/2 < \eta < 0,\ \varsigma = \pi/2)$ correspond to the future null infinities $\mathcal{J}^+$. The boundary points $(0 < \varsigma < \pi/2,\ \eta = -\pi/2)$ and $(0 < \eta < \pi/2,\ \varsigma = -\pi/2)$ correspond to the past null infinities $\mathcal{J}^-$. One can verify that the conformal metric $d\tilde{s}^2$ remains regular at $\mathcal{J}^\pm$. In the coordinates (ς, η) the singularities $r = 0$ are determined by the following implicit equation

$$\sinh[\tan(\varsigma)]\, \sinh[\tan(\eta)] = e^{-1}. \qquad (10.1.21)$$

The corresponding conformal Carter–Penrose diagram is shown in Figure 10.6.

The coordinate and conformal transformations with the required properties are not unique. For example, one can make an additional transformation $\varsigma \to \tilde{\varsigma} = f(\varsigma),\ \eta \to \tilde{\eta} = f(\eta),$

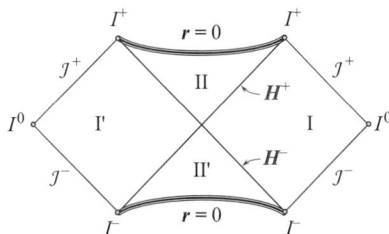

Fig. 10.6 Carter–Penrose conformal diagram of the Schwarzschild 'eternal' black and white holes.

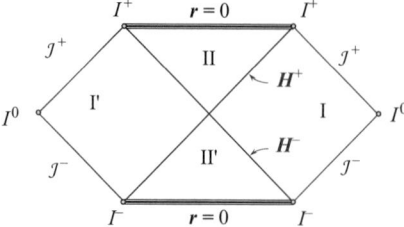

Fig. 10.7 'Canonical form' of the Carter–Penrose conformal diagram. On this diagram the lines representing the singularities $r = 0$ are straightened by the additional coordinate transformation given in the text.

where f is a regular monotonic function on the interval $(-\pi/2, \pi/2)$. One can use this ambiguity to straighten the lines representing the singularity $r = 0$. Let $\eta = g(\varsigma)$ be a solution of Eq. (10.1.21). It evidently possesses the property $\varsigma = g(\eta)$. The coordinate transformation that straightens the singularity is given by the function $f(x) = (1/2)[g(x) - x] + \pi/4$. The corresponding 'canonical form' of the Penrose–Carter conformal diagram with the straightened singularity lines is shown in Figure 10.7.

10.1.4 Asymptotically flat spacetime

The properties of the Schwarzschild black hole at infinity are similar to the properties of the flat spacetime. The existence of the conformal transformation that 'brings' the infinity to the finite distance, and regularity of the 'null infinite boundary' surface in the conformal spacetime is a common property of both metrics. One can use this property to provide a general covariant description of the *asymptotically flat spacetime*. This was done by Penrose (1963, 1964, 1965, 1968) who suggested the following definitions. A so-called *asymptotically simple spacetime* is to be defined first (Frauendiener 2004).

A spacetime M with metric $g_{\mu\nu}$ is said to be *asymptotically simple* if there exists another ('unphysical') space $\tilde{M}$ with boundary $\partial\tilde{M} \equiv \mathcal{J}$ and a regular metric $\tilde{g}_{\mu\nu}$ on it such that:

1. $\tilde{M} \setminus \partial\tilde{M}$ is conformal to M, and $g_{\mu\nu} = \Omega^{-2}\tilde{g}_{\mu\nu}$ in M;
2. $\Omega|_M > 0$, $\Omega|_{\partial\tilde{M}} = 0$, $\Omega_{,\mu}|_{\partial\tilde{M}} \neq 0$;
3. each null geodesic in M begins and ends on $\partial\tilde{M}$.

We call $\tilde{M}$ the *conformal Penrose space*. Penrose proved that if the metric $g_{\mu\nu}$ satisfies Einstein's vacuum equations in the neighborhood of $\mathcal{J}$ (or Einstein's equations with the energy-momentum tensor that decreases at infinity sufficiently fast) and the natural conditions of causality and spacetime orientability are satisfied, then an asymptotically simple space has the following properties:

1. The topology of the space M is $\mathbb{R}^4$; its boundary $\mathcal{J}$ is light-like and consists of two disconnected components $\mathcal{J} = \mathcal{J}^+ \cup \mathcal{J}^-$, each with topology $S^2 \times \mathbb{R}^1$.
2. The generators of the surfaces $\mathcal{J}^\pm$ are null geodesics in $\tilde{M}$; tangent vectors to these geodesics coincide with $\tilde{g}^{\mu\nu}\Omega_{,\mu}|_{\mathcal{J}}$.
3. The curvature tensor in the physical space M decreases as one 'moves' along a null geodesic to infinity, and the so-called *peeling off property* holds. Here, we do not focus on this property as its detailed analysis can be found in the literature (Sachs 1964; Penrose 1968).

Property 1 signifies that the global structure of the asymptotically simple space is the same as that of Minkowski space, and it does not include black holes. In order to include the possibility of the existence of localized regions of strong gravitational fields, which do not alter the asymptotic properties of spacetime, it is sufficient to analyze the class of spaces that can be converted into asymptotically simple spaces by 'cutting out' certain inner regions with singularities of some kind (due to the strong gravitational field) and by subsequent smooth 'patching' of the resultant 'holes'. Such spaces are said to be *weakly asymptotically simple spacetimes*.

To be more rigorous, a space M is said to be weakly asymptotically simple if there exists an asymptotically simple space $\tilde{M}$ with an open subset K such that the space $\tilde{M} \cap K$ is isometric to a subset of M. One says that a weakly asymptotically simple space is *asymptotically flat* if its metric in the neighborhood of $\mathcal{J}$ satisfies Einstein's vacuum equations (or Einstein's equations with an energy-momentum tensor that decreases sufficiently fast).

10.1.5 Asymptotically predictable spacetime

To study a *global causal structure* of a spacetime it is convenient to use the notions of causal curves and causal future and past, introduced in Section 3.2.5. In the asymptotically flat spacetime the set of events visible to distant observers coincides with $J^-(\mathcal{J}^+)$. For the Minkowski spacetime $J^-(\mathcal{J}^+)$ includes the entire space. This means that any event that occurs in this space can be registered by a distant observer, and there are no 'hidden' regions. In the presence of gravitating objects this is not always true. The spacetime may have 'hidden' regions, from where causally propagated signals cannot reach $\mathcal{J}^+$. This may happen when there exists a singularity visible from the infinity, or a black hole has been formed.

To separate these two cases let us introduce the following definition. Let $(M, g_{\mu\nu})$ be an asymptotically flat spacetime with associated conformal space $(\tilde{M}, \tilde{g}_{\mu\nu})$. We say that $(M, g_{\mu\nu})$ is *strongly asymptotically predictable* if in the conformal space there is an open region $\tilde{V} \subset \tilde{M}$ with $M \cap J^-(\mathcal{J}^+) \subset \tilde{V}$ such that $(\tilde{V}, \tilde{g}_{\mu\nu})$ is globally hyperbolic. An asymptotically flat spacetime that fails to be strongly asymptotically predictable is said to possess a *naked singularity*.

In 1969 Roger Penrose conjectured that no naked singularities other than the Big Bang singularity exist in the universe. In other words, the future evolution of the regular initial state would never produce a singularity visible from future null infinity. Simply speaking this *weak cosmic censorship hypothesis* states that singularities are always hidden from the observer at infinity by the event horizon of a black hole. There are some counterexamples of this hypothesis but they look physically unrealistic and probably will be excluded by more refined versions of the hypothesis. But the rigorous definition of the *cosmic censorship* conjecture is yet to be formulated.

10.2 Black Holes: General Definition and Properties

10.2.1 Black hole definition

Now we can give a rigorous definition of the concept of a black hole: *A strongly asymptotically predictable spacetime M is said to contain a* black hole *if M is not contained in $J^-(\mathcal{J}^+)$. The* black hole region, *B, of such a spacetime is defined to be $B = [M - J^-(\mathcal{J}^+)]$, and the*

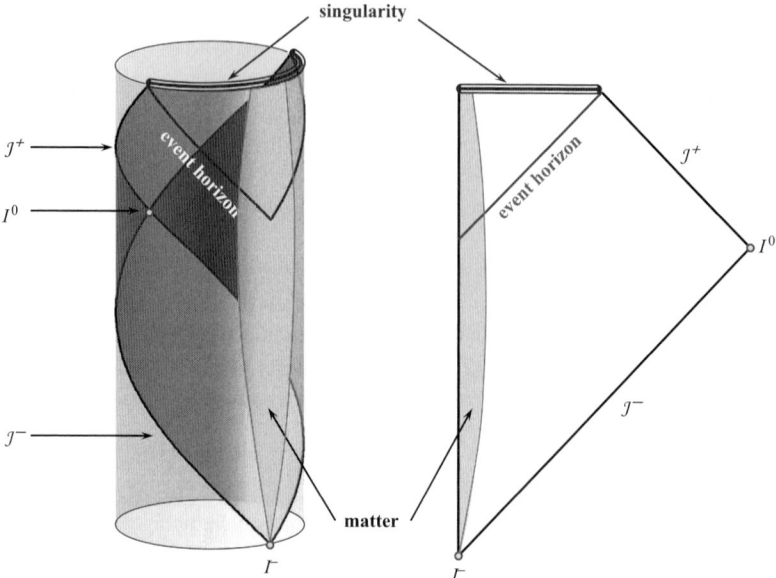

Fig. 10.8 Carter–Penrose diagrams for the spacetime of a collapsing spherical star that forms a black hole. The part of the spacetime that is depicted as the triangle with the singularity at the upper side and the event horizon as another side, is invisible from the future null infinity $\mathcal{J}^+$. It describes the black hole interior region.

boundary[2] *of B in M, $H^+ = \dot{J}^-(\mathcal{J}^+) \cap M$, is called the* event horizon. In other words, a *black hole* in asymptotically flat spacetime is defined as a region such that no causal signal (i.e. a signal propagating at a velocity not greater than that of light) from it can reach future null infinity $\mathcal{J}^+$ (see Figure 10.8). The region outside the black hole is called the *domain of outer communications*. An observer in this domain can send causal signals to $\mathcal{J}^+$ and receive such signals from $\mathcal{J}^-$.

Obviously, not just one but several black holes may exist in a bounded region of spacetime; new black holes may be created, and those already existing may interact and merge. In such cases, $\dot{J}^-(\mathcal{J}^+)$ is the union of the boundaries of all black holes. If an asymptotically flat space contains no horizon, then all events taking place in this space can be recorded, after an appropriate time, by a distant observer. If an event horizon appears it means that a black hole has been born, and the strongly enhanced gravitational field has produced qualitative changes in the causal structure of spacetime. The enhanced gravitational field prevents signals from escaping to infinity so that the observer can never find out about the events inside the black hole unless he chooses to cross the event horizon and fall into the black hole. (For a comprehensive review of global approach and conformal structure of a spacetime see (Penrose 1968).)

---

[2]The boundary $\dot{Q}$ of a set Q is defined as $\dot{Q} = \overline{Q} \cap \overline{(M - Q)}$, where M is the complete spacetime, and $\overline{Q}$ is the closure of Q.

10.2.2   The topology of black holes

Before discussing the geometry of the event horizon, let us first answer the question what is its topology. Let Σ_t be a one-parameter family of 3D space-like Cauchy surfaces. Such surfaces extend to infinity and cross the horizon. We assume that a surface at given parameter t_0 is in the future with respect to a surface with the smaller value of the time parameter $t < t_0$. We denote by $\mathcal{B}_t$ a two-dimensional space-like surface of intersection of Σ_t and the event horizon. One can consider $\mathcal{B}_t$ as a horizon surface at the 'given moment of time t'. When one speaks about the topology of a black hole one has in mind the topology of these 2D sections $\mathcal{B}_t$. In the general case, there may exist several black holes, so that $\mathcal{B}_t$ has several disconnected components. For simplicity, we assume that only one black hole is present.

The following theorem was proved by Gannon (1976): *The horizon of the black hole must be either spherical or toroidal, provided the dominant energy condition*[3] *is satisfied and the horizon is smooth to the future of some slice.* The latter condition entails in particular that no new generators enter in the future of that slice. If the topology of the horizon is toroidal it can occur only during short enough time. This follows from the following result (Friedman *et al.* 1993):

Topological censorship theorem. *If an asymptotically flat, globally hyperbolic spacetime satisfies the averaged null-energy condition, then every causal curve from $\mathcal{J}^-$ to $\mathcal{J}^+$ can be continuously deformed to a causal line γ_0 totally located in the black hole exterior in the region where the gravitational field is weak.*

The topological censorship theorem states that general relativity does not allow an observer to probe the topology of spacetime: any topological structure collapses too quickly to allow even the light to traverse it. More precisely, in a globally hyperbolic, asymptotically flat spacetime satisfying the null-energy condition, every causal curve from past null infinity to future null infinity is fixed-end point homotopic to a curve in a topologically trivial neighborhood of infinity.

If the toroidal topology of the horizon exists for long enough time, a causal line from $\mathcal{J}^-$ to $\mathcal{J}^+$ can pass through the inner region of the torus. Such a curve cannot be deformed smoothly to γ_0. This is in contradiction with the topological censorship theorem.

The described result is a generalization of the famous **Hawking theorem** (Hawking 1972a, 1972b): *Cross-sections of the event horizon in 4D asymptotically flat stationary black hole spacetime obeying the dominant energy condition are topologically two-spheres.*

It should be emphasized that in the time-dependent case, 2D sections $\mathcal{B}_t$ depend on the choice of the foliation Σ_t. It is plausible that the toroidal topology of the horizon during a short time is a result of this choice, and by proper change of the foliation one would get only spherical topology of the black hole surface.

10.2.3   Geometrical properties of the event horizon

The following theorem proved by Penrose (1968) describes the geometry of an event horizon: *The event horizon is formed by null geodesics (generators) that have no end points in the future.*

---

[3]See Section 5.2.5 for definitions of this and other energy conditions used in this section.

Thus, at the regular points the horizon is a *null surface*. If we monitor a generator of the horizon to the future it never leaves the horizon and never intersects another generator. If we monitor a generator of the horizon to the past, two alternatives are possible:

1. either it always lies on the event horizon; or

2. it leaves the horizon at some point.

In the second case, the null generator enters the horizon at some point having left the region of the black hole exterior. Such an entry point is the point of its intersection with other generators (the *caustics*). Only one generator passes through each point of the horizon beyond the caustics.

Hawking (1971) proved the following remarkable *black hole area theorem*: *Consider a strongly asymptotically predictable spacetime satisfying the null-energy condition and containing a black hole. Let Σ_i and Σ_f be 'initial' and 'final' space-like Cauchy surfaces, and $\mathcal{B}_i$ and $\mathcal{B}_f$, are the corresponding cross-sections of the black hole. Then, the area of $\mathcal{B}_f$ is greater than or equal to the area of $\mathcal{B}_i$.*

To put it simpler, if the asymptotically flat spacetime does not have a *naked singularity* the Hawking area theorem guarantees that the surface area of any black hole either remains the same or monotonically grows provided the matter and fields respect the null-energy condition.

10.2.4   Apparent horizon

It should be noted that the definition of the event horizon is intrinsically non-local and depends on the entire future of the spacetime.

> *To illustrate this property consider the collapse of a body of the mass M_0 with a black hole formation at some time t_0. For simplicity, we assume that the body is spherical, so that after the collapse the Schwarzschild black hole is formed. Its gravitational radius is $r_{S,0} = 2M_0$. If the black hole is isolated and no matter or field fall into it later, then $r_{S,0}$ is the size of the event horizon. But this situation is certainly an idealization. What happens if at some later time $t_1 > t_0$ some additional matter falls into the black hole. We again assume that the matter density and velocity are spherically symmetric and as a result of the accretion the mass of the black hole becomes $M_1 > M_0$. If nothing happens later, the null rays propagating along $r_{S,1} = 2M_1$ form the horizon. But being traced back to the time t_0 these rays remain in the exterior of the sphere $r_{S,0}$. The matter accretion increases the gravitational field, so that the null rays originally propagating at $r_{S,0}$, after t_1 would be focused and move to the center. This simple example illustrates the general situation. In order to decide whether a given ray is a generator of the event horizon one needs to know all the future history of the spacetime evolution.*

Another problem arises when one tries to extend the notion of the event horizon to non-asymptotically flat spacetimes, for example to define a black hole in a closed Friedman–Robertson–Walker universe. In this case, the universe collapses to a singularity and there is no asymptotic region to escape to.

To deal with these kinds of problems it is convenient to introduce a notion of an *apparent horizon*. Consider a Cauchy surface Σ and let B be a compact, two-dimensional, smooth space-like submanifold of Σ. Suppose B emits light. The light emitted outside has the light front Γ_+. Its generators are null geodesics orthogonal to B. Similarly, the light emitted inside has the light front Γ_- with null generators orthogonal ro B. Denote by $k_\pm$ the corresponding tangent null vectors to the outgoing and ingoing ray congruences, and their expansion is

$$\theta_\pm = k^\alpha_{\pm;\alpha}. \tag{10.2.1}$$

In the absence of the gravitational field the *expansion parameter* for the outgoing rays, θ_+, is positive, while for the ingoing rays it is negative, $\theta_- < 0$. Suppose inside B there is matter. Its gravitational field attracts the null rays. In particular, one can expect that θ_+ becomes smaller, and for a strong enough field it may become negative. A compact smooth surface B, having the property that the expansions $\theta_\pm$ for both, ingoing and outgoing, future-directed null geodesic congruences orthogonal to B are everywhere negative is called a *trapped surface*. A region of Σ inside the trapped surface B is called a *trapped region*. The total trapped region $\mathcal{T}$ of a Cauchy surface Σ is the union of all trapped regions on Σ. The boundary of $\mathcal{T}$ is called the *apparent horizon* $\mathcal{A}$ on Σ.

The presence of the apparent horizon indicates that the gravitational field is strong. For a stationary black hole the apparent horizon coincides with the cross-section of the event horizon H^+ and the Cauchy surface Σ. In the general case, the apparent horizon differs from the *event horizon*. In a spacetime obeying the null-energy condition the apparent horizon lies inside (or coincides with) the true event horizon, $H \cap \Sigma$, on Σ.

One can prove the following property: *If the total trapped region, $\mathcal{T}$, on a Cauchy surface Σ has the structure of a manifold with a boundary, then the apparent horizon, $\mathcal{A}$, is an outer marginally trapped surface with vanishing expansion, $\theta_+ = 0$.*

The theorems proved by Penrose and Hawking imply that if the proper energy conditions are satisfied and there is no naked singularity the existence of a trapped surface indicates the geodesic incompleteness of the spacetime. In other words, inside the black holes in the classical gravity there exist singularities. The discussion of these theorems and related material can be found in the books (Penrose 1968; Hawking and Ellis 1973; Misner *et al.* 1973; Wald 1984).

10.2.5   Uniqueness and stability

The exterior of a black hole has two disconnected boundaries: (i) asymptotic infinity, and (ii) the event horizon. In a stationary spacetime a force required to keep a body at rest grows infinitely near the horizon. Only for a very special configuration of the matter is such an equilibrium possible. A similar conclusion is valid for the gravitational field itself. This means that for the stationary gravitational field the condition of its regularity at the horizon is rather restrictive. There were obtained a number of important results confirming this conclusion. According to the *uniqueness theorems* a stationary vacuum solution for a black hole in the asymptotically flat spacetime is characterized by two constants M and J and coincides with the *Kerr metric*. The corresponding constants are the mass and the angular momentum of the black hole. A generalization of this result to the electrovacuum black holes is also known. The corresponding solution is the *Kerr–Newman metric*, which besides the mass and angular momentum also has an electric charge (and/or a magnetic monopole charge).

The first *black hole uniqueness theorem* was presented by Werner Israel (1967) at a King's College meeting in London, in February 1967. This theorem proves that *the only vacuum solution for a static black hole in an asymptotically flat spacetime is the Schwarzschild metric with positive mass.* This result initiated study of the black hole uniqueness that continues till now. In 1968, Israel generalized this result to the case of electrovac[4] static black holes

---

[4]That is, static solutions of the Einstein–Maxwell equations.

and demonstrated that such a solution coincides with the *Reissner–Nordström metric*, with a natural constraint $|Q| \leq M$ (Israel 1968).

Israel's theorems, besides some other technical assumptions, contained an assumption that the horizon cross-section is a regular 2D surface with the topology of S^2. The papers by (Hawking 1972a, 1972b), initiated a general study of four-dimensional, asymptotically flat, stationary black holes. Such spaces have a time-like-at-infinity Killing vector ξ. The main results of this research were summarized in the book (Hawking and Ellis 1973). Hawking (1972a) proved that in a stationary spacetime the surface topology of a black hole must be a two-sphere.

As we have already mentioned the *Killing horizon* is a null surface whose null generators coincide with the orbits of a one-parameter isometries, generated by a Killing vector. Hawking also proved the *strong rigidity theorem*. According to this theorem the event horizon of a stationary black hole is the Killing horizon of some Killing vector l. Denote by ξ a Killing vector that generates the time translations and that is time-like at infinity. If a black hole is rotating (that is stationary but not static), the vector field l differs from ξ so that at least in the vicinity of the horizon there must exist a second Killing vector m. Using an assumption of analyticity of the metric, Hawking proved that this second Killing vector field exists everywhere in the *domain of outer communications*, and the metric of a stationary rotating black hole must be axially symmetric. Later, it has been shown that this result remains valid when the metric is non-analytic in isolated regions outside the horizon.

> *The following two assumptions must be made in order to prove the basic propositions concerning general properties of stationary black holes:*
>
> *1. The spacetime is regularly predictable.*
> *2. The spacetime is either empty or contains fields described by hyperbolic equations and satisfying the energy-dominance condition Eq. (5.2.43).*
>
> *Assumption 1, concerning the general causal structure of the space-time and thoroughly discussed in the preceding section is largely technical. The energy-dominance condition (which implies, e.g., the weak-energy condition) signifies that an arbitrary observer finds the local energy to be non-negative and the local energy flux to be non-space-like.*
> *Assumption 2 definitely holds, for example, for the electromagnetic field (for additional details, see (Hawking and Ellis 1973)). Throughout this chapter we assume that the constraints set above are satisfied.*

The Hawking theorem has the following very important consequence. Consider a rotating black hole surrounded by a non-axially symmetric distribution of matter. Since the gravitational field of the system is not axially symmetric, the black hole in the final state cannot have an angular momentum. This means that there exists a non-Newtonian gravitational interaction of a black hole with the surrounding matter. As a result of this interaction the black hole gradually transfers its angular momentum to the matter (Ipser 1971; Hawking 1972a; Press 1972; Hawking and Hartle 1972).

Having two commutative Killing vector fields for a rotating black hole, one can reduce the Einstein equations to a simpler form. Carter (1973, 1979, 1987) demonstrated that the reduced 2D problem for vacuum and electrovacuum stationary black holes can be formulated as a boundary-value problem for a system of partial differential equations. In his paper, Carter (1971) proved that stationary axisymmetric vacuum black hole solutions must belong to discrete sets of continuous families with one or two parameters. The unique solution with angular momentum is the Kerr metric with $|J| < M^2$. Later, this result was generalized to the electrovacuum solutions.

For more detail see the books (Heusler 1996, 1998). Comprehensive reviews of the uniqueness and 'no-hair' results and historical remarks can be found in (Robinson 2009; Israel 1987; Thorne 1994; Carter 1997).

It is generally adopted that for a 'reasonable' class of 'regular' initial data the vacuum spacetime outside a collapsed object will settle down to a stationary Kerr black hole. This means that at late time the metric is just the regularly perturbed Kerr metric, and the perturbations decay. As was formulated by John Wheeler: 'Black holes have no hair'. A stationary solution can describe a final state of the dynamical evolution of a system only if this solution is stable. The problem of stability in all its generality for non-linear equations is quite complicated. It is much easier to study linear stability, that is the stability with respect to small perturbations. It is sufficient to show that perturbations with proper boundary conditions are not growing with time. There exist a number of results confirming the stability of the Kerr spacetime. In particular, in 1989 Whiting proved a so-called *mode stability of the Kerr black hole*. This work studied mode decomposition of the gravitational field perturbation in the Kerr geometry of the form $h \sim \exp(-i\omega t + im\phi)R(r)\Theta(\theta)$, and demonstrated that there are no exponentially growing in time modes that are regular both at the horizon and infinity. For further discussion of the black hole stability problem see, e.g., (Dafermos and Rodnianski 2010; Chrućiel *et al.* 2010; Finster *et al.* 2009).

10.2.6   Primordial black holes

The Einstein theory allows the existence of black holes of arbitrary mass. Why do we not observe formation of small mass black holes in our everyday life? The answer is evident. If one has a body of mass M, in order to obtain a black hole one must compress it to the radius $\sim r_S = 2M$, so that before the black hole is formed the density of the matter must reach the value

$$\rho(M) \sim \frac{M}{r_S^3} \sim \rho_{Pl} \left(\frac{m_{Pl}}{M}\right)^2. \qquad (10.2.2)$$

Only at such a density would the gravity attraction be a dominant force. To compress a body in the laboratory to such a density one needs to overcome the repulsive force of the pressure. This means that there exists a huge potential barrier separating laboratory-size objects and their 'black hole' phase.[5]

Black holes of small mass can be created at the early stages of the universe evolution, when the matter density was high. Such black holes are called the *primordial black holes*. The density of the matter in the early universe (at the radiation-dominated stage) is

$$\rho \sim t^{-2}, \qquad (10.2.3)$$

where t is the time after the *Big Bang*. Thus, the expected mass of *primordial black holes* that could be formed at this time t is

---

[5]There exists the probability of quantum tunnelling through this potential barrier, but it is negligibly small.

$$M_{PBH} \sim \frac{c^3 t}{G} \sim 10^{15} \left(\frac{t}{10^{-23} \text{s}} \right) \text{g}. \tag{10.2.4}$$

The mass spectrum of the primordial black holes could span an enormous mass range. Primordial black holes formed at the Planckian time $\sim 10^{-43}$s have the Planckian mass $M_{PBH} \sim 10^{-5}$ g, while the typical black hole mass formed at $t \sim 1$s is $M_{PBH} \sim 10^5 M_{\odot}$.

The high density of the matter is a necessary condition for primordial black holes creation. But this condition is not sufficient. To make formation of primordial black holes possible there must exist large inhomogeneities that play the role of the seeds. Besides this, a sudden reduction of the pressure must occur in order to allow the overdense regions to collapse. This may happen for example during phase transitions.

At the moment, there are no observational evidences indicating the existence of the primordial black holes. At the same time the hypothesis of the primordial black holes sometimes is used in connection with the *dark matter* problem. It should be emphasized that for small-mass black holes the Hawking evaporation process is important. The characteristic time of the black hole evaporation is

$$t_{evap} \sim 10^{64} \left(\frac{M}{M_{\odot}} \right)^3 \text{yr}. \tag{10.2.5}$$

Primordial black holes with mass $M \lesssim 10^{15}$ g would decay by the present time. This gives the following natural classification of the primordial black holes (PBH):

1. **PBH with $M < 10^{15}$ g.** These black holes would have completely evaporated by now, but their radiation could affect many processes in the early universe. Namely, they may generate entropy, produce gravitons, neutrinos and other particles, change the details of baryogenesis. If there exist stable relics of the Planck mass, they may contribute to the dark matter.

2. **PBH with $M \sim 10^{15}$ g.** These black holes would have evaporated by now. Their radiation may contribute to the Galactic γ-ray background and to antiprotons and positrons in the cosmic rays. Observation of the process of their decay would give important information on the very high energy physics.

3. **PBH with $M > 10^{15}$ g.** These black holes would survive till today. They may give a contribution to the dark matter. Large PBHs might influence the details of the large-scale structure formation in the universe, for example serve as seeds for the supermassive black hole formation in the galactic nuclei.

Let us repeat that there is no evidence of the PBH existence. But even the fact that we do not see them or their remnants can be used to obtain important information concerning the properties of the matter in the early universe. For a detailed discussion of primordial black holes, their properties and results of their search see the review article (Carr 2010).

10.3   Black Holes and Search for Gravitational Waves

10.3.1   Black holes in close binary systems

As we already stressed, astrophysical black holes are the most powerful engines in the universe. Up to 42% of the proper mass of the matter accreting onto the extremal Kerr black hole can be transformed into the energy of radiation. Only annihilation of matter and anti-matter produces more energy. So far we discussed mainly the action of black holes on particle motion and light propagation. Let us discuss now gravitational waves generation by systems that contain a black hole. This subject is of high interest in connection with the search of gravitational waves by recently constructed gravitational-wave detectors and exciting plans to develop gravitational-wave astrophysics.

Close binaries are natural objects for the search of stellar-mass black holes. The main 'signal', indicating that such a system contains a compact object with strong gravity, comes from observation of X- and gamma-radiation. A more massive star in a stellar binary has faster evolution. If its mass is large enough it ends its 'stellar life' as a neutron star or a black hole. If both of the stars of the binary are massive enough, one can expect formation of a close binary system, with neutron stars and black holes as its components. These systems are very interesting for the gravitational-wave search. Since both the neutron stars and the black holes have very small radii and their self-gravity is large, the tidal effects in BH/NS binaries are small and practically do not affect its evolution. The leading effect is the *gravitational-wave radiation* emission by the system.

In the observations one can trace the effect of the gravitational-wave emission on the period of the binary system. This gives information about energy and angular momentum loss because of the gravitational radiation. Standard linear approximation and the quadrupole formula are sufficient for the calculation of this effect. The amplitude of the gravitational waves produced by these systems falls as [distance from the source]$^{-1}$. When the gravitational waves reach the Earth they become so weak that it is practically impossible to detect them. For this reason, much more promising is the radiation from BH/NS binaries at the final stage of their evolution. In fact, because of the energy and angular momentum loss, the components of the binary become closer and closer to one another, until they coalesce and form a black hole. The gravitational-wave emission during the coalescence is one of the most promising sources for detection of the gravitational waves.

It is quite difficult to predict the rate of such events in the universe. Inevitable uncertainty of such predictions is connected with unknown details of the neutron star and black formation in the binaries. Estimations of the rate of coalescence events give (see, e.g., (Sathyaprakash and Schutz 2009) and references therein)

Neutron-star–neutron-star (NS–NS) coalescence. The best estimation for the rate density gives: 5×10^{-6} yr^{-1} Mpc^{-3}. Uncertainty is a factor of the order of 0.01–10;

Neutron-star–black-hole (NS–BH) coalescence. The rate density is about 3% of the NS–NS rate density;

Black-hole–black-hole (BH–BH) coalescence. The rate density is about 0.4% of the NS–NS rate density.

NS–NS and NS–BH coalescence events might be responsible for short *gamma-ray bursts* (see Section 10.3.8). These events are also of high interest since they in principle can give us information concerning properties of the matter at nuclear densities.

10.3.2 Gravitational radiation from binary systems

The gravitational radiation from a binary system can be estimated by using results of the linearized gravity presented in Section 5.4. We suppose again that the rotation plane of the binary coincides with the (X, Y)- plane, and denote by θ the inclination angle between the direction to a distant observer and the Z-axis. Then, one has

$$h_+ = -2(1 + \cos^2 \theta) \frac{\mu(R\Omega)^2}{r} \cos[2\Omega(T - r)],$$

$$h_\times = -4 \cos \theta \frac{\mu(R\Omega)^2}{r} \sin[2\Omega(T - r)].$$

(10.3.1)

The dependence on the angle θ in these relations is the result of the TT-projection. Energy loss by the binary system is

$$\frac{dE}{dT} = -\frac{32}{5} \frac{\mu^2 M^3}{R^5},$$

(10.3.2)

where $M = M_1 + M_2$ is the total mass of the system and $\mu = M_1 M_2/M$ is its *reduced mass*. As the result of the radiation the distance between the masses decreases

$$R = R_0(1 - T/T_0)^{1/4}, \qquad T_0 = \frac{5R_0^4}{256\mu M^2} = \frac{5M}{256(M\Omega)^{8/3}}.$$

(10.3.3)

Here, T_0 is the time to merger and

$$\mathcal{M} = \mu^{3/5} M^{2/5}$$

(10.3.4)

is a so-called *chirp mass*.

The frequency of the gravitational waves

$$f = \Omega/\pi$$

(10.3.5)

depends on time

$$df/dT = -f/T_{\text{chirp}},$$

(10.3.6)

where

$$T_{\text{chirp}} = \frac{8}{3} T_0 = \frac{5M}{96} v^{-1} (R_0/M)^4, \qquad v = \mu/M = M_1 M_2/M^2,$$

(10.3.7)

is the *chirp time*.

By observing h_+ and $h_\times$ one can determine the gravitational-wave frequency. The change of the frequency with time allows one to find the time to merger and the chirp mass. By measuring the amplitudes h_+ and $h_\times$ one can obtain information about the distance to the source and the orbital inclination angle.

10.3.3   Gravitational radiation from pulsar binaries

The most famous example of a system emitting gravitational waves is a *binary pulsar* PSR B1913+16 discovered by Hulse and Taylor in 1974. The radio waves emitted by this system were identified with the emission of a pulsar with the pulse period 59 ms. By observing this object Hulse and Taylor found that the radio pulses are modulated with the period of 7.75h. This is the period of the orbital motion in the binary system. Its companion is a neutron star. The pulsar provides an extremely accurate clock that allows demonstration of the post-Newtonian effects and as a result determination with a very high accuracy of the parameters of this binary system. Both neutron stars have nearly equal masses, about $1.4M_\odot$. The mass of the companion is $1.387M_\odot$. The orbital period is 7.751939106h. The orbit has the following parameters: eccentricity 0.617131, semi-major axis 1 950 100 km, periastron and apoastron separation 746 600 km and 3153 600 km, respectively.

Because of the gravitational radiation the period of the binary P_b changes with time as

$$\dot{P}_b = -\frac{192\pi}{5}\left(\frac{2\pi\mathcal{M}}{P_b}\right)^{5/3}. \tag{10.3.8}$$

Thus, measuring the period and the rate of its change one can find the *chirp mass* of the system. The observed rate of decrease of the orbital period is about 76.5 ms per year. More accurately, the observed change of the period is $\dot{P}_b = -(2.4184 \pm 0.0009) \times 10^{-12}$. The theoretical prediction is $\dot{P}_b^{\text{theor}} = -(2.4056 \pm 0.0051) \times 10^{-12}$. This result is obtained by using the quadrupole formula for the gravitational-wave radiation and takes into account a small correction connected with acceleration of the binary system toward the center of our galaxy. This confirms the existence of the gravitational waves. For this discovery Hulse and Taylor received the 1993 Nobel Prize in Physics.

There are several confirmed double-neutron-star binaries in which it is possible to test the gravitational-wave radiation (Stairs 2003). The most remarkable of them is a *double pulsar* PSR J0737-3039 discovered in 2003. This system consists of two pulsars (A and B). Their masses are 1.337 and 1.250 solar mass, for the pulsars A and B respectively. The spin period is 23 ms (for A) and 2.8 s (for B). The orbital period is 2.4h. What makes this system unique is that the orientation of the magnetic field of the pulsars is such that one can see radio pulses from both of them. As a result of the gravitational-wave radiation the common orbit shrinks by 7 mm per day. The lifetime T_0 is about 85 million years.

10.3.4 Gravitational radiation from black hole binaries

The evolution of a black hole binary has several phases:

Inspiral: As the binary evolves the frequency and amplitude of the gravitational wave grow. One can expect that the most powerful radiation comes from the binary at the moment just before the coalescence.

Collision: After reaching the last stable orbit black holes fall into one another.

Merger: As a result of the black hole collision a single black hole surrounded by a horizon is formed.

Ringdown: The newly born black hole is in its 'excited' state. Relaxation to a final stationary configuration is also combined with emission of gravitational waves. The characteristic frequencies of these waves are determined by the spectrum of the quasinormal modes (see Section 9.4.4).

Tails: After exponential decay of the ringing radiation there still remain power-law tails of the gravitational radiation.

The characteristic form of the time dependence of the gravitational-wave amplitude is shown in Figure 10.9. The amplitude and frequency increase at the inspiral stage reflects the

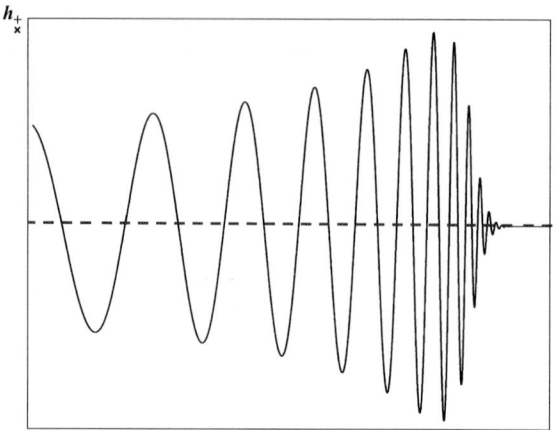

Fig. 10.9 A typical waveform of the gravitational radiation from two coalescent black holes.

effect of the shrinking of the orbit radius because of the gravitational energy loss. To estimate the maximal amplitude at the end of this stage one can use the quadrupole formula (see Eq. (5.4.33)). More accurate estimations give for the effective amplitude of the gravitational waves the following expression (Sathyaprakash 1999; Sathyaprakash and Schutz 2009)

$$h_{\text{eff}} \sim \frac{4\beta\nu M}{\pi r}. \tag{10.3.9}$$

Here, $\nu = M_1 M_2/(M_1 + M_2)^2$ is a *symmetric mass ratio* and M is the total mass. For equal mass black holes $\nu = 0.25$, while for an extreme mass ratio, when $M_2 \ll M_1$ one has $\nu = M_2/M_1$. β is a parameter that depends on the dimensionless rotation parameter $\alpha = a/M$ that takes values between 0.7 (for a non-rotating black hole) and 0.4 (for an extremely rotating one). If one chooses the radius of the *innermost stable circular orbit* $R = 6M$, one has (Sathyaprakash 1999; Sathyaprakash and Schutz 2009)

$$
\begin{aligned}
h_{\text{eff}} &\sim 10^{-21} \left(\frac{\nu}{0.25}\right) \left(\frac{M}{20 M_\odot}\right) \left(\frac{200\,\text{Mpc}}{r}\right), \\
&\sim 10^{-21} \left(\frac{\nu}{10^{-4}}\right) \left(\frac{M}{10^6 M_\odot}\right) \left(\frac{4\,\text{Gpc}}{r}\right), \\
&\sim 3 \times 10^{-17} \left(\frac{\nu}{0.25}\right) \left(\frac{M}{2 \times 10^6 M_\odot}\right) \left(\frac{6.5\,\text{Gpc}}{r}\right).
\end{aligned}
\tag{10.3.10}
$$

The first of these relations shows that the characteristic amplitude of the gravitational waves coming from a binary system of stellar-mass black holes with equal masses $10 M_\odot$. located at the distance of 200 Mpc is about 10^{-21}. These waves have a characteristic frequency of around 500 Hz.

The second of the relations demonstrates that a 100 $M_\odot$. black hole plunging into a $10^6 M_\odot$. black hole at the distance 4 Gpc produces gravitational waves with an amplitude of about

10^{-21}. In this case the total mass practically coincides with the mass of the supermassive black hole, while the parameter ν is about 10^{-4}. The characteristic frequency is about 15 mHz.

For a binary system of two supermassive black holes of mass $10^6 M_\odot$. and located at a distance of 6.5 Gpc, the effective amplitude of the gravitational waves is about 3×10^{-17}. Their characteristic frequency is of about 7.5 mHz.[6]

10.3.5 Gravitational-wave detectors

Gravitational waves that reach the Earth induce relative motion between separated 'free' masses. To trace these tiny changes of distances it is convenient to use laser interferometers. There exist several operational ground-based gravitational-wave detectors. The most sensitive is LIGO – the Laser Interferometer Gravitational Wave Observatory. LIGO has 3 detectors: one in Livingston (Luisiana) and the other two at Hanford (Richmond, Washington). Initial operation of the LIGO detectors began in 2001. Each detector consists of two storage arms, which are vacuum tubes of 1 m diameter and 4 km length at the right angle to each other. At Hanford the detector besides the main 4 km interferometer contains also a half-length interferometer using the same vacuum tubes. In November 2005 the sensitivity of the LIGO detectors reached the projected value of 10^{-21} over a 100 Hz frequency bandwidth.

There are other working ground-based detectors. GEO 600 is the German–British laser interferometer located near Hannover. It has two arms of length 600 m each. It is working in the frequency range 50Hz–1.5kHz. Its sensitivity is $10^{-20} - 10^{-21}$. The GEO600 detector reached the designed sensitivity in 2006. VIRGO is a gravitational-wave detector at Cascina in Italy. It has two arms of 300 m each. It commenced operations in 2007. TAMA 300 is a gravitational-wave detector constructed in Japan. It has two arms of 300 m each.

In November 2005, LIGO and GEO 600 began a joint science run. In 2007 Virgo joined this collaboration. At the present time there exist a few months of 3-detector data and 3 years of 2-detector data. These detectors are effective in the frequency bandwidth required for observation of the gravitational waves from NS–NS, BH–NS, and stellar mass BH–BH binaries.

It is planned to upgrade LIGO. The advanced LIGO, which should be fully operational in 2016, will have sensitivity exceeding the initial LIGO by more than a factor of 10. It is expected that this new instrument will detect gravitational waves from compact relativistic sources as often as 1 per a day. The joint work of several existing and newly constructed detectors will give detailed information on the strength and the form of the wave front, as well as a position of the source of the gravitational waves in the sky.

The study of the gravitational waves generated by supermassive black holes requires operation in the low-frequency window. The laser Interferometer Space Antenna (LISA) is the most developed project. LISA is planned as a space-based laser interferometer with three test masses forming an equilateral triangle. The frequency window of LISA is $10^{-4} - 10^{-1}$ Hz. It would be able to see such events as massive ($10^6 M_\odot$) black hole mergers in distant galaxies at $z = 1$, formation of the earliest black holes with $10^4 M_\odot$ at $z = 15$, and stellar-mass black hole capture by the massive black holes.

---

[6]As it is adopted in astronomy we give distances in parsecs. 1 pc$\approx 3 \times 10^{13}$ km=3×10^{18} cm; 1 Mpc= 10^6 pc, and 1 Gpc= 10^9 pc. The cosmological redshift at the distance 6.5Gpc is $z \sim 1$. The presented estimation of the frequency for the binary system of supermassive black holes takes into account this redshift effect.

First detection of the gravitational waves would be able to test the predictions of the Einstein general relativity. Further observations will allow scientists to study the processes in the strong gravitational field, and, in particular, to 'see' such dramatic events as black hole merging. This will open up to us a new 'gravitational' channel for study of the universe and black holes.

10.3.6   Numerical relativity

The results in the previous subsection gave an order of magnitude estimation of the amplitudes of the gravitational waves. These estimations were based on the following assumptions:

- components of the binary moved along the Keplerian orbits calculated in the Newtonian approximation;
- the motion is slow, $v/c \ll 1$;
- gravitational field is weak.

For real gravitational-wave data analysis it is necessary to know the templates for gravitational-wave profiles with rather high precision. Since the expected signal is very weak one needs to use the detailed information not only about the amplitudes but also about the phase of the wave sample. The required cumulative phase accuracy is 0.1 rad for LIGO/GEO/VIRGO searches. The accuracy must be 0.01 rad for the information extraction. For planning observations with LISA it must be much higher.

For the early stage of the BH–BH inspiral one can use the post-Newtonian approximation. For late-time ringdown radiation the black hole perturbation theory can be applied. For the radiation of the stellar-mass black hole plunging into the massive black hole one can use the perturbation theory with the ratio of the black hole masses as a small parameter. In all other cases (late inspiral, collision, merger, and early ringdown phases) the final resort is the numerical relativity. Remarkable progress was achieved in this direction recently (see, e.g., a review (Hinder 2010) and references therein).

Besides an accurate calculation of the amplitude of the gravitational waves from merging BH–BH, BH–NS, and NS–NS binaries, and for a massive black hole capture of compact relativistic objects, the numerical relativity allows one to calculate the total energy release. In particular, it is shown that enormous energy, up to 8% of the rest-mass energy, can be released in such processes in the form of gravitational waves.

10.3.7   Kick effect

Another exciting discovery of the numerical relativity is a prediction that the merger of two black holes can result in a kick of the newly born black hole up to very high velocity. The *kick effect* in a binary system is a result of asymmetry in the emission of the momentum carried by the gravitational waves. As a result of this emission the center-of-mass of the binary gets a *recoil velocity*.

Consider a system of two black holes revolving on the circular orbit. If the masses of the black holes are the same, the center-of-mass motion remains at rest. This conclusion follows from the symmetry. When the masses are different, the motion of the black hole of smaller mass is faster, and its radiation is more collimated. The net momentum is ejected in the plane of rotation along the velocity direction of this mass. The recoil motion of the center-of-mass is periodic. Averaged over the period it vanishes. In the black hole binaries two effects are important. At the late inspiral phase the radius of the orbit changes significantly with time, so that there will be a non-vanishing accumulative effect in the center-of-mass recoil. What is more important, there exists a final kick, which arises when a binary approaches the plunge, and the averaging is not effective.

> *The total energy emitted during the coalescence of the black hole binary $\sim Mf(M_1/M_2)$, where M is the total mass of the system. If we multiply both masses by some factor λ, the total energy release also changes by the same factor. A similar rescaling property is valid for the momentum $\Delta \vec{P}$ radiated by the black hole. Since the recoil velocity is $\vec{V} = \Delta \vec{P}/M$, it remains invariant under rescaling of the masses.*

To describe the last stage of the binary evolution numerical relativity is required. The numerical calculations performed demonstrated the existence of the *kick effect* and determined its characteristics. In particular, it was demonstrated that the gravitational radiation during the plunge and early ring-down phase can decrease the remnant black hole velocity (*antikick effect*). Another important result is that the kick velocity essentially depends on the black holes spins and their orientation. If the masses of black holes are equal and spin orientation is the same, the symmetry argument shows again that the kick effect vanishes.

Numerical calculations for non-spinning black hole binary inspiral shows that the maximal kick of 175.2 ± 11 km/s is achieved for the symmetric mass ratio parameter $\nu = M_1 M_2/(M_1 + M_2)^2 = 0.195 \pm 0.005$ (Gonzaález *et al.* 2007a). Much larger kick velocities exist in the binary system of two rotating black holes. It was demonstrated that the maximal kick velocity can be 2500 km/s for equal-mass black holes with antialigned spins $S/M^2 \approx \pm 0.8$ in the orbital plane Gonzaález *et al.* 2007b).

> *This kick velocity is so high that the final black hole formed in the merger of two massive black holes would escape from dwarf elliptic and spheroidal galaxies (with typical escape velocity of below 300 km/s) and from giant elliptic galaxies with the escape velocity ~ 2000 km/s (Tichy and Marronetti 2007).*

Kick velocites of more than 1000 km/s are quite usual (Tichy and Marronetti 2007). For certain configurations of black hole spins and masses the predicted kick velocities are up to a maximum of 4000 km/s (Campanella *et al.* 2007). Much higher recoil velocity of up to 10000 km/s (*superkick*) are expected for the hyperbolic encounter of two black holes (Healy *et al.* 2009),

There are indications that some of the quasars may have unusually large velocities. Two such objects, quasars E1821+643 (at $z = 0.287$) and SDSS J1054.35+345631.3 (at $z = 0.272$) have velocities 2100 km/s and 3000 km/s, respectively. They are discussed as possible candidates for black hole recoil after merging supermassive black holes (Robinson *et al.* 2010; Shields *et al.* 2009).

10.3.8 Gamma-ray bursts

Gamma-ray bursts (GRB) are the brightest objects in the universe. They are observed once or twice a day, as highly energetic explosions. They are registed as lower-energy gamma-ray bursts, which last typically a few seconds. The initial burst is usually followed by a longer-lived *afterglow* in the X-ray, optical and radio frequencies. The distribution of GRBs over the sky is homogeneous. The sources of most GRBs are billions of light years away from the

Earth. The energy released during a burst is about 10^{51} erg within a few seconds. Their power is only a few orders of magnitude less than the power of the rest of the universe. It is only 8 orders of magnitues less than the maximal theoretical luminosity possible in the relativistic gravitating systems $c^5/G \sim 10^{59}$ erg/s. GRBs were first detected in 1967 by Vela satellites, but the discovery was reported only in 1973.

The GRBs are classified as *long gamma-ray bursts* with duration greater that 2 s, and *short gamma-ray bursts* with duration less than 2 s. The long gamma-ray bursts are associated with a rapidly star forming galaxies. It is believed that the central engine in them is a black hole formed by the collapse of massive rapidly rotating stars. A part of matter, left after the collapse, forms an unstable accretion disk. This matter finally, during the time of order of tens of seconds, falls into the black hole. The radiation of the black hole is a highly focused narrow jet travelling at speeds exceeding 99.995% of the speed of light. Afterglows are produced by interaction of the jet with surrounding matter. *Short gamma-ray bursts* are usually associated with a black hole formation in a binary system of two neutron stars.

Gamma-ray bursts open for us an exciting opportunity to study directly black hole formation processes. In future, it may even become possible to detect the gravitational waves emitted during these cataclysmic events (Rees 2000; Mészáros 2002, 2006; Vedrenne and Atteia 2009; Mundell *et al.* 2010).

10.4 'Black Holes' in Laboratories

10.4.1 Dumb hole

Hawking evaporation of black holes is one of the most exciting predictions of theoretical physics. Unfortunately, till now we did not find evidence of the existence of mini-and micro-black holes for which this effect is important. An interesting idea is to try to construct in the laboratory a device in which the analog of the Hawking effect can be observed. Of course, the analog is never a perfect copy, but it can accurately reflect some important features of the analyzed phenomenon. In this section we discuss some of such analog models that can be used to study the Hawking effect. It should be emphasized from the very beginning that usually one tries to model the classical or quantum field propagation in the background spacetime in the presence of the horizon. The restriction of such models is that the background does not obey the equations similar to the Einstein equation in the theory of gravity. A basic idea of such modelling is that the geometry can effectively arise as an emergent phenomenon (see, e.g., the book by Volovik (2003)). One can find a review of the *analog gravity models* in (Barceló *et al.* 2005)

It is quite surprising that some of the effects of the Einstein theory of general relativity have analogs in everyday physics. The best known of them is the *acoustic analog model*. The equations for propagation of sound waves (phonons) in a moving fluid have a form similar to that of photons in curved spacetime. In this analogy the sound velocity plays the role of the speed of light. The acoustic analog of a black hole, proposed by Bill Unruh, is the so-called '*dumb hole*' (from deaf and dumb). In the experiment it is possible to construct a system in which the velocity of the fluid depends on the point and in some region it becomes higher than the sound speed. A surface separating two regions, with subsonic and supersonic motion is an analog of the black hole horizon. A sound signal emitted in the supersonic region will be dragged away by the fluid flow and will not be heard outside the 'horizon'. This analogy is quite precise and is valid not only classically but for quantum effects too.

The origin of the effective curved metric in the acoustic analog model comes from the equation for propagation of sound. Let us consider the flow of fluid that is potential $\nabla \times \mathbf{v} = 0$, i.e. without rotation and viscosity. In the laboratory frame the vector field of the 3-velocity of the fluid $\mathbf{v}$ may depend on the space and time coordinates. For the irrotational flow the sound waves, which correspond to the perturbations of the velocity $\delta\mathbf{v}$ above the background flow $\mathbf{v}$, can be described by a potential ϕ such that $\delta\mathbf{v} = \nabla\phi$. The wave equation for the propagation of phonons is

$$\left(\frac{\partial}{\partial t} + \nabla \cdot \mathbf{v}\right) \frac{\rho}{c_s^2} \left(\frac{\partial}{\partial t} + \nabla \cdot \mathbf{v}\right) \phi = \nabla \cdot (\rho \nabla \phi), \tag{10.4.1}$$

where ρ is the fluid density and c_s is the speed of sound, both of them being space and time dependent in a generic case. One can rewrite this equation in a 'covariant' form

$$\Box_{(s)}\phi = \frac{1}{\sqrt{-g_{(s)}}} \partial_\mu \left(\sqrt{-g_{(s)}} g_{(s)}^{\mu\nu} \partial_\nu\right) \phi = 0 \tag{10.4.2}$$

by introducing the effective *acoustic metric*

$$g_{(s)\mu\nu} = \frac{\rho}{c_s} \begin{pmatrix} v_i v^i - c_s^2 & -v_i \\ -v_j & \delta_{ij} \end{pmatrix}, \quad g_{(s)}^{\mu\nu} = \frac{1}{\rho c_s} \begin{pmatrix} -1 & -v^i \\ -v^j & c_s^2 \delta^{ij} - v^i v^j \end{pmatrix}. \tag{10.4.3}$$

This metric belongs to the *Painlevè–Gullstrand* class of metrics. For a special case where the background matter density ρ is homogeneous and constant in time and the liquid flow is spherically symmetric with the following velocity profile

$$v^i = \frac{x^i}{r} v_r, \quad v_r = \sqrt{\frac{2GM}{r}}, \quad r^2 = \delta_{ij} x^i x^j, \tag{10.4.4}$$

this metric coincides with the Schwarzschild metric in the Painlevè–Gullstrand coordinates Eq. (6.5.1). In this analogy

$$c_s = c = \text{const.} \tag{10.4.5}$$

So, if one could create the fluid flow of this type, then phonons would perceive their world as the curved spacetime of the Schwarzschild black hole, with the speed of sound playing the role of the speed of light and the horizon being at the surface where the velocity of the fluid reaches the speed of sound $v^2 = c_s^2$. Equation (10.4.2) for the sound is identical to the massless scalar field equation in the curved spacetime. This means that both classical and quantum effects related to the field propagation on the fixed background spacetime are correctly reproduced in this analog model. In particular, one can expect that in the above-discussed model there exists an analog of the Hawking effect and the sonic horizon must emit thermal phonons.

In fact, the continuity equation $\nabla \cdot (\rho\mathbf{v}) = \mathbf{0}$ contradicts the assumption $\rho =$const. For the adopted radial velocity profile, the continuity equation gives $\rho \sim r^{-3/2}$. For this choice the acoustic metric is conformal to the Schwarzschild metric. The corresponding 'acoustic' spacetime has also the horizon, surface gravity, Hawking temperature, and other black hole features, which can be in principle tested in the laboratory experiments.

Certainly the *dumb hole* model, as any other analog model of gravity, has restrictions. We have already mentioned that the effective metric is determined by the laws of hydrodynamics, and it is not a solution of the gravity equations. Another important restriction is that the sound wave in the fluid is an approximation for the description of the media excitations that works at the scales much larger than atomic ones. The sound is in fact a collective motion of the media constituents. The nature of the constituents becomes important when the sound wavelength becomes comparable with the atomic distances. In this high-energy regime the notion of sound wave has not much sense and the atomic properties become important.

> *It is plausible that the 'real' gravity is in fact the emergent phenomenon. For example, this approach naturally arises in the string theory, in which the strings play the role of the constituents. It is interesting that such fundamental effects as the quantum radiation of black holes and quantum effects in the early universe, responsible for its large-scale structure formation, formally require for their calculation to plead to the physics at the Planckian and trans-planckian energies. This is certainly the 'quantum gravity land'. The analog models of gravity can be used to demonstrate that similar effects that are determined by the large-scale parameters, in fact do not depend on the details of the theory in this superhigh-energy regime.*

10.4.2 Other black hole analog models

The acoustic analog gravity is only one example among a variety of analog gravity models. Other models can be either classical or quantum. An incomplete list of them includes

- classical acoustics;
- water waves;
- emergent spacetime in quantum liquids like superfluid ^{3}He and ^{4}He;
- emergent spacetime in Bose–Einstein condensates (BEC);
- classical and quantum optics;
- slow light.

Shallow water waves

Shallow water waves (gravity waves) present another example of the emergence of an effective metric in a very simple and purely classical system. Assuming that the flow of the liquid is irrotational and neglecting the viscosity one can show that the propagation of long waves on the surface of shallow water can be presented in the form of the relativistic scalar wave equation in the effective (2+1)-dimensional curved spacetime. Because the velocity of surface waves depends on the depth of the basin filled with liquid and the effective metric is the function of this velocity, it is very easy to model different effective spacetimes in experiments (see, e.g., Weinfurtrer *et al.* 2011).

Quantum liquids

Liquid helium is the only substance known, that has the property to remain liquid at temperatures down to the absolute zero temperature. *Quantum liquids* do not solidify even at the absolute zero temperature due to large zero-point motion of atoms. The latter property is essentially quantum. This explains their name. There are two kinds of such liquids: helium-4 and helium-3. From the chemical point of view the atoms ^{3}He and ^{4}He are identical since they have the same structure of the electron shells. However, the properties of ^{3}He and ^{4}He quantum liquids are quite different. The reason is that ^{4}He nuclei have zero spin and are bosons, while ^{3}He nuclei have spin 1/2 and are fermions. At the temperature of the order of several kelvins the statistics becomes important. ^{4}He is the *quantum Bose liquid*, while ^{3}He is the *quantum Fermi liquid*. In both cases the liquids in the corresponding phases can be described by hydrodynamics. However, there exist two important differences that make the quantum liquids attractive for modelling gravity: (i) They do not have viscosity, so that the ideal fluid description works well, and (ii) they have very low temperature. Because the dissipation is zero for superfluids and their temperature is close to zero, these liquids are very promising for the study of thermal properties of the Hawking radiation from the black hole analogs. At the same time, the superfluid helium has many different phases with quite rich structures. This opens up an opportunity to model a variety of effects in curved spacetimes including quantum physics. One can find a detailed discussion of the *emergent gravity models* in the superfluids ^{3}He and ^{4}He in the remarkable book by Volovik (2003).

^{4}He quantum liquid

Atoms of ^{4}He are bosons. At a temperature below $T_c = 2.17$ K a macroscopic number of atoms condense in the superfluid state, which displays the phenomenon of the *Bose–Einstein condensate*. Its ground state is characterized by the scalar complex function $\Psi = |\Psi| \exp(i\Phi)$. The motion of the liquid ^{4}He below T_c is frictionless with the irrotational velocity

$$\vec{v} = \hbar/m_{(^4\mathrm{He})}\vec{\nabla}\Phi, \tag{10.4.6}$$

where $m_{(^4\mathrm{He})}$ is the mass of the helium atoms. Its sound excitations are collective modes related to perturbations of the order parameter Φ. The propagation of phonons on this background is described by the scalar wave equation on the effective (3+1)-dimensional spacetime. Phonons (the quanta of sound waves) have linear spectrum $E_k = \hbar c k$ which is the analog of the Lorentz-invariant spectrum of massless scalar particles. For the moving superfluid background they obey an equation similar to Eq. (10.4.2) for the effective metric determined by the fluid motion. The interaction of phonons leads to deviation from the linear dispersion relation. There are two characteristic scales connected with this effect:

- The first scale is of the order of the $E_1 = m_{(^4\mathrm{He})}\, c_s^2 \sim 30$ K for realistic conditions corresponding to the liquid ^{4}He, where the speed of sound $c_s \sim 2.5 \times 10^4$ cm s^{-1}. At this scale the dispersion relation considerably deviates from the relativistic one.

- The second scale $E_2 = \hbar c_s/a \sim 5$K describes the energies where the discreteness of the matter becomes important. Here, a is the average interatomic distance. In solids this scale would correspond to the Debye temperature.

From an experimental point of view, though, there are problems in the study of quantum effects in the analog black holes in ^{4}He. When the velocity of the superfluid reaches the speed of sound, what is necessary for modelling of the black hole horizons, the conditions for the existence of superfluidity break down.

^{3}He quantum liquid

The superfluid ^{3}He is a more promising candidate for the laboratory probing of black hole analogs. Atoms of the ^{3}He are fermions. The transition temperature to the superfluid state is very low $T_c \sim 2.7$ mK. This huge (three orders of magnitude) difference with the transition temperature for ^{4}He has a simple explanation. The superfluid transition is a natural process for Bose systems, where the Bose condensate naturally occurs. Superfluidity of ^{3}He is a much more complicated process. The helium-3 atoms must first form bosons, by means of the mechanism similar to the Cooper-pair creation in superconductors. The Bose condensate of the Cooper pairs is responsible for the superfluidity property. The superfluid ^{3}He has several phases with quite different physical properties. In the superfluid phase ^{3}He-A isotropy is spontaneously broken. In anisotropic superfluids the 'speed of light', i.e. the velocity of quasiparticles, depends on the direction of propagation. In ^{3}He-A the fermionic excitations, which propagate along the anisotropy vector field $\mathbf{l(x)}$, have the speed $c_\parallel \sim 6000$ cm s^{-1} while in the transverse direction they propagate at the speed $c_\perp \sim 3$ cm s^{-1}. The effective metric for excitations depends on:

- the flow of the superfluid component of the quantum liquid, exactly as in the case of the acoustic metric;
- the spatial distribution of anisotropy parameter $\mathbf{l(x)}$.

After proper constant rescaling of the coordinates the effective metric of the emergent space-time can be written as

$$g^{\mu\nu}_{(s)} = \begin{pmatrix} -1 & -v^i \\ -v^j & g^{ij}_{(anis)} - v^i v^j \end{pmatrix}, \tag{10.4.7}$$

where

$$g^{ij}_{(anis)} = c^2_\parallel l^i l^j + c^2_\perp (\delta^{ij} - l^i l^j). \tag{10.4.8}$$

Note that the corresponding Riemann tensor for this emergent geometry is not trivial. By manipulating with the anisotropy field one can create the situation when the velocity of the flow of the superfluid accedes the $c_\perp$, while the property of superfluidity is still preserved. This provides one with the laboratory analog model of a black hole.

Bose–Einstein condensates

The other example of the condensed-matter systems revealing the effective geometry for quasiparticles are the *Bose–Einstein condensates* (BEC). A Bose-Einstein condensate is a state of matter of a dilute gas of weakly interacting bosons. When bosons are confined in an external potential and cooled to temperatures close to absolute zero, a macroscopically significant fraction of them condense to the ground quantum state. BEC systems have a high degree of quantum coherence, low temperature, and low speed of sound. In the hydrody-namical approximation for BEC one can introduce the acoustic metric for the propagation of quasiparticles (phonons) of the condensate. Phonons can be consistently quantized on the background acoustic metric. Their dynamics mimics that of the quantum scalar fields in a curved spacetime. This combination of properties makes BECs very attractive candidates for testing quantum effects in the analog models of curved spacetime.

In practice, a variety of isotopes, which are bosons, of alkaline atoms at ultralow temper-atures $\sim 10^{-7}$ K or below are mainly used to study BECs experimentally.

Optical analog models

When the light is propagating in an inhomogeneous media with a refractive index $n > 1$ its velocity c/n is less than the velocity of light in the vacuum. One can describe this propagation by introducing an effective metric[7] dependent on $n(x)$. When the optical media is moving, the corresponding metric is modified by the media velocity similarly to the case of the sound propagation we discussed earlier in the section.

In the case where the velocity of the media motion depends on spatial coordinates, so that in some regions it becomes higher than the effective light speed c/n, one gets a system that is similar to a black or a white hole. The surface separating domains with 'subluminal' and 'superluminal' motion of the media plays a role of the horizon. In such a system the large velocity gradients within a few wavelengths of the light can be achieved. Such gradients would correspond to the Hawking temperature of the order of 10^3 K. The first experimental results on classical effects of horizons were recently obtained (Philbin *et al.* 2008) (see also a review by (Leonhardt and Philbin 2008)).

Another realization of a similar idea is possible for the light propagation in fiber optics. In the fiber-optics communication, the signals carrying the information are propagated in the media. But because of the *Kerr effect* they modify the refractive index n. The corresponding change δn is proportional to the intensity of the signal. This modification of the refractive parameter moves together with the signal with the velocity equal to its group velocity. Consider a test electromagnetic signal propagating in such a system. Equations for the test field can be obtained by the linearization of the corresponding non-linear electromagnetic equations, and they are effectively equivalent to the equation in the moving media with non-trivial refractive index. Each signal pulse generates two artificial horizons, the black hole horizon at the front of the pulse and the white hole horizon at its end. However, in the performed experiments the gradient of the refractive index at these horizons, and hence the effective Hawking temperature, is quite small (for discussion see (Leonhardt and Philbin 2008) and references therein).

The speed of light in most dielectric media is comparable to the speed of light in vacuum. In practice, it would be very convenient to somehow slow down the speed of light. Fortunately, the contemporary technology of electromagnetically induced transparency provides us with such exotic media. Using the laser beam coupled to excited states of atoms of these media one can strongly affect their optical properties. In such experiments the electromagnetic perturbations propagate in the effective curved metric, while both the group and the phase velocities can be made well below the vacuum speed of light.

We mentioned a few possible laboratory black hole analogues (a detailed review can be found in (Barceló *et al.* 2005; Leonhardt and Philbin 2008)). One can expect that in future the accuracy achieved in these experiments will increase. It may also happen that new materials will be discovered where the classical and quantum analogs of black holes would be observed. This makes the 'search for black holes' in laboratories a very exciting direction of the research.

---

[7] Strictly speaking, this is correct in the geometric-optics approximation, or for the Maxwell equation in the media, when the magnetic permeability $\mu(x)$ is equal to the electric permittivity $\varepsilon(x)$. In the general case, the analogy of the light propagation in an inhomogeneous media and in an external gravitational field is not exact (see discussion in . (Barceló *et al.* 2005))

10.5 Black Holes in Colliders?

10.5.1 Minimal mass of black holes

Modelling black holes in laboratory experiments is, however, rather restrictive. Such experiments might give us confirmation on only some aspects of the black hole physics, mainly connected with their quantum radiation. However, such modelling does not look promising for study of the strong gravity regime of general relativity and black hole fundamental laws. Intriguing questions are:

- Can we in any future experiments observe a 'real' black hole creation in the 'laboratory'?
- Can, for example, mini-black holes be created in the high-energy collision experiments?

To answer these questions, let us note that there exists a natural lowest-mass limit for mini-black holes. The standard arguments are as follows. Consider an object of mass M. Its *Compton wavelength*

$$\lambda_M = \frac{\hbar}{Mc} \tag{10.5.1}$$

is comparable with the gravitational radius $r_S \propto GM/c^2$ when

$$M \propto m_{Pl} = \sqrt{\frac{\hbar c}{G}}. \tag{10.5.2}$$

This result indicates that one cannot use the classical theory of gravity in this regime. One can also interpret this result by saying that the quantum spreading of a wave packet for a particle with mass $M \lesssim m_{Pl}$ does not allow a black hole to be formed. Any attempt to 'compress' a particle wavepacket into the region with a size less than λ_M would result in increase of its energy and mass beyond m_{Pl}.

Usually, adopted conclusions are:

- Black holes of mass smaller than m_{Pl} do not exist.
- Black holes with the mass slightly larger than m_{Pl} would require quantum gravity for their description.
- If there is no mechanism that makes the black hole of mass $\gtrsim m_{Pl}$ stable, they would decay in the Planck time.
- If there exists a mechanism stabilizing such black holes, there might exist black hole remnants of Planckian mass. Such hypothetical objects were called *maximons* (Markov 1965) or *elementary black holes* (Hawking 1971).

We shall call a *micro-black hole* a black hole with mass of the order of m_{Pl}, or greater than this value only by a few orders. Consider collision of two identical particles and suppose that their energies in the center-of-mass frame are $\mu = E/2$. For creation of a micro-black hole the energy must be of the order of $E_{Pl} = m_{Pl}c^2 \sim 10^{28}$ eV or even greater. This energy scale is far beyond any realistic energy available in any future collider. That is why, at least in the framework of the standard gravitational theory one cannot expect production of micro-black holes in the future collider experiments.

10.5.2 Mini-black holes and extra dimensions

Recent 'reincarnation' of the idea of possible micro-black hole creation in the collision of two ultrarelativistic particles is related to now intensively discussed models with *large extra dimensions*. What makes this new development interesting is that the proposed scale of the micro-black holes in these models is on the order of TeV, that is many orders of magnitude smaller than the Planck energy. Before discussing these ideas we should emphasize that this micro-black hole creation in colliders is not an unavoidable prediction of any existent fundamental theory. This is a feature of the brane models with large extra dimensions. In these models the fundamental scale of the order of several TeVs is introduced mainly 'by hands'. Namely, for such a value of the fundamental scale the brane models do not contradict the observations and, at the same time, allow experimental tests. There are several nice reviews of the different aspects of this problem (Kanti 2004; Landsberg 2006). We refer the readers to these reviews and references therein. In this section we consider only a few rather simple questions related to the black hole creation in the collision of two ultrarelativistic particles. As earlier, we try to focus on those aspects that can be illustrated by simple calculations.

In the models with large extra dimensions it is assumed that besides our three spatial dimensions there exist one or more extra dimensions. In a so-called *ADD model* (Arkani-Hamed *et al.* 1998) our 4D world is embedded in a larger space with k flat, space-like dimensions. In the *Randall–Sundrum models* (Randall and Sundrum 1999) it is assumed that the bulk spacetime, which has anti-de Sitter (AdS) geometry, has one additional spatial dimension.

In both cases, all the usual matter can propagate only within the 4D brane, representing our world, while gravity can propagate in extra dimensions as well. In both models, in order to be consistent with the laboratory test of the Newtonian law, one needs to assume that the characteristic scale of extra dimensions is of the order of 0.1 mm or smaller. To simplify the presentation we restrict ourselves by considering the ADD model. Namely, we assume that there exist flat extra dimensions, which are compactified. We denote by L the compactification length.

> *In four dimensions the size 0.1 mm corresponds to the gravitational radius of the mass $M_1 \sim 10^{26}$ g. This is approximately 1/60 of the mass of Earth $\simeq 5.97 \times 10^{27}$ g, and of the same order as the mass of the Moon $\simeq 7.35 \times 10^{25}$ g. For black holes with mass much less than M_1 the gravitational radius is much smaller than the size of the extra dimensions.*
> *This conclusion remains valid in the higher-dimensional theory. In the general case, black holes in a space with compact extra dimensions would be distorted, but if their gravitational radius is much smaller than the compactification scale, the distortion is small and its effects can be neglected. We shall use this approximation and consider black holes as isolated in the spacetime with the flat topology. The dimensionality of the bulk spacetime is $D = 4 + k$, where k is the number of extra dimensions.*

First, we need to explain why large extra dimensions imply such a dramatic change of the fundamental scale for the quantum gravity. For this purpose let us repeat the calculations of the previous subsection. We again make 'order of magnitude' estimations and omit all the numerical coefficients of order of 1.

Consider a particle of mass M. Its Compton length does not depend on the number of dimensions and is again[8]

---

[8] We put $c = \hbar = 1$ but keep the gravitational coupling constant G.

$$\lambda_M = M^{-1}. \tag{10.5.3}$$

On the other hand, the gravitational radius of $D = 4 + k$ dimensional black hole is

$$r_{S,k} \sim (G^{(k+4)}M)^{1/(k+1)}. \tag{10.5.4}$$

In a spacetime with k compact extra dimensions of size L, the higher-dimensional gravitational coupling constant $G^{(k+4)}$ and the Newtonian 4D constant G are related as follows (see Eq. (5.5.57))

$$G^{(k+4)} = L^k G. \tag{10.5.5}$$

Denote by $m_\star$ the mass of the particle for which the Compton length coincides with its gravitational radius. Using Eq. (10.5.5) one obtains

$$m_\star^{k+2} L^k = m_{\mathrm{Pl}}^{k+2} l_{\mathrm{Pl}}^k. \tag{10.5.6}$$

Here, as earlier, m_{Pl} is a standard 4D Planck mass. Since $G = m_{\mathrm{Pl}}^{-2}$ one also has

$$G^{(k+4)} = m_\star^{-(k+2)}. \tag{10.5.7}$$

In the higher-dimensional theory $m_\star$ plays the role of the fundamental scale. The minimal mass of black holes is also equal to $m_\star$. The gravitational radius of $(k + 4)$-dimensional black hole of mass M is

$$r_{S,k} \sim m_\star^{-1}(M/m_\star)^{1/(k+1)}. \tag{10.5.8}$$

In particular, for the black hole of the fundamental mass $m_\star$ this radius $r_{S,k} \sim m_\star^{-1}$ in any number of dimensions. So, after restoring the dimensionful constants we have

$$r_\star \sim l_{\mathrm{Pl}}(m_{\mathrm{Pl}}/m_\star) = l_{\mathrm{Pl}} \left(\frac{L}{l_{\mathrm{Pl}}}\right)^{\frac{k}{k+2}}, \qquad m_\star \sim m_{\mathrm{Pl}} \left(\frac{L}{l_{\mathrm{Pl}}}\right)^{-\frac{k}{k+2}}. \tag{10.5.9}$$

We use the name *micro-black hole* to refer to small black holes of mass comparable with $m_\star$ or slightly greater than it.

For description of black holes of mass comparable with $m_\star$ the quantum gravity is required. For two extra dimensions ($k = 2$) and $L = 0.1$ mm the fundamental scale is $m_\star \sim 10^{12}$ eV. The corresponding gravitational radius $r_\star \sim 10^{-17}$ cm. It is many orders of magnitude smaller than the scale L of extra dimensions. This verifies the approximation we use: distortion of micro-black holes is negligible. The case of $k = 2$ and $L = 0.1$ mm is the most optimistic and phenomenologically most interesting scenario. In this case the fundamental scale is of order of a few TeV, and collider experiments with $E > m_\star$ might create black holes with the mass $\propto E$.

10.5.3   Cross-section of black hole creation

Suppose a model with large extra dimension is valid and the 'new' fundamental scale is about several TeV. A natural question is: Can one produce a micro-black hole in the collider experiment? To answer this question the following simplifying assumptions are usually made:

- Elementary particles, which are quantum objects, are approximated by classical point-like masses.
- The energy of the collision is high enough to create a black hole with the mass $\gg m_\star$, so that such a black hole is a classical object and quantum gravity corrections are not important.
- The cross-section of the black hole formation is estimated by the area of the apparent horizon. The null-energy condition and the Penrose hypothesis of *cosmic censorship* are assumed, so that a trapped surface is inside the black hole. The surface area of the trapped surface gives the lower limit for the cross-section.
- The tension of the brane, representing our 4D world, is small. Its gravitational field and its influence on the micro-black holes are neglected.

Let us use these simplifying assumptions and estimate the *cross-section of black hole creation* for collision of two ultrarelativistic particles. Consider two particles of equal mass m that have velocity $\mathbf{v}$ and collide with each other. In the center-of-mass frame their momenta are

$$p_1^\mu = m(\gamma, \mathbf{v}\gamma), \qquad p_2^\mu = m(\gamma, -\mathbf{v}\gamma). \tag{10.5.10}$$

Here, $\mathbf{v}$ is $(D-1)$-dimensional velocity, and $\gamma = (1 - \mathbf{v}^2)^{-1/2}$. If in the collision of these two particles a single object is created, it will have the momentum

$$P^\mu = (M, \mathbf{0}), \qquad M = 2\mu \equiv 2m\gamma. \tag{10.5.11}$$

We assume that the gamma-factor γ is big.[9] This means that the velocity of each of the particles in the center-of-mass frame is very close to the speed of light.

Denote by b an impact parameter between colliding particles. It is natural to assume that when $b \lesssim r_S(M)$ a black hole might be formed in the collision of particles with the total energy in the center-of-mass $M = 2\mu$. This gives a 'natural' estimation for the cross-section of the black hole formation

$$\sigma = B_D \pi \, r_S^2(M). \tag{10.5.12}$$

Here, B_D is a dimensionless factor of order of one, which depends on the number of dimensions and

$$r_S = \left(\frac{16\pi\, G^{(D)}(2\mu)}{(D-2)\Omega_{D-2}} \right)^{1/(D-3)}, \qquad \Omega_{D-2} = \frac{2\pi^{(D-1)/2}}{\Gamma[(D-1)/2]}. \tag{10.5.13}$$

Here, r_S is the gravitational radius of the D-dimensional Tangherlini black hole of mass 2μ (see Eq. (6.8.5)), and Ω_{D-2} is the surface area of a $(D-2)$-sphere.

Let us emphasize that in the ultrarelativistic particle collision (with and without black hole formation) gravitational waves will be produced. This is an important and robust prediction of models with large extra dimensions. It has two consequences:

---

[9]The expected values of the gamma-factor for the proton–proton collision in the LHC is about 7500.

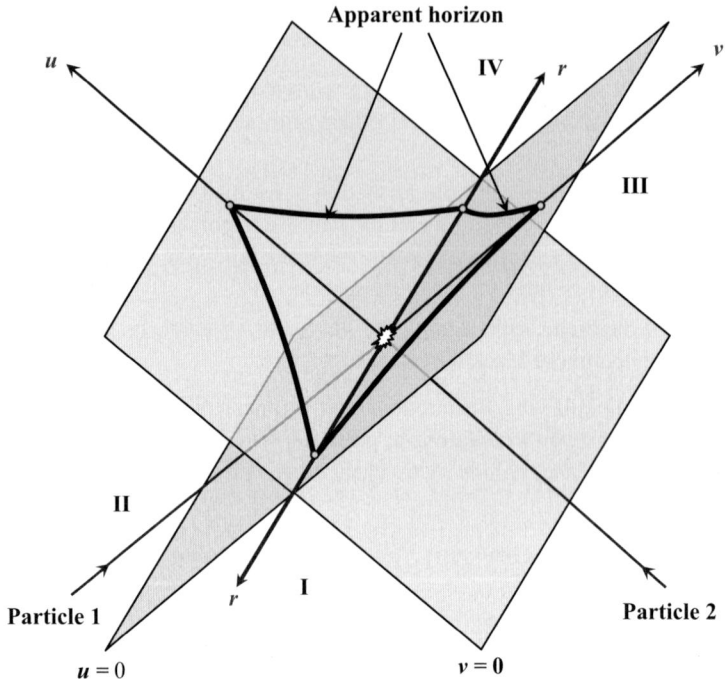

Fig. 10.10 A schematic picture of apparent horizons in two instantaneous slices $u = 0$ and $v = 0$ for a head-on collision. Particles collide at the spacetime point $u = v = r = 0$, where the coordinate $r = \sqrt{x^2}$.

- The mass of the formed black hole would be smaller than M.
- Since gravitational radiation can propagate in the bulk, a brane observer would register energy non-conservation.

The estimation Eq. (10.5.12) can be improved and one can obtain a quite accurate lower bound of the prefactor in this relation. For this purpose one uses an approximate expression for the gravitational field of an ultrarelativistic particle. In the center-mass reference frame both of the particles move practically with the speed of light. Their gravitational field can be approximated by the *Aichelburg–Sexl metric*.[10] Figure 10.10 illustrates the setup of this problem. Two relativistic particles do not interact before the collision. In regions I, II, and III the spacetime is flat except for the null planes $u = 0$ and $v = 0$. After the collision (in region IV) the non-linearity of the Einstein equations becomes important.

To estimate from below the cross-section of the black hole formation in the collision it is sufficient to find the surface of the apparent horizon. In the generic non-stationary case it always lies inside the event horizon provided the stress-energy tensor of matter satisfies the null-energy condition. So, if we calculate the cross-section corresponding to the size of the apparent horizon, we are guaranteed that the real cross-section of the black hole creation is

[10]This is a special case of a gyraton metric that was discussed in Section 5.5.4.

larger. The *Hawking area theorem* states that the area of the event horizon always increases. Therefore, the area of the apparent horizon at the moment of collision of particles gives a restriction from below on the cross-section of the creation of the black hole.

The size of the apparent horizon can be calculated at the moment of the collision, that is at the last moment, when one can still use the superposition of the Aichelburg–Sexl metrics. Penrose (1974, unpublished) studied the head-on collision and constructed an apparent horizon on the Λ-shaped slice formed by two null surfaces, which is a future boundary of the region I. This result can be generalized to the *D*-dimensional case, where the total area of the apparent horizon is (Eardley and Giddings 2002)

$$\mathcal{A}_{D-2} = \frac{2\Omega_{D-3}}{D-2} \left(\frac{8\pi G^{(D)}\mu}{\Omega_{D-3}} \right)^{(D-2)/(D-3)}. \tag{10.5.14}$$

On can use for the calculation of the apparent horizon another 'natural' slice, namely a *V*-shaped slice, which is the past boundary of the region IV. For the head-on collision the answer is the same Eq. (10.5.14), while for the collision with non-vanishing impact parameter, the calculations at the *V*-shaped slice give a better approximation. Such calculations were done in (Yoshino and Rychkov 2005). The corresponding value for the factor B_D in Eq. (10.5.12) was obtained in this work for different number of dimensions. In particular one has

$$B_D = 0.71 \text{ (for } D = 4 \text{)} \text{ and } 3.20 \text{ (for } D = 11 \text{)}. \tag{10.5.15}$$

The exact cross-section of the black hole formation in such experiments can be calculated only numerically. Unfortunately, numerical calculations for the collision with a large value of the γ-factor are very difficult.

10.5.4 'Birth, life, and death' of micro-black holes

It is convenient to describe the formation of a micro-black hole and its further evolution as a sequence of the following phases (Giddings 2007):

Formation: As a result of collision of two ultrarelativistic particles with small enough impact parameter a micro-black hole is formed. In the context of hadron colliders, the cross-section Eqs. (10.5.12)–(10.5.15) must be applied to partons, which are point-like constituents of the proton. To calculate the total cross-section of the proton–proton collision the parton distribution functions must be used.

Balding phase: When a micro-black hole is formed it is very asymmetric. Moreover, in higher dimensions black objects might have non-trivial topology of the horizon (see the next section). For $E \gg m_\star$ the formed black hole is a classical object, and its behavior can be described in classical gravitational theory. One can expect that all the multipoles that can be radiated, are finally radiated during this stage. The characteristic time of this process is $t \sim r_S$. Ultraspinning black holes may lose their angular momentum as a result of classical instability (see the next section). During the balding phase the gravitational waves are emitted. This results in a visible 'brane energy non-conservation'. It is usually assumed that the final result at the end of this stage will be a spinning uncharged stationary black hole.

Spindown phase: After the balding phase the micro-black holes are classically stable, but due to the Hawking effect they will lose their energy and momentum. The characteristic time of this stage is

$$t_H \sim m_\star^{-1}(E/m_\star)^{(D-1)/(D-3)}. \tag{10.5.16}$$

The rate of spin loss is faster than the rate of the mass loss. Estimations show that the black hole will lose over 50% of its mass during the spindown phase.

Schwarzschild phase: This is a phase of thermal radiation. The temperature of the black hole is

$$\Theta_H = \frac{D-3}{4\pi\, r_S} \sim M^{-1/(D-3)}. \tag{10.5.17}$$

The thermal spectrum of radiation is deformed by the graybody factors. During this and the spindown phase normal particles are emitted into the brane, while the gravitons are emitted in the bulk as well. Since the number of gravitational degrees of freedom grows fast with the number of spacetime dimensions, for a large enough number of extra dimensions the emission of the gravitons to the bulk might be comparable with the emission within the brane.

Planck phase: When the mass of the evaporating micro-black hole reaches the fundamental scale $m_\star$ one cannot use the classical theory and predictions can be made only after one constructs a consistent quantum theory of gravity. One can expect production of a few very energetic particles or formation of a stable remnant (if it exists). Due to recoil connected with emission of high-energy gravitons, a small-mass micro-black hole can leave the brane.

The most optimistic rate of micro-black hole production at the nominal Large Hadronic Collider (LHC) luminosity of $10^{34}\,\mathrm{cm}^{-2}\mathrm{s}^{-1}$ is one black hole every one or ten minutes (Giddings 2007). There exist some rather striking qualitative signatures indicating that an observed event might be connected with micro-black hole formation. Let us mention that a possiblity of micro-black hole formation in the cosmic-ray events is also widely discussed.

10.6 Higher-Dimensional Black Holes

10.6.1 Introduction

In the higher-dimensional gravity black hole can be used as natural probes of extra dimensions. The study of possible black hole creation in colliders stimulated high interest in higher-dimensional black holes and their properties (Cavaglià 2003; Kanti 2004).

But this is only one of several reasons for such interest. In string theory, which provides us with the most developed theory of quantum gravity, the spacetime time is naturally higher dimensional. The consistency of this theory requires that the total number of dimensions is ten (for the superstring theory). Six of these dimensions have small size, the other four dimensions describe our observable world. In the string theory the gravity is an *emergent phenomenon*. Successful statistical counting of black hole entropy in the string theory was first performed for higher-dimensional black holes. The microscopic origin and statistical mechanical explanation of black hole entropy remains one of the most important problems of black hole physics. Higher-dimensional black holes play the role of a natural 'laboratory' for microstate counting in the string theory (see, e.g., review (Sen 2008) and references therein).

Another exciting direction of the research, generated by the string theory, is a so-called *AdS/CFT correspondence*. This correspondence establishes relations between the classical gravitational theory in the D-dimensional bulk spacetime, and the quantum conformal field theory on its $(D-1)$-dimensional boundary. D-dimensional black hole solutions play an important role in this correspondence (see, e.g., review (Aharony 2000)).

There is one more reason to study higher-dimensional black hole objects. Namely, it is both interesting and important to understand which properties of solutions of the Einstein equations are specific for the four-dimensional spacetime, and which are of a general nature, valid in any number of dimensions. This approach may help one to find a simpler and more general proof of the already known results in 4D gravity. It also might help to look at the known problems from a new point of view.

Many higher-dimensional black hole solutions have been already obtained. One can find their discussion, e.g., in the book by Ortín (2004), and in a wonderful review by Emparan and Reall (2008). The definition of the black hole object in the higher dimensions remains basically the same as in the 4D spacetime. Namely, *a black hole is a spacetime region from where no information-carrying signals can escape to infinity*. Such regions have a boundary, the *event horizon*. As in the 4D case the horizon of a D-dimensional black hole is a $(D-1)$-dimensional null surface. A space-like slice crossing the horizon determines a $(D-2)$-dimensional boundary of the black hole object at a given 'moment of time', determined by this slice.

The main lesson of this study is that higher dimensions have 'more room' for black hole objects and there exists a wide variety of black objects in the higher dimensions. For simplicity, we discuss here only vacuum black holes. The main new features of higher-dimensional black holes are:

- The topology of the horizon of 4D black holes is always S^2. In the higher dimensions there exist different classes of the black hole solutions that differ by the topology of the horizon;
- In 4D besides the mass of the black hole there is only one additional parameter, the angular momentum. In higher dimensions the number of independent components of the angular momentum is $[(D-1)/2]$. This makes the structure of the space of parameters for higher-dimensional black holes more complicated.
- Some aspects of rotating black holes are determined by 'competition' between the Newtonian attraction force $\sim r^{2-D}$, and the centrifugal force of repulsion $\sim r^{-3}$. The conditions that follow from this 'competition' are different in $D=4$, $D=5$, and $D \geq 6$.
- For any value of the rotation parameter $0 \leq a/M \leq 1$ the size of the 4D horizon is mainly determined by the mass of the black hole. The ratio of the lengths of its equator and a closed circle along the 'meridian' is of the order of 1. The scale of the size of the higher-dimensional black hole is also determined by its mass, but the horizon may have quite different sizes in different 'directions'.

10.6.2 Black branes and black strings

We already know a vacuum solution for a spherically symmetric *higher-dimensional black hole*. This is the *Tangherlini metric*. The topology of its horizon is S^{D-2}. Starting with this solution it is quite easy to obtain a new vacuum black object solution with a different topology of the horizon.

Let ds_N^2 be a Ricci-flat metric in N-dimensional spacetime. Let $D > N$ and consider the following metric

$$ds_D^2 = ds_N^2 + \sum_{i=1}^{D-N} dz_i^2.$$ (10.6.1)

After adding of $D - N$ flat dimensions the metric remains Ricci flat, so that Eq. (10.6.1) is a vacuum solution of the D-dimensional Einstein equations.

If we start with the Tangherlini solution for an N-dimensional black hole with the topology of the horizon S^{N-2}, the metric Eq. (10.6.1) is a black hole with the topology of the horizon $S^{N-2} \times \mathbb{R}^{D-N}$. Such a solution is called a *black brane*. A special case of a black brane solution with $D = N + 1$ is known as a *black string*.

*Special names, such as black branes, black strings and so on, are commonly used in the litera-
ture in order to distinguish higher-dimensional black holes with different topology of the horizon.
A 'nickname' black hole is used only for solutions with the spherical topology of the horizon. It should
be emphasized that all these black objects, according to the definition, are black holes. To escape
possible confusion, we shall use a name* black object *as a common name for the higher-dimensional
black holes, and keep using the adopted 'nicknames' for special classes of the black objects with the
topology different from the spherical one.*

For an isolated N-dimensional Tangherlini black hole the spacetime is asymptotically flat. A surface of $r =$ constant at large r has the topology $S^{N-2} \times \mathbb{R}$. The *Komar mass* calculated at the spatial infinity is finite and proportional to the black hole mass. The properties of black branes are quite different. A spacetime with a black brane is not asymptotically flat. The D-dimensional Komar mass is a product of the mass of the N-dimensional black hole and the volume of flat extra dimensions. Since the latter is infinite, the D-dimensional mass of the black brane diverges.

A 'natural' way to make the mass of the black brane finite is to impose periodic boundary conditions for flat extra dimensions. This makes the volume of extra dimensions finite. The topology of the horizon of black branes with periodic boundary conditions is $S^{N-2} \times \mathbb{T}^{D-N}$, that is a product of $(N - 2)$-dimensional sphere and $(D - N)$-dimensional torus.

10.6.3   Boosted black strings

To illustrate important features of black branes we focus on the simplest case when the 4D solution is the Schwarzschild metric. We write the corresponding 5D black string metric in the form

$$ds^2 = -g dt^2 + \frac{dr^2}{g} + r^2 d\omega^2 + dz^2, \qquad g = 1 - \frac{r_S}{r}.$$ (10.6.2)

Let us perform the following transformation

$$t = c\tilde{t} + s\tilde{z}, \qquad z = s\tilde{t} + c\tilde{z},$$
$$c = \cosh\alpha, \qquad s = \sinh\alpha.$$ (10.6.3)

Here, α is a boost parameter. This transformation generates the motion in the z-direction with the velocity $v = s/c$, and c is the corresponding Lorentz γ-factor. Applying this transformation to the black string metric Eq. (10.6.2) one gets

$$ds^2 = -d\tilde{t}^2 + d\tilde{z}^2 + (1 - g)(c\,d\tilde{t} + s\,d\tilde{z})^2 + \frac{dr^2}{g} + r^2 d\omega^2. \qquad (10.6.4)$$

To compactify the metric Eq. (10.6.2) we identify

$$z + L \leftrightarrow z \quad \text{at} \quad t = \text{const}. \qquad (10.6.5)$$

Similarly, for the boosted string we use the identification

$$\tilde{z} + L \leftrightarrow \tilde{z} \quad \text{at} \quad \tilde{t} = \text{const}. \qquad (10.6.6)$$

It is easy to see that the compactified spacetimes Eqs. (10.6.5) and (10.6.6) are different. This means that the compactification procedure singles out a reference frame, and the motion of the black string with respect to this frame has a well-defined physical meaning.

In the general case of a black brane one can also consider a Lorentz transformation in (t, z_i) sector. By compactifying boosted $\tilde{z}_i$ coordinates one obtains a *boosted black brane* solution.

10.6.4 Gregory–Laflamme instability

Spherically symmetric black holes in four and higher dimensions are stable. However, the black branes constructed from them are generally unstable. Let us discuss this instability. For simplicity, we consider again a five-dimensional black string solution Eq. (10.6.2). Let M be the mass of the 4D Schwarzschild metric and L be the size of the fifth dimension. One can use $r_S = 2M$ as a scale parameter and write the metric Eq. (10.6.2) in the form

$$ds^2 = r_S^2 dS^2. \qquad (10.6.7)$$

The dimensionless metric, written in the dimensionless coordinates t/r_S and r/r_S, depends on only one dimensionless parameter

$$\beta = L/r_S. \qquad (10.6.8)$$

This parameter is the ratio of two scales, characterizing the form of the horizon, its length in the z-direction and the radius of the 2-sphere. In principle, this ratio can be arbitrary. This makes 5D black string solutions quite different from 4D black holes. In fact, black strings with large values of β are dynamically unstable. This instability was discovered by Gregory and Laflamme (1993) and it is known as the *Gregory–Laflamme instability*.

Simple arguments show why one can expect such an instability. These arguments are based on the comparison of the entropy of a 5D black string and a 5D black hole of the same mass M. The entropy of the black string is proportional to the 3D volume of the horizon

$$\mathcal{A}_{BS} \sim L \times r_S^2, \qquad r_S = 2G^{(4)}M. \qquad (10.6.9)$$

The gravitational radius $r_S^{(5)}$ of the 5D black hole of the mass M is

$$r_S^{(5)} \sim (G^{(5)}M)^{1/2}. \qquad (10.6.10)$$

Its horizon 3D volume is

$$\mathcal{A}_{BH} \sim (r_S^{(5)})^3 \sim (G^{(5)}M)^{3/2}. \qquad (10.6.11)$$

In a spacetime with a compact extra dimension of size L the 5D and 4D gravitational coipling constants are related as follows (see Eq. (5.5.56))

$$G^{(5)} \sim LG^{(4)}. \tag{10.6.12}$$

Using Eqs. (10.6.8), (10.6.9), (10.6.11), and (10.6.12) one obtains

$$\frac{\mathcal{A}_{\text{BH}}}{\mathcal{A}_{\text{BS}}} \sim \beta^{1/2}. \tag{10.6.13}$$

This relation shows that for large β the 5D black hole has higher entropy than the black string of the same mass M. In other words, the 5D black hole is entropically more preferable than a long (with large β) black string.

It was shown by Gregory and Laflamme (1993) that this 'entropic' argument is correct, and there exists a *dynamical instability* of long black strings. To prove this instability they considered perturbations of the black string metric. The perturbations can be decomposed into scalar, vector and tensor components according to their transformations under spatial rotations. The scalar and vector perturbations have been proved to be stable, while there exist unstable modes of the tensor perturbations. To show this, one considers modes with the dependence $\sim \exp[-i(\omega t - kz)]$. The frequency ω acquires a positive imaginary part for wavelengths $k^{-1} > k_{\text{GL}}^{-1} \sim r_{\text{S}}$. For these wavelengths the amplitude of the mode exponentially grows. The maximal value of this imaginary part of ω occurs for $k \approx k_{\text{GL}}/2$. It corresponds to the fastest growing mode of the instability. The instability creates inhomogeneity in the z-direction. At the moment, it is not clear what is the final state of an unstable black string. It is plausible that a decay of the long black string results in growth of the spacetime curvature until a singularity forms.

The *Gregory–Laflamme instability* is a common feature of black branes. It generically occurs when the compactification size becomes large enough with respect to the 'transverse' size of the brane. Boosted black branes also have a similar instability. More details about black branes can be found in the reviews (Kol 2006; Harmark *et al.* 2007).

10.6.5 Higher-dimensional rotating black holes

Myers–Perry metric with one rotation parameter

A higher-dimensional generalization of the Kerr metric to an arbitrary number of dimensions $D \geq 5$ was found by Myers and Perry (1986). It is an asymptotically flat vacuum solution of the Einstein equations. In the D-dimensional spacetime it contains $[(D+1)/2]$ free parameters: the mass M and $[(D-1)/2]$ components J_i of the angular momentum. This metric was originally obtained in the Kerr–Schild form, similar to Eq. (8.1.15).

Let us discuss first this solution for a special case when all J_i except one vanish, that is the black hole has a single rotation plane. The *Myers–Perry metric* in this case is

$$ds^2 = -dt^2 + \frac{m}{r^{D-5}\Sigma}(dt - a\sin^2\theta\, d\phi)^2 + \frac{\Sigma}{\Delta}dr^2$$
$$+ \Sigma\, d\theta^2 + (r^2 + a^2)\sin^2\theta\, d\phi^2 + r^2\cos^2\theta\, d\Omega_{(D-4)}^2. \tag{10.6.14}$$

Here,

$$\Sigma = r^2 + a^2 \cos^2 \theta\,, \qquad \Delta = r^2 + a^2 - \frac{m}{r^{D-5}}. \qquad (10.6.15)$$

From the asymptotic form of this metric one finds that the parameters m and a are related to the mass M and the angular momentum J as follows

$$M = \frac{(D-2)\,\Omega_{(D-2)}}{16\pi\,G^{(D)}} m\,, \qquad J = \frac{2Ma}{D-2}. \qquad (10.6.16)$$

Here, $\Omega_{(D-2)} = 2\pi^{(D-1)/2}/\Gamma((D-1)/2)$ is the area of a unit $(D-2)$-dimensional sphere.

> **Problem 10.3:** *Show that in four dimensions the metric Eq. (10.6.14) and Eq. (10.6.15) coincides with the Kerr metric Eq. (8.1.4) with a properly defined mass M and rotation parameter a. (Note that in this case the last term in the expression Eq. (10.6.14) containing $d\Omega^2_{(D-4)}$ must be omitted.)*

> **Problem 10.4:** *Show that in the absence of rotation, $a = 0$, the metric Eq. (10.6.14) reduces to the Tangherlini metric Eqs. (6.8.2)–(6.8.4).*

The *event horizon* for the solution Eq. (10.6.14) is at the largest real root r_+ of the equation $\Delta = 0$, i.e.

$$r_+^2 + a^2 - \frac{m}{r_+^{D-5}} = 0. \qquad (10.6.17)$$

Its horizon area is

$$\mathcal{A}_{\mathrm{H}} = \Omega_{(D-2)} r_+^{D-4}(r_+^2 + a^2). \qquad (10.6.18)$$

For $D = 5$ the solution of Eq. (10.6.17) is $r_+ = \sqrt{m - a^2}$. As for the Kerr metric the rotation parameter cannot be larger than some maximal value, connected with mass, $|a| \le \sqrt{m}$. However, in the extremal case, when $|a| = \sqrt{m}$ the surface area of the horizon vanishes and it becomes singular.

For $D \ge 6$ the function $\Delta(r)$ grows monotonically from $-\infty$ at $r = 0$ to ∞ at $r \to \infty$. Hence, it always has a single root in this domain. In particular, it means that there exists an event horizon for any value of the rotation parameter a. When $a \gg r_+$ one can neglect the term r_+^2 in Eq. (10.6.17) and get

$$r_+ \sim (m/a^2)^{1/(D-5)}. \qquad (10.6.19)$$

In this limit the horizon flattens along the plane of rotation. Its size along the plane is $\sim a$, while in the transverse direction the size is $\sim r_+$.

Myers–Perry metric: general case

The form of the Myers–Perry (MP) metric is slightly different for even and odd numbers of the dimension. To deal with both cases simultaneously we denote as earlier

$$D = 2n + \varepsilon, \qquad (10.6.20)$$

where $\varepsilon = 0$ and $\varepsilon = 1$ for even and odd dimensions, respectively. In such a spacetime there exists $N = n - 1 + \varepsilon$ independent rotation planes. We denote the corresponding rotation parameters by a_i.

The *Myers–Perry metric* for arbitrary rotation parameters in each of the rotation planes written in the *Boyer–Lindquist coordinates* has the form

$$ds^2 = -dt^2 + \sum_{k=1}^{N} \left[(r^2 + a_k^2)(d\mu_k^2 + \mu_k^2 \, d\phi_k^2) \right] + (1 - \varepsilon) \, r^2 \, d\alpha^2$$

$$+ \frac{m \, r^{1+\varepsilon}}{\Pi F} \left[dt + \sum_{k=1}^{N} a_k \, \mu_k^2 \, d\phi_k \right]^2 + \frac{\Pi F}{\Pi - m r^{1+\varepsilon}} \, dr^2,$$

(10.6.21)

where

$$F = 1 - \sum_{k=1}^{N} \frac{a_k^2 \, \mu_k^2}{r^2 + a_k^2}, \qquad \Pi = \prod_{k=1}^{N} (r^2 + a_k^2).$$

(10.6.22)

Coordinates in this metric are $(t, r, \mu_i, \phi_i, \alpha)$. The last of these coordinates, α, is present only in the even-dimensional case. ϕ_k are angles with the period 2π in each plane $x^{2k-1} - x^{2k}$, and μ_k are direction cosines with respect to these planes having the range $0 \leq \mu_k \leq 1$. Simple counting shows that the total number of these coordinates is $D + 1 = 2n + 1 + \varepsilon$. In fact, the coordinates are not independent. They obey a constraint

$$\sum_{k=1}^{N} \mu_k^2 + (1 - \varepsilon) \, \alpha^2 = 1.$$

(10.6.23)

The metric contains $n + \varepsilon$ parameters: m and a_i. These parameters are related to the mass M and the angular momentum components J_i as follows

$$m = \frac{8\Gamma\left(\frac{D-1}{2}\right)}{(D-2)\pi^{\frac{D-3}{2}}} \, G^{(D)} M, \qquad a_k = \frac{D-2}{2} \frac{J_k}{M}.$$

(10.6.24)

The metric does not depend on t and $N = n - 1 + \varepsilon$ angle variables ϕ_i. Thus, it admits $N + 1$ commuting Killing vectors ∂_t and ∂_{ϕ_k}.

> *The group of isometries of this solution is* $\mathbb{R} \times U(1)^N$. *If p rotation parameters are equal and non-vanishing, the subgroup* $U(1)^p$ *for the corresponding rotation planes is enhanced to a non-Abelian group* $U(p)$. *If p rotation parameters vanish the corresponding subgroup* $U(1)^p$ *is enhanced and becomes* $SO(2p + 1 - \varepsilon)$.

The outer boundary of the ergosphere, where $\xi_{(t)}^2 = 0$, is the surface of infinite redshift or the *ergosurface*. The equation of this surface is

$$\Pi F - m \, r^{1+\varepsilon} = 0.$$

(10.6.25)

Denote by

$$\eta^A = \xi_{(t)}^A + \sum_{k=1}^{N} \Omega_k \, \xi_{(k)}^A.$$

(10.6.26)

Outside the event horizon one can always choose the coefficients Ω_k in such a way that η is a time-like Killing vector. This means that a stationary reference frame is possible in this region. At the event horizon the linear combination (10.6.26) is a space-like or null vector The vector η is null for $\Omega_k = \Omega_k^H$, where Ω_k^H are the angular velocities of the black hole. The value of Ω_k^H as well as the position of the horizon are defined by the following system of equations

$$\left[\frac{\partial \eta^2}{\partial \Omega_k}\right]_{H,\Omega_k=\Omega_k^H} = 0, \qquad \left[\eta^2\right]_{H,\Omega_k=\Omega_k^H} = 0. \tag{10.6.27}$$

By solving these equations, one gets

$$\Pi - m \, r^{1+\varepsilon} = 0, \qquad \Omega_k^H = \frac{a_k}{r_+^2 + a_k^2}. \tag{10.6.28}$$

The first of these equations determines the position r_+ of the horizon. The second relation determines components of the *angular velocity* of the higher-dimensional black hole. The number of these components, $n - 1 + \varepsilon$, coincides with the number of rotation planes.

Both the event horizon and the ergosurface have the topology $S^{D-2} \times \mathbb{R}$. The ergosurface is outside the event horizon. It touches the horizon at the point where $\mu_k = 0$ for each $a_k \neq 0$.

Equation (10.6.28) is a polynomial in r of the order $D - 2 + \varepsilon$. For odd dimensions it can be simplified. In this case, the polynomial of the order $2n$ contains only even powers of r. Introducing a new variable $y = r^2$ one obtains an equation with a polynomial in y of the power n. In the general case, the MP metric either has a naked singularity or an event horizon that hides it. The critical case that separates these two different types of solutions is an *extremal black hole*. For a given mass the condition of extremality can be written as an equation $f(a_i) = 0$, which defines a surface in the parameter space of a_i. The general analysis of these conditions can be found in (Emparan and Reall 2008; Doukas 2010).

The surface gravity κ on the horizon is

$$\kappa = \begin{cases} \left.\dfrac{\partial_r \Pi - m}{2mr}\right|_{r=r_+} & \text{even } D, \\[3mm] \left.\dfrac{\partial_r \Pi - 2mr}{2mr^2}\right|_{r=r_+} & \text{odd } D, \end{cases} \tag{10.6.29}$$

while the area of the horizon $\mathcal{A}$ reads

$$\mathcal{A} = \frac{8\pi \, G^{(D)} M}{(D-2)\kappa}\left(D - 3 - 2\sum_i \frac{a_i^2}{r_+^2 + a_i^2}\right). \tag{10.6.30}$$

The laws of black hole thermodynamics can be easily extended to higher dimensions (Myers and Perry 1986).

5D rotating black holes are quite similar to the Kerr metric. In the dimensions $D \geq 6$ there are important differences:

- For $D \geq 6$ and fixed black hole mass there exist solutions with arbitrary large angular momentum. Such black holes are called *ultra–spinning*.

- For odd D there exist black hole solutions with the negative values of the mass parameter M. The solutions with negative mass involve causality violation and, hence, are pathological. Spinning black hole solutions with causality violation in their exterior are not unknown in 4D.

- Ultra–spinning black holes are expected to be dynamically unstable (Emparan and Myers 2003).

Separation of variables and complete integrability of geodesic equations

In the general case when all the rotation parameters are non-vanishing and different, the symmetry of the MP-metric is $\mathbb{R} \times U(1)^N$. This, together with a trivial conserved quantity $g_{\mu\nu}\dot{x}^\mu \dot{x}^\nu = m^2$, gives $I_D = N + 2 = n + 1 + \varepsilon$ integrals for geodesic motion. This number is 3 for the 4D Kerr spacetime. As we know, the complete integrability of the geodesic equation in the Kerr metric is a result of the existence of an additional, quadratic in momentum, integral of motion connected with *hidden symmetry*. For $D = 5$ one has $I_5 = 4$. One more integral of motion is required for the complete integrability. This integral of motion can be easily found by separating variables in the corresponding *Hamilton–Jacobi equation* (Frolov and Stojkovic 2003). The required fifth integral of motion is quadratic in the momentum and is generated by the rank-two *Killing tensor*.

It is evident that in higher dimensions the problem of the integration of geodesic equation becomes more complicated. Besides I_D integrals of motion, 'naturally' existing in the MP metric, one needs additional $n - 1$ integrals of motion to provide the complete integrability. Certainly, this can be achieved in special cases, when some of the rotation parameters coincide and the symmetry is enhanced. But the attempts to find the hidden symmetries by separating the Hamilton–Jacobi equations in the Myers–Perry metric in the Boyer–Lindquist coordinates a general case for $D \geq 6$ were not successful.

> *It should be emphasized that the property of the separability of the Hamilton–Jacobi equation depends on the choice of the coordinates that are used. If this equation is not separable in the chosen coordinates it does not mean that there do not exist other coordinates in which these equations allow the separation of variables. In Section 4.3 we have mentioned that the separability of the Hamilton–Jacobi equation is closely related to the existence of explicit and hidden symmetries. The latter can be characterized by the corresponding Killing vectors and Killing tensors that have a covariant definition. For this reason the special coordinates in which the separation takes place are determined by these tensor objects. We discuss this subject in more detail in Appendix D.*

It 'miraculously' occurs that the additional integrals of motion, quadratic in momentum, do exist for the black hole solutions with an arbitrary number of dimensions. Moreover, there always exist a sufficient number of irreducible independent Killing tensors of the second rank, which make the geodesic equations completely integrable.

The reason for this remarkable property is the existence of a so-called *principal conformal Killing–Yano tensor* (PCKYT). This is a non-degenerate closed conformal Killing–Yano 2-form $\boldsymbol{h}$. It has the form $\boldsymbol{h} = d\boldsymbol{b}$, where 1-form $\boldsymbol{b}$ in the Boyer–Lindquist coordinates is (Frolov and Kubizňák 2007; Kubiznak and Frolov 2007)

$$b = \frac{1}{2}\left[\left(r^2 + \sum_{i=1}^N a_i^2 \mu_i^2\right) dt + \sum_{i=1}^N a_i \mu_i^2 (r^2 + a_i^2)\, d\phi_i \right]. \tag{10.6.31}$$

This principal conformal Killing–Yano tensor generates a 'tower' of independent second-rank Killing tensors, which provide a set of the integrals of motion required for the complete integrability of geodesic equations. The PCKYT allows one to define special *canonical coordinates*. This set of coordinates consists of 'eigenvalues' of the PCKYT and a number of Killing coordinates. These coordinates differ from the MP coordinates:

1. They are unconstrained.
2. They are directly determined by the hidden symmetries of the metric
3. The Hamilton–Jacobi equation is completely separable in these coordinates.

The related material can be found in Appendix D.9 (see also a review (Frolov and Kubizňák 2008)).

> The existence of a PCKYT is a rather restrictive property. Solutions of the Einstein equations with a cosmological constant that possess this property form a class of so-called Kerr-NUT-(A)dS metrics. Besides MP metric and its generalization to asymptotically de Sitter or anti-de Sitter spacetimes, this metric has some additional, so-called NUT parameters (see Appendix D.9).

The MP metric and its Kerr-NUT-(A)dS generalizations in any number of dimensions possess the following properties:

1. Hamilton–Jacobi and Klein–Gordon equations allow complete separation of variables (Frolov *et al.* 2007).
2. Massive Dirac equation is separable (Oota and Yasui 2008).
3. Stationary string equations are completely integrable (Kubiznak and Frolov 2008).
4. Tensorial gravitational perturbation equations are separable (Oota and Yasui 2010).
5. Equations of the parallel transport along time-like and null geodesics can be integrated (Connell *et al.* 2008; Kubiznak *et al.* 2009).
6. Equations for charged-particle motion in such a spacetime in the presence of a test electromagnetic field are completely integrable, provided this field is generated by the *primary Killing vector* (Frolov and Krtous 2011).

10.6.6 Black rings and their 'relatives'

We have already mentioned that higher dimensions open up wide opportunities for black objects that are quite different from 4D black holes. Black branes, discussed in Section 10.6.1, and their compactified versions are examples of black objects with non-spherical topology of the horizon. However, the spacetime in these solutions is not asymptotically flat. We discuss now 5D vacuum black objects with the non-spherical topology of the horizon in an asymptotically flat spacetime. Known solutions in the class are stationary and have two additional rotational Killing vectors.

> In D-dimensional spacetime with $D - 2$ commuting non-null Killing vectors the metric depends essentially on 2 variables. The Einstein equations effectively reduce to a nonlinear sigma model with group $GL(D - 2, \mathbb{R})$. This is a completely integrable system. Only in $D = 4, 5$ are the corresponding regular solutions asymptotically flat (for more details see (Emparan and Reall 2008; Harmark et al. 2007)).

The first class of such solutions was found by Emparan and Reall (2002). They are called *black rings*. These solutions describe a black object in 5D asymptotically flat spacetime with the topology of the horizon $S^1 \times S^2$. Besides the mass M it contains two dimensionless parameters, that are related to the angular momentum of the black ring, and the radius of the contractible direction S^1 of the horizon. The black ring rotates in the S^1 direction. This rotation provides the centrifugal repulsion that helps to 'compensate' the attractive self-gravity force. As a result, one has an equilibrium configuration. One can define a 'thickness' parameter ν that characterizes the shape of the horizon, and that is roughly equal to the ratio of S^2 and S^1 radii. The black ring solutions have two branches:

Thin black rings with $0 < \nu < 1/2$. The horizon has the shape of a long narrow cylinder bent into a circle. The angular momentum is large, while the surface area is small;
Fat black rings with $1/2 \leq \nu < 1$. The horizon resembles a doughnut.

Denote $j^2 = 27\pi J^2/(32G^{(5)}M^3)$. Then, for fixed mass M and dimensional angular momentum parameter j in the range $\sqrt{27/32} \leq j < 1$ there exist three different vacuum black objects: a MP black hole and thin and fat black rings. This is a direct indication that the *black hole uniqueness theorem* proved in the 4D spacetime is not valid in the higher dimensions.

We know that in 5D spacetime there is 'enough room' for two independent components of the angular momentum. For black rings the second independent plane can contain rotation of the S^2 of the ring. The black ring solution with two angular momenta was found in (Pomeransky and Senkov 2006).

In 4D it is believed that there are no stationary solutions for vacuum black holes with horizons consisting of several disconnected components. Such solutions were discovered in five dimensions. The first of them was called *black Saturn* (Elvang and Figueras 2007). It describes a MP black hole surrounded by a concentric black ring. Both of these black objects have a common 'equatorial plane'. Another solution, called a *black di-ring*, describes two concentric black rings with the common plane of rotations. It was found in (Iguchi and Mishima 2007). There are known solutions with several such rings. There also exists

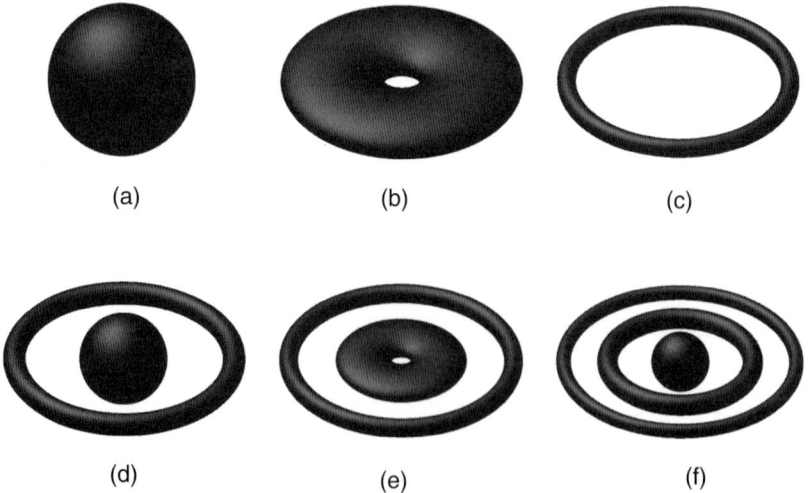

(a) (b) (c)

(d) (e) (f)

Fig. 10.11 Schematical illustration of 5D black objects: (a) MP black hole; (b) Fat black ring; (c) Thin black ring; (d) Black Saturn; (e) Di-ring; (f) Black Saturn with several rings.

a solution, called *orthogonal black di-rings* or *bicycling black rings*, with two rings lying and rotating in two independent orthogonal planes (Izumi 2008). Some of the black object configurations are schematically shown in Figure 10.11. It is expected that black objects with non-spherical topology exist in $D \geq 6$, but at the moment they are not known. Comprehensive discussion of higher-dimensional objects and their properties can be found in the review (Emparan and Reall 2008).

To summarize this brief review of the higher-dimensional black holes let us emphasize that this field of research already gave quite interesting results and much more is still to be done. One of the most important open problems is a question concerning the stability of higher-dimensional black objects. We focused on the vacuum higher- dimensional black holes. There is plenty of interesting black hole solutions in the higher-dimensional supergravity, which besides gravity contains an additional set of other fields. The book (Ortín 2004) gives a wonderful review of this subject.

10.7 Wormholes

10.7.1 Topology change in classical evolution

An important lesson of general relativity is the following: As soon as we accept that a spacetime is not flat, we can no longer restrict ourselves by considering only spacetimes with the Euclidean topology. The spacetime model in general relativity is a *differential manifold* with the metric, points of which correspond to events. The topology of this manifold does not have to be trivial. An interesting question is: What is the topology of our space and how can we probe it?

> *Consider a flat space* $\mathbb{R}^3$ *and compactify it by identifying the points* X^i *and* $X^i + L^i$, *where* X^i *are the Cartesian coordinates. In such a model the 3D space is a 3-dimensional torus* $\mathbb{T}^3$. *Choose the metric to be*
>
> $$ds^2 = -dT^2 + \sum_{i=1}^{3}(dX^i)^2. \tag{10.7.1}$$
>
> *It is evident that this flat metric is a solution of the vacuum Einstein equations. That is, from the point of view of general relativity, this is a possible spacetime.*

This simple example shows that in the framework of general relativity it is possible to have non-trivial topology of space. The natural questions are:

- If space has a certain topology at some initial moment of time, can the topology be different at a later time?

- If the region with non-trivial topology is compact, can it be static or at least living long enough to make it possible to travel through it?

We briefly discuss the first question now, and consider the second question in the next subsection.

To discuss the topology change we used the standard paradigm of the classical evolution. Consider two hypersurfaces $\mathcal{S}_1$ and $\mathcal{S}_2$, and assume that they are joined by a connected interpolating manifold $\mathcal{M}$ with a Lorentzian metric on it, such that its boundaries $\mathcal{S}_1$ and $\mathcal{S}_2$ are space-like in this metric. We call $\mathcal{S}_1$ an initial surface, if no points of $\mathcal{M}$ lie to the past of it. Similarly, S_2 is a final surface if no point of M lies to the future of it. These definitions have sense only when the spacetime is time-orientable, that is a global choice of future and past is possible. We assume this property to be valid. Now, the question about the topology change can be formulated as follows: Is it possible that the topology of the final surface S_2 is different from the topology of the initial surface S_1?

One can prove that if the interpolating spacetime M can be foliated, that is presented as a one-parameter family of space-like surfaces $S(t)$, such that $S_1 = S(t_1)$ and $S_2 = S(t_2)$, then for the topology change M must contain cuts or holes of some kind (Borde 1994).

There are two ways to 'rescue' the classical evolution. One is to allow the metric to be singular at some points, for example to become degenerate. Another way is not to assume that the foliation is possible. There exist results indicating that anyway there are problems in a spacetime with the topology change. The best-known examples are the classical theorems proved by Geroch and Tipler. Geroch (1967) has shown that if S_1 and S_2 are compact, the topology change may be obtained only at the price of causality violation. Tipler (1977) demonstrated that for the topology change situation the Einstein equations cannot hold with the source that has non-negative energy density. These results can be generalized to a wide class of spacetimes, even when S_1 and S_2 are not compact (see, e.g., (Borde 1994)). These and similar results are usually interpreted as follows: *During the classical evolution described by the Einstein equations, with physically reasonable energy conditions, the spatial topology cannot change without dramatic consequences, such as the causality violation or a singularity formation.*

10.7.2 Wormholes

A special class of spacetimes with the non-trivial topology attracts a lot of attention. This is a class of so-called *traversable wormholes* in asymptotically flat spacetime. Characteristic features of such spaces are:

1. A region with the non-trivial topology is localized inside a compact space domain.
2. A particle initially moving in the exterior region outside the wormhole can propagate through a topological handle and return to the exterior region.

Under special conditions the motion through a handle between two separated points in the external space can take less time than the usual travel between these points, so that such a handle can provide one with a 'short-cut' option. In other words, one could use such handles to travel in space. Namely, this possibility attracts much attention to the wormholes.

An example of a wormhole is shown in Figure 10.12. The spatial geometry of the wormhole is multiply connected. This means that there exist non-trivial closed pathes that cannot be contracted to a point by a continious transformation. An example of such a path is a line that enters one of the mouths, goes through the throat, and after leaving the other mouth returns to the initial point. This means that between two points A and B there exist different classes of connecting these point paths. A path may be totally outside the wormhole. We say that its winding number is 0. A path may also pass through the throat m times in a chosen direction. We say that such a path has winding number m (or $-m$, if it goes in the opposite to the chosen direction). It is impossible to change the winding number of the path by its continuous transformation. This means that the winding number is a *topological invariant*.

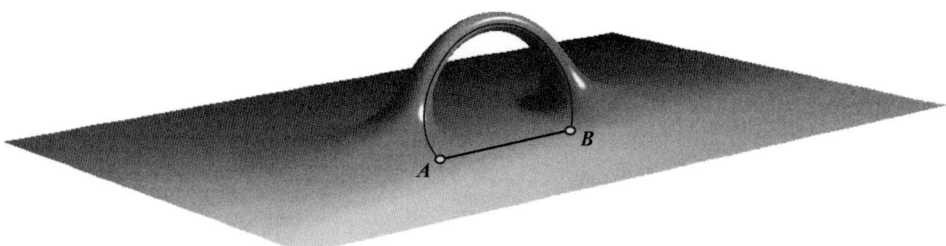

Fig. 10.12 To illustrate the wormhole geometry the figure shows the two-dimensional space embedded as a surface in the three-dimensional flat space. It is assumed that 'physical objects' can propagate along the surface and the flat extra dimension is unphysical. To reach from the point A to the point B one can either 'travel' in the external space, or 'go' through the handle.

> *In the space of paths it is possible to introduce the group structure. The corresponding group is called a* first homotopy group *and is denoted by π^1. In our case this group is equivalent to the group of integer numbers. This group may be more complicated. It happens for example, when there exist several wormholes. A space is called* simply connected, *if the first homotopy group is trivial. This means that every closed path can be contracted to a point by a continuous transformation. A space with non-trivial group π^1 is called* multiply connected. *The presence of a wormhole means that the space is multiply connected.*

The space shown in Figure 10.13 has the same topology as the space depicted in Figure 10.12 but its metric properties are different. Consider points A and B located near the mouths of the wormhole, and denote by L_1 and L_2 the length of the paths connecting A and B in the external space and passing through the throat, respectively. For the wormhole in Figure 10.12 $L_2 > L_1$, while for the wormhole in Figure 10.13 one has $L_2 < L_1$. In principle, it is possible to have $L_2 \ll L_1$. In such a case, the path through the throat gives a short-cut and can be used to 'travel through the space'.

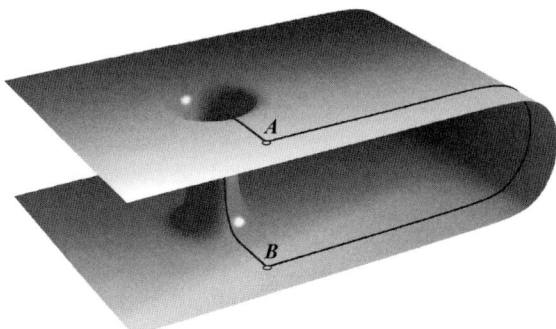

Fig. 10.13 This is another illustration of the wormhole geometry. The topology of the 2D surface, representing a space with a wormhole, is the same as in the previous figure. The difference is in the 'metric' properties of these two spaces. Now, the path through the handle is shorter than the 'external' path.

Fig. 10.14 A spherically symmetric wormhole connecting two asymptotically flat spaces.

It is instructive to discuss first a wormhole connecting two different asymptotically flat spaces. Figure 10.14 illustrates such a geometry. An observer from one infinity can enter the mouth, go through the handle, and appear in the other asymptotically flat domain.

> *When we discussed spherically symmetric black holes we mentioned that the spatial slice of the eternal Schwarzschild black hole is the Einstein–Rosen bridge (see Figure 6.4), which has the topology* $\mathbb{R} \times S^2$ *similar to the topology shown in Figure 10.14. This topology is certainly different from, for example, the spatial topology* $\mathbb{R}^3$ *of a static star. In particular, the difference in the topology can be illustrated as follows. Consider a sphere of constant radius r. In the flat space every 2D sphere can be shrunk to a point by a continuous transformation. One says that its homotopy group* π^2 *is trivial. In the space of the Einstein–Rosen bridge the spheres r =const are non-contractible. Moreover, there exists a sphere of the minimal radius* r_S, *which evidently has the minimal surface area* $4\pi r_S^2$. *This sphere is inside the throat connecting two asymptotically flat spaces. This is the simplest example of a wormhole.*

Can one use the *Einstein–Rosen bridge* in order to travel from one asymptotically flat space to another through the throat? Such a journey is evidently impossible in the eternal Schwarzschild geometry. Let us look at the Kruskal diagram (Figure 6.2). The inner (R_-) and outer (R_+) wedges of the spacetime are causally disconnected. This means that no causal curve that starts at R_- can reach R_+. Thus, the Einstein–Rosen bridge is an example of a spacetime with a *non-traversable wormhole*. The evident explanation of this is the following. Assume that at the initial slice one has the Einstein–Rosen geometry. Consider a slice located to its future and still connecting regions R_- and R_+. The size of the minimal surface in the throat becomes smaller than r_S. One can say that the throat shrinks to zero size faster than one can use it to penetrate into another flat subspace. As a result, future-directed causal curves, leaving R_+, enter the region T_- and hence inevitably reach the singularity $r = 0$. But they can never appear in R_- domain.

10.7.3 Traversable wormholes

Let us return to the discussion of wormholes connecting two regions in the same space. In order to make the 'space travel' possible, the wormhole must survive for quite a long time. Such wormholes are called traversable. Do there exist *traversable wormholes?*

First, one can conclude that such a wormhole must contain matter violating the *averaged null-energy condition* (see page 135). This conclusion follows from the *topological censorship theorem* (see page 357). Indeed, suppose we have a static traversable wormhole in an asymptotically flat spacetime. Then there exists a causal curve passing through the throat from $\mathcal{J}^-$ to $\mathcal{J}^+$ that cannot be continuously deformed to a causal line lying in a region where the geometry is practically flat. This conclusion also remains valid when the traversable wormhole is time dependent.

> *For a static spherically symmetric traversable wormhole the violation of the averaged null-energy condition also follows from the following simple observation (Morris et al. 1988; Thorne 1994).*
>
> *Consider a spherically symmetric pulse of radiation onto the wormhole. It is initially convergent, but after passing the throat it becomes expanding. This is impossible if the matter in the wormhole obeys the null-energy condition, since such matter always focuses the beams of radiation. Thus, the matter violating the null-energy condition must exist inside the wormhole.*
>
> *It is possible to obtain more detailed information concerning the location of the 'exotic matter' violating the weak-energy condition without an additional assumption of spherical symmetry (Hochberg and Visser 1997).*

Let us summarize. In classical general relativity it is impossible to create a wormhole without either violating causality, creating singularities, or using a matter that violates physically reasonable energy conditions. Usually, one assumes that either wormholes exist from the beginning of the universe evolution, or they might be created as micro-objects as a result of the topology-changing quantum gravity effects, and further they are inflated to macroscopic scales by the presence of some exotic matter. Even if such a wormhole already exists, to support it and to make it traversable, violation of the energy conditions is also required. For this reason the status of wormholes is quite different from the status of black holes. Black holes are the result of natural evolution. Their formation and existence is a corollary of the theory that is confirmed by astrophysical observations. Wormholes are hypothetical objects, which for their existence require far-reaching modification of the theory.

10.7.4 Clock synchronization in a multiply connected space

Let ξ be the Killing vector in a static spacetime, and $u = \xi/|\xi^2|^{1/2}$ the velocity of the Killing observer. A set of events $x^\mu + dx^\mu$ that are simultaneous with the event x^μ in the reference frame of the Killing observer is defined by the equation

$$u_\mu(x)dx^\mu = 0. \tag{10.7.2}$$

This condition determines a set of infinitesimal 3-planes at each of the point. In a simply connected region these simultaneity planes can be 'integrated'. This means that there exists such a function t for which $t =$const are surfaces Σ_t everywhere tangent to the local simultaneity planes (see Eq. (3.7.11)). In other words, in a simply connected region of a static spacetime it is possible to introduce the time t, which has the property that all the events with a fixed value of t are simultaneous. Let p and p' be two points on Σ_t. Let γ_1 and γ_2 be two paths connecting these points. In the simply connected region they can be continuously transformed into each other, and the result of the clock's synchronization for the points p and p' does not depend on which of these paths is used to synchronize the clocks.

In the presence of a wormhole, when the space is not simply connected, the situation is quite different. For simplicity, we assume that there exists one static traversable wormhole. Let us discuss a problem that arises in connection with the *clock synchronization* in such a spacetime. This problem allows the following 'elegant' formulation. Let C be a closed path in this multiply-connected static spacetime. Denote by m its *winding number*, which characterizes how many times the contour passes through the wormhole in the chosen direction. The contours with the same winding number can be continuously transformed into each other.

Let ξ be the Killing vector in a static spacetime and denote $\eta = \xi/|\xi^2|$. We know that $\eta_{[\mu;\nu]} = 0$ (see Section 3.7.4), that is the 1-form η is closed, $d\eta = 0$. Denote by J_C the following integral

$$J_C = \oint_C \eta_\mu dx^\mu. \tag{10.7.3}$$

Since the form η is closed, $d\eta = 0$, the value of J_C remains the same for a continuous deformation of the contour C. This means that the value of this integral is the same for each of the contours with a given winding number. If the contour has the winding number m, one has

$$J_m = J_{C_m} = mJ_1. \tag{10.7.4}$$

In particular, if the contour can be contracted to a point, $J_C = 0$.

Let us calculate the value of J_1. The static spacetime can be foliated by the 3D family S of the Killing trajectories. Let us use a closed pass Γ in S and try to synchronize the clocks along it. Let p_0 be a starting point in the spacetime and γ_0 is its projection on S (see Figure 10.15). Points simultaneous with p_0 along Γ are shown by the curve $\sigma_\Gamma[p_0]$ on this figure. When

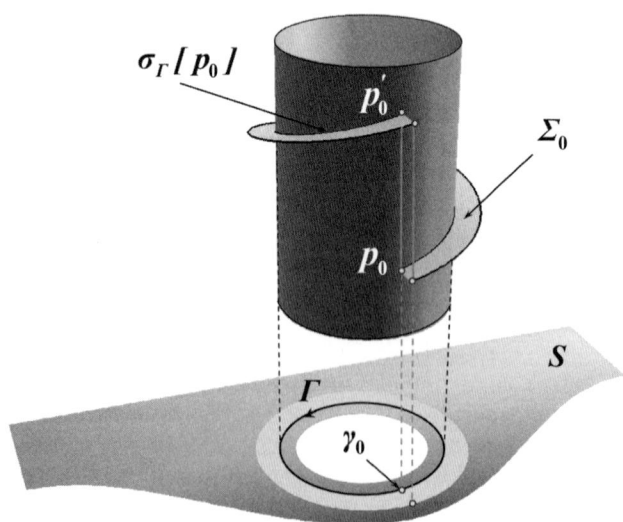

Fig. 10.15 The clock synchronization in the multiply-connected spacetime.

one returns to the same spatial point γ_0 in S, the point p'_0 in the spacetime is at the same Killing trajectory as the original point p_0, but does not necessarily coincide with it. Let us construct a closed path in the spacetime passing through the wormhole as follows. We add to the path $\sigma_\Gamma[p_0]$ a segment along the Killing trajectory connecting the points p_0 and p'_0. It is easy to calculate the invariant J_1 for this pass. Namely, γ_0 is everywhere orthogonal to η, so the invariant J_1 is equal to the time difference of points p_0 and p'_0 along the Killing trajectory

$$J_1 = \Delta t. \tag{10.7.5}$$

Here, Δt is the *time gap for the clock synchronization*.

10.8   'Time Machine' Problem

10.8.1   What's the problem?

General relativity has opened Pandora's box: *Not only can the topology of space in this theory be different from the flat one, but the global causal structure can also be non-trivial.*

In general, causality means that the cause precedes the effect. *Causality* is one of the most fundamental principles of contemporary physics. In general relativity the structure of the spacetime locally is the same as in the Minkowski space. The information-carrying signals from the 'cause' event that can produce a given effect propagate with the velocity not higher than the velocity of light, that is within the local null cones. In other words 'the cause' and 'the effect' can be connected by a causal curve. This means that general relativity 'respects' local causality. At the same time we saw that the global causal structure of the spacetime may be rather non-trivial. Black holes are an example.

In this section we discuss a so-called *time machine problem* that has attracted a lot of attention. Usually, by a 'time machine' one understands some compact macroscopic device, which allows one to return to his/her own past. This situation inevitable leads to a number of well-known paradoxes. Remarkably, the structure of general relativity is rich enough to address many questions, that earlier were quite difficult to formulate as a well-posed physical problems. The 'time machine problem' belongs to this category. In this framework it has been intensively discussed in the physical literature. There exist nice books and review articles where the reader can find comprehensive discussion of different aspects of the 'time machine' problem (see, e.g., Thorne 1994; Visser 1996).

If we neglect the size of the object sent to the past, its 'travel' is described by a worldline. In order to 'respect' local causality we assume that this worldline is time-like, that is it is always inside the future-directed null cones. Suppose the time traveller starts at some point p and reaches the event p' in the past of p. This means that there exists a future-directed time-like (or causal) curve connecting p' and p. Thus, a possibility for the 'time travel' is connected with the existence of closed time-like (or causal) curves in the spacetime.

Does general relativity allow solutions with *closed time-like curves* (CTCs)? Yes, it does. Solutions of Einstein's equations with closed time-like curves have been known for a long time. The earliest example of such a spacetime is a solution obtained by Van Stockum, which describes an infinitely long cylinder of rigidly and rapidly rotating matter. Another well-known example is Gödel's solution representing a stationary homogeneous universe with non-zero cosmological constant, filled with rotating dust. Closed time-like curves are also present in the interior of the Kerr black hole.

> As first pointed out by Carter (1968) such closed time-like curves exist inside the inner horizon in the vicinity of the ring where the Kerr solution has a singularity. In this particular case it is not clear whether such CTCs are physical. In the presence of the inner horizon the solution might be unstable.

Besides the general problem of stability of solutions with CTCs, there is another 'more practical' question. In the discussion of a time machine one usually assumes that "the causality violations appear in a finite region of spacetime without curvature singularities" (Hawking 1992). We also would prefer to have regions with such violations macroscopically large, to be able to deal with classical systems.[11]

10.8.2   Wormholes and a time machine

In order to create a time machine one can use a *traversable wormhole*. A wormhole allows one to journey through the space. If its internal size is small, after such a journey through the wormhole throat, a traveller would appear at a point that is separated by a space-like interval in the external geometry. But in relativity theory time travel and faster-than-light travel are closely related. How a traversable wormhole can be transformed into a 'time machine' was first demonstrated by Morris *et al.* (1988). Namely, the authors of this paper showed that closed time-like curves can arise when one mouth of the traversable wormhole is moving with respect to the other.

> It is instructive to use a simplified toy model of the wormhole. Let (X, Y, Z) be Cartesian coordinates in the 3D flat space. Take two points $X = -L/2$ and $X = L/2$ along the X-axis and consider two spheres of the same radius $r_0 < L/2$ with the centers at these points (see Figure 10.16). Cut the interior of these spheres and identify the points of the spherical boundaries of the cut regions. This identification is not unique, the ambiguity is connected with possible rigid rotation of the spheres

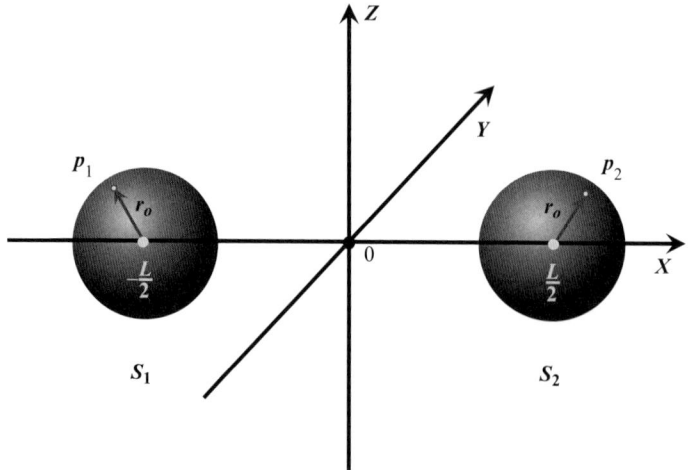

Fig. 10.16 Toy model of a wormhole

---

[11] We do not consider here possible microscopic causality violation, which might occur, for example, at the Planckian scales as the quantum gravity effects.

before their identification. To fix this ambiguity we 'glue' together points of the two spheres that are connected by the reflection in the X = 0 plane.

If a curve enters a point of one of the spheres, S_1, it has continuation that goes out of the corresponding symmetric point of the other sphere S_2. It is evident that the performed 'surgery' creates a non-simply connected space. An example of a non-contractible closed path is an interval $[-L/2 + r_0, L/2 - r_0]$ of the X-axis. According to our prescription the boundary points of this interval are identified.

For this toy model of the wormhole the spacetime outside the mouths is flat and it is a solution of the vacuum Einstein equations in this region. The mass of the mouths is zero. The vacuum Einstein equations cannot be valid in the throat. The extrinsic curvature has a jump at the identified spheres S_1 and S_2. This jump implies the existence of δ-like distribution of matter. It is possible to show that the stress-energy tensor for this distribution violates the *null-energy condition.*

Figure 10.17 shows a spacetime with a toy wormhole. We assume that the spheres S_1 and S_2 are at rest in a chosen reference frame. The parallel cylindrical tubes show the time evolution of the spherical mouths. We assume that a worldline that enters one of the mouths appears from the other one *at the same time T*. Figure 10.18 schematically shows the same wormhole as in Figure 10.17 for the case when the size r_0 of the mouths is much smaller than the distance L between the mouths. The points on the lines representing the mouths correspond to the 'equal time' moments for the 'inner' synchronization. For such an identification the time gap between external and internal synchronization vanishes. In principle, this time gap can be arbitrary. If this time gap $\Delta T > L$ there exist closed time-like curves passing through the wormhole. In this case the *wormhole* is in fact the *'time machine'*.

The time gap ΔT remains constant only for completely symmetric static wormholes. Relative motion of the mouths, or asymmetry of the matter distribution in their vicinity

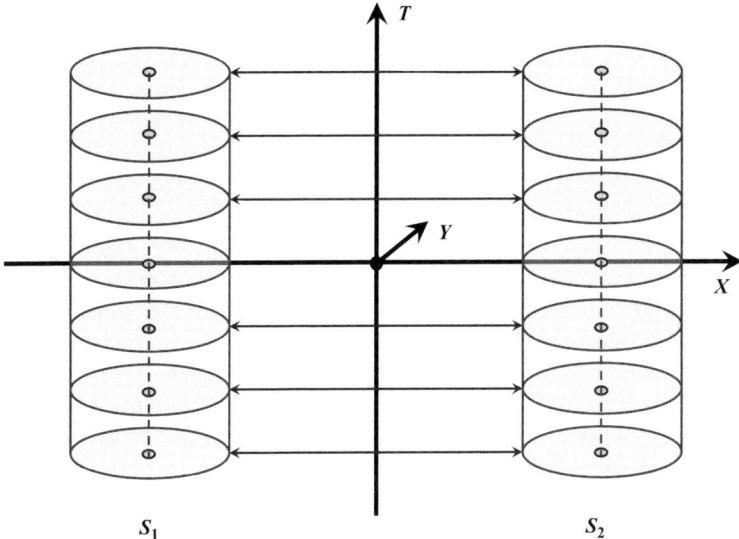

Fig. 10.17 Time evolution of a static wormhole. Equal time sections (synchronized through the handle) are connected by arrows. This is a case when the time gap for the clock synchronization vanishes.

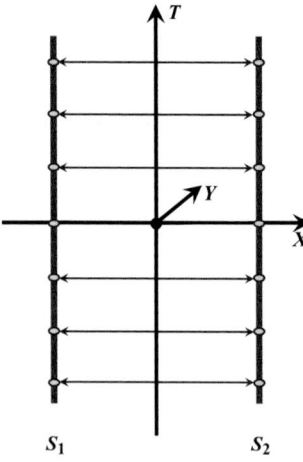

Fig. 10.18 The wormhole with zero time gap. The radius r_0 of the mouths is much smaller than the distance between them. Points representing mouths with the same 'internal' time are shown by dots connected by horizontal arrows.

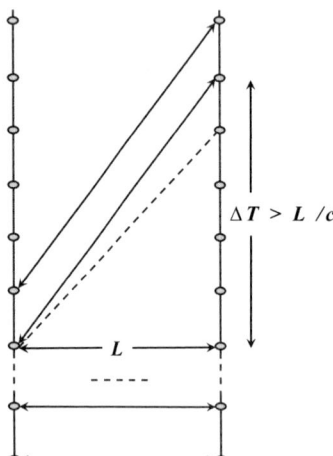

Fig. 10.19 The same diagram of the 'dynamical' wormhole. A missing part of its evolution is shown by dots. After the evolution the time gap ΔT becomes larger than the distance L. A 'time machine' is formed.

changes the time gap. It was first demonstrated in (Morris *et al.* 1988) for the relative motion of the mouths. For example, one can accelerate one of the mouths up to some velocity, and after bring it back to its initial position. One can also revolve one mouth with respect to the other for some time. One usually assumes that the interior of the traversable wormhole is rigid enough, so that its deformation, generated by the acceleration, can be done negligibly small. In principle, by such non-inertial motion the time gap can be made arbitrary large. Figure 10.19 illustrates a situation when after the motion the time gap ΔT becomes larger than L, and one obtains a 'time machine'.

10.8.3 Non-potential gravitational field

We show now that there exists a simpler way to transform a traversable wormhole into a 'time machine' without moving its mouths (Frolov and Novikov 1990).

Let us consider again a static traversable wormhole connecting two asymptotically flat regions, shown in Figure 10.14. Let $\boldsymbol{\xi}$ be a Killing vector. There is an ambiguity in its normalization: after multiplication $\boldsymbol{\xi}$ by a constant one again has a Killing vector. To fix this ambiguity we impose the normalization condition $\boldsymbol{\xi}^2 = -1$ at the 'upper infinity' of the wormhole. After this the Killing vector is uniquely defined everywhere, but its norm at the 'down infinity' is not necessarily equal to one. Moreover, this is a generic situation. A static gravitational field, generated by a non-symmetric with respect to the 'neck' of the throat matter distribution, produces an asymmetry in the normalization of the Killing vector at the infinities.

One can approximate wormholes 10.12 and 10.13 by gluing properly parts of the asymptotic regions of the wormhole shown in Figure 10.14. In this process one glues two infinite regions of this wormhole that are practically flat, keeping the spacetime smooth. What happens after this 'reconstruction' of the space with the 'non-symmetric' Killing vector field ξ^μ? Locally everything remains the same, and ξ^μ in any simply connected region is uniquely defined after choosing its normalization at a single point of this region. But since generically the norm of ξ^μ is different at the two flat infinities, which we identify to construct a non-simply connected traversable wormhole, the Killing vector field is not globally uniquely defined any longer. We call such a gravitational field a *non-potential* static field.

Let us note that the time-like Killing vector $\boldsymbol{\xi}$ is well defined in any simply connected region and it obeys the following equations

$$\xi_{(\mu;\nu)} = 0, \qquad \xi_{[\mu;\nu} \xi_{\lambda]} = 0. \tag{10.8.1}$$

Denote by $\boldsymbol{u}$ the four-velocity of the Killing observer, $\boldsymbol{u} = \boldsymbol{\xi}/|\xi^2|^{1/2}$. Then, this vector obeys the following relations

$$u_\mu u^\mu = -1, \qquad u_{\mu;\nu} = -w_\mu u_\nu, \qquad w_{[\mu;\nu]} = 0, \tag{10.8.2}$$

where $w_\mu = u^\alpha u_{\mu;\alpha}$ is the vector of 4-acceleration. In any simply connected region these equations are equivalent to Eq. (10.8.1). At the same time, the vectors $\boldsymbol{u}$ and $\boldsymbol{w}$ do not depend on the normalization of $\boldsymbol{\xi}$. We define a *locally static spacetime* as a spacetime that admits two globally defined vector fields u^μ and w^μ obeying the relations (10.8.2).

It is convenient to consider a locally static spacetime M as the collection S of Killing trajectories. We assume that a unique Killing trajectory passes through each point of the spacetime M, and that $M = \mathbb{R}^1 \times S$. One may refer to S as the three-dimensional space under consideration.

In the presence of a wormhole this space has a non-trivial topology. A non-contractible loop passing through a wormhole m times in the chosen direction can be characterized by its winding number m. Two loops with the same winding number m are connected by a continuous transformation (are homotopic). These loops are also homological, i.e. they are boundaries of some two-dimensional surface. In other words, the first fundamental group π_1 and the first homotopy group H^1 calculated for S (as well as for the spacetime M itself) coincide with the group of integer numbers $\mathbb{Z}$.

The last equation in Eq. (10.8.2) shows that the acceleration 1-form $w \equiv w_\mu dx^\mu$ is closed, $dw = 0$. According to Stokes' theorem, the integral of this form over any closed path C_m

$$I_m[w] = \oint_{C_m} w \tag{10.8.3}$$

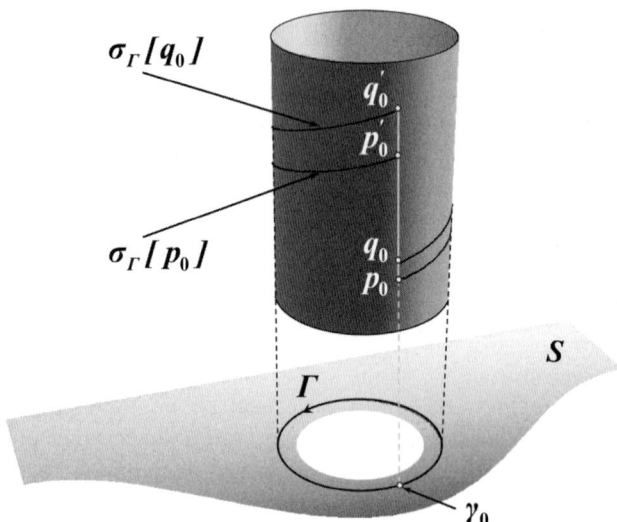

Fig. 10.20 Clock synchronization in a multiply-connected spacetime with a non-potential gravitational field. Points p_0 and q_0 are located on the same Killing trajectory. The lines $\sigma_\Gamma[p_0]$ and $\sigma_\Gamma[q_0]$ represent sequences of events, along the closed spatial path Γ, which are simultaneous with the initial events p_0 and q_0. These lines intersect the initial Killing trajectory at points p'_0 and q'_0. If the gravitational field is non-potential, the difference of time between p'_0 and q'_0 is not equal to the difference of time between p_0 and q_0.

depends only on the winding number m of the path. One has $I_m = mI$. We choose the direction of the path with $m = 1$ in such a way that $I[w] \equiv I_1[w] \le 0$. We call $I[w]$ a period associated with the closed 1-form w.

If the period $I[w]$ vanishes, then the form w is exact, i.e. there exists a globally defined scalar function φ (gravitational potential) such that $w = d\varphi$. For such a *potential gravitational field* the vector field $\xi^\mu = e^\varphi u^\mu$ is a global Killing vector field.

In the general case (when $I[w] \ne 0$), the gravitational potential φ cannot be defined as a single-valued function. Such a gravitational field is called *non-potential*. The period $I[w]$ may be considered as the measure of non-potentiality of the field. The important physical property of a *non-potential gravitational field* is that the work done by such a field on a particle moving along a closed contour C_m passing through a wormhole does not vanish. This work is proportional to $e^{-I_m[w]}$.

In a *non-potential gravitational field*, the time gap $\Delta t \equiv t_0 - t_1$ for clock synchronization is no longer time independent (see Figure 10.15). It can be shown (Frolov and Novikov 1990) that it satisfies the equation

$$\frac{d\Delta t}{dt} = 1 - e^{I[w]}. \tag{10.8.4}$$

This relation shows that the time gap for clock synchronization along any closed path with a winding number $n = 1$ grows linearly with time, and $|I[w]|$ is the measure of this growth. As soon as the gap Δt becomes greater than the minimal time t required for a light signal to propagate along a closed path passing through the wormhole and return back, a *closed time-like curve* can arise.

Frolov and Novikov (1990) demonstrated that the interaction of a wormhole with classical matter generically generates a non-potential gravitational field. In other words, a locally static wormhole is generically unstable with respect to the processes that transform it into a 'time machine'.

10.8.4   Chronology horizon

In the general case, a spacetime can be divided into *chronal regions*, without closed time-like curves, and *achronal regions* that contain closed time-like curves. The boundaries between the chronal and achronal regions are formed by *chronology horizons*. Chronal regions end and an achronal region begins at a *future chronology horizon*. Achronal regions end and a chronal region begins at the *past chronology horizon*. Thus, achronal regions are intersections of the regions bounded by both of these horizons. A future chronology horizon is a special type of future Cauchy horizon, and as such it is subject to all the properties of such horizons. In particular, it is generated by null geodesics that have no past end points but can leave the horizon when followed into the future. If the generators, monitored into the past, enter one or more compact regions of spacetime and never thereafter leave them, the future chronology horizon is said to be *compactly generated*. In a wormhole model with closed time-like curves, the future chronology horizon is compactly generated. A compactly generated chronology horizon cannot form in a spacetime developed from a space-like non-compact surface without boundary if the null-energy condition (Hawking 1992).

The past-directed generators of the compactly generated future chronology horizon have no past end points. They will enter and remain in a compact region C. Hawking (1992) showed that there exists a non-empty set E of generators, each of which remains in the compact set C in the future direction, as well as in the past direction. The sets E generically contain at least one closed null geodesic. More exactly, Hawking (1992) proved that:

1. *If E contains such a closed null geodesic, small variations of the metric preserve this property.*
2. *If E does not contain a closed null geodesic, then in geometries obtained by small variation of the metric such curves do exist.*

The generators traced into the past either wander ergodically around C or they asymptote to one or more smoothly closed geodesics. In the latter case, followed forward in time they are seen to originate in fountains and spew out of them. That is why Thorne (1993) proposed to name such closed null geodesics as *fountains*. Hawking's result indicates that in the generic case, C will contain such fountains, and it is likely that generically almost all the horizon generators will emerge from them (Thorne 1993).

10.8.5   Problems. Chronology protection

It is well known that the possibility to travel in time and violate the chronology results in many paradoxes. The best known one is the *grandfather paradox*. Namely, such a time traveller would be able to kill his grandfather at a time before the latter had met his grandmother. As a result his father or mother would have not been born, so that the 'time traveller' would not be born either. Another version of this paradox is a situation when the 'time traveller' returns to the time when he was a child and kills himself. Proposed models of the 'time machine', for example based on the traversable wormholes, allow one to formulate these and similar paradoxes in the form the can be addressed as physical problems and analyzed by means of the tools of theoretical physics. In this approach, the main question can be formulated as follows: *To what extent can one develop the structure of the theoretical physics in a spacetime that admits closed time-like curves?*

There were many interesting results obtained in this way. However, the problem is still open and continues to attract a lot of attention. Here, we make only a few general remarks.

> As a simple illustration of possible problems in the presence of CTCs let us consider a 2D locally flat spacetime, with the time coordinate T being periodic with the period ΔT. The lines $X =$const are closed time-like geodesics. Consider a massless scalar field φ in this spacetime that obeys the equation
>
> $$(-\partial_T^2 + \partial_X^2)\varphi = 0. \tag{10.8.5}$$
>
> If one chooses an initial date $\varphi = \varphi_0$ and $\dot\varphi = \dot\varphi_0$ at the slice $T = 0$, then the field evolution is uniquely determined by Eq. (10.8.5). This gives the values φ and $\dot\varphi$ at $t = \Delta T$ that in the general case do not coincide with the initial conditions at $T = 0$. This contradicts the assumed periodicity of the time.

In this particular case the way to rescue the situation is evident. One can decompose a solution into modes $\sim \exp[-i\omega(T \pm X)]$ and to construct a solution by using only those modes that obey the time-periodicity condition. This makes the spectrum ω discrete. As a result, the initial conditions at $T = 0$ cannot be chosen arbitrarily, but they must obey special *consistency conditions*. The classical free-field theory obeying this condition is well defined. The situation becomes much more complicated if the field has self-interaction or there are several interacting fields.

The origin of this problem can be illustrated by a so-called *billiard ball model*. In this model, instead of the fields one considers a classical theory of motion of the billiard balls. For the toy model of the 'time machine' a ball can enter one of the mouths and appear from the other mouth at the earlier time. Its motion can be chosen in such a way that the outgoing ball strikes its own 'version' before the latter has entered the mouth. This is a 'grandfather killing' situation. In a general case a stroke would change the motion of the 'earlier version' of the ball so that it would never pass the wormhole. This is an evident contradiction. To solve this paradox, Novikov proposed a *self-consistency principle*: *The only solutions to the laws of physics that can occur locally in the real universe are those that are globally self-consistent.* In other words it is forbidden to change the past. All events happen only once, and cannot be changed. (See (Friedman *et al.* 1990).)

The natural questions are:

1. Does such a consistent solution to this system exist for any initial conditions?

2. Is this consistent solution unique?

For the billiard problem these questions were analyzed by Echeverria *et al.* (1991). They demonstrated that for the same initial data that give self-inconsistent 'solutions', there also exist self-consistent solutions. They tried to find initial data that do not allow any self-consistent solutions at all, but none were found. They demonstrated also that in the general case for the same initial conditions there could be an infinite number of self-consistent solutions.

More severe problems arise when one passes from a single-body motion to the statistical mechanics. One of the evident problems is that in standard physics the entropy of a system is a non-decreasing function of time. For the spacetime with closed time-like curves this cannot be always true: The increase of the entropy for some period of the 'time loop' must be compensated by its decrease during the other period.

Quantum aspects of the 'time machine' problem seem to be very important. To start with, in order to change the topology of space and create a wormhole, the quantum gravity effects somehow must be involved. Quantum effects might violate the energy conditions, which are an obstacle for the topology change. They may also be useful to keep the wormhole traversable, or even to force the size of the wormhole throat to grow.

At the same time, the quantum effects might dramatically change the classical 'picture'. There is a big difference between the properties of the vacuum in the classical and quantum theory. The classical vacuum is just an empty spacetime. In the quantum field theory the vacuum is the ground state in which the field-average value vanishes. But, similarly to the quantum oscillator, there exist non-vanishing *zero-point fluctuations*. In the flat spacetime under normal conditions the contribution of the fluctuations to the energy vanishes, as a result of the renormalization. In the case when a 'time machine' is created in an asymptotically flat spacetime a compactly generated Cauchy horizon is formed. According to Hawking (1992) such a spacetime contains one or more closed null geodesics. Modes of the zero-point fluctuations trapped in the region pass again and again through the throat, and they become more and more blueshifted. As a result, their contribution to the average vacuum stress-energy tensor becomes non-vanishing. Moreover, it grows infinitely until the backreaction effects of the vacuum on the background geometry become important. This result was first obtained in the papers (Kim and Thorne 1991; Frolov 1991a, 1991b, Hawking 1992).

This is one of the possible mechanisms that might protect our world from the creation of 'time machines'. Hawking (1992) formulated this as a general *chronology protection conjecture*: *"The laws of physics do not allow the appearance of closed timelike curves."*

For further discussion and references on this interesting subject we address a reader to the books (Thorne 1994, Visser 1996).

Appendix A
Fundamental Constants and Units

A.1 Fundamental Constants

Speed of light $\qquad\qquad\qquad c = 2.997\,924\,58 \times 10^{10}\ \frac{\text{cm}}{\text{s}}$

Gravitational constant $\qquad\quad G = 6.674\,28(67) \times 10^{-8}\ \frac{\text{cm}^3}{\text{g·s}^2}$

Reduced Planck constant $\quad \hbar = h/(2\pi) = 1.054\,571\,628(53) \times 10^{-27}\ \frac{\text{g·cm}^2}{\text{s}}$

A.2 Planck Units

Planck length $\qquad\qquad l_{\text{Pl}} = \sqrt{\dfrac{\hbar G}{c^3}} \qquad = 1.616\,252(81) \times 10^{-33}\ \text{cm}$

Planck time $\qquad\qquad\quad t_{\text{Pl}} = \sqrt{\dfrac{\hbar G}{c^5}} \qquad = 5.391\,24(27) \times 10^{-44}\ \text{s}$

Planck mass $\qquad\qquad\quad m_{\text{Pl}} = \sqrt{\dfrac{\hbar c}{G}} \qquad = 2.176\,44(11) \times 10^{-5}\ \text{g}$

Planck momentum $\qquad m_{\text{Pl}}\,c = \dfrac{\hbar}{l_{\text{Pl}}} = \sqrt{\dfrac{\hbar c^3}{G}} \quad = 6.524\,85 \times 10^5\ \frac{\text{g·cm}}{\text{s}}$

Planck energy $\qquad\qquad m_{\text{Pl}}\,c^2 = \dfrac{\hbar}{t_{\text{Pl}}} = \sqrt{\dfrac{\hbar c^5}{G}} \quad = 1.961 \times 10^{16}\ \text{erg} = 1.22 \times 10^{28}\ \text{eV}$

Planck density $\qquad\qquad \rho_{\text{Pl}} = \dfrac{m_{\text{Pl}}}{l_{\text{Pl}}^3} = \dfrac{c^5}{\hbar G^2} \quad = 5.155 \times 10^{93}\ \frac{\text{g}}{\text{cm}^3}$

Planck temperature $\qquad \Theta_{\text{Pl}} = \dfrac{m_{\text{Pl}}\,c^2}{k_{\text{B}}} = \sqrt{\dfrac{\hbar c^5}{k_{\text{B}}^2 G}} = 1.416\,785(71) \times 10^{32}\ \text{K}$

A.3 Conversion Factors

Distance $\qquad\qquad\qquad\qquad$ 1 pc $= 3.0856 \times 10^{18}\ \text{cm} = 3.2616\ \text{lt-yr}$

$\qquad\qquad\qquad\qquad\qquad$ 1 lt-yr $= 9.4605 \times 10^{17}\ \text{cm} = 0.30659\ \text{pc}$

$\qquad\qquad\qquad\qquad\qquad$ 1 A.U. $= 1.495\,98 \times 10^{13}\ \text{cm} = 1.581\,28 \times 10^{-5}\ \text{lt-yr}$

Time $\qquad\qquad\qquad\qquad\qquad$ 1 year $= 3.155\,692\,6 \times 10^7\ \text{s}$

$\qquad\qquad\qquad\qquad\qquad$ 1 day $= 86\,400\ \text{s}$

$\qquad\qquad\qquad$ 1 sidereal day $= 86\,164.091\ \text{s}$

Energy $\qquad\qquad\qquad\qquad\quad$ 1 eV $= 1.160\,48 \times 10^4\ \text{K} = 1.602\,192 \times 10^{-12}\ \text{erg}$

Boltzmann's constant $\qquad\quad k_{\text{B}} = 1.380\,65 \times 10^{-16}\ \frac{\text{erg}}{\text{K}} = 8.617\,38 \times 10^{-5}\ \frac{\text{eV}}{\text{K}}$

A.4   Various Scales of Masses

Electron mass	m_e	=	$9.109\,382\,15(45) \times 10^{-28}$ g
	$m_e\,c^2$	=	$0.510\,998\,910(13)$ MeV
Proton mass	m_p	=	$1.672\,621\,637(83) \times 10^{-24}$ g
	$m_p\,c^2$	=	$938.272\,013(23)$ MeV
Neutron mass	m_n	=	$1.674\,927\,29(28) \times 10^{-24}$ g
	$m_n\,c^2$	=	$939.565\,560(81)$ MeV
Earth mass	$M_\oplus$	=	$5.973\,6 \times 10^{27}$ g
Solar mass	$M_\odot$	=	$1.989\,1 \times 10^{33}$ g

A.5   Milky Way Galaxy Observational Data

Mass	$\sim 5.8 \times 10^{11}\ M_\odot = 1.2 \times 10^{45}$ g
Radius	$\sim 50\,000$ lt-yr
Thickness	~ 1000 lt-yr
Number of stars	$\sim 1 - 4 \times 10^{11}$
Number of stellar-mass black holes	$\sim 10^8 - 10^9$
Mass of the central black hole	$\sim 4 \times 10^6\ M_\odot$
Radius of the central black hole	$\sim 1.2 \times 10^{12}$ cm ≈ 40 lt-s
Sun's distance to galactic center	$\sim 25\,000$ lt-yr
Sun's galactic rotation period	~ 250 million years

A.6   Universe Observational Data

Age	$\sim 13.75 \times 10^9$ years		
Number of galaxies	$\sim 8 \times 10^{10}$		
Number of stars	$\sim 3 - 7 \times 10^{22}$		
Number of atoms	$\sim 10^{80}$		
Density of matter	$\sim 9.9 \times 10^{-30}\ \frac{\text{g}}{\text{cm}^3}$	$\approx 4.56\%^{+0.16\%}_{-0.16\%}$	(normal matter)
		$+22.7\%^{+1.4\%}_{-1.4\%}$	(dark matter)
		$+72.8\%^{+1.5\%}_{-1.6\%}$	(dark energy)

A.7   Dimensionless Entropy (S/k_B)

Single stellar mass $(M_{BH} = 10\,M_\odot)$ **black hole**	$\sim 10^{79}$
Single supermassive $(M_{BH} = 4 \times 10^6\ M_\odot)$ **black hole**	$\sim 1.7 \times 10^{90}$
All stars in the observable Universe	$\sim 10^{79}$
Relic gravitons	$\sim 10^{86}$
CMB photons	$\sim 10^{88}$
Relic neutrinos	$\sim 10^{88}$

Appendix B
Gauss–Codazzi Equations

B.1 Gauss–Codazzi Equations

Let Σ be a submanifold of codimension 1 in the D-dimensional spacetime. We denote by x^μ coordinates in the bulk manifold, ($\mu = 0, 1, \ldots, D - 1$), and by y^i coordinates in Σ, ($i = 1, \ldots, D - 1$). The equation of embedding is $x^\mu = x^\mu(y^i)$. Denote by n^μ a unit normal to Σ and

$$\varepsilon(n) \equiv n_\mu n^\mu = \pm 1 . \tag{B.1.1}$$

Σ is space-like for $\varepsilon(n) = -1$ and time-like if $\varepsilon(n) = 1$. The case of null Σ is special and we do not consider it here.

The *induced metric* on Σ is given by

$$h_{ij} = \frac{\partial x^\mu}{\partial y^i} \frac{\partial x^\nu}{\partial y^j} g_{\mu\nu} . \tag{B.1.2}$$

The D-dimensional Riemann tensor $R^\alpha{}_{\mu\beta\nu}$ for the metric g and the $(D-1)$-dimensional Riemann tensor $\mathcal{R}^m{}_{ijk}$ for the metric h are related as follows

$$R^m{}_{ijk} = \mathcal{R}^m{}_{ijk} + \varepsilon(n) \left[K_{ij} K^m{}_k - K_{ik} K_j{}^m \right], \tag{B.1.3}$$

$$n_\alpha R^\alpha{}_{ijk} = \varepsilon(n) \left[K_{ik|j} - K_{ij|k} \right], \tag{B.1.4}$$

$$n^\alpha n^\beta G_{\alpha\beta} = -\frac{1}{2} \mathcal{R} + \frac{1}{2} \varepsilon(n) \left[K^2 - K_{ij} K^{ij} \right], \tag{B.1.5}$$

$$n^\alpha G_{\alpha i} = -\varepsilon(n) \left[K_i{}^m{}_{|m} - K_{|i} \right], \tag{B.1.6}$$

$$n_\alpha n^\beta R^\alpha{}_{i\beta k} = \varepsilon(n) \left[n^\alpha K_{ik,\alpha} + K_{im} K^m{}_k \right]. \tag{B.1.7}$$

Here,

$$K = h^{ij} K_{ij} , \qquad \mathcal{R} = h^{ij} \mathcal{R}_{ij} , \qquad \mathcal{R}_{ij} = \mathcal{R}^m{}_{imj}, \tag{B.1.8}$$

and $(\ldots)_{|i}$ denotes a covariant derivative with respect to the metric h_{ij}.

These equations are simplified if one uses the Gaussian coordinates Eq. (3.8.13), in which the bulk metric has the form

$$ds^2 = \varepsilon(n)\, d\tau^2 + h_{ij}(\tau, y)\, dy^i\, dy^j, \tag{B.1.9}$$

and the *extrinsic curvature* is

$$K_{ij} = -\frac{1}{2}\,\varepsilon(n)\,\partial_\tau h_{ij}.\tag{B.1.10}$$

B.2 Static Surface in a Static Spacetime

Consider a static spacetime with the metric Eq. (3.7.14)

$$ds^2 = -e^{2U}\,dt^2 + h_{ij}\,dy^i\,dy^j,\tag{B.2.1}$$

and let Σ be a surface $t = 0$ in it. Since the metric is invariant under the reflection $t \to -t$, the extrinsic curvature for this surface vanishes

$$K_{ij} = 0.\tag{B.2.2}$$

The *Gauss–Codazzi equations* (B.1.3)–(B.1.7) give

$$
\begin{aligned}
R^i{}_{jkl} &= \mathcal{R}^i{}_{jkl},\\
R^t{}_{jkl} &= 0,\\
R^t{}_{itj} &= -U_{|ij} - U_{|i}U_{|j}.
\end{aligned}
\tag{B.2.3}
$$

Here, $\mathcal{R}^i{}_{jkl}$ is the corresponding $(D{-}1)$-dimensional curvature tensor. Using these relations one can also derive the *Ricci tensor*

$$
\begin{aligned}
R^i{}_j &= \mathcal{R}^i{}_j - U^{|i}{}_{|j} - U^{|i}U_{|j},\\
R^t{}_j &= 0,\\
R^t{}_t &= -U^{|k}{}_{|k} - U^{|k}U_{|k},
\end{aligned}
\tag{B.2.4}
$$

the *scalar curvature*

$$R = \mathcal{R} - 2(U^{|k}{}_{|k} + U^{|k}U_{|k}),\tag{B.2.5}$$

and the *Einstein tensor*

$$
\begin{aligned}
G^i{}_j &= \mathcal{G}^i{}_j - U^{|i}{}_{|j} - U^{|i}U_{|j} + \delta^i_j(U^{|k}{}_{|k} + U^{|k}U_{|k}),\\
G^t{}_j &= 0,\\
G^t{}_t &= -\frac{1}{2}\mathcal{R}.
\end{aligned}
\tag{B.2.6}
$$

On the right-hand side of these equations all the tensors and covariant derivatives are defined with respect to the metric h_{ij}, while on the left-hand side all tensors correspond to the D-dimensional metric Eq. (B.2.1) .

Appendix C
Conformal Transformations

Here, we give without derivation some useful formulas for *conformal transformations* of the Ricci tensor and the scalar curvature in D-dimensional spacetime. Consider two *conformally related metrics* g and $\bar{g}$

$$
\begin{aligned}
g_{\alpha\beta} &= \Omega^2 \, \bar{g}_{\alpha\beta} = \exp[\, 2\sigma \,] \bar{g}_{\alpha\beta}, \\
g^{\alpha\beta} &= \Omega^{-2} \, \bar{g}^{\alpha\beta} = \exp[-2\sigma] \, \bar{g}^{\alpha\beta}, \\
g^{1/2} &= \Omega^D \bar{g}^{1/2} = \exp[\, D\sigma \,] \, \bar{g}^{1/2}.
\end{aligned}
\tag{C.0.1}
$$

The conformal factor is presented in two forms $\Omega^2(x) = \exp[\, 2\sigma(x)\,]$, both of which may be convenient in different applications. We mark by a bar objects and operators related to the metric $\bar{g}$. The *Christoffel symbols* for the conformally related spaces are related as follows

$$
\begin{aligned}
\Gamma^\lambda_{\alpha\beta} &= \bar{\Gamma}^\lambda_{\alpha\beta} + \gamma^\lambda_{\alpha\beta}, \\
\gamma^\lambda_{\alpha\beta} &= \Omega^{-1} \left(\delta^\lambda_\alpha \Omega_{,\beta} + \delta^\lambda_\beta \Omega_{,\alpha} - \bar{g}_{\alpha\beta} \bar{g}^{\lambda\epsilon} \Omega_{,\epsilon} \right) \\
&= \delta^\lambda_\alpha \sigma_{,\beta} + \delta^\lambda_\beta \sigma_{,\alpha} - \bar{g}_{\alpha\beta} \bar{g}^{\lambda\epsilon} \sigma_{,\epsilon}.
\end{aligned}
\tag{C.0.2}
$$

The *conformal transformation of the scalar curvature* is

$$
\begin{aligned}
R &= \frac{1}{\Omega^2} \left[\bar{R} - 2(D-1)\frac{\bar{\Box}\,\Omega}{\Omega} - (D-1)(D-4) \left(\frac{\bar{\nabla}\Omega}{\Omega} \right)^2 \right] \\
&= \exp[-2\sigma] \left[\bar{R} - 2(D-1)\,\bar{\Box}\,\sigma - (D-1)(D-2)\,(\bar{\nabla}\sigma)^2 \right]. \\
\bar{R} &= \Omega^2 \left[R + 2(D-1)\frac{\Box\Omega}{\Omega} - D(D-1) \left(\frac{\nabla\Omega}{\Omega} \right)^2 \right] \\
&= \exp[\, 2\sigma \,] \left[R + 2(D-1)\,\Box\,\sigma - (D-1)(D-2)\,(\nabla\sigma)^2 \right].
\end{aligned}
\tag{C.0.3}
$$

Here, $(\nabla\sigma)^2 = \nabla^\alpha \sigma \nabla_\alpha \sigma$ and $\Box\sigma = \nabla^\alpha \nabla_\alpha \sigma$. The corresponding objects with 'bar' are defined in a similar way with respect to the metric $\bar{g}$.

The *conformal transformation of the Ricci tensor* is

$$R_{\alpha\beta} = \bar{R}_{\alpha\beta} - (D-2)\frac{\bar{\nabla}_\alpha\bar{\nabla}_\beta\Omega}{\Omega} + 2(D-2)\frac{\bar{\nabla}_\alpha\Omega\,\bar{\nabla}_\beta\Omega}{\Omega\;\Omega}$$

$$- \bar{g}_{\alpha\beta}\left[\frac{\bar{\Box}\,\Omega}{\Omega} + (D-3)\left(\frac{\bar{\nabla}\Omega}{\Omega}\right)^2\right]$$

$$= \bar{R}_{\alpha\beta} + (D-2)\left[-\bar{\nabla}_\alpha\bar{\nabla}_\beta\sigma + \bar{\nabla}_\alpha\sigma\bar{\nabla}_\beta\sigma\right]$$

$$- \bar{g}_{\alpha\beta}\left[\bar{\Box}\,\sigma + (D-2)(\bar{\nabla}\sigma)^2\right], \tag{C.0.4}$$

$$\bar{R}_{\alpha\beta} = R_{\alpha\beta} + (D-2)\frac{\nabla_\alpha\nabla_\beta\Omega}{\Omega} + g_{\alpha\beta}\left[\frac{\Box\Omega}{\Omega} - (D-1)\left(\frac{\nabla\Omega}{\Omega}\right)^2\right]$$

$$= R_{\alpha\beta} + (D-2)\left[\nabla_\alpha\nabla_\beta\sigma + \nabla_\alpha\sigma\nabla_\beta\sigma\right] + g_{\alpha\beta}\left[\Box\sigma - (D-2)(\nabla\sigma)^2\right].$$

Transformation of the box-operator $\Box = \nabla^\alpha\nabla_\alpha$ acting on a scalar Φ is

$$\Box\,\Phi = \frac{1}{\Omega^2}\left[\bar{\Box} + (D-2)\frac{\bar{\nabla}^\epsilon\Omega}{\Omega}\bar{\nabla}_\epsilon\right]\Phi = \exp[-2\sigma]\left[\bar{\Box} + (D-2)\bar{\nabla}^\epsilon\sigma\bar{\nabla}_\epsilon\right]\Phi. \tag{C.0.5}$$

$$(\Box - \xi_D R)\,\Phi = \frac{1}{\Omega^2}\,\Omega^{-(\frac{D}{2}-1)}(\bar{\Box} - \xi_D\bar{R})\,\Omega^{(\frac{D}{2}-1)}\,\Phi,$$

$$(\Box - \xi_D R)\,\Phi = \exp\left[\left(-\frac{D}{2}-1\right)\sigma\right](\bar{\Box} - \xi_D\bar{R})\,\exp\left[\left(\frac{D}{2}-1\right)\sigma\right]\Phi, \tag{C.0.6}$$

where $\xi_D = \dfrac{D-2}{4(D-1)}$.

In the four dimensions $(D=4)$ the scalar curvature and the Ricci tensor transform as follows:

$$R = \frac{1}{\Omega^2}\left[\bar{R} - 6\frac{\bar{\Box}\,\Omega}{\Omega}\right] = \exp[-2\sigma]\left[\bar{R} - 6\,\bar{\Box}\,\sigma - 6\,(\bar{\nabla}\sigma)^2\right]$$

$$\bar{R} = \Omega^2\left[R + 6\frac{\Box\Omega}{\Omega} - 12\left(\frac{\nabla\Omega}{\Omega}\right)^2\right] = \exp[\,2\sigma\,]\left[R + 6\,\Box\sigma - 6\,(\nabla\sigma)^2\right], \tag{C.0.7}$$

$$R_{\alpha\beta} = \bar{R}_{\alpha\beta} - 2\frac{\bar{\nabla}_\alpha\bar{\nabla}_\beta\Omega}{\Omega} + 4\frac{\bar{\nabla}_\alpha\Omega\,\bar{\nabla}_\beta\Omega}{\Omega\;\Omega} - \bar{g}_{\alpha\beta}\left[\frac{\bar{\Box}\,\Omega}{\Omega} + \left(\frac{\bar{\nabla}\Omega}{\Omega}\right)^2\right]$$

$$= \bar{R}_{\alpha\beta} + 2\left[-\bar{\nabla}_\alpha\bar{\nabla}_\beta\sigma + \bar{\nabla}_\alpha\sigma\bar{\nabla}_\beta\sigma\right] - \bar{g}_{\alpha\beta}\left[\bar{\Box}\,\sigma + 2(\bar{\nabla}\sigma)^2\right],$$

$$\bar{R}_{\alpha\beta} = R_{\alpha\beta} + 2\frac{\nabla_\alpha\nabla_\beta\Omega}{\Omega} + g_{\alpha\beta}\left[\frac{\Box\Omega}{\Omega} - 3\left(\frac{\nabla\Omega}{\Omega}\right)^2\right] \tag{C.0.8}$$

$$= R_{\alpha\beta} + 2\left[\nabla_\alpha\nabla_\beta\sigma + \nabla_\alpha\sigma\nabla_\beta\sigma\right] + g_{\alpha\beta}\left[\Box\sigma - 2(\nabla\sigma)^2\right].$$

Appendix D
Hidden Symmetries

D.1   Conformal Killing Tensor

The Kerr metric is an example of a spacetime with hidden symmetries. This property is discussed in Section 8.7. In this appendix we collect additional information, concerning spacetimes with hidden symmetries. We discuss first hidden symmetries in the 4D spacetime. Brief remarks on the hidden symmetries in the higher dimensions are given in Sections D.9 and D.10. Additional material can be found in the review article (Frolov and Kubizňák 2008).

By definition, a spacetime has a *hidden symmetry* if the geodesic equations possess conserved quantities of higher than the first order in momentum. The geometric structure responsible for such a conservation law is the *Killing tensor*. We define first a *conformal Killing tensor*. This is a symmetric tensor $K_{\mu_1\ldots\mu_p}$ of rank p that obeys the equation

$$K_{(\mu_1\ldots\mu_p;\nu)} = g_{\nu(\mu_1}\tilde{K}_{\mu_2\ldots\mu_p)}, \tag{D.1.1}$$

where $\tilde{K}_{\mu_2\ldots\mu_p}$ is a symmetric tensor of rank $p-1$. For *null geodesics* $x^\mu(\lambda)$, where λ is the affine parameter, in a spacetime with the conformal Killing tensor the following quantity is conserved

$$K_{\mu_1\ldots\mu_p} u^{\mu_1} \ldots u^{\mu_p}, \qquad u^\mu = \frac{dx^\mu}{d\lambda}. \tag{D.1.2}$$

A *Killing tensor* is a special type of a conformal Killing tensor for which the right-hand side of Eq. (D.1.1) vanishes. For the Killing tensor the quantity Eq. (D.1.2) is conserved for any (not necessary null) geodesics. For non-null geodesics $u^\mu = dx^\mu/d\tau$, where τ is a proper time parameter for the time-like geodesics and the proper distance for the space-like one.

D.2   Killing–Yano Tensors

D.2.1   Conformal Killing–Yano tensor

A Killing tensor is a natural symmetric generalization of the Killing vector. There exists also an antisymmetric generalization, known as a *Killing–Yano tensor*. We introduce first a *conformal Killing–Yano tensor*. This is an anti–symmetric tensor $h_{\mu_1\mu_2\ldots\mu_p}$ of the rank p (p-form) that obeys the equation

$$\nabla_\mu h_{\mu_1\mu_2\ldots\mu_p} = \nabla_{[\mu} h_{\mu_1\mu_2\ldots\mu_p]} + p\, g_{\mu[\mu_1}\tilde{h}_{\mu_2\ldots\mu_p]}. \tag{D.2.1}$$

The tensor $\tilde{h}_{\mu_2...\mu_p}$ can be obtained by tracing both sides of Eq. (D.2.1) with respect to the indices μ and μ_1, and in 4D it has the form

$$\tilde{h}_{\mu_2...\mu_p} = \frac{1}{5-p} \nabla^\mu h_{\mu\mu_2...\mu_p}. \tag{D.2.2}$$

Another equivalent form of Eq. (D.2.1) is

$$\nabla_{(\mu_1} h_{\mu_2)\mu_3...\mu_{p+1}} = g_{\mu_1\mu_2} \tilde{h}_{\mu_3...\mu_{p+1}} - (p-1) g_{[\mu_3(\mu_1} \tilde{h}_{\mu_2)...\mu_{p+1}]}. \tag{D.2.3}$$

When $\tilde{h} = 0$, and hence the right-hand side of Eq. (D.2.3) vanishes, the corresponding antisymmetric tensor is called a *Killing–Yano tensor*. If a spacetime has a Killing–Yano tensor $k_{\mu_1\mu_2...\mu_p}$ then

- $k_{\mu_1\mu_2...\mu_p} u^{\mu_p}$ is parallelly propagated along a geodesic with a tangent vector u^{μ_p}.
- $K_{\mu\nu} = (k \bullet k)_{\mu\nu} \equiv k_{\mu\mu_2...\mu_p} k_\nu^{\mu_2...\mu_p}$ is a *Killing tensor*.

The latter property means that in a spacetime with a Killing–Yano tensor there always exists a Killing tensor that is a 'square' of the Killing–Yano tensor. Note, that in the general case, the 'square root' of the Killing tensor does not exist, that is one cannot present the Killing tensor as a 'square' of some Killing–Yano tensor.

D.2.2 Closed conformal Killing–Yano tensor

When the first term on the right-hand side of Eq. (D.2.1) vanishes

$$\nabla_{[\mu} h_{\mu_1\mu_2...\mu_p]} = 0, \tag{D.2.4}$$

the conformal Killing–Yano tensor is closed. Such a p-form h can be written (at least locally) as

$$h = db. \tag{D.2.5}$$

The $(p-1)$-form b is called a potential-generating closed conformal Killing–Yano tensor.

Consider a rank-2 skew-symmetric tensor $h_{\mu\nu}$ in a four-dimensional spacetime obeying the equation

$$h_{\mu\nu;\lambda} = g_{\lambda\nu} \xi_\mu - g_{\lambda\mu} \xi_\nu. \tag{D.2.6}$$

The contraction of this relation gives

$$\xi_\mu = \frac{1}{3} h_\mu{}^\lambda{}_{;\lambda}. \tag{D.2.7}$$

This tensor obeys the relations

$$\nabla_{(\lambda} h_{\mu)\nu} = g_{\nu(\lambda} \xi_{\mu)} - g_{\lambda\mu} \xi_\nu,$$
$$\nabla_{[\lambda} h_{\mu\nu]} = 0, \tag{D.2.8}$$

and, hence, it is a rank-2 closed conformal Killing–Yano tensor. This tensor can be written in the form

$$h_{\mu\nu} = \nabla_\mu b_\nu - \nabla_\nu b_\mu = -2\, b_{[\mu;\nu]}, \tag{D.2.9}$$

where b_μ is its potential.

D.2.3 Duality relations

There exists a *duality relation* between a closed conformal Killing–Yano tensor and a Killing–Yano tensor. To obtain this relation we use the following properties of the *totally skew-symmetric tensor*

$$e_{\mu_1\mu_2\mu_3\mu_4} e^{\nu_1\nu_2\nu_3\nu_4} = -24\, \delta^{[\nu_1}_{\mu_1}\delta^{\nu_2}_{\mu_2}\delta^{\nu_3}_{\mu_3}\delta^{\nu_4]}_{\mu_4},$$

$$e_{\mu_1\mu_2\mu_3\lambda} e^{\nu_1\nu_2\nu_3\lambda} = -6\, \delta^{[\nu_1}_{\mu_1}\delta^{\nu_2}_{\mu_2}\delta^{\nu_3]}_{\mu_3},$$

$$e_{\mu_1\mu_2\epsilon\lambda} e^{\nu_1\nu_2\epsilon\lambda} = -4\, \delta^{[\nu_1}_{\mu_1}\delta^{\nu_2]}_{\mu_2}, \tag{D.2.10}$$

$$e_{\mu_1\pi\epsilon\lambda} e^{\nu_1\pi\epsilon\lambda} = -6\, \delta^{\nu_1}_{\mu_1},$$

$$e_{\sigma\pi\epsilon\lambda} e^{\sigma\pi\epsilon\lambda} = -24.$$

A *Hodge dual* tensor of an antisymmetric tensor $h_{\mu\nu}$ is defined as

$$k_{\alpha\beta} = \frac{1}{2} h^{\mu\nu} e_{\mu\nu\alpha\beta}. \tag{D.2.11}$$

Lemma 3: *If $h_{\mu\nu}$ is a closed conformal Killing–Yano tensor, then its Hodge dual $k_{\mu\nu}$ is a Killing–Yano tensor.*

Proof: One has

$$k_{\alpha\beta;\gamma} = \frac{1}{2} (\delta^\nu_\gamma \xi^\mu - \delta^\mu_\gamma \xi^\nu) e_{\mu\nu\alpha\beta} = \xi^\mu e_{\mu\alpha\beta\gamma}. \tag{D.2.12}$$

Hence, $k_{\alpha(\beta;\gamma)} = 0$ and k is a Killing–Yano tensor.

Lemma 4: *If $f_{\mu\nu}$ is a Killing-Yano tensor, then its Hodge dual $h_{\mu\nu}$ is a closed conformal Killing–Yano tensor.*

Proof: According to the definition of the Killing-Yano tensor $f_{\mu\nu;\lambda} = -f_{\mu\lambda;\nu}$. Since $f_{\mu\nu}$ is antisymmetric with respect to its indices, $f_{\mu\nu;\lambda}$ is also an antisymmetric tensor. Let us denote

$$\xi^\nu = -\frac{1}{6} f_{\mu_1\mu_2;\mu_3} e^{\mu_1\mu_2\mu_3\nu}. \tag{D.2.13}$$

Multiplying both sides of this relation by $e_{\nu_1\nu_2\nu_3\nu}$ and using Eq. (D.2.10) we get

$$f_{\mu_1\mu_2;\mu_3} = -\xi^\nu e_{\nu\mu_1\mu_2\mu_3}. \tag{D.2.14}$$

Consider now a Hodge dual tensor of $f_{\mu\nu}$

$$h_{\mu\nu} = \frac{1}{2} f_{\alpha\beta} e^{\alpha\beta}{}_{\mu\nu}. \tag{D.2.15}$$

For its covariant derivative we have

$$h_{\mu\nu;\lambda} = \frac{1}{2} f_{\alpha\beta;\lambda} \, e^{\alpha\beta}{}_{\mu\nu}. \tag{D.2.16}$$

Substituting the representation Eq. (D.2.14) in this relation and using Eq. (D.2.10) we get

$$h_{\mu\nu;\lambda} = g_{\nu\lambda}\xi_\mu - g_{\mu\lambda}\xi_\nu. \tag{D.2.17}$$

Hence, $h_{\mu\nu}$ is a closed conformal Killing–Yano tensor.

Note that, because $**h = -h$, we have the relation $f_{\mu\nu} = -k_{\mu\nu}$.

D.3   Primary Killing Vector

Solutions of the Einstein equations with the cosmological constant Λ

$$R_{\mu\nu} = \Lambda g_{\mu\nu}, \tag{D.3.1}$$

are called *Einstein spaces*. We prove now the following statement:

Lemma 5: *Define a vector* $\boldsymbol{\xi}$ *in the Einstein space with a closed conformal Killing–Yano tensor* $\boldsymbol{h}$ *by the relation*

$$\xi_\mu = \frac{1}{3} h_\mu{}^\lambda{}_{;\lambda}. \tag{D.3.2}$$

If $\boldsymbol{\xi} \neq 0$ *it is a Killing vector.*

Proof: Using the formula for the commutator of the covariant derivatives we write

$$h_{\mu\epsilon;\lambda;\nu} - h_{\mu\epsilon;\nu;\lambda} = h_{\alpha\epsilon}R^\alpha{}_{\mu\lambda\nu} + h_{\mu\alpha}R^\alpha{}_{\epsilon\lambda\nu}. \tag{D.3.3}$$

By multiplying this relation by $g^{\epsilon\lambda}$ we get

$$h_{\mu\lambda}{}^{;\lambda}{}_{;\nu} - h_{\mu\lambda;\nu}{}^{;\lambda} = h_{\alpha\lambda}R^\alpha{}_\mu{}^\lambda{}_\nu - h_{\mu\alpha}R^\alpha{}_\nu. \tag{D.3.4}$$

Consider a vector ξ^μ defined by Eq. (D.2.7). For this vector

$$\xi_{\mu;\nu} = \frac{1}{3} h_{\mu\lambda}{}^{;\lambda}{}_{;\nu}, \tag{D.3.5}$$

and Eq. (D.3.4) implies

$$2\xi_{\mu;\nu} + g_{\mu\nu}\xi_\lambda{}^{;\lambda} = h_{\alpha\lambda}R^\alpha{}_\mu{}^\lambda{}_\nu - h_{\mu\alpha}R^\alpha{}_\nu. \tag{D.3.6}$$

Contraction of this equation gives

$$\xi_\lambda{}^{;\lambda} = 0. \tag{D.3.7}$$

After symmetrization of the relation Eq. (D.3.6) we get

$$\xi_{(\mu;\nu)} = -\frac{1}{2} R^\alpha{}_{(\nu}h_{\mu)\alpha}. \tag{D.3.8}$$

For the Einstein space Eq. (D.3.1) the right-hand side of this relation vanishes. It implies that ξ^μ is a Killing vector. We call this vector a *primary Killing vector*.

D.4 Properties of the Primary Killing Vector

Lemma 6: *The following relation is valid:*

$$\xi^\alpha h_{\mu\nu;\alpha} = 0. \tag{D.4.1}$$

Proof: Using Eq. (D.2.17) we get

$$\xi^\alpha h_{\mu\nu;\alpha} = -2\xi_{[\mu}\xi_{\nu]} = 0 . \tag{D.4.2}$$

We denote $\zeta_\alpha = -h_{\alpha\beta}\xi^\beta$. This vector obeys the property $\zeta_\mu\xi^\mu = 0$.

Lemma 7: *The following relation is valid:*

$$\zeta_\mu = \frac{1}{4}(h_{\alpha\beta}h^{\alpha\beta})_{;\mu}. \tag{D.4.3}$$

Proof: One has

$$(h_{\alpha\beta}h^{\alpha\beta})_{;\mu} = 2h^{\alpha\beta}h_{\alpha\beta;\mu} = 2h^{\alpha\beta}(g_{\mu\beta}\xi_\alpha - g_{\mu\alpha}\xi_\beta) = 4\zeta_\mu. \tag{D.4.4}$$

Lemma 8: *Let ξ be a primary Killing vector for the closed conformal Killing–Yano tensor h. Then, the following relation is valid:*

$$\mathcal{L}_\xi h_{\mu\nu} = 0. \tag{D.4.5}$$

Proof: Using the definition of the *Lie derivative* we have

$$\mathcal{L}_\xi h_{\mu\nu} = \xi^\alpha h_{\mu\nu;\alpha} + \xi_{\mu}{}^{;\alpha}h_{\alpha\nu} + \xi_{\nu}{}^{;\alpha}h_{\mu\alpha}. \tag{D.4.6}$$

We also have

$$\xi_{\mu}{}^{;\alpha}h_{\alpha\nu} = -\xi^\alpha{}_{;\mu}h_{\alpha\nu} = -(\xi^\alpha h_{\alpha\nu})_{;\mu} + \xi^\alpha h_{\alpha\nu;\mu} = g_{\mu\nu}\xi^2 - \zeta_{\nu;\mu} - \xi_\mu\xi_\nu. \tag{D.4.7}$$

Using Eq. (D.4.1) we obtain

$$\mathcal{L}_\xi h_{\mu\nu} = 2\zeta_{[\mu;\nu]}. \tag{D.4.8}$$

Finally, we use the relation Eq. (D.4.3) to prove Eq. (D.4.5).

D.5 Secondary Killing Vector

Let $h_{\mu\nu}$ be a closed conformal Killing–Yano tensor and $k_{\mu\nu}$ is its Hodge dual Killing–Yano tensor. Then, the Killing tensor associated with $k_{\mu\nu}$

$$K_{\mu\nu} = k_{\mu\lambda}k^\lambda{}_\nu \tag{D.5.1}$$

can be written as follows:

$$K_{\mu\nu} = H_{\mu\nu} - \frac{1}{2}g_{\mu\nu}H_\lambda{}^\lambda, \qquad H_{\mu\nu} = h_{\mu\lambda}h^\lambda{}_\nu. \tag{D.5.2}$$

The relation Eq. (D.5.2) can be checked directly by using the expression Eq. (D.2.11) and the relation Eq. (D.2.10).

Denote $\zeta_\mu = -h_{\mu\nu}\xi^\nu$, then direct calculations give

$$H_{\mu\nu;\lambda} = 2\left[h_{\lambda(\mu}\xi_{\nu)} - g_{\lambda(\mu}\zeta_{\nu)}\right],$$

$$K_{\mu\nu;\lambda} = 2\left[h_{\lambda(\mu}\xi_{\nu)} - g_{\lambda(\mu}\zeta_{\nu)} + g_{\mu\nu}\zeta_\lambda\right], \tag{D.5.3}$$

$$K_{\mu\lambda}{}^{;\lambda} = -2\,\zeta_\mu.$$

Lemma 9: *If $K_{\mu\nu}$ is a Killing tensor associated with the closed conformal Killing–Yano tensor $h_{\mu\nu}$ and ξ^μ is a primary Killing vector then*

$$\eta_\mu = K_{\mu\lambda}\xi^\lambda \tag{D.5.4}$$

is a Killing vector. We call this Killing vector a secondary Killing vector.

Proof: One has

$$\eta_{\mu;\nu} = K_{\mu\lambda;\nu}\xi^\lambda + K_{\mu\lambda}\xi^\lambda{}_{;\nu}. \tag{D.5.5}$$

Thus,

$$\eta_{(\mu;\nu)} = K_{\lambda(\mu;\nu)}\xi^\lambda + \frac{1}{2}[K_{\mu\lambda}\xi^\lambda{}_{;\nu} + K_{\nu\lambda}\xi^\lambda{}_{;\mu}]. \tag{D.5.6}$$

Using the definition of the Killing tensor we get $K_{\lambda(\mu;\nu)} = -\frac{1}{2}K_{\mu\nu;\lambda}$. The expression in the squared brackets in Eq. (D.5.6) can be rewritten as

$$K_{\mu\lambda}\xi^\lambda{}_{;\nu} + K_{\nu\lambda}\xi^\lambda{}_{;\mu} = \mathcal{L}_\xi K_{\mu\nu} - \xi^\lambda K_{\mu\nu;\lambda}. \tag{D.5.7}$$

Combining these results we obtain

$$\eta_{(\mu;\nu)} = \frac{1}{2}\mathcal{L}_\xi K_{\mu\nu} - \xi^\lambda K_{\mu\nu;\lambda}. \tag{D.5.8}$$

The first term vanishes because $K_{\mu\nu}$ is constructed in terms of $h_{\mu\nu}$ and $g_{\mu\nu}$, for which $\mathcal{L}_\xi h_{\mu\nu} = 0$ and $\mathcal{L}_\xi g_{\mu\nu} = 0$. Using Eq. (D.5.3) one can check that the second term also vanishes. Thus, we proved that

$$\eta_{(\mu;\nu)} = 0, \tag{D.5.9}$$

and η_μ is a Killing vector.

Lemma 10: *Killing vector fields ξ^μ and η^μ commute with one another.*

Proof: The following relations are valid:

$$[\xi, \eta]^\mu = \mathcal{L}_\xi \eta^\mu = \left(\mathcal{L}_\xi K^\mu_\nu\right)\xi^\nu + K^\mu_\nu\,\mathcal{L}_\xi\xi^\nu = 0. \tag{D.5.10}$$

D.6 Darboux Basis

Consider a symmetric operator

$$H_{\mu\nu} = h_{\mu\lambda} h^{\lambda}_{\ \nu}, \tag{D.6.1}$$

constructed from a closed conformal Killing–Yano tensor h. Consider the following eigenvalue problem:

$$H^{\mu}_{\ \nu} n^{\nu} = A n^{\mu}. \tag{D.6.2}$$

We assume that a vector n^{μ} is normalized

$$(n, n) = \varepsilon_n, \tag{D.6.3}$$

where $\varepsilon_n = 1$ for a space-like vector and $\varepsilon_n = -1$ for a time-like one.

Lemma 11: *If n^{ν} is a normalized eigenvector of $H_{\mu\nu}$ with $A \neq 0$, then*

$$\bar{n}^{\mu} = \frac{1}{\sqrt{|A|}} h^{\mu}_{\ \lambda} n^{\lambda} \tag{D.6.4}$$

is also a normalized eigenvector of the same operator with the same eigenvalue. One also has

$$n^{\mu} = \frac{sign(A)}{\sqrt{|A|}} h^{\mu}_{\ \lambda} \bar{n}^{\lambda}. \tag{D.6.5}$$

Proof: Denote $\bar{n}^{\mu} = \beta h^{\mu}_{\ \lambda} n^{\lambda}$. It is easy to see that $(n, \bar{n}) = 0$, and one has

$$H^{\mu}_{\ \nu} \bar{n}^{\nu} = \beta h^{\mu}_{\ \epsilon} h^{\epsilon}_{\ \nu} h^{\nu}_{\ \lambda} n^{\lambda} = \beta h^{\mu}_{\ \epsilon} H^{\epsilon}_{\ \lambda} n^{\lambda} = A \bar{n}^{\mu}. \tag{D.6.6}$$

Thus, $\bar{n}^{\mu}$ is an eigenvector of $H^{\mu}_{\ \nu}$ and its eigenvalue is A. For $\beta = 1/\sqrt{|A|}$ one also has

$$(\bar{n}, \bar{n}) = \bar{\varepsilon}_n, \qquad \bar{\varepsilon}_n = -sign(A)\varepsilon_n. \tag{D.6.7}$$

The relation Eq. (D.6.5) can be easily checked by substituting Eq. (D.6.4) into this relation and using Eq. (D.6.2). We call the vector $\bar{n}$ *conjugated* to n. Two mutually conjugated orthonormal vectors $\{n, \bar{n}\}$ span a two-dimensional eigenspace of $H^{\mu}_{\ \nu}$.

We call the closed conformal Killing–Yano tensor $h_{\mu\nu}$ in the 4D spacetime *non-degenerate* if the eigenvalue problem Eq. (D.6.2) has two different eigenvalues A_1 and A_2. In this case one has two linearly independent eigenspaces. Let us show that these spaces are orthogonal.

Lemma 12: *Any two eigenvectors n_1 and n_2 of H with different eigenvalues are orthogonal.*

Proof: Denote by A_1 and A_2 eigenvalues for the eigenvectors n_1 and n_2, respectively. Then, one has

$$A_2(n_1, n_2) = (n_1, Hn_2) = n_1^{\mu} h_{\mu}^{\ \lambda} h_{\lambda}^{\ \nu} n_{2\ \nu} = (Hn_1, n_2) = A_1(n_2, n_1). \tag{D.6.8}$$

Since $(\boldsymbol{n}_1, \boldsymbol{n}_2) = (\boldsymbol{n}_2, \boldsymbol{n}_1)$, for $A_1 \neq A_2$ one has

$$(\boldsymbol{n}_1, \boldsymbol{n}_2) = 0. \tag{D.6.9}$$

Thus, at each point of a 4D spacetime, a non-degenerate closed conformal Killing–Yano tensor defines 2 mutually orthogonal planes. We call a plane *space-like* if all its vectors are space-like, and call it *time-like* if it contains a time-like vector. We assume that at any spacetime point the 4D tangent space is spanned by one time-like and one space-like 2D eigenspace.

Consider a space-like eigenspace and denote the conjugated normalized vectors that span it by $\{\boldsymbol{n}_2, \bar{\boldsymbol{n}}_2\}$. For these vectors $\varepsilon_n^{(2)} = \bar{\varepsilon}_n^{(2)} = 1$, and the relation Eq. (D.6.7) implies that $A_2 < 0$. Similar arguments for the vectors $\{\boldsymbol{n}_1, \bar{\boldsymbol{n}}_1\}$, which span the timelike 2D eigen-subspace, show that $A_1 > 0$. In what follows we choose $\boldsymbol{n}_1$ to be a time-like vector, so that $\bar{\boldsymbol{n}}_1$ as well as $\boldsymbol{n}_2$ and $\bar{\boldsymbol{n}}_2$, are space-like vectors. We also denote

$$A_1 = r^2, \qquad A_2 = -y^2, \qquad r > 0, \qquad y > 0. \tag{D.6.10}$$

Thus, we have the following result:

Lemma 13: *In each point of a spacetime with a non-degenerate closed conformal Killing–Yano tensor there exists a canonical basis of normalized vectors $\{\boldsymbol{n}_1, \bar{\boldsymbol{n}}_1, \boldsymbol{n}_2, \bar{\boldsymbol{n}}_2\}$ in which*

$$\begin{aligned}
g_{\mu\nu} &= -n_{1\mu}n_{1\nu} + \bar{n}_{1\mu}\bar{n}_{1\nu} + n_{2\mu}n_{2\nu} + \bar{n}_{2\mu}\bar{n}_{2\nu}, \\
h_{\mu\nu} &= 2r\,n_{1[\mu}\bar{n}_{1\nu]} - 2y\,n_{2[\mu}\bar{n}_{2\nu]}.
\end{aligned} \tag{D.6.11}$$

We call such a basis a Darboux basis.

The quantities r and y depend on the choice of a point. In the general case, they change when we move from one point to another. In other words, r and y are scalar functions on the spacetime manifold. We extend our assumption of non-degeneracy of the closed conformal Killing–Yano tensor to include into it a requirement that r and y are *functionally independent* (at least in some region U of the spacetime). This allows one to use r and y as coordinates in U. We call these coordinates *Darboux coordinates*.

We assume that the associated Killing vectors ξ^μ and η^μ do not vanish and are linearly independent of U. The integral lines $x^\mu(\tau)$ and $y^\mu(\psi)$ for these vector fields are defined by equations

$$\frac{dx^\mu(\tau)}{d\tau} = \xi^\mu, \qquad \frac{dy^\mu(\psi)}{d\psi} = \eta^\mu. \tag{D.6.12}$$

We can use τ and ψ as two coordinates. We call them *Killing coordinates*.

To summarize, in the presence of a non-degenerate closed conformal Killing–Yano tensor there exist, at least locally, coordinates (τ, r, y, ψ), which we call *canonical coordinates*, such that (r, y) are the Darboux coordinates and τ, ψ are the Killing coordinates.

D.7 Canonical Form of Metric

D.7.1 Off-shell canonical metric

Let us consider the following metric

$$ds^2 = \frac{1}{r^2 + y^2}\left[-\Delta_r(d\tau + y^2 d\psi)^2 + \Delta_y(d\tau - r^2 d\psi)^2\right]$$

$$+ (r^2 + y^2)\left[\frac{dr^2}{\Delta_r} + \frac{dy^2}{\Delta_y}\right], \tag{D.7.1}$$

where $\Delta_r = \Delta_r(r)$ and $\Delta_y = \Delta_y(y)$ are arbitrary functions.

Theorem: *The metric Eq. (D.7.1) has a closed conformal Killing–Yano tensor $h = db$. The corresponding potential b is*

$$b = \frac{1}{2}[(r^2 - y^2)\,d\tau + r^2 y^2\,d\psi]. \tag{D.7.2}$$

> *This statement can be checked by direct calculations, which are straightforward but rather long. Another, much more efficient way is to use a computer for analytical calculations. For example, the GRTensor package can be used to make the required calculations in a very short time.*

The form of the metric Eq. (D.7.1) was introduced by Debever (1971). Carter (1968b) demonstrated separation of variables in the Hamilton–Jacobi and Schrodinger equations in such a spacetime. For the discussion of the closed conformal Killing–Yano tensor for the metric Eq. (D.7.1) (see (Frolov and Kubizňák 2008; Kubiznak 2008) and references therein.)

We collect below some formulas for geometrical objects calculated for the metric Eq. (D.7.1). The required calculations are straightforward, but some of them are really long. The formulas below are obtained by using GRTensor package developed for the analytical calculations of the geometrical objects in a curved spacetime (see page 244). In particular, the determinant g of the metric Eq. (D.7.1) is

$$g = -(r^2 + y^2)^2. \tag{D.7.3}$$

The closed conformal Killing–Yano tensor h and its Hodge dual Killing tensor $k = *h$ are

$$h = h_{\mu\nu}\,dx^\mu \wedge dx^\nu = -y dy \wedge (d\tau - r^2 d\psi) + r dr \wedge (d\tau + y^2 d\psi),$$

$$k = k_{\mu\nu}\,dx^\mu \wedge dx^\nu = -r dy \wedge (d\tau - r^2 d\psi) - y dr \wedge (d\tau + y^2 d\psi). \tag{D.7.4}$$

The tensor H and the Killing tensor K are

$$H^\mu{}_\nu = \begin{pmatrix} r^2 - y^2 & 0 & 0 & r^2 y^2 \\ 0 & r^2 & 0 & 0 \\ 0 & 0 & -y^2 & 0 \\ 1 & 0 & 0 & 0 \end{pmatrix}, \tag{D.7.5}$$

$$K^{\mu}{}_{\nu} = \begin{pmatrix} 0 & 0 & 0 & r^2 y^2 \\ 0 & y^2 & 0 & 0 \\ 0 & 0 & -r^2 & 0 \\ 1 & 0 & 0 & -(r^2 - y^2) \end{pmatrix}.$$ (D.7.6)

Here, tensor indices are ordered in accordance with (τ, r, y, ψ) choice of the coordinates. The One also has $H^{\mu}{}_{\mu} = 2(r^2 - y^2)$.

D.7.2 Darboux basis for canonical metric

The characteristic equation for the eigenvalues of H is

$$\det(H - \lambda I) = 0.$$ (D.7.7)

Here, I is a unit matrix. The equation Eq. (D.7.7) has the form

$$(\lambda - r^2)^2 (\lambda + y^2)^2 = 0.$$ (D.7.8)

Thus, the eigenvalues of H are r^2 and $-y^2$, and (r, y) are the *Darboux coordinates*.
A corresponding *normalized Darboux basis* is

$$n_{1\mu} dx^{\mu} = \sqrt{\frac{\Delta_r}{r^2 + y^2}} (d\tau + y^2 d\psi), \qquad \bar{n}_{1\mu} dx^{\mu} = -\sqrt{\frac{r^2 + y^2}{\Delta_r}} dr,$$

$$n_{2\mu} dx^{\mu} = \sqrt{\frac{\Delta_y}{r^2 + y^2}} (d\tau - r^2 d\psi), \qquad \bar{n}_{2\mu} dx^{\mu} = -\sqrt{\frac{r^2 + y^2}{\Delta_y}} dy.$$ (D.7.9)

The primary and secondary Killing vectors are

$$\xi^{\mu} \partial_{\mu} = \partial_{\tau}, \qquad \eta^{\mu} \partial_{\mu} = \partial_{\psi}.$$ (D.7.10)

The coordinates τ and ψ are corresponding Killing coordinates for these vectors.
Note that Eqs (D.7.2)–(D.7.6) do not contain the functions $\Delta_r(r)$ and $\Delta_y(y)$. That is, these equations are valid even if $\Delta_r(r)$ and $\Delta_y(y)$ describe the metrics that do not fulfill the Einstein equations. We call the metric Eq. (D.7.1) with arbitrary functions Δ_r and Δ_y the *off-shell metric*, to distinguish it from an *on-shell metric* satisfying the Einstein equations.

D.7.3 On-shell canonical metric

We discuss now conditions on the metric functions $\Delta_r(r)$ and $\Delta_y(y)$ imposed by the vacuum Einstein equations with the cosmological constant

$$R_{\mu\nu} = \Lambda g_{\mu\nu}.$$ (D.7.11)

We use the GRTensor program to calculate the components of the Ricci tensor for the metric Eq. (D.7.1). We consider first the trace equation

$$R = 4\Lambda,$$ (D.7.12)

which takes a very simple form

$$\partial_r^2 \Delta_r + \partial_y^2 \Delta_y = -4\Lambda(r^2 + y^2). \tag{D.7.13}$$

This equation allows a separation of variables

$$\partial_r^2 \Delta_r + 4\Lambda r^2 = C, \qquad \partial_y^2 \Delta_y + 4\Lambda y^2 = -C. \tag{D.7.14}$$

A solution of each of these two equations contains 2 independent integration constants. Thus, together with C one has 5 integration constants. Note that the metric Eq. (D.7.1) remains invariant under the following rescaling

$$r \to pr, \quad y \to py, \quad \tau \to p^{-1}\tau, \quad \psi \to p^{-3}\psi, \quad \Delta_r \to p^4 \Delta_r, \quad \Delta_y \to p^4 \Delta_y. \tag{D.7.15}$$

This means that one of the 5 integration constants can be excluded by means of these transformations. We write the answer in the following standard form

$$\Delta_r = (r^2 + a^2)(1 - \Lambda r^2/3) - 2Mr,$$
$$\Delta_y = (a^2 - y^2)(1 + \Lambda y^2/3) + 2Ny. \tag{D.7.16}$$

The four parameters in these functions are Λ, M, N, and a. For $\Lambda = 0$ and $N = 0$ this metric coincides with the Kerr metric, M and a being the mass and the rotation parameter, respectively. In addition to these two parameters, a general solution Eq. (D.7.16) contains the cosmological constant Λ, and a so-called NUT parameter N. Solutions with a non-trivial N contain singularities in the black hole exterior. For this reason we do not consider them here. In the general case, a solution with parameters M, a, and Λ describes a rotating black hole in the asymptotically de Sitter (for $\Lambda > 0$), anti-de Sitter (for $\Lambda < 0$), or flat (for $\Lambda = 0$) spacetime. A similar solution containing the NUT parameter N is known as *Kerr-NUT-(A)dS spacetime*. Direct calculation shows that for the choice Eq. (D.7.16) of the metric functions the Einstein equations are satisfied.

The metric for the Kerr-NUT-(A)dS spacetime can be written in a more symmetric form. Let us denote $r = iz$, $N = Ny$, $M = iN_z$,

$$\Delta_z = (a^2 - z^2)(1 + \Lambda z^2/3) + 2N_z z,$$
$$\Delta_y = (a^2 - y^2)(1 + \Lambda y^2/3) + 2N_y y. \tag{D.7.17}$$

Then, the Kerr-NUT-(A)dS metric Eq. (D.7.1) takes the form

$$ds^2 = \frac{1}{y^2 - z^2} \left[\Delta_y (d\tau + z^2 d\psi)^2 - \Delta_z (d\tau + y^2 d\psi)^2 \right]$$
$$+ (y^2 - z^2) \left[\frac{dy^2}{\Delta_y} - \frac{dz^2}{\Delta_z} \right]. \tag{D.7.18}$$

This metric is symmetric with respect to the formal substitution $z \leftrightarrow y$.

D.8 Separation of Variables in Canonical Coordinates

A massive *Klein–Gordon equation* and the *Hamilton–Jacobi equation* allow a separation of variables in the off-shell canonical metric Eq. (D.7.1). To demonstrate this we calculate the inverse metric to Eq. (D.7.1). Using this metric one can write

$$\sqrt{-g}g^{\mu\nu}\partial_\mu\partial_\nu = \Delta_r(\partial_r)^2 + \Delta_y(\partial_y)^2 + \Delta_y^{-1}(y^2\partial_\tau - \partial_\psi)^2 - \Delta_r^{-1}(r^2\partial_\tau + \partial_\psi)^2. \quad (D.8.1)$$

The *Klein–Gordon equation*

$$\Box\Phi - \mu^2\Phi = 0 \quad (D.8.2)$$

in the canonical coordinates takes the form

$$\sqrt{-g}(\Box - \mu^2)\Phi = \partial_r(\Delta_r\partial_r\Phi) + \partial_y(\Delta_y\partial_y\Phi)$$
$$-\frac{1}{\Delta_r}(r^2\partial_\tau + \partial_\psi)^2\Phi + \frac{1}{\Delta_y}(y^2\partial_\tau - \partial_\psi)^2\Phi - \mu^2(r^2 + y^2)\Phi = 0. \quad (D.8.3)$$

To demonstrate that this equation allows a separation of variables we write

$$\Phi = e^{-i\omega\tau}e^{i\Psi\psi}R(r)Y(y). \quad (D.8.4)$$

Substitution of this expression into Eq. (D.8.3) gives

$$\partial_r(\Delta_r\partial_r R) + \frac{1}{\Delta_r}(r^2\omega - \Psi)^2 R - \mu^2 r^2 R + \lambda R = 0,$$
$$\partial_y(\Delta_y\partial_y Y) - \frac{1}{\Delta_y}(y^2\omega + \Psi)^2 Y - \mu^2 y^2 Y - \lambda Y = 0. \quad (D.8.5)$$

The parameter λ is a separation constant.
Similarly, the *Hamilton–Jacobi equation*

$$g^{\mu\nu}\partial_\mu S\partial_\nu S + \mu^2 = 0 \quad (D.8.6)$$

can be separated. Indeed, let us write

$$S = -\omega\tau + \Psi\psi + \mathcal{R}(r) + \mathcal{Y}(y). \quad (D.8.7)$$

After multiplying Eq. (D.8.6) by $\sqrt{-g}$ and substituting Eq. (D.8.7) one obtains

$$\Delta_r(\partial_r\mathcal{R})^2 - \Delta_r^{-1}(r^2\omega - \Psi)^2 + \mu^2 r^2 + \lambda = 0,$$
$$\Delta_y(\partial_y\mathcal{Y})^2 + \Delta_y^{-1}(y^2\omega + \Psi)^2 + \mu^2 r^2 - \lambda = 0. \quad (D.8.8)$$

D.9 Higher-Dimensional Generalizations

D.9.1 Principal conformal Killing–Yano tensor

In the models with large extra dimensions there exist a variety of *black objects*, which are generalizations of the four-dimensional black holes. We discuss them in Section 10.6.2. These black objects differ by the topology of the event horizon. As in four dimensions, the horizon

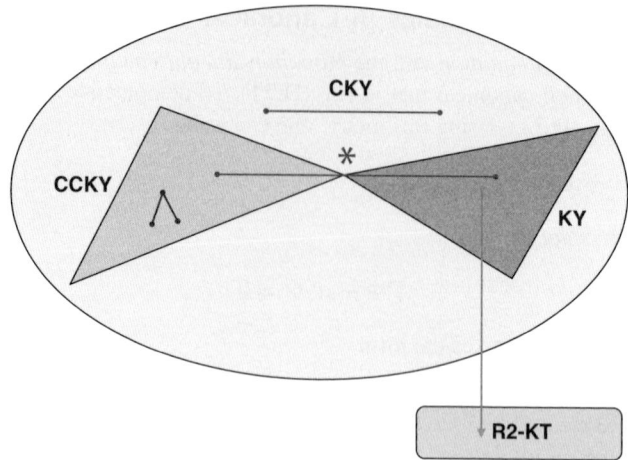

Fig. D.1 Schematical illustration of the properties of conformal Killing-Yano tensors. Points inside a large oval correspond to conformal Killing-Yano tensors. Horizontal lines connect two conformal Killing-Yano tensors, related by the Hodge–duality transformation. Two triangles represent closed conformal Killing–Yano tensors and Killing-Yano tensors. The Hodge, duality transformation gives a map between these objects. An *exterior product* maps two closed conformal Killing-Yano tensors to another closed conformal Killing–Yano tensor. A 'square' of a Killing-Yano tensor gives a rank-2 Killing tensor.

of a higher-dimensional black object is a null surface. A *black hole surface*, that is a spatial slice of the horizon, is a compact space-like surface of the dimension $D-2$, where D is the dimension of the spacetime. We focus here on the black objects with the topology of the horizon surface S^{D-2}. We call them *black holes*. The most general known solution of the higher-dimensional Einstein equations for rotating higher-dimensional black holes was obtained in (Chen *et al.* 2007). This, so-called, higher-dimensional *Kerr-NUT-(A)dS metric* is quite complicated. The remarkable fact is that this solution in many aspects is quite similar to its four-dimensional 'cousin'. The main reason for this is that it possesses a *non-degenerate closed conformal Killing–Yano tensor* (Frolov and Kubizňák 2007; Kubiznak and Frolov 2007). Here, we briefly summarize important properties of these solutions. More detailed information and references can be found, e.g., in (Frolov and Kubizňák 2008).

We denote the number of dimensions of the spacetime by $D = 2n + \varepsilon$, where $\varepsilon = 0$ for even dimensions, and $\varepsilon = 1$ for odd dimensions. The basic objects responsible for the hidden symmetry are, Killing–Yano tensor and conformal Killing-Yano tensor. They are antisymmetric tensors, or forms. To describe their properties it is very convenient to use the 'language' of differential forms (see Section 3.1.4). The exterior (or wedge) product of p-form $\boldsymbol{\alpha}_p$ and q-form $\boldsymbol{\beta}_q$ is a $(p+q)$-form $\boldsymbol{\alpha}_p \wedge \boldsymbol{\beta}_q$. The *Hodge dual* of a p-form $\boldsymbol{\alpha}_p$ is a $(D-p)$-form $(\ast\boldsymbol{\alpha})_{D-p}$. An *exterior derivative* of the p-form $\boldsymbol{\alpha}_p$ is a $(p+1)$-form $d\boldsymbol{\alpha}_p$.

A conformal Killing–Yano tensor in D dimensions is defined by the same equation Eq. (D.2.1) as in, 4-dimensional case. For the *closed conformal Killing–Yano tensor* the first term in the right-hand side of this equation vanishes. The conformal Killing–Yano tensor in the D-dimensional spacetime possesses the following properties:

1. The Hodge dual of a conformal Killing–Yano tensor is a conformal Killing-Yano tensor.
2. The Hodge dual of a closed conformal Killing–Yano tensor is a Killing-Yano tensor.
3. The Hodge dual of a Killing–Yano tensor is a closed conformal Killing–Yano tensor.
4. An exterior product of two closed conformal Killing–Yano tensors is a closed conformal Killing–Yano tensor.

Consider a 2-form h obeying the equation

$$h_{\mu\nu;\lambda} = g_{\lambda\nu}\xi_{\mu} - g_{\lambda\mu}\xi_{\nu}\,. \tag{D.9.1}$$

This equation implies

$$h_{[\mu\nu;\lambda]} = 0\,, \qquad \xi_{\mu} = \frac{1}{D-1}h_{\mu}{}^{\lambda}{}_{;\lambda}. \tag{D.9.2}$$

This object is a closed conformal Killing–Yano tensor. We assume that it is non-degenerate, that is it matrix rank is $2n$. Since this object plays a fundamental role in study of the hidden symmetries, we call it a *principal conformal Killing–Yano tensor* .

D.9.2 Killing–Yano tower

Starting with a principal conformal Killing–Yano tensor h one can construct a set of new closed conformal Killing–Yano tensors

$$h^{\wedge j} = \underbrace{h \wedge h \wedge \ldots \wedge h}_{j\text{times}}\,. \tag{D.9.3}$$

The index j enumerates how many hs are in the exterior product, so that $h^{\wedge 1} = h$. For $j = n$ the object $h^{\wedge n}$ is either proportional to the totally antisymmetric tensor (for $\varepsilon = 0$), or its dual is a vector (for $\varepsilon = 1$). Excluding these 'trivial' cases, one has $(n-1)$ non-trivial closed conformal Killing–Yano tensors. The Hodge dual of these tensors

$$k^{j} = *h^{\wedge j}, \tag{D.9.4}$$

are $(n-1)$ Killing–Yano tensors, which can be used to construct $(n-1)$ Killing tensors

$$K^{j} = k^{j} \bullet k^{j}. \tag{D.9.5}$$

The latter notation means that we use the operation described on page 415. Since the metric g is a (trivial) Killing tensor, this construction shows that in a spacetime with a *principal conformal Killing–Yano tensor* there exist n integrals of motion for geodesic equations, that are quadratic in momentum.

Let us show that in a general case this spacetime also has $(n + \varepsilon)$ Killing vectors. The first of this set is a *primary Killing vector* ξ defined in Eq. (D.9.2). This vector obeys the relation

$$\xi_{(\mu;\nu)} = -\frac{1}{D-2}R^{\alpha}{}_{(\nu}h_{\mu)\alpha}. \tag{D.9.6}$$

In an *Einstein space*, which is a solution of the equation

$$R_{\mu\nu} = \frac{2}{D-2}\Lambda\, g_{\mu\nu},$$

(D.9.7)

the relation Eq. (D.9.6) implies that $\xi_{(\mu;\nu)} = 0$ is a Killing vector.[1]

It is also possible to show that the following objects constructed from the primary Killing vector

$$\xi_\mu^j = K_{\mu\nu}^j \xi^\nu$$

(D.9.8)

are, in fact, again Killing vectors. In the odd-dimensional spacetime there exists one more Killing vector that can be obtained as a Hodge dual of $h^{\wedge n}$. We call these objects *secondary Killing vectors*. The total number of the (primary plus secondary) Killing vectors is $n + \varepsilon$. This gives $(n + \varepsilon)$ additional integrals for geodesic motion, that are of the first order in momentum.

Thus, in a spacetime with a *principal conformal Killing–Yano tensor* geodesic equations have

$$n + (n + \varepsilon) = 2n + \varepsilon = D$$

(D.9.9)

integrals of motion. It is possible to show that they are in involution and (generically) they are functionally independent. This provides the complete integrability of geodesic equations in a spacetime with a principal conformal Killing-Yano tensor. This is a direct generalization of the similar result for the four-dimensional Kerr metric. It should be emphasized that this gives a new important example of a physically interesting dynamical system, which allows the *complete integrability*.

D.10   Higher-Dimensional Kerr-NUT-(A)dS Metric

In a spacetime with symmetry generated by Killing vectors one can find special coordinates, in which the form of the metric is simplified. In such coordinates the metric coefficients are functions of fewer variables and their number is less than in the general case. Spacetimes with a *hidden symmetry* have a similar property. In particular, the existence of a principal conformal Killing–Yano tensor h imposes constraints on the geometry.

To describe a canonical form of the metric, in which the hidden symmetries generated by the *Killing–Yano tower* induced by h become most transparent, we define first a *Darboux basis*. Consider an orthonormal basis $(e_a, e_{\bar{a}}, e_0)$. The last vector, e_0, is present only when $\varepsilon = 1$, that is the number of spacetime dimensions is odd. The indices a and $\bar{a}$, which enumerate the vectors, take values $1, 2, \ldots, n$. Denote by $(\omega^a, \omega^{\bar{a}}, \omega^0)$ a dual basis of 1-forms. It is possible to choose the basis in such a way, that

$$g_{\mu\nu} = \sum_{a=1}^{n}(\omega_\mu^a \omega_\nu^a + \omega_\mu^{\bar{a}}\omega_\nu^{\bar{a}}) + \varepsilon\omega_\mu^0\omega_\nu^0,$$

$$h_{\mu\nu} = 2\sum_{a=1}^{n} x_a \omega_{[\mu}^a \omega_{\nu]}^{\bar{a}}.$$

(D.10.1)

[1] This property is also valid 'off-shell', that is even if the Einstein equations are not satisfied. The proof of this is, however, much more complicated (see, e.g, (Krtous *et al.* 2008; Houri *et al.* 2009)).

This is a special choice of the Darboux basis, which possesses an additonal property: in this basis the metric is also diagonal. We define $H^\mu{}_\nu = h^{\mu\lambda} h_{\lambda\nu}$. For fixed a two vectors e_a and $e_{\bar{a}}$ span a two-dimensional eigenspace of H, and the corresponding coordinate x_a is determined by the eigenvalue of H. Since h is non-degenerate, its eigenvalues x_a are functionally independent and can be used as Darboux coordinates on the spacetime manifold. By adding $(n + \varepsilon)$ Killing coordinates ψ_b one obtains the canonical coordinates. It is possible to show (Krtous *et al.* 2008; Houri *et al.* 2009), that in such (canonical) coordinates the metric takes the form

$$ds^2 = \sum_{a=1}^{n} \left[\frac{dx_a^2}{Q_a} + Q_a \left(\sum_{b=0}^{n-1} A_a^{(b)} d\psi_b \right)^2 \right] - \frac{\varepsilon c}{A^{(n)}} \left(\sum_{b=0}^{n} A^{(b)} d\psi_b \right)^2. \tag{D.10.2}$$

Here, $A^{(b)}$ and $A_a^{(b)}$ are polynomials in the Darboux coordinates x_a, defined by the following relations

$$\prod_{a=1}^{n} (1 + \lambda x_a^2) = \sum_{b=0}^{n} \lambda^b A^{(b)},$$

$$(1 + \lambda x_a^2)^{-1} \prod_{a'=1}^{n} (1 + \lambda x_{a'}^2) = \sum_{b=0}^{n-1} \lambda^b A_a^{(b)}, \tag{D.10.3}$$

$$Q_a = \frac{X_a}{U_a}, \qquad U_a = \prod_{a' \neq a} (x_{a'}^2 - x_a^2).$$

Quantities X_a are functions of one variable, $X_a = X_a(x_a)$. Equations (D.10.2) and (D.10.3) give the most general form of the metric that allows a *principal conformal Killing–Yano tensor*.

In the case when the canonical metric is a solution of the D-dimensional Einstein equations, the functions X_a take the form

$$X_a = \sum_{k=\varepsilon}^{n} c_k x_a^{2k} - 2b_a x_a^{1-\varepsilon} + \frac{\varepsilon c}{x_a^2}. \tag{D.10.4}$$

With this choice of X_a the metric obeys Eq. (D.9.7) with

$$\Lambda = \frac{(D - 1)(D - 2)}{2} (-1)^n c_n. \tag{D.10.5}$$

The asymptotically flat spacetime is recovered when $c_n = 0$

Time is denoted by ψ_0, azimuthal (Killing) coordinates by ψ_k, $k = 1, \ldots, n - 1 + \varepsilon$, and x_a, $a = 1, \ldots, n$, stand for Darboux coordinates. The metric with proper Lorentzian signature is recovered from Eq. (D.10.2) after the analytical continuation of the radial coordinate $r = -ix_n$ and the redefinition of the mass parameter $M = (-i)^{1+\epsilon} b_n$.

The total number of constants that enter the solution is $2n + 1$: ε constants c, $n + 1 - \varepsilon$ constants c_k and n constants b_a. However, the form of the metric is invariant under a 1-parameter scaling coordinate transformations, thus the total number of independent parameters is $2n$. These parameters are related to the cosmological constant, mass, angular momenta, and NUT parameters. One of them, say Λ, may be used to define a scale, while the other $2n - 1$ parameters can be made dimensionless (Chen *et al.* 2007). Similar to the 4D case, the signature of the symmetric form of the metric depends on the domain of x_as and the signs of X_as.

In the four-dimensional spacetime the number of independent parameters in the solution is 4. They are the cosmological constant, mass, angular momentum and NUT parameter. When the cosmological constant and the NUT parameter vanish, after a proper analytical continuation one recovers the Kerr metric.

Appendix E
Boundary Term for the Einstein–Hilbert Action

E.1 An Example Illustrating the Problem

Let us consider a massless scalar field φ in a D-dimensional region V of the spacetime with the boundary $\partial V \equiv \Sigma$. The action that describes the dynamics of the field φ is of the form

$$S[\varphi] = -\frac{1}{2} \int_V (\nabla \varphi)^2 \sqrt{-g}\, d^D x. \tag{E.1.1}$$

Its variation is

$$\delta S[\varphi] = \int_V \delta \varphi \,\Box \varphi \sqrt{-g}\, d^D x - \int_V \nabla(\delta \varphi \,\nabla \varphi) \sqrt{-g}\, d^D x$$

$$= \int_V \delta \varphi \,\Box \varphi \sqrt{-g}\, d^D x - \int_\Sigma \delta \varphi \,\varphi_{;\mu}\, d\sigma^\mu. \tag{E.1.2}$$

If we fix the value of φ on the boundary, i.e. $\delta \varphi|_\Sigma = 0$, then the variational problem is well defined because the boundary term in Eq. (E.1.2) vanishes. Since the variation $\delta \varphi$ in the bulk is arbitrary we obtain the equation for the scalar field in the bulk

$$\Box \varphi = 0. \tag{E.1.3}$$

However, it is possible to modify the action by adding an extra boundary term to it. This does not change the variation of the action in the bulk. The difference will only be in the boundary variations. To give an example, let us consider the following action

$$S_\alpha[\varphi] = -\frac{1}{2} \int_V (\nabla \varphi)^2 \sqrt{-g}\, d^D x + \alpha \int_\Sigma \varphi \,\varphi_{;\mu}\, d\sigma^\mu. \tag{E.1.4}$$

The second term in the modified action can be written as a bulk integral of a total derivative. This gives another (equivalent) form of the action

$$S_\alpha[\varphi] = -\frac{1}{2} \int_V [(1 - 2\alpha)(\nabla \varphi)^2 - 2\alpha \varphi \Box \varphi] \sqrt{-g}\, d^D x. \tag{E.1.5}$$

The variation of Eq. (E.1.4) gives

$$\delta S_\alpha[\varphi] = \int_V \delta\varphi \, \Box\varphi \, \sqrt{-g} \, d^D x + (\alpha - 1) \int_\Sigma \delta\varphi \, \varphi_{;\mu} \, d\sigma^\mu + \alpha \int_\Sigma (\delta\varphi)_{;\mu} \, \varphi \, d\sigma^\mu. \qquad (E.1.6)$$

For $\alpha = 0$ the action Eq. (E.1.4) reproduces Eq. (E.1.1) and the variational problem is well defined for the *Dirichlet boundary condition* $\delta\varphi|_\Sigma = 0$. For $\alpha = 1$ the variational problem is also well defined, but for the Neumann boundary condition $n^\mu (\delta\varphi)_{;\mu}|_\Sigma = 0$. In the general case, in order to make the variation procedure self-consistent one needs to impose the following condition

$$[(\alpha - 1)\delta\varphi + \alpha n^\mu (\delta\varphi)_{;\mu}]_\Sigma = 0. \qquad (E.1.7)$$

The lesson of this example is that the boundary term in the action and the boundary conditions are related to each other and have to be consistent. *Natural boundary conditions* are those that imply the vanishing of boundary terms in the first variation of the action functional for a free problem, that is when functions are not subject to constraints on the boundary. As stated by Courant (1943) "Now the dominant fact is: *appropriate boundary conditions for differential equations are obtained as natural boundary conditions of corresponding variational problems.*"

E.2   Boundary Term for the Einstein–Hilbert action

The Einstein–Hilbert action contains second derivatives of the metric, and thus it is similar to the action (E.1.5) for the scalar field. It is more involved but the algorithm for treatment of the boundary terms is quite similar to the scalar case. In most physical problems the metric on the boundary of the manifold is assumed to be given, i.e. the Dirichlet problem for the metric is usually considered. In this case the variation of the metric at the boundary is set to be zero $\delta g_{\alpha\beta}|_\Sigma = 0$. To be more accurate it is the intrinsic $(D{-}1)$-metric on the boundary that is to be fixed rather than all the components of $g_{\alpha\beta}$. In order for the Dirichlet boundary conditions to be natural boundary conditions one has to determine the form of the boundary term in the action. This means that the normal derivatives of the metric variations at the boundary should not appear in the variation of the total action, including the surface contribution, because otherwise the variational problem will not be consistent.

In the variation of the Einstein action Eq. (5.2.11) only the term $\sqrt{-g}g^{\alpha\beta}\delta R_{\alpha\beta}$ contains second derivatives of the metric variation. After integration they will be boiled down to a surface integral that contains not higher than the first derivative variations. Written explicitly these surface terms are

$$\delta S[g] = \frac{c^3}{16\pi G} \int_\Sigma g^{\alpha\beta} \left(\delta g_{\mu\alpha;\beta} - \delta g_{\alpha\beta;\mu}\right) d\sigma^\mu + \dots. \qquad (E.2.1)$$

Here, ... denote terms proportional to $\delta g_{\alpha\beta}$ in the bulk and on the boundary. We have to add to the Einstein action Eq. (5.1.6) a surface integral, variation of which cancels the first term in Eq. (E.2.1). Let us determine this integral. On the surface Σ we have at hand the metric itself, the unit normal vector n^μ, and their first derivatives. Variations of other possible terms are of higher order in derivatives of $\delta g_{\alpha\beta}$, so we do not need to include them.

The obvious candidate for the surface integral, which is linear in derivatives, has the form

$$\int_\Sigma n^\alpha{}_{;\alpha} n^\mu \, d\sigma_\mu. \tag{E.2.2}$$

The orientation is defined by the requirement that n^μ be the unit outward-pointing vector field at the boundary. The surface element is

$$d\sigma^\mu = \varepsilon(n) \, n^\mu \sqrt{|h|} \, d^{D-1} y. \tag{E.2.3}$$

Here, y^i are the coordinates on the boundary, h_{ij} is the induced $(D-1)$ metric on the boundary, and $h = \det(h_{ij})$. We exclude a special case when the boundary surface is null, so that the normal vector n^μ can be either space-like $n^\alpha n_\alpha = +1$ or time-like $n^\alpha n_\alpha = -1$

$$\varepsilon(n) = n^\alpha n_\alpha = \pm 1. \tag{E.2.4}$$

The variations of n^μ and $d\sigma_\mu$ are

$$\delta n^\mu = \left(\frac{1}{2} \varepsilon(n) \, n^\mu \, n^\alpha - g^{\mu\alpha} \right) n^\beta \, \delta g_{\alpha\beta}, \qquad \delta(d\sigma_\mu) = \frac{1}{2} g^{\alpha\beta} \, \delta g_{\alpha\beta} \, d\sigma_\mu. \tag{E.2.5}$$

The first relation here trivially follows from the expression

$$n^\mu = g^{\mu\nu} f_{,\nu} / \sqrt{\varepsilon(n) \, g^{\alpha\beta} f_{,\alpha} f_{,\beta}}, \tag{E.2.6}$$

provided the equation $f(x) = \text{const}$ defines the boundary surface. Now consider the variation of $n^\alpha{}_{;\alpha}$

$$\delta(n^\alpha{}_{;\alpha}) = (\delta n^\alpha)_{;\alpha} + \delta \left(\Gamma^\alpha_{\alpha\beta} \right) n^\beta. \tag{E.2.7}$$

Taking into account that $d\sigma_\mu$ is proportional to n^μ one can show that

$$\begin{aligned}
\int_\Sigma \delta \left(n^\alpha{}_{;\alpha} n^\mu \, d\sigma_\mu \right) &= -\frac{1}{2} \varepsilon(n) \int_\Sigma \left[g^{\alpha\beta} \delta g_{\alpha\mu;\beta} + (g^{\alpha\beta} - \varepsilon(n) n^\alpha n^\beta) \delta g_{\alpha\mu;\beta} \right] d\sigma^\mu \\
&\quad + \varepsilon(n) \int_\Sigma \delta(\Gamma^\alpha_{\alpha\beta}) \, d\sigma^\beta + \dots \\
&= -\frac{1}{2} \varepsilon(n) \int_\Sigma (g^{\alpha\beta} - \varepsilon(n) n^\alpha n^\beta) \delta g_{\alpha\mu;\beta} \, d\sigma^\mu + \dots \\
&\quad - \frac{1}{2} \varepsilon(n) \int_\Sigma g^{\alpha\beta} (\delta g_{\mu\alpha;\beta} - \delta g_{\alpha\beta;\mu}) \, d\sigma^\mu.
\end{aligned} \tag{E.2.8}$$

One can see that the last term in this expression, when taken with the proper coefficient, is exactly the same as the variation Eq. (E.2.1). Thus, by subtracting it one can 'kill' the dangerous terms containing normal derivatives of the metric. The other term in this expression, which accompanies it, survives on the boundary, and its vanishing requires that tangential derivatives of the metric variation have to be zero. This is evident, since the combination

$$p_{\alpha\beta} = g_{\alpha\beta} - \varepsilon(n) \, n_\alpha \, n_\beta \tag{E.2.9}$$

is the *projection operator* onto the boundary

$$p_{\alpha\epsilon} n^{\epsilon} = 0, \qquad p_{\alpha\epsilon} p^{\epsilon}{}_{\beta} = p_{\alpha\beta}. \qquad (\text{E.2.10})$$

We can rewrite Eq. (E.2.2) in terms of the extrinsic curvature using the relation

$$n^{\alpha}{}_{;\alpha} = -K. \qquad (\text{E.2.11})$$

Here, K is the trace of the *extrinsic curvature*

$$K_{\alpha\beta} = -p^{\mu}{}_{\alpha} p^{\nu}{}_{\beta} n_{\nu;\mu}, \qquad K = p^{\alpha\beta} K_{\alpha\beta}. \qquad (\text{E.2.12})$$

Combining all these formulas together we finally obtain the proper *Einstein action with boundary terms* (York 1972, 1986; Gibbons and Hawking 1977), corresponding to the variational problem when the metric variations and the tangential derivatives of the metric variations are zero on the boundary, while normal derivatives of the metric variations are not fixed

$$
\begin{aligned}
S[\,g\,] &= \frac{c^3}{16\pi G} \left(\int_V d^D x \sqrt{-g}\,[R - 2\Lambda] - 2\sum_B \varepsilon_B(n) \int_\Sigma K\,n^\mu\,d\sigma_\mu \right) \\
&= \frac{c^3}{16\pi G} \left(\int_V d^D x \sqrt{-g}\,[R - 2\Lambda] - 2\sum_B \varepsilon_B(n) \int_\Sigma d^{D-1} y \sqrt{|h|}\,K \right).
\end{aligned}
\qquad (\text{E.2.13})
$$

Here, $\sum_B$ means the sum over all time-like and space-like boundaries of the region V with the appropriate $\varepsilon_B(n) = \pm 1$. The surface integral in this action is called the *Gibbons–Hawking–York boundary term*.

Note that in the literature one can find different signs for the boundary term in Eq. (E.2.13). This apparent discrepancy comes from different sign conventions in the definitions of the extrinsic curvature (see Eqs. (3.8.9), and (E.2.11)). We accepted here the (Misner *et al.* 1973) sign conventions.

E.3   Boundary Term for the Euclidean Einstein–Hilbert Action

In the Euclidean spacetime the calculations are practically the same. The only differences are:

- The normal to Σ is always space-like and hence $\varepsilon(n) = 1$.
- The total sign in the Euclidean bulk action is opposite.

The *Euclidean Einstein–Hilbert action* with the boundary term is

$$
\begin{aligned}
S_E[\,g\,] &= -\frac{c^3}{16\pi G} \left(\int_V d^D x \sqrt{g}\,[R - 2\Lambda] - 2\int_\Sigma K\,n^\mu\,d\sigma_\mu \right) \\
&= -\frac{c^3}{16\pi G} \left(\int_V d^D x \sqrt{g}\,[R - 2\Lambda] - 2\int_\Sigma d^{D-1} y \sqrt{h}\,K \right).
\end{aligned}
\qquad (\text{E.3.1})
$$

Appendix F
Quantum Fields

F.1 Classical Oscillator

F.1.1 Lagrangian formalism

The action for a classical oscillator of mass m and frequency ω is

$$S = \int L \, dt, \qquad L = \frac{m}{2} \left(\dot{q}^2 - \omega^2 q^2 \right). \tag{F.1.1}$$

The variation of the action with the fixed end points gives the equation of motion

$$\ddot{q} + \omega^2 q = 0. \tag{F.1.2}$$

A solution of this equation for the initial conditions $q(0) = q_0$, $\dot{q}(0) = \dot{q}_0$ is

$$q(t) = q_0 \cos \omega t + \frac{\dot{q}_0}{\omega} \sin \omega t. \tag{F.1.3}$$

In the comparison of the quantum and classical theory an important role is played by the action as a function of its end points. Suppose the initial point at the time t_1 is q_1 and the final point at the time t_2 is q_2. The value of the action calculated for the solution Eq. (F.1.3) obeying these boundary conditions is

$$S[q_1, q_2] = \frac{m\omega}{2 \sin(\omega \Delta t)} \left[\left(q_1^2 + q_2^2 \right) \cos(\omega \Delta t) - 2 q_1 \, q_2 \right], \tag{F.1.4}$$

where $\Delta t = t_2 - t_1$.

F.1.2 Hamiltonian formalism

Denote

$$p = \frac{\partial L}{\partial \dot{q}} = m \dot{q}, \tag{F.1.5}$$

then the Hamiltonian of the system is

$$H = \left[\dot{q} p - L \right]_{\dot{q} = \dot{q}(p)} = \frac{p^2}{2m} + \frac{m\omega^2 q^2}{2}. \tag{F.1.6}$$

In this formalism $\{p, q\}$ are considered as coordinates in a 2D *phase space*. For any two functions on the phase space, $A = A(p, q)$ and $B = B(p, q)$, one defines the *Poisson brackets*

$$\{A, B\} = \frac{\partial A}{\partial p} \frac{\partial B}{\partial q} - \frac{\partial A}{\partial q} \frac{\partial B}{\partial p}. \tag{F.1.7}$$

It is easy to check that

$$\{p, q\} = 1, \quad \{q, q\} = \{p, p\} = 0. \tag{F.1.8}$$

The oscillator equations in the Hamiltonian form are

$$\dot{q} = \{H, q\} = p/m, \qquad \dot{p} = \{H, p\} = -m\omega^2 q. \tag{F.1.9}$$

This system is equivalent to Eq. (F.1.2) and its solution, with the initial data (p_0, q_0), is

$$q = q_0 \cos(\omega t) + \frac{p_0}{m\omega} \sin(\omega t),$$

$$p = -m\omega q_0 \sin(\omega t) + p_0 \cos(\omega t). \tag{F.1.10}$$

F.1.3   Euclidean oscillator

In the quantum theory the Euclidean version of quantum oscillators is quite often used. In particular, this allows one to establish general relations between quantum observables in the thermal state and the Euclidean expectation values. To perform a transformation to the *Euclidean space* we use the following *Wick's rotation* of time t

$$t = i\tau. \tag{F.1.11}$$

After making this analytical continuation one considers τ as a real parameter, or the *Euclidean time*. We define the action for the *Euclidean oscillator* as follows:

$$S_{\rm E} = iS|_{t=i\tau} = \frac{m}{2} \int d\tau \left[\left(\frac{dq}{d\tau} \right)^2 + \omega^2 q^2 \right]. \tag{F.1.12}$$

The variation of this action with the fixed end points gives the equation

$$\frac{d^2 q}{d\tau^2} - \omega^2 q = 0. \tag{F.1.13}$$

A solution for the 'initial' data $q(\tau_0) = q_0$, $\frac{dq}{d\tau}(\tau_0) = \dot{q}_0$ is

$$q(\tau) = q_0 \cosh[\omega(\tau - \tau_0)] + \frac{\dot{q}_0}{\omega} \sinh[\omega(\tau - \tau_0)]. \tag{F.1.14}$$

The Euclidean action as a function of end points is

$$S_{\rm E}(q_1, q_2) = \frac{m\omega}{2} \left[\frac{\cosh(\omega \Delta \tau)}{\sinh(\omega \Delta \tau)} \left(q_1^2 + q_2^2 \right) - \frac{2 q_1 q_2}{\sinh(\omega \Delta \tau)} \right]. \tag{F.1.15}$$

Here, $\Delta \tau = \tau_2 - \tau_1$.

F.2 Quantum Oscillator

F.2.1 Heisenberg picture

For a *quantum oscillator*, q and p are operators obeying the following *canonical commutation relations*

$$[\hat{q}, \hat{p}] = \hat{q}\hat{p} - \hat{p}\hat{q} = i\hbar, \qquad [\hat{q}, \hat{q}] = [\hat{p}, \hat{p}] = 0. \tag{F.2.1}$$

The *quantum Hamiltonian* $\hat{H}$ is defined as

$$\hat{H} = \frac{\hat{p}^2}{2m} + \frac{m\omega^2 \hat{q}^2}{2}. \tag{F.2.2}$$

In the *Heisenberg picture* $\hat{q}$ and $\hat{p}$ are time-dependent operators and their evolution is determined by the following equations of motion:

$$\dot{\hat{q}} = \frac{i}{\hbar}[\hat{H}, \hat{q}] = \frac{\hat{p}}{m}, \qquad \dot{\hat{p}} = \frac{i}{\hbar}[\hat{H}, \hat{p}] = m\omega^2 \hat{q}. \tag{F.2.3}$$

These relations imply that any operator $\hat{A}$ that is a function of $\hat{q}$ and $\hat{p}$ obeys a similar equation

$$\dot{\hat{A}} = \frac{i}{\hbar}[\hat{H}, \hat{A}]. \tag{F.2.4}$$

Since $\hat{H}$ does not depend on time one can write a solution of this equation in the form

$$\hat{A}(t) = e^{i\hat{H}t/\hbar} \hat{A}(0) \, e^{-i\hat{H}t/\hbar}. \tag{F.2.5}$$

It is convenient to rescale q and p

$$q = \tilde{q}/\sqrt{m}, \qquad p = \sqrt{m}\tilde{p}. \tag{F.2.6}$$

Working with the new variables $\tilde{q}$ and $\tilde{p}$ is equivalent to taking the mass parameter m equal to 1. From now on we shall use the rescaled variables and omit the tilde. In what follows we shall also put the Planck constant $\hbar$ equal to 1.

F.2.2 Operators of creation and annihilation

Let us write the operators $\hat{q}(t)$ and $\hat{p}(t)$ in the form

$$\hat{q} = \frac{1}{\sqrt{2\omega}}(\hat{a}(t) + \hat{a}^\dagger(t)), \qquad \hat{p} = -i\sqrt{\frac{\omega}{2}}(\hat{a}(t) - \hat{a}^\dagger(t)). \tag{F.2.7}$$

Since $\hat{q}$ and $\hat{p}$ are *Hermitian operators* the operator $\hat{a}^\dagger$ is the *Hermitian conjugated operator* to $\hat{a}$. Using Eq. (F.1.10) and Eq. (F.2.5) one gets

$$\hat{a}(t) = e^{i\hat{H}t} \hat{a} \, e^{-i\hat{H}t} = e^{-i\omega t}\,\hat{a}, \quad \hat{a}^\dagger(t) = e^{i\hat{H}t} \hat{a}^\dagger \, e^{-i\hat{H}t} = e^{+i\omega t}\,\hat{a}^\dagger. \tag{F.2.8}$$

Here, constant operators $\hat{a}$ and $\hat{a}^\dagger$ are the corresponding initial data. These operators obey the following commutation relations:

$$[\hat{a}, \hat{a}^\dagger] = 1, \qquad [\hat{a}, \hat{a}] = 0, \qquad [\hat{a}^\dagger, \hat{a}^\dagger] = 0. \tag{F.2.9}$$

They are known as the *operator of creation*, $\hat{a}^\dagger$, and the *operator of annihilation*, $\hat{a}$. The Hamiltonian $\hat{H}$ of the quantum oscillator, written in terms of the operators of creation and annihilation, takes the form

$$\hat{H} = \omega\,\hat{a}^\dagger\hat{a} + \omega/2. \tag{F.2.10}$$

In the *occupation number representation* the basis in the *Hilbert space* of states is

$$|n\rangle = \frac{1}{\sqrt{n!}}\left(\hat{a}^\dagger\right)^n |0\rangle. \tag{F.2.11}$$

This basis is orthonormal

$$\langle n|m\rangle = \delta_{nm}. \tag{F.2.12}$$

One also has

$$\hat{a}|n\rangle = \sqrt{n}\,|n-1\rangle, \qquad \hat{a}^\dagger|n\rangle = \sqrt{n+1}\,|n+1\rangle,$$
$$\hat{N}|n\rangle = n|n\rangle, \qquad \hat{N} = \hat{a}^\dagger\hat{a}. \tag{F.2.13}$$

The state $|0\rangle$ has the lowest possible energy $\omega/2$. It is called a *ground state* or a *vacuum*.

F.2.3 Thermal quantum oscillator

Let Θ be a temperature. We denote by $\beta = 1/\Theta$ the *inverse temperature* of the system. A thermal state in quantum mechanics is described by a *thermal density matrix*

$$\hat{\rho} = Z(\beta)^{-1}e^{-\beta\hat{H}}. \tag{F.2.14}$$

The *partition function* $Z(\beta)$ is determined by the normalization condition

$$\mathrm{Tr}\,\hat{\rho} = 1, \tag{F.2.15}$$

that is by the following expression

$$Z(\beta) = \mathrm{Tr}\,e^{-\beta\hat{H}}. \tag{F.2.16}$$

The partition function for the quantum oscillator can be easily calculated using the fact that $|n\rangle$ are eigenvectors of the Hamiltonian $\hat{H}$ with the eigenvalue equal to $(n + 1/2)\omega$. One has

$$\mathrm{Tr}\,e^{-\beta\hat{H}} = \sum_{n=0}^{\infty}\langle n|e^{-\beta\hat{H}}|n\rangle = e^{-\beta\omega/2}\sum_{n=0}^{\infty}e^{-\beta\omega n} = e^{-\beta\omega/2}\left(1 - e^{-\beta\omega}\right)^{-1}. \tag{F.2.17}$$

The partition function determines the *free energy* F defined by the relation

$$Z(\beta) = e^{-\beta F}. \tag{F.2.18}$$

All the thermodynamical characteristics of the *canonical ensemble* can be obtained from its free energy. For a thermal quantum oscillator the free energy is

$$F = \beta^{-1}\ln\left(1 - e^{-\beta\omega}\right) + \omega/2. \tag{F.2.19}$$

The temperature-independent term $\omega/2$ in the free energy is a result of the presence of the *vacuum energy* $\omega/2$ term in the expression for the Hamiltonian Eq. (F.2.10). We can rewrite $\hat{H}$ in the form

$$\hat{H} = \hat{H}_0 + \omega/2, \qquad \hat{H}_0 = \omega\,\hat{a}^\dagger\hat{a}. \qquad \text{(F.2.20)}$$

We use the following standard notation for the *thermal average* of an operator $\hat{A}$

$$\langle\hat{A}\rangle_\beta = Z(\beta)^{-1}\mathrm{Tr}\,(e^{-\beta\hat{H}}\hat{A}). \qquad \text{(F.2.21)}$$

The expression for the thermal average can also be written in the form

$$\langle\hat{A}\rangle_\beta = Z_0(\beta)^{-1}\mathrm{Tr}\,(e^{-\beta\hat{H}_0}\hat{A}),$$

$$Z_0(\beta) = e^{-\beta F_0} = \mathrm{Tr}\,e^{-\beta\hat{H}_0}, \qquad F_0 = F - \omega/2. \qquad \text{(F.2.22)}$$

F.2.4   Green functions

There is a variety of different Green functions or two-point correlators, which proved to be useful in quantum physics. Usually, they have a form of the average value of bilinear combinations of $\hat{q}(t)$ and $\hat{q}(t')$. They differ by the choice of the state and the concrete form of the bilinear form, which may involve, for example, special ordering of the operators. We focus now our attention on the following thermal correlators

$$\mathcal{G}_\beta^+(t,t') = \langle\hat{q}(t)\,\hat{q}(t')\rangle_\beta,$$

$$\mathcal{G}_\beta^-(t,t') = \langle\hat{q}(t')\,\hat{q}(t)\rangle_\beta = \mathcal{G}_\beta^+(t',t). \qquad \text{(F.2.23)}$$

Let us calculate $\mathcal{G}_\beta^+(t,t')$. One has

$$\mathcal{G}_\beta^+(t,t') = \frac{1}{Z_0}\mathrm{Tr}\left[e^{-\beta\hat{H}_0}\hat{q}(t)\hat{q}(t')\right] = \frac{1}{Z_0}\sum_{n=0}^{\infty}e^{-n\beta\omega}\langle n|\hat{q}(t)\hat{q}(t')|n\rangle. \qquad \text{(F.2.24)}$$

Using Eq. (F.2.7), Eqs. (F.2.8), and (F.2.13) one obtains

$$\langle n|\hat{q}(t)\hat{q}(t')|n\rangle = \frac{1}{2\omega}\left[e^{-i\omega(t-t')}(n+1) + e^{i\omega(t-t')}n\right]. \qquad \text{(F.2.25)}$$

It is easy to check that

$$\sum_{n=0}^{\infty}n e^{-n\beta\omega} = -\frac{1}{\omega}\partial_\beta Z_0. \qquad \text{(F.2.26)}$$

Using Eq. (F.2.25) and Eq. (F.2.26) one obtains

$$\mathcal{G}_\beta^+(t,t') = \frac{e^{-i\omega(t-t')}}{2\omega} + \frac{1}{\omega}\frac{\cos(\omega(t-t'))}{e^{\beta\omega}-1}. \qquad \text{(F.2.27)}$$

In the zero-temperature limit, when $\beta \to \infty$, the second term in the right-hand side vanishes. Thus[1]

$$\mathcal{G}^+(t, t') = \langle 0|\hat{q}(t)\hat{q}(t')|0\rangle = \frac{e^{-i\omega(t-t')}}{2\omega}. \tag{F.2.28}$$

From the representation Eq. (F.2.27) it is evident that $\mathcal{G}^+_\beta(t, t')$ depends only on the time difference $t - t'$ and, hence, it is a function of only one variable. We denote this function by the same symbol $\mathcal{G}^+_\beta(t)$, so that one has

$$\mathcal{G}^+_\beta(t, 0) \equiv \mathcal{G}^+_\beta(t) = \frac{e^{-i\omega t}}{2\omega} + \frac{1}{\omega} \frac{\cos(\omega t)}{e^{\beta\omega} - 1},$$
$$\mathcal{G}^-_\beta(t, 0) \equiv \mathcal{G}^-_\beta(t) = \frac{e^{i\omega t}}{2\omega} + \frac{1}{\omega} \frac{\cos(\omega t)}{e^{\beta\omega} - 1}. \tag{F.2.29}$$

This property is generic for any two-time thermal correlator $\langle \hat{A}(t)\hat{B}(t')\rangle_\beta$. In order to prove this we use the property that for any bounded operators $\hat{A}$ and $\hat{B}$ one has

$$\text{Tr}\,(\hat{A}\,\hat{B}) = \text{Tr}\,(\hat{B}\hat{A}). \tag{F.2.30}$$

Let us substitute the following expressions

$$\hat{A}(t) = \hat{U}_t^{-1}\hat{A}(0)\,\hat{U}_t, \quad \hat{B}(t) = \hat{U}_t^{-1}\hat{B}(0)\,\hat{U}_t, \quad \hat{U}_t = e^{-i\hat{H}t}, \tag{F.2.31}$$

into the average $\langle \hat{A}(t)\hat{B}(t')\rangle_\beta$. Then, we apply Eq. (F.2.30) and the property that $e^{-\beta\hat{H}}$ commutes with $\hat{U}_t$ and $\hat{U}_t^{-1}$. As a result of simple manipulations we obtain

$$\langle \hat{A}(t)\hat{B}(t')\rangle_\beta = \langle \hat{A}(t - t')\hat{B}(0)\rangle_\beta. \tag{F.2.32}$$

One can use the expression Eq. (F.2.29) for the functions $\mathcal{G}^+_\beta(t)$ and $\mathcal{G}^-_\beta(t)$ to define them in the complex plane of the variable t. Then, it is easy to check that the following relation is valid

$$\mathcal{G}^+_\beta(t) = \mathcal{G}^-_\beta(t + i\beta). \tag{F.2.33}$$

This result is known as the Kubo–Martin–Schwinger relation, or briefly the *KMS relation*. Similar KMS relations are valid for any thermal two-point correlator

$$\langle \hat{A}(t)\hat{B}(0)\rangle_\beta = \langle \hat{B}(0)\hat{A}(t + i\beta)\rangle_\beta. \tag{F.2.34}$$

To prove this result we use relations Eq. (F.2.31) for a complex time $t + i\beta$ to obtain

$$\hat{A}(t + i\beta) = e^{-\beta\hat{H}}\hat{A}(t)\,e^{\beta\hat{H}}, \tag{F.2.35}$$

---

[1] We use the standard notation $\mathcal{G}$ for the zero-temperature Green functions, so that $\mathcal{G} = \mathcal{G}_\infty$.

so that

$$\langle \hat{B}(0)\hat{A}(t+i\beta)\rangle_\beta = Z^{-1}\mathrm{Tr}\left(e^{-\beta\hat{H}}\hat{B}(0)\,e^{-\beta\hat{H}}\hat{A}(t)\,e^{\beta\hat{H}}\right)$$

$$= Z^{-1}\mathrm{Tr}\left(e^{-\beta\hat{H}}\hat{A}(t)\hat{B}(0)\right) = \langle \hat{A}(t)\hat{B}(0)\rangle_\beta.$$

(F.2.36)

Starting with $\mathcal{G}_\beta^\pm$ functions one can construct other Green functions for the quantum oscillator in the thermal state. In particular, for the commutator $[\hat{q}(t),\hat{q}(t')]$ one has

$$i\mathcal{G}(t) = \langle [\hat{q}(t),\hat{q}(0)]\rangle_\beta = \mathcal{G}_\beta^+(t) - \mathcal{G}_\beta^-(t) = -i\frac{\sin(\omega t)}{\omega}.$$

(F.2.37)

This correlator in fact does not depend on the temperature, and it coincides with the vacuum expectation value. The reason is evident: The commutator $[\hat{q}(t),\hat{q}(t')]$ is a c-number. A similar thermal Green function with the anticommutator $\{\ldots,\ldots\}$ being substituted for the commutator is known as the *Hadamard function*[2]

$$\mathcal{G}_\beta^{(1)}(t) = \langle\{\hat{q}(t),\hat{q}(0)\}\rangle_\beta \equiv \langle \hat{q}(t)\hat{q}(0) + \hat{q}(0)\hat{q}(t)\rangle_\beta$$

$$= \mathcal{G}_\beta^+(t) + \mathcal{G}_\beta^-(t) = \frac{\cos(\omega t)}{\omega}\coth\left(\frac{\beta\omega}{2}\right).$$

(F.2.38)

Another Green function that plays an important role in the quantum theory is the *causal Green function* or the *Feynman propagator*. It is defined as follows

$$-i\mathcal{G}_{F,\beta}(t) = \langle T(\hat{q}(t)\hat{q}(0))\rangle_\beta,$$

(F.2.39)

where the T-product of the operators $\hat{q}(t)$ and $\hat{q}(0)$ is defined as

$$T(\hat{q}(t)\hat{q}(0)) = \theta(t)\,\hat{q}(t)\hat{q}(0) + \theta(-t)\,\hat{q}(0)\hat{q}(t).$$

(F.2.40)

Here, $\theta(t)$ is a Heaviside step function. The definition Eq. (F.2.39) implies that

$$-i\mathcal{G}_{F,\beta}(t) = \theta(t)\,\mathcal{G}_\beta^+(t) + \theta(-t)\,\mathcal{G}_\beta^-(t) = \frac{e^{-i\omega|t|}}{2\omega} + \frac{1}{\omega}\frac{\cos(\omega t)}{e^{\beta\omega} - 1}.$$

(F.2.41)

This function is continuous at $t = 0$, but its derivative has a jump. $\mathcal{G}_{F,\beta}$ can be defined as a solution of the following equation

$$\left(\frac{d^2}{dt^2} + \omega^2\right)\mathcal{G}_{F,\beta}(t) = \delta(t).$$

(F.2.42)

It obeys the following boundary conditions: At zero temperature this function propagates positive frequency modes to the future and negative frequency modes to the past.

Equation (F.2.35) can be used to determine the evolution of the operators in the purely imaginary or Euclidean time

$$\hat{A}(i\beta) = e^{-\beta\hat{H}}\hat{A}(0)\,e^{\beta\hat{H}}.$$

(F.2.43)

---

[2] Sometimes, it is also called the *Hadamard elementary solution*.

The following object is known as the *Matsubara propagator*

$$\Delta(\tau) = \langle T_\tau(\hat{q}(-i\tau)\hat{q}(0))\rangle_\beta. \tag{F.2.44}$$

The operator T_τ is the *Euclidean time ordering*

$$T_\tau(\hat{q}(i\tau)\hat{q}(0)) = \theta(\tau)\,\hat{q}(i\tau)\hat{q}(0) + \theta(-\tau)\,\hat{q}(0)\hat{q}(i\tau). \tag{F.2.45}$$

One has

$$\Delta(\tau) = \theta(\tau)\,\mathcal{G}_\beta^+(-i\tau) + \theta(-\tau)\,\mathcal{G}_\beta^-(-i\tau). \tag{F.2.46}$$

For $\tau \in [0, \beta]$ the Matsubara propagator $\Delta(\tau)$ coincides with $\mathcal{G}_\beta^+(-i\tau)$ and is of the form

$$\Delta(\tau) = \frac{\cosh\left[(\beta/2 - \tau)\omega\right]}{2\omega\sinh(\beta\omega/2)}. \tag{F.2.47}$$

One can easily check that

$$\Delta(0) = \Delta(\beta) = \frac{1}{2\omega}\coth(\beta\omega/2). \tag{F.2.48}$$

In other words, the Matsubara propagator is periodic in the *imaginary time* with the period β. It can also be uniquely defined as the Green function for the *Euclidean oscillator* obeying the differential equation on a circle of length β

$$\left(\frac{d^2}{d\tau^2} - \omega^2\right)\Delta(\tau) = -\delta_\beta(\tau). \tag{F.2.49}$$

Here, $\delta_\beta(\tau)$ is a delta-function defined on a circle with the period β

$$\delta_\beta(\tau) = \sum_{m=-\infty}^{\infty} \delta(\tau + \beta m). \tag{F.2.50}$$

F.3   Quantum Field in Flat Spacetime

F.3.1   Classical scalar field

As an example of the field theory we consider a *scalar massive field* φ in a flat $(3 + 1)$-dimensional spacetime. The action is

$$S[\varphi] = \int dt\, L, \qquad L = \frac{1}{2}\int d\mathbf{X}\left[\dot{\varphi}^2 - (\partial_\mathbf{X}\varphi)^2 - m^2\varphi^2\right]. \tag{F.3.1}$$

Here, $t = T$ and X are Cartesian coordinates, and m is the mass of the field. The variation of the action gives the field equation. For the scalar field it is the *Klein–Gordon equation*

$$(\Box - m^2)\varphi = (-\partial_t^2 + \Delta - m^2)\varphi = 0. \tag{F.3.2}$$

It is convenient to consider a system in a box with the sizes $\{L_1, L_2, L_3\}$ and to impose the following boundary conditions

$$\varphi|_{X_i=-L_i/2} = \varphi|_{X_i=L_i/2} = 0, \qquad i = 1, 2, 3. \tag{F.3.3}$$

Momentum π conjugated to the field φ is

$$\pi = \frac{\delta L}{\delta \dot\varphi} = \dot\varphi. \tag{F.3.4}$$

The Hamiltonian of the system is

$$H = \frac{1}{2} \int d\mathbf{X} \left[\pi^2 + (\partial_{\mathbf{X}}\varphi)^2 + m^2\varphi^2 \right]. \tag{F.3.5}$$

Denote

$$\mathbf{k} = (k_1, k_2, k_3), \qquad k_i = \pi n_i/L_i, \tag{F.3.6}$$

where n_i are positive integer numbers. Then, a complete set of eigenfunctions of the Laplace operator $\triangle$ with the imposed boundary conditions is formed by functions

$$\Phi_{\mathbf{k}} = \sqrt{\frac{8}{V}} \sin(k_1 X^1) \sin(k_2 X^2) \sin(k_3 X^3), \tag{F.3.7}$$

where $V = L_1 L_2 L_3$ is the volume of the box. These functions obey the equation

$$\triangle \Phi_{\mathbf{k}} = -\mathbf{k}^2 \Phi_{\mathbf{k}}, \tag{F.3.8}$$

and the normalization condition

$$\int_V d\mathbf{X} \, \Phi_{\mathbf{k}}(X) \Phi_{\mathbf{k}'}(X) = \delta_{\mathbf{k},\mathbf{k}'} \equiv \delta_{n_1,n_1'} \delta_{n_2,n_2'} \delta_{n_3,n_3'}. \tag{F.3.9}$$

A general solution of Eq. (F.3.2) can be written as

$$\varphi = \sum_{\mathbf{k}} q_{\mathbf{k}}(t) \Phi_{\mathbf{k}}(X), \tag{F.3.10}$$

where the amplitudes of the field $q_{\mathbf{k}}(t)$ satisfy the equation

$$\left(\frac{d^2}{dt^2} + \omega_{\mathbf{k}}^2 \right) q_{\mathbf{k}}(t) = 0, \qquad \omega_{\mathbf{k}}^2 = m^2 + \mathbf{k}^2. \tag{F.3.11}$$

Thus, the amplitude of the field for a given wave number $\mathbf{k}$ obeys the oscillator equation. The reduction of the field theory Eq. (F.3.1) to the infinite set of oscillators Eq. (F.3.11) enumerated by $\mathbf{k}$ can be made more transparent by substituting the field decomposition in harmonics Eq. (F.3.10) into the Lagrangian Eq. (F.3.1). This substitution gives the following result

$$L = \frac{1}{2} \sum_{\mathbf{k}} \left[\dot{q}_{\mathbf{k}}^2 - \omega_{\mathbf{k}}^2 q_{\mathbf{k}}^2 \right]. \tag{F.3.12}$$

F.3.2 Quantum field

To quantize the field φ it is sufficient to quantize a decoupled set of oscillators described by the Lagrangian Eq. (F.3.12). To do this one considers the amplitude q_k and a conjugated momenta $p_k = \dot{q}_k$ as Hermitian operators obeying the canonical commutation relations

$$[\hat{q}_k, \hat{p}_{k'}] = i\hbar\, \delta_{k,k'}, \quad [\hat{q}_k, \hat{q}_{k'}] = [\hat{p}_k, \hat{p}_{k'}] = 0.$$

Using the completeness of the orthonormal basis Eq. (F.3.7) it is possible to show that these relations are equivalent to following *canonical commutation relations* for the field variables

$$[\hat{\varphi}(t, \mathbf{X}), \hat{\pi}(t, \mathbf{X}')] = i\hbar\, \delta\left(\mathbf{X} - \mathbf{X}'\right),$$

$$[\hat{\varphi}(t, \mathbf{X}), \hat{\varphi}(t, \mathbf{X}')] = 0, \qquad [\hat{\pi}(t, \mathbf{X}), \hat{\pi}(t, \mathbf{X}')] = 0. \tag{F.3.13}$$

One may describe the quantum field as an infinite set of oscillators. But now each oscillator is 'living' not in the 'physical space', but in a 'space of amplitudes'. Spatial modes Φ_k describe an amplitude of the probability for a given mode to be in the vicinity of the space point X. The quantum nature of the field is connected with discrete levels of its quantum amplitudes. Since the free quantum field is equivalent to a set of decoupled oscillators, one can easily apply all previous results concerning the quantum oscillator to such a set. In particular, the operators of creation and annihilation for each mode $\mathbf{k}$ are related to the amplitudes $\hat{q}_k$ and their momenta $\hat{p}_k = \dot{\hat{q}}_k$ as follows

$$\hat{q}_k = \frac{1}{\sqrt{2\omega_k}}(\hat{a}_k(t) + \hat{a}_k^\dagger(t)), \qquad \hat{p}_k = -i\sqrt{\frac{\omega}{2}}(\hat{a}_k(t) - \hat{a}_k^\dagger(t)), \tag{F.3.14}$$

where

$$\hat{a}_k(t) = e^{-i\omega_k t}\hat{a}_k, \qquad \hat{a}_k^\dagger(t) = e^{i\omega_k t}\hat{a}_k^\dagger. \tag{F.3.15}$$

The operators $\hat{a}_k$ and $\hat{a}_k^\dagger$ obey the commutation relations

$$[\hat{a}_k, \hat{a}_{k'}^\dagger] = \delta_{k,k'}, \qquad [\hat{a}_k, \hat{a}_{k'}] = 0, \qquad [\hat{a}_k^\dagger, \hat{a}_{k'}^\dagger] = 0. \tag{F.3.16}$$

The Hamiltonian Eq. (F.3.5) written in terms of the operators of creation and annihilation takes the form

$$\hat{H} = \hat{H}_0 + E_0, \qquad \hat{H}_0 = \sum_k \omega_k \hat{a}_k^\dagger \hat{a}_k, \qquad E_0 = \frac{1}{2}\sum_k \omega_k. \tag{F.3.17}$$

A *vacuum* is a state $|0\rangle$ with the lowest possible energy. It is defined by the condition

$$\hat{a}_k|0\rangle = 0. \tag{F.3.18}$$

The quantity E_0 is the energy of the vacuum *zero-point fluctuations*. The vacuum energy E_0 is unobservable. However, in the presence of an external field and/or for non-trivial boundary conditions the vacuum energy E_0' differs from its value E_0 in the empty Minkowski spacetime. The difference $E_0' - E_0$ can be measured and hence it has a well-defined physical meaning. This phenomenon is known as the *Casimir effect*.

Denote by $n_{\mathbf{k}}$ a non-negative integer number for a mode $\mathbf{k}$ and by $\{n_{\mathbf{k}}\}$ a set of such integer numbers. We assume that only a finite number of them do not vanish. In the *occupation number representation* the basis in the *Hilbert space* of states is

$$|\{n_{\mathbf{k}}\}\rangle = \prod_{\mathbf{k}} \left[\frac{1}{\sqrt{n_{\mathbf{k}}!}} \left(\hat{a}_{\mathbf{k}}^{\dagger} \right)^{n_{\mathbf{k}}} \right] |0\rangle. \tag{F.3.19}$$

This basis is orthonormal

$$\langle \{n_{\mathbf{k}}\} | \{m_{\mathbf{k}'}\} \rangle = \prod_{\mathbf{k},\mathbf{k}'} \left[\delta_{\mathbf{k},\mathbf{k}'} \delta_{n_{\mathbf{k}},m_{\mathbf{k}'}} \right]. \tag{F.3.20}$$

F.3.3 Thermal fields

A thermal state of the quantum field with the inverse temperature β is a state when each of its modes is in a thermal state described by the density matrix Eq. (F.2.14). The corresponding density matrix for the quantum field is

$$\hat{\rho} = Z^{-1}(\beta) e^{-\beta \hat{H}} = Z_0^{-1}(\beta) e^{-\beta \hat{H}_0}. \tag{F.3.21}$$

Here, $\hat{H}$ is a complete Hamiltonian Eq. (F.3.17) for the quantum field and $\hat{H}_0$ is its version with zero-point energy subtracted. The trace operator is the trace over the complete Hilbert space of the field states. Using the representation of the field as a set of decoupled oscillators, one obtains

$$Z_0 = e^{-\beta F_0}, \qquad F_0 = \frac{1}{\beta} \sum_{\mathbf{k}} \ln \left(1 - e^{-\beta \omega_{\mathbf{k}}} \right). \tag{F.3.22}$$

This is a *free energy* of the thermal quantum field at the temperature $\Theta = \beta^{-1}$.

F.3.4 Continuous spectrum

The imposed boundary conditions, that is vanishing of the field at the boundary, are somehow artificial. However, if the quantization box is large and one is studying observables in a domain far away from the boundary, the result does not depend on the particular boundary conditions. They affect the state of the field only close to the boundary. Usually, for large volume the surface effects can be neglected.

In order to exclude effects connected with the spatial location of the boundary it is convenient to use the other boundary conditions, known as the *periodicity conditions*. Let us again take a cube with boundaries at $|X_i| = L_i/2$ and impose the following periodic boundary conditions

$$\varphi(X)|_{X_i=-L_i/2} = \varphi(X)|_{X_i=L_i/2},$$
$$\partial_{X_i}\varphi(X)|_{X_i=-L_i/2} = \partial_{X_i}\varphi(X)|_{X_i=L_i/2}, \qquad i = 1, 2, 3. \tag{F.3.23}$$

In other words, the opposite boundaries of the cube are identified with one another and we have a space that is 3-torus $\mathbb{T}^3$.

The complete set of eigenfunctions of the Laplace operator with given periodicity conditions is well known. For each of three orthogonal directions one can use $\sin(k_i X_i)$ and $\cos(k_i X_i)$
functions. The periodicity conditions imply that

$$k_i = \frac{2\pi n_i}{L_i},$$

(F.3.24)

where n_i are non-negative integer numbers. The 'doubling' of the basic functions is 'compensated' by the fact that the corresponding wavelengths k_i are twice 'rare' than before. Very
often it is more convenient instead of real eigenfunctions of the Laplace operator to use the
complex eigenfunctions $\exp(\pm i k_i X_i)$. These functions also obey the periodicity conditions
when Eq. (F.3.24) is satisfied. We use the following solutions

$$\Phi_{\mathbf{k}}(X) = \frac{1}{\sqrt{V}} e^{i\mathbf{k}X}, \qquad V = L_1 L_2 L_3.$$

(F.3.25)

Here, $\mathbf{k} = (\frac{2\pi n_1}{L_1}, \frac{2\pi n_2}{L_2}, \frac{2\pi n_3}{L_3})$, where n_i are the integer numbers, which can be both positive
and negative. Under this condition the real basic functions $\sin(k_i X_i)$ and $\cos(k_i X_i)$ can be
obtained as linear combinations of solutions $\Phi_{\mathbf{k}}$ and their complex conjugated. The basis
functions obey the following normalization conditions

$$\int dX \, \Phi_{\mathbf{k}}(X)\Phi_{\mathbf{k}'}^*(X) = \delta_{\mathbf{k},\mathbf{k}'} \equiv \delta_{n_1,n_1'}\delta_{n_2,n_2'}\delta_{n_3,n_3'}.$$

(F.3.26)

A mode decomposition of the field operator $\hat{\phi}(t, X)$ now takes the form

$$\hat{\varphi}(t, X) = \sum_{\mathbf{k}} \left[\hat{a}_{\mathbf{k}} \frac{e^{-i\omega_{\mathbf{k}}t}}{\sqrt{2\omega_{\mathbf{k}}}} \Phi_{\mathbf{k}}(X) + \hat{a}_{\mathbf{k}}^\dagger \frac{e^{i\omega_{\mathbf{k}}t}}{\sqrt{2\omega_{\mathbf{k}}}} \Phi_{\mathbf{k}}^*(X) \right].$$

(F.3.27)

Making the size of the quantization box infinitely large one reduces the distance between
the nearby wavelength, so that finally in the infinite-size limit one obtains a continuous
spectrum of the Laplace operator.

The continuous spectrum formulas can be obtained from the discrete ones by the following substitution

$$\Phi_{\mathbf{k}} \rightarrow A\,\Phi_{\mathbf{k}}, \qquad \hat{a}_{\mathbf{k}} \rightarrow A\,\hat{a}_{\mathbf{k}}, \qquad \hat{a}_{\mathbf{k}}^\dagger \rightarrow A\,\hat{a}_{\mathbf{k}}^\dagger,$$

$$\sum_{\mathbf{k}} \rightarrow A^{-2} \int d\mathbf{k}, \qquad \delta_{\mathbf{k},\mathbf{k}'} \rightarrow A^2\,\delta(\mathbf{k} - \mathbf{k}'),$$

(F.3.28)

$$\left[\hat{a}_{\mathbf{k}}, \hat{a}_{\mathbf{k}'}^\dagger\right] = \delta_{\mathbf{k},\mathbf{k}'} \quad \rightarrow \quad \left[\hat{a}_{\mathbf{k}}, \hat{a}_{\mathbf{k}'}^\dagger\right] = \delta(\mathbf{k} - \mathbf{k}'),$$

*where $A = \sqrt{V/(2\pi)^3}$. The discrete and continuous δ-functions are defined as usual to fulfill the
relations $\sum_{\mathbf{k}} \delta_{\mathbf{k},\mathbf{k}'} = 1$ and $\int d\mathbf{k}\,\delta(\mathbf{k} - \mathbf{k}') = 1$.*

The continuous spectrum basis functions are

$$\Phi_{\mathbf{k}}(X) = \frac{1}{\sqrt{(2\pi)^3}} e^{i\mathbf{k}X}.$$

(F.3.29)

They obey the normalization condition

$$\int dX \, \Phi_{\mathbf{k}}(X)\Phi^*_{\mathbf{k}'}(X) = \delta(\mathbf{k} - \mathbf{k}'). \tag{F.3.30}$$

The field representation in the limit of an infinite box is

$$\hat{\varphi}(t, X) = \int \frac{d\mathbf{k}}{\sqrt{2\omega_{\mathbf{k}}}} \left[\hat{a}_{\mathbf{k}} e^{-i\omega_{\mathbf{k}}t} \Phi_{\mathbf{k}}(X) + \hat{a}^{\dagger}_{\mathbf{k}} e^{i\omega_{\mathbf{k}}t} \Phi^*_{\mathbf{k}}(X) \right], \tag{F.3.31}$$

where the operators of creation and annihilation with continuous index $\mathbf{k}$ obey the commutation relations

$$[\hat{a}_{\mathbf{k}}, \hat{a}^{\dagger}_{\mathbf{k}'}] = \delta(\mathbf{k} - \mathbf{k}'), \qquad [\hat{a}_{\mathbf{k}}, \hat{a}_{\mathbf{k}'}] = 0, \qquad [\hat{a}^{\dagger}_{\mathbf{k}}, \hat{a}^{\dagger}_{\mathbf{k}'}] = 0. \tag{F.3.32}$$

F.3.5 Green functions

A *Green function* for a quantum field $\hat{\varphi}$ is an average value of a bilinear combination of $\hat{\varphi}(t, X)$ and $\hat{\varphi}(t', X')$. Again, as in the case of a quantum oscillator, there exists a variety of Green functions that differ by the choice of a state used for averaging and by a concrete form of the bilinear form, which may include special operator ordering.

Let us consider the following thermal correlators

$$G^+_{\beta}(t, X; t', X') = \langle \hat{\varphi}(t, X)\hat{\varphi}(t', X') \rangle_{\beta},$$
$$G^-_{\beta}(t, X; t', X') = \langle \hat{\varphi}(t', X')\hat{\varphi}(t, X) \rangle_{\beta} = G^+_{\beta}(t', X'; t, X), \tag{F.3.33}$$

where

$$\langle \hat{A}\hat{B} \rangle_{\beta} = Z_0^{-1} \text{Tr} \left(e^{-\beta \hat{H}_0} \hat{A}\hat{B} \right). \tag{F.3.34}$$

These functions are known as *thermal Wightman functions*. Since the spacetime is static one has

$$\langle \hat{A}(t)\hat{B}(t') \rangle_{\beta} = \langle \hat{A}(t - t')\hat{B}(0) \rangle_{\beta}. \tag{F.3.35}$$

This property can be proved by using the relations

$$\hat{A}(t) = e^{i\hat{H}_0 t}\hat{A}(0)e^{-i\hat{H}_0 t}, \qquad \hat{B}(t') = e^{i\hat{H}_0 t'}\hat{B}(0)e^{-i\hat{H}_0 t'}, \tag{F.3.36}$$

and the fact that cyclic permutations of operators within the Tr-operation do not change the result.

The property Eq. (F.3.35) means that G^+_{β} and G^-_{β} functions depend only on the difference $t - t'$. Moreover, since the system is invariant under spatial translations and rotations, these fuctions depend only on $r = |X - X'|$. Thus, we have

$$G^+_{\beta}(t, X; t', X') = G^+_{\beta}(t - t', r),$$
$$G^-_{\beta}(t, X; t', X') = G^-_{\beta}(t - t', r). \tag{F.3.37}$$

To calculate these quantities it is sufficient to substitute Eq. (F.3.31) in their definition Eq. (F.3.33), and to use the relations

$$\langle \hat{a}_{\mathbf{k}} \hat{a}_{\mathbf{k}'}^{\dagger} \rangle_{\beta} = \frac{\delta(\mathbf{k} - \mathbf{k}')}{1 - e^{-\beta \omega_{\mathbf{k}}}}, \qquad \langle \hat{a}_{\mathbf{k}}^{\dagger} \hat{a}_{\mathbf{k}'} \rangle_{\beta} = \frac{\delta(\mathbf{k} - \mathbf{k}')}{e^{\beta \omega_{\mathbf{k}}} - 1}. \tag{F.3.38}$$

The calculations give

$$G_{\beta}^{\pm}(t,r) = \frac{1}{2\pi^2 r} \int_0^{\infty} dk\, k \sin(kr)\, G_{\beta,\omega_k}^{\pm}(t). \tag{F.3.39}$$

Here, $G_{\beta,\omega}^{\pm}(t)$ is the corresponding Green function for a thermal oscillator with the frequency ω given by Eq. (F.2.29). Similar representations are valid for the other Green functions of the quantum field.

Let G, $G_{F,\beta}$, and $G_{\beta}^{(1)}$ be the thermal average of the commutator, the *Feynman propagator* and the *Hadamard function* for the quantum field (Takagi 1986; Birrell and Davies 1982), respectively

$$iG(t,r) \equiv \langle [\hat{\varphi}(t,X), \hat{\varphi}(0,X')] \rangle_{\beta},$$

$$-iG_{F,\beta}(t,r) \equiv \langle T(\hat{\varphi}(t,X)\hat{\varphi}(0,X')) \rangle_{\beta}, \tag{F.3.40}$$

$$G_{\beta}^{(1)}(t,r) \equiv \langle \{\hat{\varphi}(t,X)\hat{\varphi}(0,X')\} \rangle_{\beta}.$$

Here,

$$T(\hat{\varphi}(t,X)\hat{\varphi}(0,X')) = \theta(t)\,\hat{\varphi}(t,X)\hat{\varphi}(0,X') + \theta(-t)\,\hat{\varphi}(0,X')\hat{\varphi}(t,X). \tag{F.3.41}$$

To obtain any of these functions it is sufficient to substitute in the relation similar to Eq. (F.3.39) a corresponding Green function for the oscillator

$$G_{\beta}^{\bullet}(t,r) = \frac{1}{2\pi^2 r} \int_0^{\infty} dk\, k \sin(kr)\, G_{\beta,\omega_k}^{\bullet}(t). \tag{F.3.42}$$

Here, the $\bullet$-symbol specifies a concrete choice of the Green function.

The thermal oscillatory Green functions consists of two parts. The first, which coincides with the vacuum average, differs for different Green functions, while the second (which vanishes at the zero-temperature limit and may be absent in some cases) is a universal expression

$$B(t) = -\frac{1}{\omega} \frac{\cos(\omega t)}{e^{\beta \omega} - 1}. \tag{F.3.43}$$

In a special case of a massless field, $m = 0$, the expressions for the Green functions can be written in terms of the elementary functions. One has $\omega = k$, and the integral in Eq. (F.3.42) can be easily calculated. For $B_{\beta}(t,r)$ related to $B(t)$ by Eq. (F.3.42) one finds

$$B_{\beta}(t,r) = \frac{\coth[\pi(r+t)/\beta] + \coth[\pi(r-t)/\beta]}{8\pi\beta r} + \frac{1}{4\pi^2(t^2 - r^2)}. \tag{F.3.44}$$

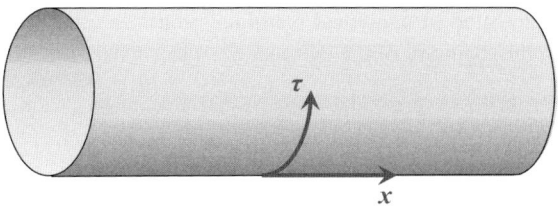

Fig. F.1 Euclidean space for the thermal field. It has the topology of the cylinder $S^1 \times \mathbb{R}^3$. The directions of the Euclidean time τ and spatial dimensions are schematically shown.

Using this result and the following relation

$$\frac{1}{x \pm i\epsilon} = \mathcal{P}\frac{1}{x} \mp i\pi \delta(x), \tag{F.3.45}$$

where $\epsilon > 0$ is an infinitesimal quantity and $\mathcal{P}$ is the principal value, one obtains the real-time thermal Green functions for massless fields

$$G_\beta^\pm(t, r) = -\frac{1}{4\pi^2}\frac{1}{(t \mp i\epsilon)^2 - r^2} + B_\beta(t, r),$$

$$iG(t, r) = \frac{1}{4\pi r}[\delta(r + t) - \delta(r - t)],$$

$$-iG_{F,\beta}(t, r) = -\frac{1}{4\pi^2}\frac{1}{t^2 - r^2 - i\epsilon} + B_\beta(t, r),$$

$$G_\beta^{(1)}(t, r) = -\frac{1}{2\pi^2}\frac{1}{t^2 - r^2} + 2B_\beta(t, r).$$

$$\tag{F.3.46}$$

For imaginary time $t = -i\tau$ the Wightman functions are equal, $G_\beta^+(-i\tau, r) = G_\beta^-(-i\tau, r)$ and the *Matsubara propagator* is

$$\Delta(\tau, r) = G_\beta^\pm(-i\tau, r) = \frac{1}{4\pi\beta r}\frac{\sinh(2\pi r/\beta)}{\cosh(2\pi r/\beta) - \cos(2\pi \tau/\beta)}. \tag{F.3.47}$$

It is periodic in the imaginary time with the period β

$$\Delta(\tau + \beta, r) = \Delta(\tau - \beta, r) = \Delta(\tau, r), \tag{F.3.48}$$

and obeys the following equation

$$(\partial_\tau^2 + \Delta)\Delta(\tau, r) = -\delta(\tau)\delta(\mathbf{X} - \mathbf{X}'). \tag{F.3.49}$$

For $r = 0$

$$\Delta(\tau, 0) = \frac{1}{2\beta^2}\frac{1}{1 - \cos(2\pi \tau/\beta)}. \tag{F.3.50}$$

Thermal Green functions for massive fields and fields of other spins can be obtained in a similar manner. In all cases the general property remains valid: The quantum field at the finite temperature β^{-1} is connected by analytical continuation to the *Euclidean field theory* on the Euclidean space with the topology of the cylinder $S^1 \times \mathbb{R}^3$, where the length of S^1 is β.

F.3.6   Measurement of temperature

Let us discuss now how the temperature of the radiation can be measured. In a general setup, a thermometer is a system with internal degrees of freedom that can interact with thermal radiation. We assume that this interaction is weak and does not disturb the state of the radiation. We expect that the internal degrees of freedom of the thermometer will be excited as a result of an interaction and, after some time, they will be distributed thermally. By 'measuring' observables depending on this distribution one can 'measure' temperature of the radiation. Usually, it is assumed that the thermometer has small size. This is important when the thermal radiation is affected by an external field. For example, in the presence of a static gravitational field the local temperature depends on a point. In order to measure this local temperature the size of the thermometer must be much smaller than a characteristic scale at which the temperature changes.

For illustration let us consider a simple model. We choose a quantum oscillator as a thermometer and consider its interaction with a scalar massless field φ. Let the 'thermometer' be located at a point X_0 of a flat spacetime. The system, that is the quantum field φ and the oscillator q, is described by the following action

$$S[\varphi, q] = S_0[\varphi] + S_0[q] + S_{\text{int}}[\varphi, q]. \tag{F.3.51}$$

Here, $S_0[\varphi]$ is the field action Eq. (F.3.1), $S_0[q]$ is the oscillator action Eq. (F.1.1). The interaction action is

$$S_{\text{int}}[\varphi, q] = \int dt\, V, \qquad V(t) = \lambda \varphi(t)\, q(t), \tag{F.3.52}$$

where $\varphi(t) = \varphi(t, X_0)$. We denote by $\hat{H}$ the Hamiltonian of the complete system. It has the form

$$\hat{H} = \hat{H}_0 + \hat{V}, \qquad \hat{H}_0 = \hat{H}_0^{\text{field}}[\varphi] + \hat{H}_0^{\text{osc}}[q]. \tag{F.3.53}$$

In the *Schrödinger picture* the evolution of the complete system is described by the state vector $|\Psi(t)\rangle$ obeying the *Schrödinger equation*

$$i\partial_t |\Psi(t)\rangle = \hat{H}|\Psi(t)\rangle. \tag{F.3.54}$$

In this picture the operators do not evolve, so that

$$\hat{V} = \hat{V}_0 = \lambda \hat{\varphi}(0)\hat{q}(0). \tag{F.3.55}$$

Let us write

$$|\Psi(t)\rangle = \exp\left(-i\hat{H}_0 t\right)|\Psi(t)\rangle. \tag{F.3.56}$$

Substitution of Eq. (F.3.56) into Eq. (F.3.54) gives

$$i\partial_t |\Psi(t)\rangle = \hat{V}(t)|\Psi(t)\rangle,$$

$$\hat{V}(t) = \exp(i\hat{H}_0 t)\,\hat{V}_0\,\exp(-i\hat{H}_0 t) = \lambda\hat{\varphi}(t)\hat{q}(t). \tag{F.3.57}$$

Denote by $|A\rangle$ the vectors of the orthonormal basis in the *Hilbert space* of the states of the system without interaction, and A is an index enumerating the vectors of this set. We additionally assume that $|A\rangle$ are eigenstates of the free Hamiltonian

$$\hat{H}_0 |A\rangle = E_A |A\rangle. \tag{F.3.58}$$

Let us write

$$|\Psi(t)\rangle = \sum_A b_A(t)|A\rangle. \tag{F.3.59}$$

Substituting this relation into Eq. (F.3.57) and multiplying the obtained relation by $\langle B|$ we get

$$i\partial_t b_B(t) = \sum_A V_{BA}(t) b_A(t), \qquad V_{BA}(t) = \langle B|\hat{V}(t)|A\rangle. \tag{F.3.60}$$

We assume that the interaction constant λ is small and use the perturbation theory to solve the Eq. (F.3.60). In the first-order approximation, the amplitude of the probability of the transition from the initial state $|A\rangle$ to the final state $|B\rangle$ at time t is

$$\mathcal{A}_{BA} = -i\int_{-\infty}^{t} dt'\, V_{BA}(t'). \tag{F.3.61}$$

The probability of the transition $A \to B$ is

$$w_{BA} = |\mathcal{A}_{BA}|^2 = \int_{-\infty}^{t} dt' \int_{-\infty}^{t} dt''\, V_{BA}(t') V_{BA}^*(t''). \tag{F.3.62}$$

Till now the consideration was quite general and we did not use specific properties of the interaction. Basically, the relation Eq. (F.3.62) reproduces the standard quantum mechanical result of the perturbation theory for the probability of transitions under a general perturbation $\hat{V}$. In order to apply this result to our system of a quantum oscillator interacting with the thermal radiation we proceed as follows. First, note that a state $|A\rangle$ of the complete system can be written as

$$|A\rangle = |n\rangle|N\rangle, \tag{F.3.63}$$

where $|n\rangle$ is a complete set of the states of a non-interacting quantum oscillator and $|N\rangle$ is a complete set of the field states. Accordingly, we write the index $A = \{n, N\}$. For the index B we write $B = \{m, M\}$. We choose the states $|n\rangle$ and $|N\rangle$ to be the eigenstates of the corresponding free Hamiltonian

$$\hat{H}_0^{\text{osc}}|n\rangle = \omega_n|n\rangle, \qquad \hat{H}_0^{\text{field}}|N\rangle = E_N|N\rangle. \tag{F.3.64}$$

In this basis one has

$$V_{Mm,Nn}(t) = \lambda e^{i(\omega_m - \omega_n)t} \langle m|\hat{q}(0)|n\rangle \langle M|\hat{\varphi}(t)|N\rangle, \quad q_{mn} = \langle m|\hat{q}(0)|n\rangle. \tag{F.3.65}$$

We used here that $\langle m|\hat{q}(t)|n\rangle = e^{i(\omega_m - \omega_n)t} q_{mn}$. These matrix elements can be found by using the expression Eq. (F.2.7)

$$\hat{q}(0) = \frac{1}{\sqrt{2\omega}}(\hat{a} + \hat{a}^\dagger), \tag{F.3.66}$$

where ω is the frequency of the oscillator. One has

$$q_{mn} = \frac{1}{\sqrt{2\omega}} \left(\sqrt{n}\, \delta_{m,n-1} + \sqrt{n+1}\, \delta_{m,n+1} \right). \tag{F.3.67}$$

The relations Eq. (F.3.65) and Eq. (F.3.67) show that in the lowest order of the perturbation theory the transitions are only of two types:

- either the oscillator absorbs a single 'photon' of energy ω and 'jumps' from the lower level n to the upper level $n + 1$;
- or it emits a 'photon' of the frequency ω and 'jumps' from the upper level $n + 1$ to the lower level n.

To obtain the probability for the transition w_{mn} one needs to average over unobserved final states of the field $|M\rangle$ and to use the thermal density matrix for the initial state of the field

$$\hat{\rho}_{\text{field}} = Z^{-1} \sum_N P_N |N\rangle \langle N|. \tag{F.3.68}$$

Here, P_N is a probability of a state $|N\rangle$. As a result of these operations one gets

$$w_{mn} = Z^{-1} \sum_M \sum_N P_N w_{mM,nN}$$

$$= \lambda^2 |q_{mn}|^2 \int_{-\infty}^{t} dt' \int_{-\infty}^{t} dt'' \, e^{i(\omega_m - \omega_n)(t' - t'')} G_\beta^+(t'' - t'). \tag{F.3.69}$$

Here,

$$G_\beta^+(t' - t'') = \langle \hat{\varphi}(t')\hat{\varphi}(t'') \rangle_\beta \tag{F.3.70}$$

is the positive-frequency *Wightman function*

$$G_\beta^+(t) = G_\beta^+(t, \mathbf{X}_0, 0, \mathbf{X}_0). \tag{F.3.71}$$

It is easy to show that a double integral in Eq. (F.3.69) can be rewritten as follows

$$\int_{-\infty}^{t} dt' \int_{-\infty}^{t} dt'' \, F(t' - t'') = \int_{-\infty}^{t} dt' \int_{-\infty}^{\infty} d\tau \, F(\tau). \tag{F.3.72}$$

This relation shows that w_{mn} is proportional to time. Thus, for the transision probability per unit time one has

$$\dot{w}_{mn} = \frac{dw_{mn}}{dt} = \lambda^2 |q_{mn}|^2 \tilde{G}_\beta(\omega_m - \omega_n), \tag{F.3.73}$$

where

$$\tilde{G}_\beta(\omega) = \int_{-\infty}^{\infty} d\tau\, e^{-i\omega\tau} G_\beta^+(\tau). \tag{F.3.74}$$

Let us denote $\dot{w}_{n+} = \dot{w}_{n+1,n}$ and $\dot{w}_{n-} = \dot{w}_{n,n+1}$. Equation (F.3.67) shows that

$$q_{n\pm} = q_n \equiv \frac{\sqrt{n+1}}{\sqrt{2\omega}}. \tag{F.3.75}$$

Thus,

$$\dot{w}_{n\pm} = \lambda^2 q_n^2 \tilde{G}_\beta(\pm\omega). \tag{F.3.76}$$

Simple calculations give

$$\tilde{G}_\beta(\omega) = \frac{1}{2\pi} \frac{\omega}{e^{\beta\omega} - 1}, \tag{F.3.77}$$

for either positive or negative ω. Thus,

$$\frac{\dot{w}_{n+}}{\dot{w}_{n-}} = \frac{\tilde{G}_\beta(\omega)}{\tilde{G}_\beta(-\omega)} = e^{-\beta\omega}. \tag{F.3.78}$$

The obtained relation implies that the ratio of the probability of the transition $n \to n+1$ to the probability of the inverse process $n+1 \to n$ is a universal function. It does not depend on the details of interaction and it is determined only by the temperature of the radiation β^{-1} and the energy difference of the corresponding levels. This ratio is determined only by the energy spectrum of the Wightman function $\tilde{G}(\omega)$ for the radiation. This result is quite general and it allows one to prove that the equilibrium state of the quantum oscillator interacting with the thermal bath with temperature β^{-1} is described by the thermal density matrix with the same temperature.

This can be easily shown as follows. Denote by $p_n(t)$ the probability of the oscillator to be at the level n. Then, the change in time of this probability is

$$\frac{dp_n}{dt} = \dot{w}_{n-1,+}p_{n-1} + \dot{w}_{n,-}p_{n+1} - (\dot{w}_{n,+} + \dot{w}_{n-1,-})p_n, \qquad n \geq 1. \tag{F.3.79}$$

For $n = 0$ one has

$$\frac{dp_0}{dt} = \dot{w}_{0,-}p_1 - \dot{w}_{0,+}p_0. \tag{F.3.80}$$

In the equilibrium state $\dot{p}_n = 0$. The relation Eq. (F.3.80) gives

$$p_1 = \frac{\dot{w}_{0,+}}{\dot{w}_{0,-}}p_0 = e^{-\beta\omega}p_0. \tag{F.3.81}$$

Starting with this relation and using Eq. (F.3.79) we obtain

$$p_n = e^{-n\beta\omega} p_0. \tag{F.3.82}$$

The normalization condition $\sum_{n=0}^{\infty} p_n = 1$ determines p_0

$$p_0 = \frac{1}{1 - e^{-\beta\omega}}. \tag{F.3.83}$$

Thus, the equilibrium distribution of the oscillator over the energy levels has a thermal law. The details of its interaction with the thermal bath, and especially the value of the coupling constant λ determine the rate of transitions $\dot{w}_{n\pm}$, and, in particular, how fast the equilibrium is reached. The smaller λ the longer one needs to wait when the oscillator reaches its equilibrium thermal state. In other words, if the interaction is weak the time required to 'measure' the temperature of the thermal radiation is large.

F.4   Quantum Theory in (1+1)-Spacetime

F.4.1   Field equations

Till now we have considered the case when a quantum field is freely propagating in a flat spacetime, so that the mode functions are the standard plane waves. Let us consider a slightly more complicated situation when the scalar field moves in a space with a potential. For simplicity we assume that the space has only one dimension and denote a spatial coordinate by x. The action for the field $\varphi(t, x)$ is

$$S[\varphi] = \int dt\, L, \quad L = \frac{1}{2} \int dx \left[\dot{\varphi}^2 - (\partial_x \varphi)^2 - U(x)\varphi^2 \right]. \tag{F.4.1}$$

Here, $U(x)$ is the potential of the form shown in Figure F.2. It is positive and falls rapidly enough at $|x| \to \infty$. One can also consider this action as a generalization of the $(1 + 1)$-dimensional version of Eq. (F.3.1) to the case when the mass of the field depends on x.

The field equation is

$$(\Box - U)\varphi \equiv (-\partial_t^2 + \partial_x^2 - U)\varphi = 0. \tag{F.4.2}$$

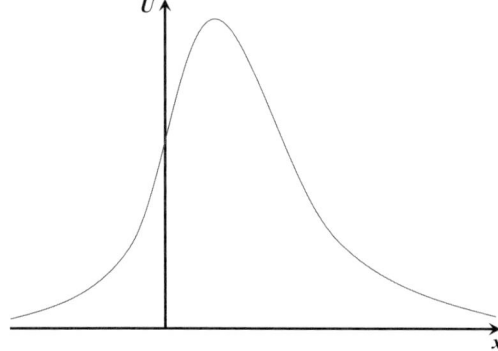

Fig. F.2 Potential $U(x)$. At far distances $|x| \to 0$ it falls rapidly enough.

The field can be decomposed into modes

$$\varphi_k \sim e^{-ikt} u_k(x), \tag{F.4.3}$$

where u_k are complex solutions of the following eigenvalue problem

$$D[u_k] \equiv \left[\frac{d^2}{dx^2} - U \right] u_k = -k^2 u_k. \tag{F.4.4}$$

The latter equation is simply a standard *Schrödinger equation* of quantum mechanics

$$\left[\frac{d^2}{dx^2} + (E - U) \right] u_k = 0. \tag{F.4.5}$$

It describes a one-dimensional motion of a particle of the energy $E = k^2$ in the presence of the potential barrier. This problem is discussed in standard books on quantum mechanics. Here, we just collect the main results that are required for our analysis.

F.4.2   Bases in the solution space

The spectrum of the problem Eq. (F.4.5) is continuous and the energy E is positive. The energy levels are degenerate, namely there exist two linear independent solutions for a given energy E. It is convenient to consider complex solutions of Eq. (F.4.5). The scalar product for any two complex solutions f_1 and f_2 is defined as follows

$$< f_1, f_2 >= \int_{-\infty}^{\infty} dx f_1^*(x) f_2(x). \tag{F.4.6}$$

A complex solution is uniquely specified by its asymptotic behavior at $x = \pm\infty$. In the general case, for a given $k > 0$ one has

$$f_k(x) = \frac{1}{\sqrt{2\pi}} \begin{cases} \gamma_k e^{-ikx} + \beta_k e^{ikx} , \ x \to -\infty , \\ \alpha_k e^{-ikx} + \delta_k e^{ikx} , \ x \to +\infty. \end{cases} \tag{F.4.7}$$

For the continious spectrum the scalar product can be written in terms of the scattering coefficients α_k, β_k, γ_k and δ_k. Since a solution of the original wave equation Eq. (F.4.2) is obtained by multiplying $f_k(x)$ by e^{-ikt}, it is evident that a term with e^{-ikx} describes a mode propagating to the left, while a mode with e^{ikx} moves to the right. This can be presented by the following *conformal diagram* (see Figure F.3).

One can interpret the solution Eq. (F.4.7) as follows. An initial wave consists of two parts: One, with the amplitude β_k, is going from $x = -\infty$ to the right, and the other, with the amplitude α_k, is propagating from $x = \infty$ to the left. After the scattering on the potential $U(x)$ the right-moving mode will have the amplitude δ_k at $x = \infty$, while the left-moving mode will have the amplitude γ_k at $x = -\infty$. Since the mode is uniquely specified by its initial data α_k and β_k, the 'future' asymptotic data, γ_k and δ_k, are functions of the initial data. These functions depend on the form of the potential $U(x)$. For any two solutions Eq. (F.4.7) one has

$$< f_k, f_k' >= (|\alpha|_k^2 + |\beta|_k^2) \delta(k - k') = (|\gamma|_k^2 + |\delta|_k^2) \delta(k - k'). \tag{F.4.8}$$

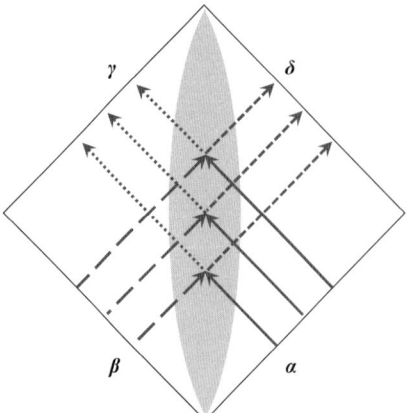

Fig. F.3 Conformal diagram for a general solution f. A shadowed region schematically shows a domain where the scattering on the potential takes place.

We shall use the normalization condition

$$|\alpha|_k^2 + |\beta|_k^2 = |\gamma|_k^2 + |\delta|_k^2 = 1. \tag{F.4.9}$$

A standard convenient choice of the complete set of normalized complex solutions consists of functions u_{in} and u_{up}. These functions are specified by the following scattering data

$$u_{\text{in},k}(x) = \frac{1}{\sqrt{2\pi}} \begin{cases} T_k e^{-ikx}, & x \to -\infty, \\ e^{-ikx} + R_k e^{ikx}, & x \to +\infty; \end{cases} \tag{F.4.10}$$

$$u_{\text{up},k}(x) = \frac{1}{\sqrt{2\pi}} \begin{cases} e^{ikx} + r_k e^{-ikx}, & x \to -\infty, \\ t_k e^{ikx}, & x \to +\infty. \end{cases} \tag{F.4.11}$$

In these solutions we assume that $k > 0$. The conformal diagrams for these functions are shown in Figure F.4.

The normalization condition Eq. (F.4.9) results in

$$|T_k|^2 + |R_k|^2 = 1, \qquad |t_k|^2 + |r_k|^2 = 1. \tag{F.4.12}$$

These solutions obey the following relations

$$< u_{\text{in},k}, u_{\text{in},k'} >\, =\, < u_{\text{up},k}, u_{\text{up},k'} >\, = \delta(k - k'),$$
$$< u_{\text{in},k}, u_{\text{up},k'} >\, =\, < u_{\text{up},k}, u_{\text{in},k'} >\, = 0. \tag{F.4.13}$$

It is well known (and easy to demonstrate) that the *Wronskian*

$$W(v_1, v_2) = v_1 (dv_2/dx) - v_2 (dv_1/dx) \tag{F.4.14}$$

of any two linearly independent solutions v_1 and v_2 to Eq. (F.4.5) is a constant. By calculating the Wronskian at both infinities, $x = \pm\infty$, for solutions Eqs. (F.4.10) and (F.4.11) and their

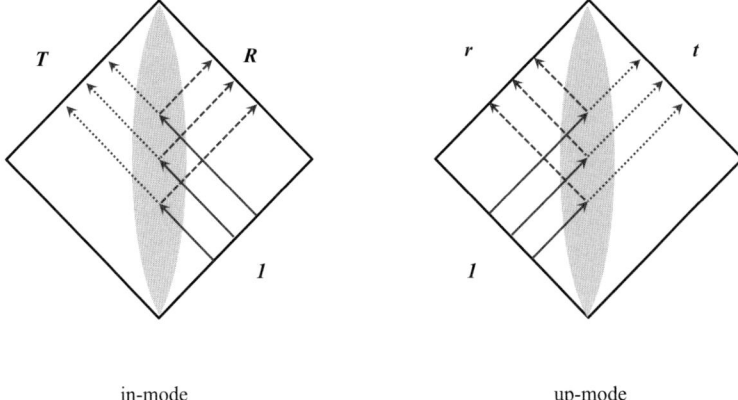

Fig. F.4 Conformal diagram for the solutions u_{in} and u_{up}.

complex conjugate, one obtains relations between the coefficients that enter the solutions. In particular, these conditions reproduce Eq. (F.4.12). Besides this they give two additional relations

$$T_k = t_k, \qquad R_k t_k^* + T_k r_k^* = 0. \tag{F.4.15}$$

A solution $u_{\text{up},k}$ describes a stationary wave propagating from $x = -\infty$. It is partially scattered back and partially passes through the potential barrier. The coefficients r_k and t_k are the reflection and transmission coefficients, respectively. Similarly, $u_{\text{in},k}(x)$ describes a stationary wave propagating from $x = \infty$, and R_k and T_k are its reflection and transmission coefficients. Equation (F.4.15) shows that the transmission amplitudes T_k and t_k are the same, while for the reflection amplitudes one has $|R_k| = |r_k|$, while their phases can be different.

Since the potential in Eq. (F.4.4) is real, one can always choose basic solutions to be real. For each of $k > 0$ one has two real solutions that we denote $v_{1,k}$ and $v_{2,k}$. The conformal diagram for these solutions are shown in Figure F.5. These solutions are special versions of a general solution Eq. (F.4.7). For $v_{1,k}$ $\alpha_k = \delta_k^* = C_{+,k}$ and $\gamma_k = \beta_k^* = C_{-,k}$. Similarly, for $v_{2,k}$ $\alpha_k = \delta_k^* = D_{+,k}$ and $\gamma_k = \beta_k^* = D_{-,k}$. The normalization and orthogonality conditions for $v_{1,k}$ and $v_{2,k}$ give

$$|C_{-,k}|^2 + |C_{+,k}|^2 = |D_{-,k}|^2 + |D_{+,k}|^2 = 1,$$
$$C_{-,k}D_{-,k}^* + C_{+,k}^*D_{+,k} = 0. \tag{F.4.16}$$

The coefficients $C_{+,k}$ and $D_{+,k}$ are uniquely determined by the coefficient $C_{-,k}$ and $D_{-,k}$. The corresponding relations include the reflection and transmission coefficient. One can use the following asymptotic data, specifying the real solutions $v_{1,k}$ and $v_{2,k}$:

$$C_{-,k} = \frac{1}{2}(z_k + r_k z_k^{-1}), \qquad C_{+,k} = \frac{1}{2}\frac{t_k^*}{|t_k|}(z_k - r_k z_k^{-1}),$$

$$D_{-,k} = \frac{i}{2}(z_k - r_k z_k^{-1}), \qquad D_{+,k} = \frac{i}{2}\frac{t_k^*}{|t_k|}(z_k + r_k z_k^{-1}). \tag{F.4.17}$$

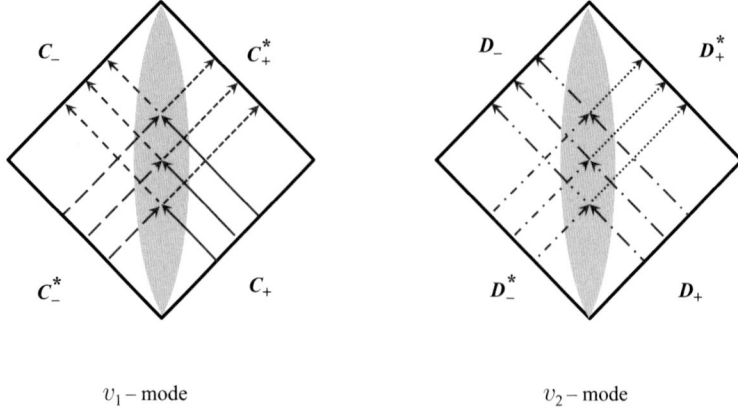

v_1 – mode v_2 – mode

Fig. F.5 Conformal diagram for a real solutions v_1 and v_2.

Here, $z_k = \sqrt{1 + |t_k|}$.

Basic solutions $u_{in,k}$ and $u_{up,k}$ can be written as linear combinations of $v_{1,k}$ and $v_{2,k}$

$$u_{in,k} = P_{1,k}v_{1,k} + P_{2,k}v_{2,k}, \qquad u_{up,k} = Q_{1,k}v_{1,k} + Q_{2,k}v_{2,k}, \qquad (F.4.18)$$

where

$$P_{1,k} = \frac{|t_k|}{2r_k t_k^*} z_k(r_k - 2 + z_k^2), \qquad P_{2,k} = -\frac{i|t_k|}{2r_k t_k^*} z_k(r_k + 2 - z_k^2),$$

$$Q_{1,k} = \frac{z_k^2 + r_k}{2z_k}, \qquad Q_{2,k} = \frac{i(z_k^2 - r_k)}{2z_k}. \qquad (F.4.19)$$

Denote

$$V_k(x, x') = v_{1,k}(x)v_{1,k}(x') + v_{2,k}(x)v_{2,k}(x'). \qquad (F.4.20)$$

It is evident that this function is real and symmetric with respect to its arguments x and x'

$$V_k(x, x') = V_k(x', x) = V_k^*(x, x'). \qquad (F.4.21)$$

Using Eqs. (F.4.18) and (F.4.19) one can check that

$$V_k(x, x') = u_{in,k}^*(x)u_{in,k}(x') + u_{up,k}^*(x)u_{up,k}(x'). \qquad (F.4.22)$$

At far distances, where the potential $U(x)$ becomes small, the function V_k has the following asymptotic form

$$V_k(x, x') = \frac{1}{\pi} \cos[k(x - x')] + \dots. \qquad (F.4.23)$$

In this expression ... denote omitted terms that either decrease at infinity or contain the fast-oscillating factor $\sim \exp[\pm ik(x + x')]$. The latter does not contribute to the integral over k.

F.4.3   Quantization

To quantize the scalar field φ one can write it in the following two equivalent forms

$$\hat{\varphi}(t,x) = \int_0^\infty dk \left[v_{1,k}(x)\hat{q}_{1,k}(t) + v_{2,k}(x)\hat{q}_{2,k}(t) \right]$$

$$= \int_0^\infty \frac{dk}{\sqrt{2k}} \left[e^{-ikt}[u_{\text{in},k}(x)\hat{a}_k + u_{\text{up},k}(x)\hat{b}_k] \right. \tag{F.4.24}$$

$$\left. + e^{ikt}[u^*_{\text{in},k}(x)\hat{a}^\dagger_k + u^*_{\text{up},k}(x)\hat{b}^\dagger_k] \right].$$

In the first line, the operators $\hat{q}_{1,k}(t)$ and $\hat{q}_{2,k}(t)$ are standard oscillatory-amplitude operators. In the second and third lines, $\hat{a}^\dagger_k$, $\hat{b}^\dagger_k$, $\hat{a}_k$, and $\hat{b}_k$ are the operators of creation and annihilation obeying the relations

$$[\hat{a}_k, \hat{a}^\dagger_{k'}] = \delta(k - k'), \qquad [\hat{b}_k, \hat{b}^\dagger_{k'}] = \delta(k - k'). \tag{F.4.25}$$

All other independent commutators vanish. The operator $\hat{a}^\dagger_k$ creates left-moving quanta in the mode $u_{\text{in},k}$, while $\hat{b}^\dagger_k$ creates right-moving quanta in the mode $u_{\text{up},k}$. The vacuum state is a state when neither left-moving nor right-moving quanta are present. It is defined by the conditions

$$\hat{a}_k|0\rangle = \hat{b}_k|0\rangle = 0. \tag{F.4.26}$$

The Hamiltonian of the system (with zero-point fluctuation contribution excluded) is

$$\hat{H}_0 = \hat{H}_{\text{in}} + \hat{H}_{\text{up}}$$

$$\hat{H}_{\text{in}} = \int_0^\infty dk\, k\, \hat{a}^\dagger_k \hat{a}_k, \qquad \hat{H}_{\text{up}} = \int_0^\infty dk\, k\, \hat{b}^\dagger_k \hat{b}_k. \tag{F.4.27}$$

F.4.4   Equilibrium thermal state

Thermal equilibrium state with temperature $\Theta = \beta^{-1}$ is described by the thermal density matrix

$$\hat{\rho} = Z_0^{-1} e^{-\beta \hat{H}_0}. \tag{F.4.28}$$

The *thermal Wightman functions* are defined as

$$G^+_\beta(t - t', x, x') = \langle \hat{\varphi}(t,x)\hat{\varphi}(t',x') \rangle_\beta,$$

$$G^-_\beta(t - t', x, x') = \langle \hat{\varphi}(t',x')\hat{\varphi}(t,x) \rangle_\beta = G^+_\beta(t' - t, x', x). \tag{F.4.29}$$

Using Eq. (F.4.24) one obtains

$$G^\pm_\beta(t, x, x') = \int_0^\infty dk\, \mathcal{G}^\pm_{\beta,k}(t)\, V_k(x, x'). \tag{F.4.30}$$

Similar expressions are valid for all other Green function

$$G_\beta^\bullet(t, x, x') = \int_0^\infty dk\, \mathcal{G}_{\beta,k}^\bullet(t)\, V_k(x, x'). \tag{F.4.31}$$

Here, the •-symbol specifies the corresponding Green function.

In particular, the *Matsubara propagator* is

$$\Delta(\tau, x, x') = G_\beta^\pm(-i\tau, x, x') = \int_0^\infty dk\, \Delta_k(\tau)\, V_k(x, x'). \tag{F.4.32}$$

This propagator is periodic in τ with the period β and obeys an equation

$$(\partial_\tau^2 + \partial_x^2 - U)\Delta(\tau, x, x') = -\delta(\tau)\delta(x - x'). \tag{F.4.33}$$

The *Matsubara propagator* can be obtained as a Green function on the Euclidean cylinder $S^1 \times \mathbb{R}^1$ in the presence of the potential U and with the size of the compact dimension equal to β (see Figure F.6).

F.4.5 Local observables

As an application of the developed formalism, let us calculate the asymptotic value of the *thermal averages* of the *stress-energy tensor* for the massless field $\hat\varphi$. To obtain the temperature-dependent part of $T_{\mu\nu}$ we shall use the renormalized *Hadamard function*

$$D(t, x, x') = \frac{1}{2}[G_\beta^{(1)}(t, x, x') - G_{\beta=\infty}^{(1)}(t, x, x')]$$

$$= \frac{1}{2\pi}\int_0^\infty \frac{dk}{k}\, \cos[k(x - x')]\cos(kt)[\coth(\beta k/2) - 1].$$

In the asymptotic region, where the potential $U(x)$ vanishes, the stress-energy tensor can be written as follows

$$T_{\mu\nu} \sim \lim_{t\to 0, x'\to x}[\partial_\mu \partial_{\nu'} - \frac{1}{2}g_{\mu\nu'}\partial_\alpha \partial^{\alpha'}]D(t, x, x'). \tag{F.4.34}$$

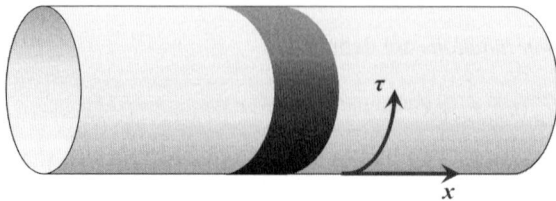

Fig. F.6 Euclidean space for the $(1 + 1)$-thermal field. It has the topology of the cylinder $S^1 \times \mathbb{R}^1$. The direction of the Euclidean time τ and the spatial dimension x are schematically shown. A shadowed region schematically shows a domain where the scattering on the potential takes place.

Simple calculations give

$$T_{\mu\nu} \sim \mathrm{diag}(\varepsilon, \varepsilon), \qquad \varepsilon = \frac{\pi}{6}\Theta^2, \qquad \Theta = \beta^{-1}. \tag{F.4.35}$$

Thus, outside the potential for the thermal state one has $p = \varepsilon$. This result is in agreement with the expected one.

F.4.6 Non-equilibrium thermal state

The density matrix Eq. (F.4.28) for the thermal equilibrium state can be written as a product of two independent density matrices

$$\hat{\rho} = \hat{\rho}_{\mathrm{in},\beta}\hat{\rho}_{\mathrm{up},\beta}, \qquad \hat{\rho}_{\mathrm{in},\beta} = Z_{\mathrm{in}}^{-1}e^{-\beta\hat{H}_{\mathrm{in}}}, \qquad \hat{\rho}_{\mathrm{up},\beta} = Z_{\mathrm{up}}^{-1}e^{-\beta\hat{H}_{\mathrm{up}}}. \tag{F.4.36}$$

Here, $\hat{H}_{\mathrm{in}}$ and $\hat{H}_{\mathrm{up}}$ are the Hamiltonians for in- and up-quanta given by Eq. (F.4.27). In the thermal equilibrium state the in- and up-quanta are emitted with the same temperature β^{-1} from $x = \infty$ and $x = -\infty$, respectively. Since $\hat{H}_{\mathrm{in}}$ and $\hat{H}_{\mathrm{up}}$ are independent, one can consider a more general state, which is described by the following density matrix

$$\hat{\rho}_{\mathrm{in},\mathrm{up}} = \hat{\rho}_{\mathrm{in},\beta_{\mathrm{in}}}\hat{\rho}_{\mathrm{up},\beta_{\mathrm{up}}}. \tag{F.4.37}$$

For $\beta_{\mathrm{in}} \neq \beta_{\mathrm{up}}$ this state is stationary, but not inequilibrium. In such a state there exists a net flux of thermal radiation from the hotter source to the colder one. We call such a state a *non-equilibrium thermal state*.

The renormalized Hadamard function for this state is

$$D(t, x, x') = D_{\mathrm{in}}(t, x, x') + D_{\mathrm{up}}(t, x, x'),$$

$$D_{\mathrm{in}}(t, x, x') = \int_0^\infty \frac{dk}{2k} n_{\mathrm{in},k}\left[e^{-ikt}u_{\mathrm{in},k}(x)u_{\mathrm{in},k}^*(x') + \{c.c.\}\right], \tag{F.4.38}$$

and D_{up} is obtained from D_{in} by changing in $\to$ up. In these relations $\{c.c.\}$ means a complex conjugated expression, and

$$n_{\mathrm{in},k} = \frac{1}{e^{\beta_{\mathrm{in}}k} - 1}, \qquad n_{\mathrm{up},k} = \frac{1}{e^{\beta_{\mathrm{up}}k} - 1}. \tag{F.4.39}$$

To calculate the renormalized stress-energy tensor at $x \to \infty$ we use the following asymptotics[3]

$$u_{\mathrm{in},k}(x)u_{\mathrm{in},k}^*(x') \sim \frac{1}{2\pi}\left(e^{-ik(x-x')} + |R_k|^2 e^{ik(x-x')} + \ldots\right),$$

$$u_{\mathrm{up},k}(x)u_{\mathrm{up},k}^*(x') \sim \frac{1}{2\pi}\left(|t_k|^2 e^{ik(x-x')} + \ldots\right). \tag{F.4.40}$$

Here, $\ldots$ denote the terms that either vanish at infinity or rapidly oscillate there, $\sim \exp[ik(x + x')]$. These terms will not contribute to the asymptotic value of the $T_{\mu\nu}$. Denote $v = t + x$ and $u = t - x$, then,

---

[3] Calculations for $x \to -\infty$ are similar.

$$D_{\rm in}(t, x, x') \sim \frac{1}{2\pi} \int_0^\infty \frac{dk}{k} n_{{\rm in},k} \left(\cos[k(v - v')] + |R_k|^2 \cos[k(u - u')] \right),$$

$$D_{\rm up}(t, x, x') \sim \frac{1}{2\pi} \int_0^\infty \frac{dk}{k} n_{{\rm up},k} |t_k|^2 \cos[k(u - u')].$$

(F.4.41)

The calculations give $T_{uv} = T_{vu} = 0$ and

$$T_{vv} = \frac{1}{2\pi} \int_0^\infty dk\, k\, n_{{\rm in},k} = \frac{\pi}{12} \beta_{\rm in}^{-2},$$

$$T_{uu} = \frac{1}{2\pi} \int_0^\infty dk\, k \left[|R_k|^2 n_{{\rm in},k} + |t_k|^2 n_{{\rm up},k} \right].$$

(F.4.42)

As expected, the ingoing flux T_{vv} is thermal with the temperature $\Theta_{\rm in} = \beta_{\rm in}^{-1}$. But the outgoing flux depends on both temperatures. It is a combination of the reflected flux with the temperature $\Theta_{\rm in} = \beta_{\rm in}^{-1}$, and the flux with the temperature $\Theta_{\rm up} = \beta_{\rm up}^{-1}$, which penetrates the potential barrier. In the limiting case, when $\Theta_{\rm in} = 0$, there is only one non-vanishing component and the flux is

$$T_{uu} = \frac{1}{2\pi} \int_0^\infty dk\, k\, |t_k|^2 n_{{\rm up},k}.$$

(F.4.43)

This is the thermal flux emitted by the thermal bath with temperature $\Theta_{\rm up}$ at $x = -\infty$, which, after scattering on the potential U, reaches $x = \infty$.

References

Abramowitz, M. and Stegun, I. A. (Eds.) (1972). Spheroidal Wave Functions. *Handbook of Mathematical Functions with Formulas, Graphs, and Mathematical Tables, 9th printing*, New York: Dover, pp. 751–759.

Aharony, O. *et al.*, (2000). Large N Field Theories, String. Theory and Gravity, *Phys. Reports*, **323**, 183–386.

Arkani-Hamed, N., Dimopoulos, S., Dvali, G. (1998). The Hierarchy problem and new dimensions at a millimeter, *Phys. Lett. B*, **436**, 263–272.

Arkani-Hamed, N., Dimopoulos, S., Dvali, G. (1999). Phenomenology, astrophysics, and cosmology of theories with submillimeter dimensions and TeV scale quantum gravity, *Phys. Rev.*, **D59**, 086004 [21 pp].

Aliev, A.N. and Gal'tsov, D.V. (1989). "Magnetized" black holes, *Sov. Phys. Usp.*, **32**, 75–92.

Arnold, V. I., (1989). *Mathematical Methods of Classical Mechanics*, Graduate Texts in Math., **60**, Springer-Verlag, New York and Berlin.

Baade, W. and Zwicky, F. (1934). On Supernovae, *Proc. Nat. Acad. Sci.*, **20**, 254–259.

Babelon O., Bernard D., and Talon M. (2006). *Introduction to Classical Integrable Systems*, Cambridge Univ. Press. Cambridge, United Kingdom.

Barceló, C., Liberati, S., Visser, M. (2005). Analogue Gravity, *Living Rev. Relativity*, **8**, 12. http://www.livingreviews.org/lrr-2005-12.

Bardeen, J.M., (1970). Kerr Metric Black Holes, *Nature*, **226**, 64–65.

Bardeen, J. M. (1973). *Timelike and Null Geodesics in the Kerr Metric*, In "Black Hole", eds. C. DeWitt and B. S. DeWitt. Gordon and Breach, New York.

Bardeen, J. M., Carter, B., and Hawking, S. W. (1973). The four laws of black hole mechanics, *Commun. Math. Phys.*, **31**, 161–170.

Bardeen, J. and Horowitz, G. T. (1999). Extreme Kerr throat geometry: A vacuum analog of $AdS_2 \times S^2$, *Phys. Rev.*, **D60**, 104030 [10 pp].

Bardeen, J. M., Press, W. H. (1973). Radiation fields in the Schwarzschild Background, *J. Math. Phys.*, **14**, 7–19.

Bardeen, J. M., Press, W. H. and Teukolsky, S. A. (1972). Rotating Black Holes: Locally Nonrotating Frames, Energy Extraction, and Scalar Synchrotron Radiation, *Astrophys. J.*, **178**, 347–369.

Barrow, I. D. and Silk I. (1983). *The Left Hand of Creation*, Basic Books, New York.

Bekenstein, J. D. (1972). Black holes and the second law, *Lett. Nuovo Cim.*, **4**, 737–740.

Bekenstein, J. D. (1973). Black Holes and Entropy, *Phys. Rev.*, **D7**, 2333–2346.

Bekenstein, J. D. (1974). Generalized second law of thermodynamics in black-hole physics, *Phys. Rev.*, **D9**, 3292–3300.

Bekenstein, J. D. (1980). Black Hole Thermodynamics, *Phys. Today*, **33**, 24–31.

Berger, M. (2003). *A Panoramic View of Riemannian Geometry*, Springer-Verlag, Berlin-Heidelberg-New York.

Berti, E., Cardoso, V., Casals, M. (2006). Eigenvalues and eigenfunctions of spin-weighted spheroidal harmonics in four and higher dimensions, *Phys. Rev.*, **D73**, 024013.

Berti, E., Cardoso, V., and Starinets, A. O. (2009). Quasinormal modes of black holes and black branes, *Class. Quantum Grav.*, **26**, 163001, 108pp.

Birrell, N. D., Davies, P. C. W. (1982). *Quantum Fields in Curved Space*, Cambridge Univ. Press. Cambridge, United Kingdom.

Blandford, R. D., Znajek, R. L. (1977). Electromagnetic extraction of energy from Kerr black holes, *Mon. Not. R. Astron. Soc.*, **179**, 433–456.

Bolton, C. T. (1972). Identification of Cygnus X-1 with HDE 226868, *Nature*, **235**, 271–273.

Bonnor, W. B. (1969). The gravitational field of light, *Commun. Math. Phys.*, **13**, 163–174.

Borde, A. (1994). Topology change in classical general relativity, arXiv: 9406053 [gr-qc].

Burrows, D. N. *et al.* (2011). Relativistic jet activity from the tidal disruption of a star by a massive black hole, Nature, **476**, 421–423.

Callon, H.B (1960). *Thermodynamics*, John Widey and Sons, New York, pp. 117–121.

Campanella, M., Lousto, C., Zlochower, Y., Merrit, D. (2007). Maximum Gravitational Recoil, *Phys. Rev. Lett.*, **98**, 231102, 4pp.

Carlip, S. (2009). Black Hole Thermodynamics and Statistical Mechanics, *Lect. Notes Phys.*, **769**, 89–123.

Carr, B. J. *et al.* (2010). New cosmological constraints on primordial black holes, *Phys. Rev.*, **D81**, 104019-1–33.

Cartan E. (1927). Sur la possibilit de plonger un espace Riemannien dans un espace Euclidien, *Annal. Soc. Polon. Math.*, **6**, 1–7.

Carter, B. (1968a). Global Structure of the Kerr Family of Gravitational Fields, *Phys. Rev.*, **174**, 1559–1571.

Carter, B. (1968b). Hamilton-Jacobi and Schrodinger separable solutions of Einstein's equations, *Commun. Math. Phys.*, **10**, 280–310.

Carter, B. (1971). Axisymmetric Black Hole Has Only Two Degrees of Freedom, *Phys. Rev. Lett.*, **26**, 331–333.

Carter, B. (1973). Black hole equilibrium states. *Black Holes, proceedings of the 1972 Les Houches Summer School*, eds. DeWitt C., and DeWitt, B. S., Gordon and Breach, New York. pp. 57–214.

Carter, B. (1976). *Proceedings of the First Marcel Grossmann Meeting on General Relativity*, ed. R. Ruffini, R., North-Holland, Amsterdam.

Carter, B. (1979). *General Relativity, an Einstein Centenary Survey*, eds. Hawking, S.W., Israel, W., Cambridge University Press, Cambridge, pp. 294–369.

Carter, B. (1987). *Gravitation in Astrophysics*, eds. Carter, B., Hartle, J., Plenum, New York, pp. 63–122.

Carter, B. (1997). *Proceedings of the Eighth Marcel Grossmann Meeting on General Relativity*, eds. Piran, T., and Ruffini, R., World Scientific, Singapore, pp. 136–165.

Cavaglià, M. (2003). Black hole and brane production in TeV gravity: A Review, *Int. J. Mod. Phys. A*, **18**, 1843–1882.

Chandrasekhar, S. (1931). The maximum mass of ideal white dwarfs, *Astrophys. J.*, **74**, 81–82.

Chandrasekhar, S. (1983). *The Mathematical Theory of Black Holes*, Oxford University Press, New York.

Chen, W., Lu, H., Pope, C.N. (2007). Kerr-de Sitter black holes with NUT charges, *Nucl. Phys. B*, **762**, 38–54.

Cherepashchuk, A.M. (1996). Masses of black holes in binary stellar systems, *Sov. Phys. Usp.*, **39**, 759–780.

Courant, R. (1943). Variational methods for the solution of problems of equilibrium and vibrations, *Bull. Am. Math. Soc.*, **49**, 1–23.

Crispino, L. C. B., Higuchi, A., Matsas, G. E. A. (2008). The Unruh effect and its applications, *Rev. Mod. Phys.*, **80**, 787–838.

Chruściel, P. T., Galloway, G. J., Pollack, D. (2010). Mathematical general relativity: A sampler, *Bull. Amer. Math. Soc.*, **47**, 567–638.

Connell, P., Frolov, V. P., and Kubiznak, D. (2008). Solving parallel transport equations in the higher-dimensional Kerr-NUT-(A)dS spacetimes, *Phys. Rev.*, **D78**, 024042, 13pp.

Dafermos, M., Rodnianski, I. (2010). The black hole stability problem for linear scalar perturbations, arXiv:1010.5137. 48 pp.

Debever, R., (1971). On type D expanding solutions of Einstein-Maxwell equations, *Bull. Soc. Math. Belg.*, **23**, 360–376.

DeWitt, B. S. (1975). Quantum field theory in curved spacetime, *Phys. Rep.*, **C19**, 296–357.

Doukas, J. (2010). Exact constraints on D≤ 10 Myers–Perry black holes, arXiv:1009.6118.

Dowker, J. S., (1977). Quantum field theory on a cone, *J. Phys. A: Math. Gen.*, **10**, 115–124.

Dubrovin, B.A., Fomenko, A.T., Novikov, S.P. (1990). *Modern Geometry - Methods and Applications*, Springer-Verlag.

Dymnikova, I.G (1986). Motion of particles and photons in the gravitational field of a rotating body (In memory of Vladimir Afanas'evich Ruban), *Sov. Phys. Usp.*, **29**, 215–237.

Eardley, D. M., and Giddings, S. B. (2002). Classical black hole production in high-energy collisions, *Phys. Rev.*, **D66**, 044011, 7pp.

Eddington, A. S. (1924). A comparison of Whitehead's and Einstein's formulas, *Nature*, **113**, 192.

Einstein, A. and Rosen, N. (1935). The particle problem in the general theory of relativity, *Phys. Rev.*, **48**, 73–77.

Eisenhart L. P. (1966). *Riemannian Geometry*, Princeton University Press, Princeton.

Echeverria, F., Klinkhammer, G., and Thorne, K. S. (1991). Billiard balls in wormhole spacetimes with closed timelike curves: Classical theory, *Phys. Rev.*, **D44**, 1077–1099.

Elvang, H., Figueras, P. (2007). Black Saturn, *JHEP*, 0705:050.

Emparan, R., Myers, R. C. (2003). Instability of ultra-spinning black holes, *JHEP*, 0309:025, 21 pp.

Emparan, R., Reall, H. S. (2002). A rotating black ring in five dimensions, *Phys. Rev. Lett.*, **88**, 101101, 4pp.

Emparan, R., Reall, H. S. (2008). Black Holes in Higher Dimensions, *Living Rev. Rel.* **11**, 6, 76 pp. http://www.livingreviews.org/lrr-2008-6.

Finkelstein, D. (1958). Past-Future Asymmetry of the Gravitational Field of a Point Particle, *Phys. Rev.*, **110**, 965–967.

Finster, F., *et al.* (2009). Linear waves in the Kerr geometry: A mathematical voyage to black hole physics, *Bull. Amer. Math. Soc.*, **46**, 635–659.

Flamm, L. (1916). Beitrge zur Einsteinischen Gravitationtheorie, *Physikalische Zeitschrift*, **17**, 448–454.

Flammer, C. (1957). *Spheroidal Wave Functions*, Stanford Univ. Press, Stanford.

Frauendiener, J.G. (2004). Conformal Infinity, *Living Reviews in Relativity*, http://relativity.livingreviews.org/open?pubNo=lrr-2004-1.

Friedman, J., Morris, M. S., Novikov, I. D., Echeverria, F., Klinkhammer, G., Thorne, K. S., Yurtsever, U. (1990). Cauchy problem in spacetimes with closed timelike curves. *Phys. Rev.*, **D42**, 1915–1930.

Friedman, J. L., Schleich, K., Witt, D. M. (1993). Topological Censorship, *Phys. Rev. Lett.*, **71**, 1486–1489.

Frolov, V. P. (1976). Black holes and quantum processes in them, *Sov. Phys. Usp.*, **19**, 244–262.

Frolov, V. P. (1991a). Vacuum polarization in a locally static multiply connected spacetime and a time-machine problem, *Phys. Rev.*, **D43**, 3878–3894.

Frolov, V. P. (1991b). Physical effects in wormholes and the "time machine" problem, *Proceedings of the sixth Marcel Grossmann meeting*, Sato, H., Nakamura, T., eds., Kyoto, 1233–1245.

Frolov, V., (2006). Embedding of the Kerr-Newman black hole surface in Euclidean space, *Phys. Rev.*, **D73**, 064021 [5 pp].

Frolov, V. P., Fursaev, D. V. (2005). Gravitational field of a spinning radiation beam pulse in higher dimensions, *Phys. Rev.*, **D71**, 104034, 16 pp.

Frolov, V. and Krtouš, P. (2011). Charged particle in higher dimensional weakly charged rotating black hole spacetime, *Phys. Rev.*, **D83**, 024016, 8 pp.

Frolov, V. P., Krtouš, P., Kubiznák, D. (2007). Separability of Hamilton-Jacobi and Klein-Gordon equations in general Kerr-NUT-AdS spacetimes, JHEP, 0702:005.

Frolov, V.P., Kubizňák, D. (2007). Hidden Symmetries of Higher Dimensional Rotating Black Holes, *Phys. Rev. Lett.*, **98**, 011101.

Frolov, V.P., Kubizňák, D. (2008). Higher-dimensional black holes: hidden symmetries and separation of variables, *Class. Quantum Grav.*, **25**, 154005, 22 pp.

Frolov, V. P., Novikov, I. D. (1990). Physical effects in wormholes and time machines, *Phys. Rev.*, **D42**, 1057–1065.

Frolov, V. and Novikov, I. (1998). *Black Hole Physics*, Kluwer Academic Publishers, Dordrecht.

Frolov, V.P., Shoom, A.A., (2007). Interior of distorted black holes, *phys. Rev.*, **D76**, 064037.

Frolov, V.P., Shoom, A.A., (2010). Motion of charged particles near a weakly magnetized Schwarzschild black hole, *Phys. Rev.*, **D82**, 084034.

Frolov, V. P., Stojkovic, D. (2003). Particle and light motion in a space-time of a five-dimensional rotating black hole. *Phys. Rev.*, **D68**, 064011.

Frolov, V. P., Israel, W., and Zelnikov, A. (2005). Gravitational field of relativistic gyratons, *Phys. Rev.*, **D72**, 084031, 11 pp.

Fronsdal, C. (1959). Completion and Embedding of the Schwarzschild Solution, *Phys. Rev.*, **116**, 778–781.

Fulton, W., (1997). *Young tableaux: with applications to representation theory and geometry*, Cambridge Monographs on Mathematical Physics, Cambridge.

Gallo, E. (2010). Radio emission and jets from microquasars, *Lecture Notes in Physics*, **794**, 85–113.

Gannon, D. (1976). On the topology of spacelike hypersurfaces, singularities, and black holes, *Gen. Rel. Grav.*, **7**, 219–232.

Geroch, R. P. (1967). Topology in general relativity, *J. Math. Phys.*, **8**, 782–786.

Gezari, S., *et al.* (2009). Luminous Thermal Flares from Quiescent Supermassive Black Holes, *Astrophys. J.*, **698**, 1367–1379.

Giacconi, R., Gursky, H., Paolini, F. R., and Rossi, B. B. (1962). Evidence for X-rays from sources outside the solar system, *Phys. Rev. Lett.*, **9**, 439–443.

Gibbons, G. W. and Hawking, S. W. (1977). Action integrals and partition functions in quantum gravity, *Phys. Rev.*, **D15**, 2752–2756.

Giddings, S. B. (2007). High-energy black hole production, arXiv:0709.1107 [hep-ph].

Ginzburg, V. L. (1985). *Physics and Astrophysics: A Selection of Key Problems*, Pergamon Press, Oxford.

Ginzburg, V. L., and Frolov V. P. (1987). Vacuum in a homogeneous gravitational field and excitation of a uniformly accelerated detector, *Sov. Phys. Uspekhi*, **30**, 1073–1095.

Goldberg, J. N., MacFarlane, A. J., Newman, E. T., Rorlich, F., Sudarshan, E. C. G. (1967). *J. Math. Phys.*, **8**, 2155–2161.

Gonzaález, J. A., Sperhake, U., Brugmann, B., Hannam, M. and Husa, S., (2007a). Maximum kick from nonspinning black-hole binary inspiral, *Phys. Rev. Lett.*, **98**, 091101. 4 pp.

Gonzaález, J. A., Hannam, M., Sperhake, U., Br ugmann, B., and Husa, S., (2007b). Super-massive Recoil Velocities for Binary Black-Hole Mergers with Antialigned Spins, *Phys. Rev. Lett.*, **98**, 231101. 4 pp.

Gou, L., *et al.* (2009). A Determination of the Spin of the Black Hole Primary in LMC X-1, *Astrophys. J.*, **701**, 1076–1090.

Gregory, R., Laflamme, R. (1993). Black Strings and p-Branes are Unstable. *Phys. Rev. Lett.*, **70**, 2837–2840.

Griffiths, J. B., Podolský, J. (2009). *Exact Space-Times in Einstein's General Relativity*, Cambridge Univ. Press, New York.

Guseinov, O. Kh. and Zel'dovich, Ya.B. (1966). Collapsed stars in binary systems. *Astron. Zh.*, **43**, 313–315.

Hansen, R.O. (1974). Multipole moments of stationary spacetimes, *JMP*, **15**, 46–52.

Harmark, T., Niarchos, V., and Obers, N. A. (2007). Instabilities of black strings and branes, *Class. Quant. Grav.*, **24**, R1-R90.

Hawking, S. W. (1971). Gravitationally collapsed objects of very low mass, *Mon. Not. Roy. Astron. Soc.*, **152**, 75–78.

Hawking, S. W. (1972a). Black holes in general relativity, *Commun. Math. Phys.*, **25**, 152–166.

Hawking, S. W. (1972b). The event horizon, *Black Holes, Les Houches lectures*, eds. DeWitt, C., DeWitt, B. S., Amsterdam, North Holland. pp. 1–55.

Hawking, S. W. (1974). Black Hole Explosions, *Nature*, **248**, 30–31.

Hawking, S. W. (1975). Particle Creation by Black Holes, *Commun. Math. Phys.*, **43**, 199–220.

Hawking, S.W., (1992). The chronology protection conjecture, *Phys. Rev.*, **D46**, 603–611.

Hawking, S. W. and Ellis, G. F. (1973). *The Large-Scale Structure of Spacetime*, Cambridge Univ. Press, Cambridge.

Hawking, S. W., Hartle, J. B. (1972). Energy and angular momentum flow into a black hole, *Commun. Math. Phys.*, **27**, 283–290.

Healy, J., Herrmann, F., Hinder, I., Shoemaker, D. M., Laguna, P., and Matzner, R. A. (2009). Superkicks in Hyperbolic Encounters of Binary Black Holes, *Phys. Rev. Lett.*, **102**, 041101, 4 pp.

Heusler, M. (1996). *Black Hole Uniqueness Theorems*, Cambridge Univ. Press., Cambridge.

Heusler, M. (1998). Stationary Black Holes: Uniqueness and Beyond, *Living Rev. Relativity*, **1** (6); http://relativity.livingreviews.org/Articles/lrr-1998-6.

Hewish, A., Bell, S. J., Pilkington, J. D. H., Scott, P. F., and Collins, R. A. (1968). Observation of a Rapidly Pulsating Radio Source, *Nature*, **217**, 709–713.

Hinder, I. (2010). The current status of binary black hole simulations in numerical relativity, *Class. Quant. Grav.*, **27**, 114004, 19 pp.

Hioki, K. and Maeda, Kei-ichi (2009). Measurement of the Kerr spin parameter by obbservation of a compact object's shadow, *Phys. Rev.*, **D80**, 024042.

Ho, L. C. (2002). On the relationship between radio emission and black hole mass in galactic nuclei, *Astrophys. J.*, **564**, 120–132.

Hochberg, D. and Visser, M. (1997). Geometric structure of the generic static traversable wormhole throat, *Phys. Rev.*, **D56**, 4745–4755.

Houri, T., Oota, T., Yasui, Y. (2009). Closed conformal Killing-Yano tensor and uniqueness of generalized Kerr-NUT-de Sitter spacetime, *Class. Quant. Grav.*, **26**, 045015, 24pp.

Iguchi, H., Mishima, T. (2007). Black di-ring and infinite nonuniqueness, *Phys. Rev.*, **D75**, 064018; Erratum-ibid. **D78**, 069903.

Izumi, K. (2008). Orthogonal black di-ring solution, *Prog. Theor. Phys.*, **119**, 757–774.

Ipser, J. R. (1971). Electromagnetic Test Fields Around a Kerr-Metric Black Hole, *Phys. Rev. Lett.*, **27**, 529–531.

Israel, W. (1967). Event Horizons in Static Vacuum Space-Times, *Phys. Rev.*, **164**, 1776–1779.

Israel, W. (1968). Event horizons in static electrovac space-times, *Commun. Math. Phys.*, **8**, 245–260.

Israel, W. (1983). Black holes, *Sci. Prog. (Oxford)*, **68**, 333–363.

Israel, W. (1986a). Third Law of Black-Hole Dynamics: A Formulation and Proof, *Phys. Rev. Let.*, **57**, 397–399.

Israel, W. (1986b). The formation of black holes in unimpherical collopse and cosmic cenocraphiys, *Can. J. Phys.*, **64**, 120–127.

Israel, W. (1987). Dark stars: the evolution of an idea. In S. W. Hawking and W. Israel, eds. *300 Years of Gravitation*, pp. 199–276. Cambridge Univ. Press, Cambridge.

Janet M. (1926). Sur la possibilit de plonger un espace Riemannien donn dans un espace Euclidien, *Ann. Soc. Polon. Math.*, **5**, 38–42.

Jordan, P., Ehlers, J., Sachs, R. (1961). Beitrage zur Theorie der reinen Gravitationsstrahlung, *Akad. Wiss. Mainz. Abh. Math. Naturwiss. Kl.*, **1**, p. 2.

Kaluza, T. (1921). Zum Unitatsproblem der Physik, *Sitz. Preuss. Akad. Wiss.*, **K1**, 966–972.

Kanti, P. (2004). Black Holes in Theories with Large Extra Dimensions: a Review, *Int. J. Mod. Phys. A*, **19**, 4899–4951.

Kapner, D.J., *et al.* (2007). Tests of the Gravitational Inverse-Square Law below the Dark-Energy Length Scale, *Phys. Rev. Lett.*, **98**, 021101.

Kastin, J. (1968). *A Course of Thermodynamics*, Vol. 2. Blaisdell, Wattham, Massachusatts.

Kim, S. W., Thorne, K. P. (1991). Do vacuum fluctuations prevent the creation of closed timelike curves? *Phys. Rev.*, **D43**, 3929–3947.

Klein, O. (1926). Quantum Theory and Five-Dimensional Theory of Relativity, *Z. Phys.*, **37**, 895–906; (1986), *Surveys High Energ. Phys.*, **5**, 241–244.

Kol, B., (2006). The phase transition between caged black holes and black holes, *Phys. Rep*, **422**, 119–165.

Kokkotas, K. D., and Schmidt, B. G. (1999). Quasi-Normal Modes of Stars and Black Holes, *Living Rev. Rel.* **2**, 2. e-Print: arXiv: gr-qc/9909058.

Komossa, S., Zhou, H., Lu, H., (2008). A Recoiling Supermassive Black Hole in the Quasar SDSS J092712.65+294344.0?, *Astrophys. J.*, **678**, L81–L84.

Konoplya, R.A., Zhidenko, A. (2011). Quasinormal modes of black holes: From astrophysics to strong theory, e-print: arxiv: 1102.4014 [gr-gc]

Kormendy, J. (1993). In *The Nearest Active Galaxies*, J. E. Beckman, H. Netzer, and L. Colina (eds.), Consejo Superior de Investigaciones Cientificas, Madrid, p. 197.

Krtous, P., Frolov, V. P., and Kubiznak, D. (2008). Hidden symmetries of higher dimensional black holes and uniqueness of the Kerr-NUT-(A)dS spacetime, *Phys. Rev.*, **D78**, 064022, 5pp.

Kruskal, M. D. (1960). Maximal Extension of Schwarzschild Metric, *Phys. Rev.*, **119**, 1743–1745.

Kubiznak, D. (2008). Hidden Symmetries of Higher-Dimensional Rotating Black Holes, arXiv:0809.2452 [gr-qc], 170pp.

Kubiznak, D., Frolov, V. P. (2007). Hidden Symmetry of Higher Dimensional Kerr-NUT-AdS Spacetimes. *Class. Quant. Grav.*, **24**, F1–F6.

Kubiznak, D., Frolov, V. P. (2008). Stationary strings and branes in the higher-dimensional Kerr-NUT-(A)dS spacetimes, *JHEP*, 0802:007.

Kubiznak, D., Frolov, V. P., Krtouš, P., and Connell, P. (2009). Parallel-propagated frame along null geodesics in higher-dimensional black hole spacetimes, *Phys. Rev.*, **D79**, 024018, 16pp.

Pelavas, N., Neary, N., Lake, K. (2001). Properties of the instantaneous Ergo surface of a Kerr black hole, *Class. Quant. Grav.*, **18**, 1319–1332.

Landau, L. D. (1932). On the theory of stars, *Physikallische Zeitschrift der Sowjetunion* **1**, 285–288.

Landau, L. D., Lifshitz, E.M. (1980). *The Classical Theory of Fields, Fourth Edition: Volume 2*, Butterworth-Heinemann, Oxford.

Landsberg, G. (2006). Black holes at future colliders and beyond, *J. Phys. G: Nucl. Part. Phys.*, **32**, R337–R365.

Lee, H. K., Wijers, R. A. M. J., Brown, G.E. (2000). The Blandford-Znajek process as a central engine for a gamma-ray burst. *Phys. Rep.*, **325**, 83–114.

Lemaître, G (1933). L'Univers en expansion, *Ann. Soc. Sci. Bruxelles*, **A53**, 51–85.

Leonhardt, U., Philbin, T. G. (2008). The case for artificial black holes, *Phil. Trans. R. Soc. A*, **366**, 2851–2857.

Lewis, G. N. and Randall, M. (1961). *Themodynamics*, 2nd edn., revised by Pitzer, K. S. and Bremer, L., Mcgraw-Hill, New York.

Lightman, A.P., Press, W. H., Price, R.H., and Teukolsky, S. A., (1975). *Problem Book in Relativity and Gravitation, with complete solutions*, Princeton University Press, Princeton.

Linet, B. (1976). Electrostatics and magnetostatics in the Schwarzschild metric, *J. Phys.*, **A9**, 1081–1087.

Lynden-Bell, D. (1969). Galactic Nuclei as Collapced Old Quazars, *Nature*, **223**, 690–694.

Markov, M. A. (1965). Can the gravitational field prove essential for the theory of elementary particles? *Suppl. Progr. Theor. Phys*, (extra number), 85–95.

Markov, M. A. (1970). The Closed Universe and Laws of Conservation of Electric Baryon and Lepton Charges, *Ann. Phys.*,**59**, 109–128.

Markov, M. A. (1974). Global properties of matter in collapsed state ("black holes"), *Sov. Phys. Uspekhi*, **16**(5), 587–599.

McClintock, J. E., Shafee, R., Narayan, R., Remillard, R.A., Davis, S. W., Li, L. X. (2006). The Spin of the Near-Extreme Kerr Black Hole GRS 1915+105, *Astrophys. J.*, **652**, 518–539.

McClintock, J. E., Narayan,R., Gou, L., Liu, J., Penna, R.F., and Steiner, J.F. (2009). Measuring the Spins of Stellar Black Holes: A Progress Report, *arXiv:0911.5408*.

Melia, F. (2009). *High-Energy Astrophysics*, Princeton University Press, Princeton.

Mészáros, P. (2002). Theory of Gamma-Ray Bursts, *Ann. Rev. Astron. Astrophys.*, **40**, 1–40.

Mészáros, P. (2006). Gamma-Ray Bursts, *Rept. Prog. Phys.*, **69**, 2259–2322.

Miller J.M., Reynolds C.S., Fabian A.C., Miniutti G., Gallo L.C. (2009). Stellar-mass black hole spin constraints from disk reflection and continuum modeling, *Astrophys. J.*, **697**, 900–912.

Misner, Charles W., Thorne, K.S., Wheeler, J.A. (1973). *Gravitation*, W.H. Freeman and Company, New York.

Morris, M. S., Thorne, K. S., and Yurtsever, U. (1988). Wormholes, Time Machines, and the Weak Energy Condition, *Phys. Rev. Lett.*, **61**, 1446–1449.

Mundell, C. G., Guidorzi, C., and Steele, I. A. (2010). Gamma-Ray Bursts in the Era of Rapid Followup, *Advances in Astronomy*, **2010**, 718468, 14pp.

Myers, R. C. and Perry, M. J. (1986). Black holes in higher dimensional space-times, *Ann. Phys.*, **172**, 304–347.

Nash, J. (1956). The imbedding problem for Riemannian manifolds, *Annals Math.*, **63**, 20–63.

Nordström, G. (1914). Über die Möglichkeit, das elektromagnetische Feld und das Gravitationsfeld zu vereinigen, *Phys. Zeits.*, **15**, 504–506.

Norris, L.K. (1997). Schouten-Nijenhuis Brackets, *J. Math. Phys.*, **38**, 2694–2709.

Novikov, I. D. (1963). On the evolution of a semiclosed universe (in Russian), *Astron. Zh.* **40**, 772.

Novikov, I. D. (1964). R- and T-regions in a spacetime with a spherically symmetric space (in Russian). *Comm. State Sternberg Astron. Inst.*, **132**, 3.

Novikov, I. D. (1990). *Black Holes and the Universe*, Cambridge Univ. Press, Cambridge.

Oota, T., Yasui, Y. (2008). Separability of Dirac equation in higher dimensional Kerr-NUT-de Sitter spacetime. *Phys. Lett. B*, **659**, 688–693.

Oota, T., Yasui, Y. (2010). Separability of Gravitational Perturbation in Generalized Kerr-NUT-de Sitter Spacetime, *Int. J. Mod. Phys. A*, **25**, 3055–3094.

Oppenheimer, J. R. and Snyder, H. (1939). On Continued Gravitational Contraction, *Phys. Rev.*, **56**, 455–459.

Oppenheimer, J. R. and Volkoff, G. (1939). On Massive Neutron Cores, *Phys. Rev.*, **55**, 374–381.

Orosz, J. A., *et al.* (2007). A 15.65-solar-mass black hole in an eclipsing binary in the nearby spiral galaxy M 33, *Nature*, **449**, 872–875.

Orosz, J. A., *et al.* (2009). A new dynamical model for the black hole binary LMC X-1, *Astrophys. J.*, **697**, 573–591.

Ortín, T. (2004). *Gravity and Strings*, Cambridge Univ. Press, Cambridge.

Page, D. N. (1983). Comment on "Entropy Evaporated by a Black Hole", *Phys. Rev. Lett.*, **50**, 1013–1013.

Penrose, R. (1963). Asymptotic Properties of Fields and Space-Times, *Phys. Rev. Lett.*, **10**, 66–68.

Penrose, R. (1964). *Relativity, Groups and Typology* eds. Dewitt, C. and Dewitt, B., Gorden and Breach, New York, London, pp. 565–584.

Penrose, R. (1965). Zero rest mass fields including gravitation, *Proc. R. Soc. London*, **A284**, 159–203.

Penrose, R. (1968). Structure of Space-Time. *Battelle Rencontres*, eds. DeWitt C. M., and Wheeler, J. A., Benjamin, New York, pp. 121–235.

Penrose, R. (1972). Black holes, *Sci. Am.*, **226**(5), 38–46.

Penrose, R., Rindler, W., (1987). *Spinors and space-time. Volume 1. Two-spinor calculus and relativistic fields*, Cambridge Univ. Press, Cambridge.

Petterson, J.A. (1974). Magnetic field of a current loop around a Schwarzschild black hole, *Phys. Rev.*, **D10**, 3166–3170.

Philbin, T. G., Kuklewicz, C., Robertson, S., Hill, S., König, F. & Leonhardt, U. (2008). Fiber-optical analog of the event horizon, *Science*, **319**, 1367–1370.

Poisson, E. (2007). *A Relativist's Toolkit: The Mathematics of Black-Hole Mechanics*, Cambridge Univ. Press, Cambridge.

Pomeransky, A. A., and Sen'kov, R. A. (2006). Black ring with two angular momenta, arXiv: 0612005 [hep-th]

Press, W. H. (1972). Time Evolution of a Rotating Black Hole Immersed in a Static Scalar Field, *Astrophys. J.*, **175**, 243–252.

Press, W. H. and Teukolsky, S. A. (1973). Perturbations of a Rotating Black Hole. 2. Dynamical Stability of the Kerr Metric, *Astrophys. J.*, **185**, 649–673.

Price, R. H. (1972a). Nonspherical Perturbations of Relativistic Gravitational Collapse. I. Scalar and Gravitational Perturbations, *Phys. Rev.*, **D5**, 2419–2438.

Price, R. H. (1972b). Nonspherical Perturbations of Relativistic Gravitational Collapse. II. Integer-Spin, Zero-Rest-Mass Fields, *Phys. Rev.*, **D5**, 2439–2454.

Punsly, B. (2008). *Black Hole Gravitohydromagnetics*, Springer-Verlag, Berlin Heidelberg.

Randall, L., Sundrum, R. (1999). Large Mass Hierarchy from a Small Extra Dimension, *Phys. Rev. Lett.*, **83**, 3370–3373.

Rees, M. J. (2000). A Review of Gamma Ray Bursts, *Nucl. Phys.*, **A663&664**, 42c–55c.

Remillard, R.A., McClintock, J.E. (2006). X-Ray Properties of Black-Hole Binaries, *Annual Review of Astronomy and Astrophysics*, **44**, 49–92.

Robinson, D. C. (2009). Four decades of black hole uniqueness theorems. *The Kerr Spacetime: Rotating Black Holes in General Relativity*, eds. Wiltshire, D. L., Visser, M., and Scott, S. M., Cambridge University Press, pp. 115–143.

Robinson, A., Young, S., Axon, D. J., Kharb, P., Smith, J. E. (2010). Spectropolarimetric Evidence for a Kicked Supermassive Black Hole in the Quasar E1821+643, *Astrophys. J. Letters*, **717**, L122-L126.

Rubin, M. A., and Ordóñez,C. R. (1984). Eigenvalues and degeneracies for n-dimensional tensor spherical harmonics, *J. Math. Phys.*, **25**, 2888–2894.

Sachs, R. K. (1961). Gravitational waves in general relativity VI: The outgoing radiation condition, *Proc. R. Soc. London*, **A264**, 309–338.

Sachs, R. K. (1964). *Relativity, Groups, and Topology*, DeWitt, C. and DeWitt, B. S. eds., N. Y.-London.

Salpeter, E. (1964). Accretion of interstellar matter by massive objects, *Astrophys. J.*, **140**, 796–800.

Sánchez, N. (1978). Absorption and emission spectra of a Schwarzschild black hole, *Phys. Rev.*, **D18**, 1030–1036.

Sandage, A. R., *et al.* (1966). On the Optical Identification of SCO X-1, *Astrophys. J.*, **146**, 316–321.

Sathyaprakash, B. S. (1999). Gravitational Waves: The Future of Black Hole Physics, *J. Astrophys. Astr.*, **20**, 211–220.

Sathyaprakash, B. S. and Schutz, B. F. (2009). Physics, Astrophysics and Cosmology with Gravitational Waves, *Living Reviews in Relativity*, http://relativity.livingreviews.org/Articles/lrr-2009-2/.

Sen, A. (2008). Black hole entropy function, attractors and precision counting of microstates, *Gen. Relativ. Gravit.*, **40**, 2249–2431.

Schreier, E., *et al.* (1972). Evidence for the Binary Nature of Centaurus X-3 from UHURU X-Ray Observations, *Astrophys. J.*, **172**, L79–L89.

Sciama, D. W. (1976). Black holes and their thermodynamics, *Vistas Astron.*, **19**, 385.

Shafee, R., McClintock, J. E., Narayan, R., Davis, S.W., Li, L. X., Remillard, R. A. (2006). Estimating the spin of stellar-mass black holes via spectral fitting of the X-ray continuum, *Astrophys. J.*, **636**, L113-L116.

Shields, G. A. *et al.* (2009). The Quasar SDSS J105041.35+345631.3: Black Hole Recoil or Extreme Double-Peaked Emitter? *Astrophys. J.*, **707**, 936–941.

Shklovsky, I. S. (1967). *Astrophys. J.*, **148**, L1.

Silverman, J. M. and Filippenko, A. V. (2008). On IC 10 X-1, The most massive known stellar-mass black hole, *Astrophys. J.*, **678**, L17–L20.

Smarr, L. (1973). Mass Formula for Kerr Black Holes, *Phys. Rev. Lett.*, **30**, 71–73.

Stairs, I. H. (2003). Testing General Relativity with Pulsar Timing, *Living reviews in relativity*, http://relativity.livingreviews.org/Articles/lrr-2003-5.

Starobinsky, A. A. and Churilov, S. M. (1974). Amplification of electromagnetic and gravitational waves scattered by a rotating black hole, *Sov. Phys. JETP*, **38**, 1–5.

Synge, J. L. (1950). The gravitational field of a particle, *Proc. Roy. Irish. Acad.*, **A53**, 83–114.

Synge, J. L. (1959). Optical observations in general relativity, *Milan Journal of Mathematics*, **30**, 271–302.

Szekeres, G. (1960). On the singularities of a Riemannian manifold, *Publ. Math. Debrecent*, **7**, 285–301.

Takagi, S. (1986). Vacuum Noise and Stress Induced by Uniform Acceleration, *Progr. Theor. Phys. Suppl.*, **88**, 1–142.

Teo, E. (2003). Spherical photon orbits around a Kerr black hole, *Gen. Rel. Grav.*, **35**, 1909–1926.

Thorne, Kip S., (1974). Disk-Accretion onto a Black Hole. II. Evolution of the Hole, *Astrophys. J.*, **191**, 507–519.

Thorne, Kip S. (1980). Multipole expansions of gravitational radiation, *Rev. Mod. Phys.*, **52**, 299–339.

Thorne, K. S. (1993). *GR13: General Relativity and Gravitation 1992 – Proceedings of the 13th International Conference on General Relativity and Gravitation, Cordoba, Argentina, 1992*, R. J. Gleiser, C. N. Kozameh, and O. M. Moreschi eds., Institute of Physics, Bristol, p. 295.

Thorne, Kip S. (1994). *Black Holes and Time Warps*, W. W. Norton. New York, London.

Thorne, K. S., Price, R. H., and Macdonald, D. A. (1986). *Black Holes: The Membrane Paradigm*, Yale Univ. Press, New Haven.

Tichy, W. and Marronetti, P. (2007). Binary black hole mergers: Large kicks for generic spin orientations, *Phys. Rev.*, **D76**, 061502, 5pp.

Tipler, F. J. (1977). Singularities and Causality Violation, *Ann. Phys,*, **108**, 1–36.

Tolman, R. C. (2010). *Relativity, Thermodynamics and Cosmology*, Dover Publications.

Vedrenne, G. and Atteia, J.-L. (2009). *Gamma-Ray Bursts: The Brightest Explosions in the Universe*, Springer/Praxis Books.

Visser, M. (1996). *Lorentzian Wormholes: From Einstein to Hawking*, American Institute of Physics Press. Woodbury, New York.

Volovik, G.E. (2003). *The Universe in a Helium Droplet*, Clarendon Press; Oxford University Press, Oxford, U.K.; New York, U.S.A.

Wald, R. M. (1974). Gedanken experiments to destroy a black hole, *Ann. Phys.*, **82**, 548–556.

Wald, R.M. (1984). *General Relativity*, Univ. Chicago Press, Chicago and London.

Webster, B. L. and Murdin, P. (1972). Cygnus X-1 – a Spectroscopic Binary with a Heavy Companion? *Nature*, **235**, 37–38.

Weinberg, S. (1972). *Gravitation and Cosmology: Principles and Applications of the General Theory of Relativity*, John Wiley & Sons, New York.

Weinfurtner, S., Tedford, E. W., Penrice, M. C. J., Unruh, W. G., and Lawrence, G. A. (2011). Measurement of Stimulated Hawking Emission in an Analogue System, *Phys. Rev. Lett.*, **106**, 02130, 4 pp.

Weyl, H. (1917). Zur Gravitationstheorie, *Ann. Physik*, **54**, 117–145.

Whiting, B, (1989). Mode stability of the Kerr black hole, *J. Math. Phys.*, **30**, 1301–1305.

Wilkins, D. C. (1972). Bound geodesics in the Kerr metric, *Phys. Rev.*, **D5**, 814–822.

Wilson, A. M. (1957). *Thermodynamics and Statistical Mechanics,* Cambridge Univ. Press, Cambridge, Chap 7.

York, J. W. (1972). Role of Conformal Three-Geometry in the Dynamics of Gravitation, *Phys. Rev. Lett.*, **28**, 1082–1085.

York, J.W. (1986). Boundary Terms in the Action Principles of General Relativity, *Foundat. Phys.*, **16**, 249–257.

Yoshino, H., Rychkov, V. S. (2005). Improved analysis of black hole formation in high-energy particle collisions, *Phys. Rev.*, **D71**, 104028, 13 pp.

Zel'dovich, Ya. B. (1964). The Fate of a Star and the Evolution of Gravitational Energy upon Accretion, *Sov. Phys. Doklady*, **9**, 195–197.

Zel'dovich, Ya. B. and Novikov, I. D. (1967). The hypothesis of cores retarded during expansion and the hot cosmological model, *Sov. Astron.*, **10**, 602.

Zel'dovich, Ya. B. and Novikov, I. D. (1971a). *Relativistic Astrophysics, Vol. 1: Stars and Relativity*, Univ. of Chicago Press, Chicago.

Zel'dovich, Ya. B. and Novikov, I. D. (1971b). *Relativistic Astrophysics, Vol. 2: The Structure and Evolution of the Universe*, Univ. of Chicago Press, Chicago.

Zurek, W. H. (1982). Entropy Evaporated by a Black Hole, *Phys. Rev. Lett.*, **49**, 1683–1686.

Index

absolute parallelism, 90
accretion
  disk, 2, 10, 32, 36, 194, 231, 242, 257, 269, 295
  of thermal radiation, 317
  spherical, 26
acoustic
  analog model, 370
  metric, 371
action
  for electromagnetic field, 133
  Einstein–Hilbert, 127, 129, 130
  for massive scalar field, 134
action-angle variables, 124
ADD model, 45, 377
AdS/CFT correspondence, 383
advanced time, 181, 212, 309
affine parameter, 60, 76, 110
afterglow, 369
Aichelburg–Sexl metric, 153, 156, 380
analog gravity models, 370
angle deficit, 65, 66, 161, 254
angular velocity of the black hole, 248, 389
annihilation operator, 438
anomalous Doppler effect, 293
antikick effect, 369
antisymmetrization, 69
  brackets, 73
apparent horizon, 358, 359
apparent shape of the black hole, 283
approximation
  geometrical optics, 276, 298, 305, 319
  semiclassical, 43
asymmetry of the black hole shadow, 283

asymptotically
  flat spacetime, 138, 348, 354, 355
  simple spacetime, 354
    weakly, 355
atlas, 68
  canonical symplectic, 120
averaged null-energy condition, 135, 397
axisymmetric magnetic field, 231

BEC, *see* Bose–Einstein condensate
Bianchi identity, 94
Big Bang, 361
billiard ball model, 406
binary pulsar, 364
binding energy, 194
Birkhoff's theorem, 168
black body, 43
black brane, 384
  boosted, 385
black di-rings, 392
  orthogonal, 393
black hole, 1, 14, 426
  angular velocity, 248, 389
  apparent shape of, 283
  binary, 22, 29
  definition, 355
  electrodynamics, 295
  elementary, 27, 376
  entropy, 44, 337
  eternal, 177
  Euclidean, 171, 183, 321
  evaporation, 44, *see* Hawking effect
  exterior, 63
  extremal, 389
  higher dimensional, 383

black hole (*cont.*)
  interior, 63, 176, 178
  internal energy, 337
  Kerr, 33, 267
  Kerr–Newman, 245, 359
  lens, 215
  magnetized, 232
  micro, 376, 378
  of stellar mass, 15
  paradigm, 26
  particle scattering by, 237
  persistent, 30
  physics, 1, 3, 42
  primordial, 40, 361
  region, 355
  Reissner–Nordström, 182, 360
  rotating, 242
  Schwarzschild, 168
  shadow, 214, 227, 278, 281, 283
  stability, 312
  surface, 426
  Tangherlini, 183, 341, 342, 383, 387
  temperature, 337
  transient, 30
  ultraspinning, 389
  uniqueness theorem, 359, 392
black object, 384, 425
black ring, 391
  bicycling, 393
  fat, 392
  thin, 392
black Saturn, 392
black string, 384
Boltzmann factor, 331
boost transformations, 49, 83
Bose–Einstein condensate, 16, 373, 374
boundary condition
  Dirichlet, 432
  natural, 432
  Neumann, 432
boundary term
  for Euclidean Einstein–Hilbert action,
    434
  for Einstein–Hilbert action, 434
  Gibbons–Hawking–York, 434
bounded motion, 233, 261

Boyer–Lindquist coordinates, 243, 245, 288,
    314, 388
bracket
  antisymmetrization, 73
  Poisson, 119, 121, 436
  Schouten–Nijenhuis, 125, 126
brane tension, 161
bulk space, 102

canonical
  commutation relations, 437, 444
  coordinates, 391, 421
  ensemble, 438
  form of the Kerr metric, 288
  symplectic atlas, 120
capture cross-section, 239
Cartan subgroup, 152
Carter's constant, 287
Carter–Penrose conformal diagrams, 175,
    348
Cartesian coordinates, 12, 47
Casimir effect, 444
Cauchy
  domain, 79
    future, 79
    past, 79
  formula, 311
  horizon, 182
  surface, 79
causal
  curve, 68
  future, 79
  Green function, *see* Feynman propagator
  past, 79
  structure, 68
    global, 6, 355
causality, 399
  local, 6
caustics, 358
chain process, 14
Chandrasekhar limit, 21, 29
chart, 68
Cherenkov cone, 293
chirp
  mass, 364, 365
  time, 364

Christoffel symbols, 80
  conformal transformation of, 412
chronological
  future, 79
  past, 79
chronology horizon, 405
  compactly generated, 405
  future, 405
  past, 405
chronology protection conjecture, 407
circular orbit, 261
clock synchronization, 398
  time gap, 399
closed conformal Killing–Yano tensor,
      426
  non-degenerate, 426
closed time-like curve, 399, 405
codimension of submanifold, 102
commutation relations
  canonical, 437, 444
commutator
  of vector fields, 81
complete integrability, 185, 428
  of geodesic equations, 286
complete integral, 272
Compton wavelength, 17, 27, 376
conformal
  anomaly, 44
  diagram, 455
  infinity, 348
  Killing tensor, 414
  Killing–Yano tensor, 414, 426
    closed, 426
    principal, 427
  Penrose space, 354
  transformation, 412
    of Christoffel symbols, 412
    of Ricci tensor, 413
    of scalar curvature, 412
conformally related metrics, 93, 412
conical singularity, 65, 66, 161,
      171, 254
consistency conditions, 406
constituents, 44
contraction of tensor, 69
coordinate system, 68

coordinates
  Boyer–Lindquist, 243, 245, 288, 314, 388
  canonical, 421
  Cartesian, 12, 47
  Darboux, 288, 421, 423
  Eddington–Finkelstein, 181
  essential, 243
  Fermi, 80, 95
  Gaussian, 104, 108
  Killing, 243, 421
  Kruskal, 172
  Painlevè–Gullstrand, 180
  Riemann, 80, 84, 89, 94
  tortoise, 181
cosmic censorship, 355, 379
cosmic string, 161
cosmological constant, 45, 128
  problem, 46
cotangent space, 73
covariance principle, 53, 54, 127, 131
covariant
  conservation law, 132
  derivative, 52, 80
covector, 73
creation operator, 438
critical null ray, 210
critical point, 25, 159
cross-section, 239
  of black hole creation, 379
curvature
  extrinsic, 103–105, 411, 434
  Gaussian, 250
  Riemann, 84
  scalar, 86, 411

Darboux
  basis, 421, 428
    normalized, 289, 423
  coordinates, 288, 421, 423
  theorem, 119
dark matter, 362
dark star, 13
Darwin's formula, 211, 240
delta function
  coordinate, 306
  invariant, 306

density matrix, 321
  thermal, 438
DeWitt approximation, 320, 323, 346
diffeomorphism, 69, 74
differential manifold, 67, 74, 393
dilaton 2D gravity, 165
dimensional reduction, 304
Dirichlet boundary condition, 432
disk dynamo, *see* unipolar generator
distribution function, 117
domain of outer communications,
      356, 360
dominant energy condition, 135, 248, 337
Doppler effect, 40
  anomalous, 293
Doppler shift, 29, 62
double pulsar, 365
dragging-into-rotation effect, 9
duality relations, 416
dumb hole, 370, 372
dynamical instability, 386
dynamo mechanism, 295

Eddington luminosity, 26, 37
Eddington–Finkelstein coordinates, 181
effective potential, 187
  for light motion, 188
Einstein
  formula for light deflection, 204, 240
  lift, 59
  space, 417, 428
  summation rule, 50
  tensor, 86, 130, 411
  universe, 349
Einstein equations, 8, 42, 127, 131
  higher dimensional, 149
Einstein principle of equivalence, *see*
      equivalence principle
Einstein–Hilbert action, 127, 129, 130, 163,
      183
  Euclidean, 434
  with boundary term, 434
Einstein–Rosen bridge, 8, 177, 178, 254, 396
  throat of, 178
electromagnetic field, 232
electrostatic potential, 12, 291
elementary black holes, 376

ellipsoid of rotation
  oblate, 244
embedding, 250
  diagram, 178, 322, 328
  problem, 170
emergent gravity models, 373
emergent phenomenon, 43, 45, 382
energy condition, 135
  averaged null, 135, 397
  dominant, 135, 248, 337
  null, 135, 215, 401
  strong, 135
  weak, 135, 338
energy-momentum tensor, *see* stress-energy
      tensor
entropy, 317
  of a black hole
    generalized, 339
  of black hole, 44, 337
equivalence principle, 5, 6, 47,
      59, 127
  strong, 5
ergosphere, 9, 247
ergosurface, 246, 253, 388
essential coordinates, 243
eternal black hole, 177
Euclidean
  black hole, 171, 183, 321
  Einstein–Hilbert action, 434
  field theory, 450
  horizon, 171, 183
  oscillator, 436, 442
  space, 436
  time, 436, 441
    ordering, 442
event, 67
event horizon, 1, 9, 41, 168, 246, 248, 293,
      356, 359, 383, 387
  future, 63
  past, 63
exotic matter, 397
expansion parameter, 359
extended body, 95
exterior
  derivative, 73, 426
  product, 73, 426
extrinsic curvature, 103–105, 411, 434

Fermi
   coordinates, 80, 95
   degenerate gas, 16, 17
   derivative, 83
   momentum, 16
   transport, 83
Feynman propagator, 322, 329, 441, 448
fiducial geodesic, 116, 117
first homotopy group, 395
first integral of motion, 121
fixed points of Killing vector, 249
foliation of spacetime, 112
free energy, 438, 445
Frobenius theorem, 98, 166, 249
fundamental theorem of Riemannian
      geometry, 171
future null infinity, 213, 352

galactic nuclei, 36
   active, 36
gamma-ray bursts, 34, 363, 369
   long, 370
   short, 370
gauge
   fixing condition, 54
   freedom, 136
   transverse traceless, 143
Gauss law, 245
Gauss–Codazzi equations, 411
Gaussian curvature, 250
Gaussian normal coordinates, 104, 108
general linear group, 70
generators
   of event horizon, 358
geodesic, 76
   deviation equation, 96
   equation, 54, 60
      complete integrability of, 286
   fiducial, 116
   null, 76
   submanifold, 108, 176
geodesic incompleteness, 63
geodesically incomplete spacetime, 170
geometrical optics approximation, 276, 298,
      305, 319
Gibbons–Hawking instanton, 171, 322, 334

Gibbons–Hawking–York boundary term,
      434
global
   Cauchy surface, 79
   causal structure, 355
   coordinates, 47
gradient, 73
grandfather paradox, 405
gravitational
   collapse, 15, 177
   equilibrium, 21
   lensing, 38, 211
   radiation, 145
   radius, 8, 168
   waves, 142, 145
      radiation, 363
      weak, 143
gravitational field
   homogeneous, 47
   non-potential, 404
   potential, 404
Green function, 140, 447
   causal, *see* Feynman propagator
   Hadamard, 329
   renormalized, 329
   retarded, 143, 306
Gregory–Laflamme instability, 385, 386
greybody factor, 323
ground state, 438
GRTensor package, 244
gyraton, 156

Hadamard
   elementary solution, 441
   function, 329, 441, 448, 460
Hamilton–Jacobi equation, 112, 126, 272,
      274, 390, 425
Hamiltonian, 114, 119
   flow, 119, 121
   quantum, 437
   reduced, 113
   vector field, 119
Hartle–Hawking state, 333
Hawking
   area theorem, 290, 358, 381
   effect, 40, 42

Hawking (*cont.*)
  evaporation, 43
  process, 321
  radiation, 40, 42, 44, 323
  temperature, 42, 171, 184, 321, 334
Heaviside step function, 320, 441
Heisenberg picture, 437
Hermitian operator, 437
  conjugated, 437
hidden symmetry, 9, 42, 101, 272, 275, 286,
      390, 414, 428
hierarchy problem, 4, 46, 158
higher-dimensional
  black hole, 383
  Einstein equations, 149
Hilbert space, 438, 445, 451
Hodge dual, 75, 416, 426
holonomic basis, 72, 73
homopolar generator, *see* unipolar generator
hook length, 71
horizon
  apparent, 358, 359
  bifurcation surface, 176, 183
  Cauchy, 182
  chronology, 405
  Euclidean, 183
  event, *see* event horizon, 248, 387
    future, 63
    past, 63
  Killing, 248, 360
hyperbolic motion, 197, 261

imaginary time, *see* Euclidean time
impact parameter, 187, 188, 238, 351
inclination angle, 23, 225, 277, 283
induced metric, 102, 103, 410
inertial reference frame, 47
infinite redshift surface, 246
information
  loss paradox, 44
  mining process, 14
innermost stable circular orbit, 28, 33, 194,
      224, 234, 237, 242, 262, 263, 269,
      366
integrability, 120
  conditions, 51
integral of motion, 121

interval, 48, 76
inverse metric, 75
inverse temperature, 438
involution, 121
irreducible
  Killing tensor, 287
  mass, 290
  representation, 70
ISCO, *see* innermost stable circular orbit

Jacobi
  action, 272, 274
  identity, 98, 119, 125
Jeans instability, 4, 27
Jost functions, 309, 310

Kaluza–Klein tower, 45
Kepler problem
  relativistic, 38
Kepler's third law, 196
Keplerian
  orbits, 147
  velocity, 197
Kerr
  black hole, 42, 267
    extremely rotating, 33
    mode stability of, 361
  metric, 42, 243, 359
    canonical form of, 288
    extremal, 256
  spacetime, 42, 277
Kerr effect, 375
Kerr–Newman metric, 245, 359
Kerr-NUT-(A)dS
  metric, 391, 426
  spacetime, 424
Kerr–Schild metric, 245
kick effect, 368, 369
Killing
  coordinates, 243, 421
  equation, 49
  horizon, 248, 360
  observer, 315
  primary vector, 391
  tensor, 101, 125, 272, 287, 390,
      414, 415
    conformal, 414

irreducible, 287
reducible, 287
vector, 49, 64, 132, 315
primary, 418, 427
secondary, 419, 428
Killing–Yano tensor, 101, 414, 415
closed conformal, 426
conformal, 414, 426
principal conformal, 427–429
Killing–Yano tower, 428
Klein–Gordon equation, 314, 425, 442
KMS relation, 440
Komar mass, 139, 151, 168, 384
Kretschman invariant, 168
Kronecker delta symbol, 69
Kruskal
coordinates, 172
metric, 173, 174
spacetime, 174

Lagrange multipliers, 109
Lagrange point, 24, 25
Lambert W-function, 304
Laplace transform, 307
large extra dimensions, 27, 377
models with, 45
Leibnitz rule, 80
Lense–Thirring force, 33
lensing effect, 38
Levi-Civita
symbol, 74, 75, 162, 165
tensor, 75
Liénard–Wiechert potential, 61
Lie
algebra, 97
derivative, 80–82, 418
group, 98
transport, 82
light propagation, 238
LIGO, 11
Liouville
integrability theorem, 122
theorem, 115, 122, 286
local
frame, 47
null cone, 48, 68

reference frame, 58
locally static spacetime, 403
Lorentz
force, 231
gamma-factor, 60
group, 48
transformation, 48

magnetar, 22
magnetohydrodynamics, 231, 297
Maldacena conjecture, 45
marginally stable circular orbit, *see*
innermost stable circular orbit, 262
maser, 39
radiation, 39
mass
function, 24, 29
Komar, 168, 384
reduced, 23
mass shell, 111
master equation, 42, 298
Matsubara propagator, 442, 449, 460
maximon, 27, 376
Maxwell equations, 56, 298
membrane formalism, 293
Messier catalogue, 34
metastable state, 310
metric, 74
acoustic, 371
Aichelburg–Sexl, 153, 156, 380
conformally related, 93
extremal Kerr, 256
induced, 102, 103, 410
inverse, 75
Kerr, 243, 359
Kerr–Newman, 245, 359
Kerr-NUT-(A)dS, 391, 426
Kerr–Schild, 245
Kruskal, 173, 174
Myers–Perry, 386, 388
off-shell, 423
on-shell, 423
Painlevè–Gullstrand, 180, 371
Reissner–Nordström, 360
Rindler, 302, 326
Schwarzschild, 168

metric (*cont.*)
  Schwarzschild–anti-de Sitter, 166, 168
  Schwarzschild–de Sitter, 166, 168
  static, 98
  Tangherlini, 183, 341, 342
micro-black hole, 376, 378
Milky Way, 1
minimal
  submanifold, 108
  surface, 108
Minkowski
  metric, 6, 48
  spacetime, 47
  vacuum, 330
mode stability of Kerr black hole, 361
Morse
  index, 25
  point, 25
  theorem, 25
  theory, 25
multiply-connected space, 395
multipole expansion, 140
Myers–Perry metric, 386, 388

naked singularity, 28, 355, 358
natural boundary conditions, 432
near critical rays, 210
Neumann boundary condition, 432
neutron star, 16, 21
Newtonian
  coupling constant, 12
  law, 45
  limit, 137
  potential, 12
no-hair theorems, 41
non-equilibrium thermal state,
        323, 461
non-potential static field, 403
non-traversable wormhole, 396
non-linear sigma model, 391
normalized Darboux basis, 289, 423
null
  cone, 6
    local, 6, 48, 68
  energy condition, 135, 215, 401
  geodesic, 76
    principal, 286

infinity
  future, 352
  past, 352
ray, 76
  critical, 210
  principal, 285
  propagation, 267
surface, 358
vector
  principal, 289
number density in phase space, 117
NUT parameters, 391

occupation number representation, 438, 445
off-shell metric, 423
on-shell metric, 423
operator of
  annihilation, 438
  creation, 438

p-forms, 73
Painlevè–Gullstrand
  coordinates, 180
  metric, 180, 371
parabolic motion, 261
particle scattering by black hole, 237
partition function, 438
past null infinity, 213, 352
peeling off property, 354
pencil of geodesics, 116
Penrose
  conformal space, 354
  diagram, 175, 348
  limit, 153
  process, 264
Penrose theorem, 357
periapsis, 197
Petrov type D, 286
phase space, 114, 115, 119, 436
phase volume, 115
photon scattering, 277
planar orbits, 186
plane monochromatic waves, 62
Poincaré group, 48, 49
Poisson
  bracket, 119, 121, 436
  equation, 12

potential function, 187, 259
primary
  Killing vector, 391, 418, 427
  ray, 216
primordial black holes, 40, 361
principal conformal Killing–Yano tensor,
    390, 427–429
principal null
  geodesics, 286
  rays, 285
  vectors, 289
projection operator, 434
proper
  distance, 76
  length, 76
  time, 76, 109
pulsar, 22
pure quantum state, 321

quadrupole
  formula, 145
  moment of mass distribution, 144
quality factor, 312
quantum
  gravity, 21, 43
  Hamiltonian, 437
  liquid, 372
    Bose, 373
    Fermi, 373
  oscillator, 437
  state
    pure, 44, 321
quark star, 22
quasars, 36
  microquasars, 33
quasinormal modes, 310, 311, 343

radial turning point, 187
radiation dominated stage, 40
Randall–Sundrum model, 45, 377
rank of tensor, 69
ray-tracing, 38
recoil velocity, 368
red-shift factor, 315
reduced Hamiltonian, 113
reduced mass, 147, 364

reducible Killing tensor, 287
reference frame, 5
Reissner–Nordström
  spacetime, 182
Reissner–Nordström
  metric, 182, 360
relativistic jets, 33, 37
renormalized Green function, 329
resonances, 310
retarded
  Green function, 143, 306
  time, 145, 181, 212, 309
revolution surface, 251
Ricci
  scalar, 86
  tensor, 86, 411
    conformal transformation of, 413
Riemann
  curvature, 84
  normal coordinates, 80, 84,
    89, 94
Riemannian
  geometry, 67, 74
  operator, 96
  space, 74, 75
Rindler
  frame, 60
  metric, 60, 302, 326
  observer, 47
  particle, 331, 334
  space, 326
ringing radiation, 312, 343
Roche lobe, 24, 25, 31, 32
rotating black hole, 242
rotation
  parameter, 324
  rapidity, 243

scalar
  curvature, 86, 411
    conformal transformation of, 412
  massive field, 442
  massless field, 299
  product, 75
Schouten–Nijenhuis bracket, 125,
    126, 287

Schrödinger
 equation, 450, 455
 picture, 450
Schur's theorem, 95
Schwarzschild
 black hole, 168
 metric, 168
 radius, 168, 177, 183
 spacetime, 168
Schwarzschild–anti-de Sitter metric, 166, 168
Schwarzschild–de Sitter metric, 166, 168
secondary Killing vector, 419, 428
self-consistency principle, 406
semi-closed world, 177
separation of variables, 135
shallow water waves, 372
Shapiro delay, *see* time delay, 210
simply connected space, 395
singularity
 naked, 28
Sokolov–Ternov effect, 335
Sommerfeld transformation, 329
space
 cotangent, 73
 Euclidean, 436
 Hilbert, 438
 multiply connected, 395
 phase, 119
 simply connected, 395
 tangent, 72, 73
spacetime
 asymptotically flat, 138, 348, 354, 355
 asymptotically simple, 354
  weakly, 355
 curvature, 5
 geodesically incomplete, 170
 Kerr, 277
 Kerr-NUT-(A)dS, 424
 Kruskal, 174
 locally static, 403
 Reissner–Nordström, 182
 Schwarzschild, 168
 Schwarzschild–anti-de Sitter, 168
 Schwarzschild–de Sitter, 168
 simply connected, 92

spherically symmetric, 162, 163
 stationary axially symmetric, 151
 Tangherlini, 183, 341, 342, 383, 387
 time-orientable, 78
 ultrastatic, 77, 352
spatial infinity, 352
specific
 angular momentum, 187, 259
 azimuthal angular momentum, 230
 energy, 187, 230, 259
 intensity, 118
 total angular momentum, 230
spherical accretion, 26
spherically symmetric spacetime,
  162, 163
spheroidal harmonics, 314
 spin-weighted, 315
spin–spin interaction, 291
spinning up effect, 242
stable circular orbit, 233
standard clock, 47
static metric, 98
stationary action principle, 127, 149
stationary axially symmetric spacetime, 151
statistical mechanics, 44
stellar
 capture, 38
 disruption, 38
Stefan–Boltzmann constant, 319
Stokes' theorem, 107, 144
strange star, 22
stress-energy tensor, 131, 164, 460
 of electromagnetic field, 134
 of fluid, 133
 of scalar field, 134
string theory, 44
strong energy condition, 135
strong rigidity theorem, 360
strongly asymptotically predictable, 355
structure constants, 97
Sturm–Liouville eigenvalue problem,
  314
submanifold, 102, 122
 codimension of, 102
 geodesic, 176
 minimal, 108

supergravity
  theory of, 45
superkick, 369
supersymmetry, 45
surface
  area S_{n-1}, 344
  gravity, 169, 171, 184, 248, 249, 322
  minimal, 108
symmetric mass ratio, 366
symmetrization, 69
symmetry transformation, 97
symplectic
  canonical atlas, 120
  manifold, 118
  structure, 118

T-product of operators, 441
tangent
  space, 72, 73
  vector, 73
Tangherlini
  metric, 183, 341, 342, 383, 387
  spacetime, 183
tensor
  antisymmetrization of, 69
  conformal Killing–Yano, 414
  contraction, 69
  density, 74
  Einstein, 86, 130, 411
  energy-momentum, *see* stress-energy
    tensor
  Hodge dual, 75, 416
  Killing, 101, 125, 272, 287, 390, 414, 415
    conformal, 414
  Killing–Yano, 101, 414, 415
  Levi-Civita, 75
  principal conformal Killing–Yano, 390,
    427
  rank of, 69
  Ricci, 86, 411
  Riemann, 84
  stress-energy, 131, 164, 460
  symmetrization of, 69
  totally skew-symmetric, 416
  Weyl, 86, 286
theorem
  of black hole uniqueness, 359

Birkhoff's, 168
Darboux, 119
Frobenius, 98, 166, 249
fundamental of Riemannian
  geometry, 171
Hawking black hole area, 358, 381
Jacobi, 135
Liouville, 115, 122, 286
Liouville integrability, 122
no-hair, 41
of strong rigidity, 360
Schur's, 95
Stokes', 107, 144
topological censorship, 397
thermal
  average, 439, 460
  density matrix, 438
  non-equilibrium state, 319, 323, 461
  Wightman functions, 447, 459
Thomson scattering cross-section, 25
tidal
  effect, 7
  force, 5, 7, 13, 95
    tensor, 13
time
  delay, 207
  machine, 401
    problem, 399
time-orientable spacetime, 78
Tolman's
  law, 325
  temperature, 316, 334
Tolman–Oppenheimer–Volkoff limit,
  22, 29
topological
  censorship theorem, 397
  invariant, 394
tortoise coordinate, 181, 304, 305,
  326, 343
totally skewsymmetric tensor, 416
transitivity surface, 162
transmission coefficient, 320
transverse traceless gauge, 143
trapped
  region, 359
  surface, 359
traversable wormhole, 394, 397, 400

turning point, 187
twin paradox, 55, 77, 265

ultraspinning black hole, 389
ultrastatic spacetime, 77, 352
uncertainty relations, 321
uniformly accelerated
    detector, 332
    frame, 47, 334
    motion, 57
unipolar
    generator, 294
    inductor, 295
unity decomposition, 106
universal interaction, 5
Unruh
    effect, 66, 334
    temperature, 328–331

vacuum, 438, 444
    energy, 439
    in quantum theory, 321
    Minkowski, 330
    polarization, 325, 326, 331
vector, 72
    Killing, 132
volume V_n, 344

volume element
    invariant, 53

wave zone, 142
weak
    energy condition, 135, 338
    field approximation, 142, 153
    gravitational wave, 143
weakly asymptotically simple spacetime, 355
wedge product, *see* exterior product, 426
Weyl tensor, 86, 286
white
    dwarf, 16, 21
    hole, 63, 177
Wick's rotation, 65, 183, 436
Wightman functions, 452
    thermal, 447, 459
winding number, 398
wormhole, 396, 401
    non-traversable, 396
    traversable, 394, 397, 400
Wronskian, 309, 456

Young tableau, 71, 86
    shape, 71

zero-point fluctuations, 321, 326, 407, 444